CLASSICAL AND MODERN REGRESSION WITH APPLICATIONS

SECOND EDITION

Raymond H. Myers
Virginia Polytechnic Institute and State University

THE DUXBURY ADVANCED SERIES IN STATISTICS AND DECISION SCIENCES

PWS-KENT Publishing Company
Boston

PWS-KENT
Publishing Company

20 Park Plaza
Boston, Massachusetts 02116

Dedicated to my wife, Sharon, and my children, Billy and Julie

PWS-KENT Publishing Company is a division of Wadsworth, Inc.

Library of Congress Cataloging-in-Publication Data

Myers, Raymond H.
Classical and modern regression with applications.
Bibliography: p.
Includes index.
1. Regression analysis. I. Title
QA278.2.M49 1989 519.5′36 89-13638
ISBN 0-534-92178-7

Printed in the United States of America

90 91 92 93 94—10 9 8 7 6 5 4 3 2 1

Sponsoring Editor: Michael Payne
Manufacturing Coordinator: Margaret Sullivan Higgens
Interior and Cover Designer: Catherine L. Johnson
Production Editor: S. London
Typesetter: H. Charlesworth & Co Ltd
Printer and Binder: The Maple-Vail Book Manufacturing Group

PREFACE

No single statistical tool has received the attention given to regression analysis in the past 25 years. Both practical data analysts and statistical theorists have contributed to an unprecedented advancement in this important and dynamic topic. Many volumes have been written by statisticians and scientists with the result being that the arsenal of effective regression methods has increased manyfold.

My intent for this second edition is to provide a rather substantial increase in material related to classical regression while continuing to introduce relevant new and modern techniques. I have included major supplements in simple linear regression that deal with simultaneous influence, maximum likelihood estimation of parameters, and the plotting of residuals. In multiple regression, new and substantial sections on the use of the general linear hypothesis, indicator variables, the geometry of least squares, and relationship to ANOVA models are added. In addition, all new topics are illustrated with the use of real-life data sets and annotated computer printout. In the area of useful modern techniques, additional types of diagnostic residual plots are developed and illustrated, including component plus residual plots and augmented partial plots. These plots are designed to provide a two-dimensional picture of the role of each regressor in the multiple regression and graphically highlight the need for nonlinearities in the regression model.

The chapter on nonstandard conditions has received considerable enhancement in order to better highlight modern techniques. The Box–Cox transformation has been expanded considerably with illustrations. Regression with a categorical response, and otherwise nonnormal error situations, are now given the consideration they deserve. Logistic regression is now expanded and is accompanied by Poisson regression. These two important techniques provide a foundation for a generalization to *generalized linear models* from which one can build models under error structures that have distributions such as gamma, negative binomial, exponential, and others. Other nonstandard conditions such as autocorrelated errors are discussed and illustrated. The text continues to emphasize applications that are selected from a wide variety of fields. The text puts emphasis on concepts and provides a blend between illustrations with real data sets and conceptual development.

Exercises in the text have been enhanced substantially. The number of exercises have been approximately doubled, and many new ones involve new data sets and problems to provide the student experience in using new techniques developed in the text. A data disk is available containing data sets in the exercises.

It is necessary that readers have been exposed to a course in basic statistical methods and thus are familiar with the use of the normal, t, χ^2, and F distributions. A basic course in calculus is required, and familiarity with matrix algebra is desirable. The text emphasizes applications with examples that illustrate nearly all techniques discussed. I have selected examples from a wide variety of fields including the physical sciences, engineering, biology, management science, and economics.

The book is designed for seniors and graduate students majoring in statistics or user subject-matter fields. There are nine chapters and three appendices. Chapter 1 introduces the general notion of regression models, and Chapter 2 deals specifically in simple linear regression with traditional material based on ordinary least squares estimation. Chapter 3 extends least squares estimation to multiple linear regression and introduces multicollinearity with illustrations, though modern methods of diagnosis and combating collinearity are relegated to Chapter 8. Both Chapters 2 and 3 have been expanded in areas previously discussed. Chapter 4 represents a blend between classical and contemporary methodology designed to choose the optimal subset model. Sequential analytic or *stepwise* model-building methods are discussed and illustrated. In addition, Mallow's C_p-statistic is developed and motivated as a criterion that represents a compromise between model overfitting and underfitting. The obvious connection is made in Chapter 4 between *cross-validation* criteria and model selection. To this end, the PRESS statistic and data splitting are presented and illustrated as criteria for discriminating among competing models. Illustrations are given of how modern software allows PRESS, C_p, and other model selection criteria to be used in an *all possible regression* type of procedure.

Chapter 5 deals solely in analysis of residuals when the user's intent is to do model criticism and detection of violation of assumptions. Modern diagnostic methods are discussed for outlier detection and plotting of residuals. The material in Chapter 6 revolves around modern methods for detecting data points that exert disproportionate influence on the regression results. Influence diagnostics, which are rapidly becoming a standard part of the analyst's arsenal, are discussed. Chapter 7 follows logically by presenting procedures that are used as alternatives to standard methodology when assumptions are violated. Transformations are presented with illustrations. Weighted regression is used as an alternative to ordinary least squares when the homogeneous variance assumption fails. Robust regression is described for the case of outliers or non-Gaussian error assumptions. As I mentioned previously, Chapter 7 has been supplemented to include autocorrelated errors, additional development in transformations, and regression with categorical responses and other non-Gaussian conditions. Chapter 8 is dedicated completely to diagnostic and analytical methods in cases of data sets that contain multicollinearity. Ridge regression and principal components regression are developed and illustrated with real data sets. Chapter 9 covers nonlinear regression. Standard nonlinear regression methods for finding least squares estimators are developed and presented.

I have already mentioned the substantial number of real data sets that are a part of an expanded number of exercises. Other exercises are more conceptual or theoretical in nature and are certainly challenging to the graduate student or high-level undergraduate who has an appreciation of linear algebra.

The three appendices are each designed for a different purpose. Appendix A supplements the reader's background in linear algebra with a treatment of items such as eigenvalues, eigenvectors, and quadratic forms. Appendix B is designed to strengthen the reader's understanding of certain statistical concepts that are not likely to be covered in prerequisite courses. For example, treatment of notions such as maximum likelihood, the generalized variance, and expected values of quadratic forms are given. This will aid the student who is mathematically more sophisticated to achieve a higher level of understanding of some of the regression concepts in the text. In addition, Appendix B contains derivations of results dealing with quadratic forms, deletion formulae associated with influence diagnostics, and several other theoretical results that were deleted from the mainstream of the text so that the reader not be deflected from the practical concepts and consequences of the regression development. Appendix C contains statistical tables.

I would like to acknowledge those who made contributions to the project. First and foremost, I would like to thank my wife, Sharon, who proofread the manuscript at each stage and whose efforts resulted in many improvements. Thanks also go to Roy Welsh and Jeffrey Birch who made helpful suggestions on how to improve the first edition. In addition, I would like to thank the reviewers for this edition:

> Fred Andrews, University of Oregon; Indra Chakavarti, University of North Carolina at Chapel Hill; Don Edwards, University of South Carolina at Columbia; Yu Sheng Hsu, Georgia State University; Jon Matta, University of Missouri; J. Michael Steel, Princeton University; R. Kirk Steinhorst, University of Idaho, and Wayne L. Winston, Indiana University.

Raymond H. Myers

CONTENTS

CHAPTER 8

DETECTING AND COMBATING MULTICOLLINEARITY 368

CHAPTER 9

NONLINEAR REGRESSION 424

APPENDIX A

SOME SPECIAL CONCEPTS IN MATRIX ALGEBRA 452

APPENDIX B

SOME SPECIAL MANIPULATIONS 461

APPENDIX C

STATISTICAL TABLES 474

INDEX 486

CHAPTER 1

INTRODUCTION: REGRESSION ANALYSIS

The term regression analysis describes a collection of statistical techniques that serve as a basis for drawing inferences about relationships among quantities in a scientific system. In the field of applied statistics, there is a plethora of modern data analysis techniques, which are branded with eye-catching names. The motivation of a particular method stems from the need to solve a specific problem. But regression analysis has been burdened with the responsibility—under one heading—of solving a variety of problems. For this reason, volumes are written about the topic, and its use continues to expand. New subtopics and subareas are hurled under the ever-growing umbrella that we continue to call *regression analysis.* The resulting analytical methodology, a product of the imagination of both professional statisticians and subject matter scientists, has become readily accessible today due to the rapid work of computer software personnel. Though it was not the intent or grand design, regression analysis is now perhaps the most used of all data analysis methods.

In 1885 Sir Francis Galton (1885) first introduced the word "regression" in a study that demonstrated that offspring do not tend toward the size of parents, but rather toward the average as compared to the parents. The upshot of the term's application was the "regression towards mediocrity" of offspring. Credit for discovery of the method of least squares generally is given to Carl Friedrich Gauss, who used the procedure in the early part of the nineteenth century. There is some controversy concerning this discovery. Apparently Adrien Marie Legendre published the first work on its use in 1805. Regression analysis and the method of least squares nearly always were linked in practice and seemed to be compatible bedfellows until the latter part of the 1960s. It became very apparent that, in many nonideal situations, ordinary least squares (OLS) is not appropriate and, indeed, can be improved upon. Much controversy has evolved and will likely continue to cloud the notion of alternatives to OLS. However, biased estimation for combating multicollinearity (Chapter 8) and robust regression (Chapter 7) have been accepted by many data analysts and will attain a niche in the heritage of regression analysis.

Often a matter of no small question to the analyst is what effect quantities (variables) have on one another. The system that generates the data may be a chemical or biological process, the nation's economy, a group of patients in a medical experiment, or perhaps a group of metal specimen in a tensile strength study. In some cases the pertinent variables are random variables and are related in a probability sense through a joint probability distribution. In other cases, the variables are mathematical quantities, and the assumption is that there exists a functional relationship linking them. From a set of data involving measurements on the variables, regression analysis is designed to shed light on certain aspects of the mechanism that relate them. An example illustrates what kind of information regression analysis can acquire from the data. Suppose we select a group of men at random to take part in an experiment. Each individual must run a specified distance. We then make the following measures on each individual:

x_1: age
x_2: weight
x_3: pulse rate at rest (rest pulse)
x_4: pulse rate immediately following run (run pulse)
x_5: time that it takes to run the distance (run time)
y: oxygen consumption rate (oxygen rate)

For specific data on this type of application, the reader is referred to the *SAS User's Guide* (1985). The y variable represents the efficiency with which the individual utilizes oxygen. We can argue that perhaps y should play a somewhat different role than x_1, x_2, x_3, x_4, and x_5. It may be of interest to visualize the x variables as quantities that actually *determine* y or rather *predict* y. Thus the x's are called *independent variables* or *regressor variables*. Given the data, the analyst certainly would hope to derive information regarding the role of each regressor variable in terms of its influence on oxygen rate. If one or more of the variables has a negligible influence on oxygen, this is valuable information. In addition, we might expect that the data should allow for estimation of the relationship that exists among the variables (assuming a relationship exists). Before any specific techniques can be developed or even discussed, the notion of a statistical model must be introduced.

1.1

REGRESSION MODELS

In the following chapters much development revolves around the use and development of *statistical models*. Simply put, all procedures used and conclusions drawn in a regression analysis depend at least indirectly on the assumption of a regression model. A model is what the analyst perceives as the mechanism that generates the data on which the regression analysis is conducted. Regression models are usually found in an algebraic form. For example, in the oxygen consumption illustration, if the experimenter is willing to assume that the relationship is well represented by a structure that is linear in the regressor variables, then a suitable model may be given by

$$y = \beta_0 + \beta_1 x_1 + \beta_2 x_2 + \beta_3 x_3 + \beta_4 x_4 + \beta_5 x_5 + \varepsilon \tag{1.1}$$

In Eq. (1.1), β_0, β_1, ..., β_5 are unknown constants called *regression coefficients.* Procedures embraced by regression analysis concern themselves with drawing conclusions about these coefficients. The investigator may be involved in a painstaking attempt to determine if an increase in x_4 (run pulse) truly does decrease or increase the efficiency of oxygen consumption. Perhaps the sign and magnitude of the coefficient is important. The ε term in the model is added to account for the fact that the model is not exact. It essentially describes the random disturbance or *model error.* When we apply (1.1) to a set of data, we can view the ε term as an aid in accounting for any variation due to the individual, that is *apart from the terms supplied by the model.* Chapter 2 gives a description that outlines the very important assumptions that often must be made on the ε's, assumptions on which the theory underlying regression analysis depends.

The model in Eq. (1.1) falls into the class of *linear models* (linear in the parameters, the β's). Any regression procedure involves *fitting the model* to a set of data, the latter defined as readings on the variables for the various experimental units being sampled (for example, the individuals sampled in the oxygen consumption situation). The term, *fit to a set of data*, actually involves estimation of the regression coefficients and the corresponding formulation of a *fitted regression model*, an empirical device that is the basis of any statistical inference made. Measures of quality of fit are important statistics that form the foundation of a statistical analysis. Clearly if the postulated model does not describe the data satisfactorily, any fundamental conclusions recovered from the fitted model are suspect.

This text is not confined to the treatment of linear models. Nonlinear models are commonplace in many of the natural sciences or engineering applications. A biochemist may postulate a growth model of the type

$$y = \frac{\alpha}{1 + e^{\beta t}} + \varepsilon \tag{1.2}$$

to represent the growth y of a particular organism as a function of time t. Here the parameters α and β are to be estimated from the data. Many of the same problems one attempts to solve by linear regression can often be handled by *nonlinear regression*. However, the computational aspects of building nonlinear models are less straightforward and thus require a special treatment (see Chapter 9).

The fitted regression model is produced by the model builder as an estimate of the functional relationship that describes the data. The type of model postulated quite often depends on the range of the regressor variables encountered in the data. For example, a chemical engineer may have knowledge of his system that necessitates the use of terms that impart curvature on the model. Suppose x_1 and x_2 represent temperature and a reactant concentration, and the response y is a simple yield of a chemical reaction. The engineer's goal is possibly two-fold:

1. To use the fitted or estimated regression (*prediction equation*) for yield estimation at locations x_1 and x_2 other than data conditions;
2. To study the relationship in the region of the data, perhaps with a view toward finding temperature and concentration conditions that provide satisfactory yield.

To reflect model curvature, a structure of the type

$$y = \beta_0 + \beta_1 x_1 + \beta_2 x_2 + \beta_{11} x_1^2 + \beta_{22} x_2^2 + \beta_{12} x_1 x_2 + \varepsilon \tag{1.3}$$

is used. Thus the ranges of the regressor variables may well dictate the model type. With narrow ranges in x_1 and x_2, the engineer may be successful at using a model that does not involve quadratic terms. Though Eq. (1.3) includes powers and products of order two in the regressor variables, it is, nevertheless, another example of a linear model since the coefficients enter linearly.

In nearly all regression applications in which linear models describe a set of data, the model formulation is an oversimplification of what occurs in the data observational process. Areas in the social and behavioral sciences represent examples where the systems from which the data are taken are far too complicated to model with an absolutely correct structure. The linear models used are approximations that hopefully *work well in the range of the data* used to build them. There is no intention here to cast stones on those who build linear models as approximations. When the sophistication of the subject matter field is not sufficient to provide a working theory, a linear and "common sense" empirical model approach can be very informative, particularly when used in conjunction with a set of data of reasonable quality.

1.2 FORMAL USES OF REGRESSION ANALYSIS

In this text much coverage is given to categories representing specific kinds of inferences, and distinctions are necessarily drawn among them. This is crucial since either the estimation procedure or even the model that is adopted may well depend on what the intended goal of the study is. This often seems counter to the ideology of certain users of the methodology. To the inexperienced analyst, it may seem that the model that apparently best describes the data should be adopted for every purpose. However, a model that gives a satisfactory solution to one problem will not necessarily provide success in solving another. The uses of regression analysis fall into three or perhaps four categories, though there is some overlap. These categories are as follows:

1. Prediction
2. Variable screening
3. Model specification (system explanation)
4. Parameter estimation

The analyst is advised to know and bear in mind what goal is vital in his or her particular endeavor. Let us initially consider goal (1). Here, parameter estimates are not sought for their own sake. We do not search for the true model specification apart from how the functional form influences prediction. It is not important that we capture the role of each regressor in the model with strict preciseness. Certainly our example with the chemical process is a prediction problem. Reaction yield is important, and the engineer needs to predict it adequately.

Goal (2) above is relevant in a greater number of real-life applications than one might suspect. The model formulation is secondary; it is used merely as an instrument to detect the degree of importance of each variable in explaining the variation in response. Variables that are found to explain a reasonable amount of variation in the response are perhaps kept for further study. Regressors that seemingly play a minor role are eliminated. This practice often precedes a more extensive study or model-building process. A tobacco chemist, for example, conducts a taste-testing experiment with a panel in which a taste-type rating is given to each of several tobacco formulations. The regressor variables are the ingredient concentrations of the many additives put into cigarette tobacco. A model is fit to the test data with the sole purpose of determining, from a fitted regression model, which ingredients influence taste. Those that appear to play a role are retained as experimental variables for future studies.

Model specification explains itself. The analyst must take a great deal of care in postulating the model. Any analyst will acknowledge, either directly or obliquely, that various candidate models are often in competition, in different functional forms. Each functional form defines a different role for the regressor. When the model is linear, this type of exercise can be frustrating unless the complexion of each regressor variable is well defined in the data.

Parameter estimation is often the sole purpose of conducting a regression analysis in certain scientific fields. In Chapter 8, a data set is shown in which an agriculture production function is fit to a set of input regressors that represent expenditures. Six expenditure type regressors and a rainfall variable are used. The sampling unit is a year, and the data have been collected for the State of Virginia for each of 25 years. These 25 data points are fit to a linear model, and prediction and variable screening are totally unimportant. However, specific ranges of the regression coefficients support (or refute) a particular economic theory. The *signs and magnitudes* of the coefficients are crucial.

1.3

THE DATA BASE

Much used (and perhaps overused) clichés in data analysis "Garbage In—Garbage Out" and "The results are only as good as the data that produced them" apply in the building of regression models. If the data do not reflect a trend involving the variables, there will be no success in model development or in drawing inferences regarding the system. Even if some type of association does exist, this does not imply that the data will reveal it in a clearly detectable fashion. Data may be produced from a designed experiment, a well-developed survey, a collection and tabulation over time, a computer simulation, or from one of many other sources. It should be clear that sample size is important. When the sample size is too small, the analyst cannot compute adequate measures of *error* in the regression results, and there can be no basis for checking model assumptions. However, the size of the data set is far from the only consideration. Many of the difficulties with data sets are obvious. For example, we cannot develop a model to produce a broad-based relationship if the sample of experimental units does not represent the population we are attempting to model. Broad generalizations regarding any of the goals of regression analysis cannot be made if the data are too specific. At times, even the ranges of the regressor variables in the data are such that the model built and conclusions drawn are *data specific*. For the oxygen consumption example, suppose all the individuals used in the experiment were well-conditioned athletes. Both the ranges of the variables and other characteristics of the inferences made would likely apply only to that population.

In many situations, difficulties with the regression analysis are a result of a failure of one or more assumptions. In particular, the multiple linear regression model of Eq. (1.1) is analyzed under the assumption that the regressor variables are measured without error. If excessive measurement error in the regressors exists, estimates of the regression coefficients can be severely affected and other inferences such as prediction, variable screening, etc. can be clouded with uncertainty. See Seber (1977).

Perhaps the most serious limitation in a regression data set is the failure to collect data on all potentially important regressors. This inadequacy may arise because the analyst isn't aware of all of the relevant regressors. Even if all or most of the quantities are identified, limitations in the data gathering process may prevent them from being measured. When this occurs, the model may be badly underspecified, resulting in poor estimates of the regression coefficients and poor predictions.

REFERENCES FOR CHAPTER 1

1. Galton, Sir Francis. 1885. Regression towards mediocrity in heredity stature. *Journal of Anthropological Institute* 15: 246–263.
2. SAS Institute Inc. 1985. *SAS User's Guide: Statistics, Version 5*. Cary, North Carolina: SAS Institute Inc. p. 68.
3. Seber, G.A.F. 1977. *Linear Regression Analysis*. New York: John Wiley & Sons.

CHAPTER 2

THE SIMPLE LINEAR REGRESSION MODEL

2.1 THE MODEL DESCRIPTION

The model that is applicable in the simplest regression structure is the simple linear regression model. Here the term *simple* implies a single regressor variable, x, and the term *linear* implies linear in x. We begin with the description

$$y = \beta_0 + \beta_1 x + \varepsilon \tag{2.1}$$

where y is the measured response variable, β_0 and β_1 are the intercept and slope respectively, and ε is the model error. We ordinarily view (2.1) in a data setting where pairs of observations $(x_1, y_1), (x_2, y_2), \ldots, (x_n, y_n)$ are taken on experimental units, and estimates of the parameters β_0 and β_1 are sought. We then write the model

$$\boxed{y_i = \beta_0 + \beta_1 x_i + \varepsilon_i \qquad (i = 1, 2, \ldots, n)} \tag{2.2}$$

In Section 2.3, the estimation of β_0 and β_1 via the method of least squares is discussed, and we present the conditions under which least squares estimators are ideal.

2.2

ASSUMPTIONS AND INTERPRETATION OF MODEL PARAMETERS

The purpose of a model formulation in regression analysis is to allow the analyst to conceptualize how the observations are generated. This formulation of statistical theory will then allow for the study of properties of estimators of the parameters. The assumptions underlying the least squares procedure are important and should be noted by the reader. First, let us assume that the x_i are nonrandom, observed with negligible error, while the ε_i are random variables with mean zero and constant variance σ^2 (homogeneous variance assumption). Hereafter we shall use the expectation operator to denote a *population mean*. Thus we can state that $E(\varepsilon_i) = 0$ and $E(\varepsilon_i^2) = \sigma^2$. In addition, we assume that the ε_i are uncorrelated from observation to observation. Of course, in most practical situations, the x_i do experience some random variation. We assume here that any random variation in x is negligible compared to the range in which it is measured. In addition, any error in the measurement of the x_i is assumed to be small compared to the range. The case where x and y are both random variables will be treated in a later section. In addition, the effect of measurement error in the x_i will be discussed in Chapter 7.

The simple linear regression model of Eq. (2.2) describes a situation in which, at a specific value of x, say x_i, the mean of the distribution of the y_i is given by

$$E(y_i) = \beta_0 + \beta_1 x_i$$

and the variance of the distribution is σ^2, independent of the level of x in question. Thus there is an underlying linear relationship implied by (2.2), which relates the mean response to x. For this reason, we may occasionally use the notation

$$E(y \mid x) = \beta_0 + \beta_1 x$$

with $E(y \mid x)$ referring to the *mean y conditional on a specific value of x.*

ADDITIONAL DESCRIPTION OF MODEL AND MODEL PARAMETERS

An important task of the data analyst is to estimate the parameters β_0, the *intercept* in the regression model, and β_1, the *slope* of the regression line. These parameters are called *regression coefficients*, and their estimates play an important role in the total analytical procedure. The parameter β_1 is the change in $E(y)$ per unit increase in x. As a result, the values of β_0 and β_1 (as well as values of the estimates) depend on the units used on y and x. Thus the analyst must take this into account in any interpretation given to the estimates.

Often the experimenter may become confused about the interpretation of β_0 or its estimate from experimental data. As we indicated earlier in this section, the linear model is the simplest empirical device for explaining how the data were produced. Often the characteristics of the model that are computed by the data *are very much dependent on the range of* x in which the data were taken. In other words, there is a presumption that the model given by Eq. (2.2) holds only in a confined region of x. If this region covers $x = 0$, then the estimate of β_0 can certainly be interpreted as the mean y at $x = 0$. But if the data coverage is far away from the origin, then β_0 is merely a regression term that supplies little in the way of interpretation. Often, in fact, the estimate of β_0 turns out to be a value that seems quite unreasonable, or even impossible in the context of the problem. The analyst must bear in mind that interpretation of β_0 is tantamount to extrapolating the model outside the range in which it was intended to be used.

Consider Figure 2.1. Suppose the data set represents measurements of the yield of a chemical reaction (y) at different levels of temperature (x). The measurements are indicated by the 15 points. Obviously in the range of temperature covered by the data, a model linear in temperature is quite reasonable. Suppose the dotted curve represents the *true expected value* of yield which obviously has a curvilinear relationship. An "extension" of the straight line portion (dashed line) back to an $x = 0$ temperature produces a y of about 17% yield. This, of course, would be a reasonable value of an estimate of β_0 for these data. However, its interpretation

FIGURE 2.1 **Reaction yield and temperature data**

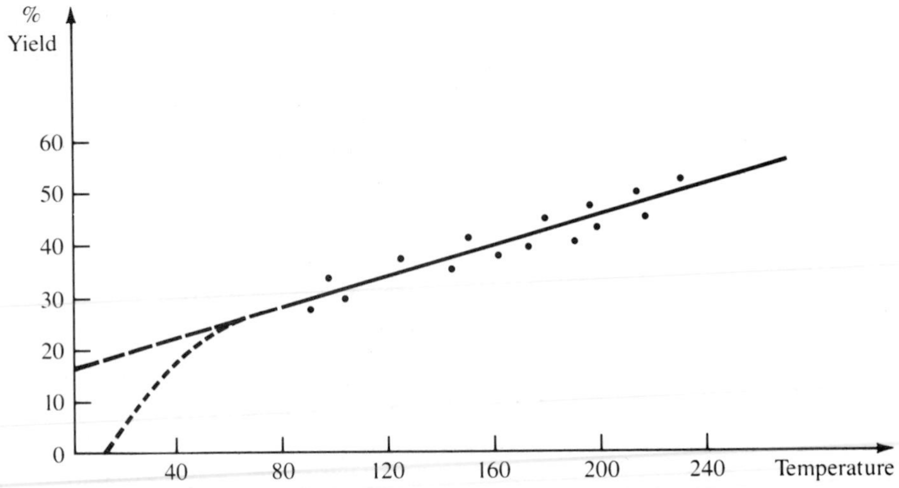

makes no scientific sense in this case because even if the scientist is not aware of the true curvilinear relationship involved, he or she may know that at a temperature below approximately 20°C no reaction will take place. Thus a 0°C projection of yield need not be made. Thus the role of the intercept becomes merely part of the mechanism that describes what transpires between 80° and 220° in temperature.

NORMAL THEORY ASSUMPTION

Clearly we should place focus on underlying assumptions that apply in any statistical analysis procedure. In order to totally understand and appreciate the regression assumptions, one must concentrate on what property or properties of the estimators are dependent on which assumptions. For example, the bias, variance, and covariance of the estimators will be discussed subsequently and each of these properties depend on one or more of the assumptions described here. In a later section, hypothesis testing will be discussed, and these inferential procedures depend on an additional assumption, namely that in the model, the ε_i are assumed to follow the Gaussian distribution. This assumption is referred to as the *normal theory assumption.*

ALTERNATIVE MODEL FORMULATIONS; THE CENTERED MODEL

The simple linear regression model given by Eq. (2.2) can be written in other forms that are used either for ease of manipulation or, in some cases, for better parameter interpretation. The reader will observe the use of the linear regression model written various ways. At this point, there is an advantage in introducing the *centered* simple linear model.

For the simple linear regression model, let $\bar{x} = \sum_{i=1}^{n} x_i/n$, i.e., the average value of the regressor variable. By centering we mean that x_i enters the model centered; i.e.,

$$y_i = \beta_0^* + \beta_1(x_i - \bar{x}) + \varepsilon_i \qquad (i = 1, 2, \ldots, n) \tag{2.3}$$

The model in (2.3) implies that the x_i have been shifted to an origin of zero. This requires a change of intercept since β_0^* is the $E(y)$ when the centered regressor, $(x_i - \bar{x})$, takes on a zero value. As a result,

$$\beta_0^* = \beta_0 + \beta_1\bar{x}$$

Then it is easy to determine that the centered and uncentered models are equivalent. In Section 2.3 the method of least squares will be used to develop the estimators for β_0 and β_1. The use of the centered model simplifies this development. Throughout the text we will use the models in (2.2) and (2.3) interchangeably. The notion of *centered regressors* will surface in a somewhat different context when we discuss collinearity diagnostics in Chapter 8.

2.3

LEAST SQUARES FORMULATION

The method of least squares is used more extensively than any other estimation procedure for building regression models. Prior to the 1970s it was employed almost exclusively. In Chapters 7 and 8, we will discuss important alternatives to least squares. The method of least squares is designed to provide estimators b_0 and b_1 of β_0 and β_1, respectively, and the *fitted value*

$$\hat{y}_i = b_0 + b_1 x_i$$

of the response so that the *residual sum of squares* $\sum_{i=1}^{n} (y_i - \hat{y}_i)^2$ is minimized. As a result, b_0 and b_1 must satisfy

$$\frac{\partial}{\partial b_0}\left[\sum_{i=1}^{n} (y_i - b_0 - b_1 x_i)^2\right] = 0 \tag{2.4}$$

and

$$\frac{\partial}{\partial b_1}\left[\sum_{i=1}^{n} (y_i - b_0 - b_1 x_i)^2\right] = 0 \tag{2.5}$$

The minimization of $\sum_{i=1}^{n} (y_i - \hat{y}_i)^2$ has good rationale, and in most instances the values of b_0 and b_1 possess good properties. The motivation is that the data points will be "close" to the *fitted regression line*

$$\hat{y} = b_0 + b_1 x$$

if the errors in fit, i.e., *the residuals*, are rendered small. Figure 2.2 illustrates the procedure, with the vertical deviations representing the residuals.

The development of the least squares estimators is easily accomplished if one uses the centered model given by Eq. (2.3). That is, we need to consider

$$\frac{\partial}{\partial b_0^*}\left\{\sum_{i=1}^{n} [y_i - b_0^* - b_1(x_i - \bar{x})]^2\right\}$$

and

$$\frac{\partial}{\partial b_1}\left\{\sum_{i=1}^{n} [y_i - b_0^* - b_1(x_i - \bar{x})]^2\right\}$$

Applying the derivatives, we obtain the pair of equations,

$$nb_0 + b_1 \sum_{i=1}^{n} (x_i - \bar{x}) = \sum_{i=1}^{n} y_i$$

$$b_0 \sum_{i=1}^{n} (x_i - \bar{x}) + b_1 \sum_{i=1}^{n} (x_i - \bar{x})^2 = \sum_{i=1}^{n} y_i(x_i - \bar{x})$$

FIGURE 2.2 **Illustration of least squares residuals**

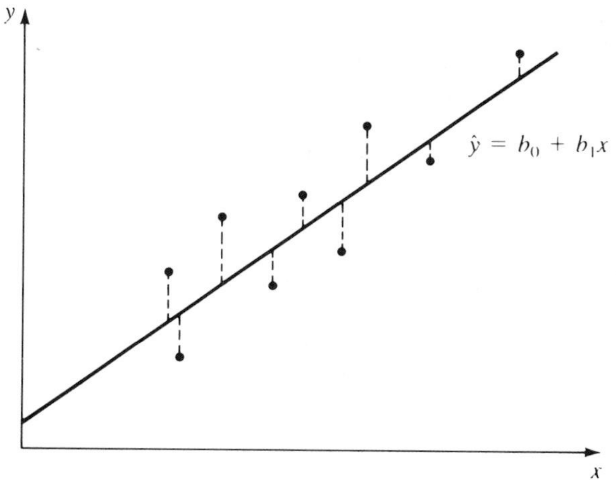

Since $\sum_{i=1}^{n} (x_i - \bar{x}) = 0$, the solutions to the *least squares normal equations* are given by the estimators

$$\boxed{\begin{aligned} b_1 &= S_{xy}/S_{xx} \\ b_0^* &= \bar{y} \end{aligned}} \tag{2.6}$$

where $S_{xy} = \sum_{i=1}^{n} (x_i - \bar{x}) y_i$ and $S_{xx} = \sum_{i=1}^{n} (x_i - \bar{x})^2$. Now, in order to obtain the estimator for β_0, the intercept in the uncentered model, we note

$$\beta_0 = \beta_0^* - \beta_1 \bar{x}$$

and hence

$$b_0 = \bar{y} - b_1 \bar{x}$$

Thus our estimators for intercept and slope are

$$\boxed{b_0 = \bar{y} - b_1 \bar{x}} \tag{2.7}$$

$$\boxed{b_1 = S_{xy}/S_{xx}} \tag{2.8}$$

In addition to the slope and intercept estimates of (2.7) and (2.8), obvious products of the least squares methodology are the residuals themselves, i.e., the $y_i - \hat{y}_i$, and the fitted least squares line given by

$$\hat{y} = b_0 + b_1 x$$

In many applications the conclusions that are drawn center directly on the estimated slope and intercept. As a result, the student of regression analysis must be aware of the statistical properties of the estimators.

PROPERTIES OF THE ESTIMATORS

It is not difficult to show that, under the assumption that the x_i are nonrandom and $E(\varepsilon_i) = 0$, the estimators are unbiased. The expectation of b_1 is given by

$$
\begin{aligned}
E\left(\frac{S_{xy}}{S_{xx}}\right) &= \sum_{i=1}^{n} \frac{(x_i - \bar{x})(E(y_i))}{S_{xx}} \\
&= \frac{1}{S_{xx}} \sum_{i=1}^{n} (x_i - \bar{x})(\beta_0 + \beta_1 x_i) \\
&= \frac{\beta_1 S_{xx}}{S_{xx}} \\
&= \beta_1
\end{aligned}
$$

For the intercept, we have

$$
\begin{aligned}
E(b_0) &= E[\bar{y} - b_1 \bar{x}] \\
&= \frac{1}{n} E\left(\sum_{i=1}^{n} y_i\right) - \beta_1 \bar{x} \\
&= \frac{1}{n} \sum_{i=1}^{n} [\beta_0 + \beta_1 x_i] - \beta_1 \bar{x} \\
&= \beta_0
\end{aligned}
$$

For the variance properties of the least squares estimators, one should first note that the homogeneous variance assumption allows that, at $x = x_i$, $\text{Var}(y_i) = \text{Var}(\varepsilon_i) = \sigma^2$, emphasizing that the model error variance is constant for fixed values of the regressor variable. Thus we can take the variance of b_1 in (2.7) and obtain

$$
\text{Var}(b_1) = \frac{1}{S_{xx}^2} \sum_{i=1}^{n} \sigma^2 (x_i - \bar{x})^2 = \frac{\sigma^2}{S_{xx}} \tag{2.9}
$$

For the intercept it is convenient to show that $\bar{y} = b_0^*$ and b_1 are uncorrelated. To show this we first write

$$
\bar{y} = \beta_0^* + \bar{\varepsilon}
$$

where $\bar{\varepsilon} = 1/n \sum_{i=1}^{n} \varepsilon_i$. Then

$$
\begin{aligned}
b_1 &= \sum_{i=1}^{n} \frac{[x_i - \bar{x}][\beta_0 + \beta_1(x_i - \bar{x}) + \varepsilon_i]}{S_{xx}} \\
&= \beta_1 + \frac{\sum_{i=1}^{n} (x_i - \bar{x})\varepsilon_i}{S_{xx}}
\end{aligned}
$$

So

$$\text{Cov}(\bar{y}, b_1) = \text{Cov}\left[\beta_0^* + \bar{\varepsilon}, \quad \beta_1 + \frac{\sum_{i=1}^{n}(x_i - \bar{x})\varepsilon_i}{S_{xx}}\right]$$

Now, if we make use of the assumption that the ε_i are uncorrelated with mean zero and variance σ^2,

$$\begin{aligned}\text{Cov}(\bar{y}, b_1) &= E[\beta_0^* + \bar{\varepsilon}]\left[\beta_1 + \frac{\sum_{i=1}^{n}(x_i - \bar{x})\varepsilon_i}{S_{xx}}\right] - \beta_0^*\beta_1 \\ &= \beta_0^*\beta_1 + \frac{\sum_{i=1}^{n}(x_i - \bar{x})\sigma^2}{nS_{xx}} - \beta_0^*\beta_1 \\ &= 0\end{aligned}$$

We can then write

$$\begin{aligned}\text{Var}(b_0) &= \text{Var}\left(\frac{\sum_{i=1}^{n} y_i}{n}\right) + \bar{x}^2\,\text{Var}(b_1) \\ &= \sigma^2\left(\frac{1}{n} + \frac{\bar{x}^2}{S_{xx}}\right) \qquad \text{(2.10)}\end{aligned}$$

It should be clear that, in the development leading to the variances in Eq. (2.9) and (2.10), the homogeneous variance assumption was used as well as the assumption that the ε_i are uncorrelated. In addition, the results depend on the condition that the x_i are nonrandom. In a later section where tests of hypotheses are considered, *estimated standard errors* of both slope and intercept will be required, at which point (2.9) and (2.10) become extremely useful.

The unbiasedness and variance properties of the least squares estimators can be developed by using a different approach. An alternative and insightful expression for b_1 is given by

$$b_1 = \beta_1 + \sum_{i=1}^{n} d_i\varepsilon_i \qquad \text{where} \quad d_i = \frac{x_i - \bar{x}}{S_{xx}}$$

(See Exercise 2.25.) One can then simply invoke $E(\varepsilon_i) = 0$ and show unbiasedness of b_1. The variance of b_1 in Eq. (2.9) follows quite easily. For the intercept, an interesting expression is given by

$$b_0 = \beta_0 + \bar{\varepsilon} - \bar{x}\sum_{i=1}^{n} d_i\varepsilon_i$$

(See Exercise 2.25.) Again, unbiasedness and the variance in Eq. (2.10) follow.

Example 2.1 ***Accounting Return Data***

There is a controversy in the accounting literature concerning the relationship between accounting rates on stocks and market returns. Theory might suggest a positive, perhaps linear relationship between the two variables. Fifty-four companies were chosen with the regressor variable (x) as the mean yearly accounting rate for the period 1959 to 1974. The dependent variable (y) is the corresponding mean market rate. The data are given in Table 2.1.

Accounting rates generally represent input into investment, and thus accounting rate is a natural explanatory or regressor variable. The model of Eq. (2.2) was fit to the data with parameter estimates given by (Eq. (2.7) and (2.8))

$$b_0 = 0.84801$$
$$b_1 = 0.61033$$

Figure 2.3 shows a plot of the data along with the least squares regression line.

TABLE 2.1 **Accounting rates and market rates from 1959 to 1974**

Company	Market Rate	Accounting Rate
McDonnell Douglas	17.73	17.96
NCR	4.54	8.11
Honeywell	3.96	12.46
TRW	8.12	14.70
Raytheon	6.78	11.90
W.R. Grace	9.69	9.67
Ford Motors	12.37	13.35
Textron	15.88	16.11
Lockheed Aircraft	−1.34	6.78
Getty Oil	18.09	9.41
Atlantic Richfield	17.17	8.96
Radio Corporation of America	6.78	14.17
Westinghouse Electric	4.74	9.12
Johnson and Johnson	23.02	14.23
Champion International	7.68	10.43
R.J. Reynolds	14.32	19.74
General Dynamics	−1.63	6.42
Colgate-Palmolive	16.51	12.16
Coca-Cola	17.53	23.19
International Business Machines	12.69	19.20
Allied Chemical	4.66	10.76
Uniroyal	3.67	8.49
Greyhound	10.49	17.70
Cities Service	10.00	9.10

TABLE 2.1
Continued

Accounting rates and market rates from 1959 to 1974

Company	Market Rate	Accounting Rate
Philip Morris	21.90	17.47
General Motors	5.86	18.45
Phillips Petroleum	10.81	10.06
FMC	5.71	13.30
Caterpillar Tractor	13.38	17.66
Georgia Pacific	13.43	14.59
Minnesota Mining & Manufacturing	10.00	20.94
Standard Oil (Ohio)	16.66	9.62
American Brands	9.40	16.32
Aluminum Company of America	.24	8.19
General Electric	4.37	15.74
General Tire	3.11	12.02
Borden	6.63	11.44
American Home Products	14.73	32.58
Standard Oil (California)	6.15	11.89
International Paper	5.96	10.06
National Steel	6.30	9.60
Republic Steel	.68	7.41
Warner Lambert	12.22	19.88
U.S. Steel	.90	6.97
Bethlehem Steel	2.35	7.90
Armco Steel	5.03	9.34
Texaco	6.13	15.40
Shell Oil	6.58	11.95
Standard Oil (Indiana)	14.26	9.56
Owens Illinois	2.60	10.05
Gulf Oil	4.97	12.11
Tenneco	6.65	11.53
Inland Steel	4.25	9.92
Kraft	7.30	12.27

Benzion Barlev and Haim Levy, "On the Variability of Accounting Income Numbers," *Journal of Accounting Research* (Autumn, 1979): 305–315.

COMPARISON TO OTHER FORMS OF ESTIMATORS

We have chosen here to develop the mean (expectation) and variance of the least squares estimators for the specific case of the simple linear regression model of Eq. (2.2). No development is given regarding properties of the least squares estimators for more general model forms. In addition, no attempt is made in this chapter to present any material regarding how the least squares estimators compare to other forms of estimation. Of particular interest is the comparison of the variances of the least squares coefficients to variances of the coefficients for, say, an alternative unbiased form of estimation. The least squares estimators for a wide class of models possess certain optimal properties under the assumption stated here regarding the model errors. These optimal properties will be discussed in Chapter 3 in conjunction with the more general linear regression model. We postpone this presentation in order to ensure that the reader understands the general nature of the models to which the properties apply.

FIGURE 2.3 **Plot of data and regression line; accounting and marketing return data of Table 2.1**

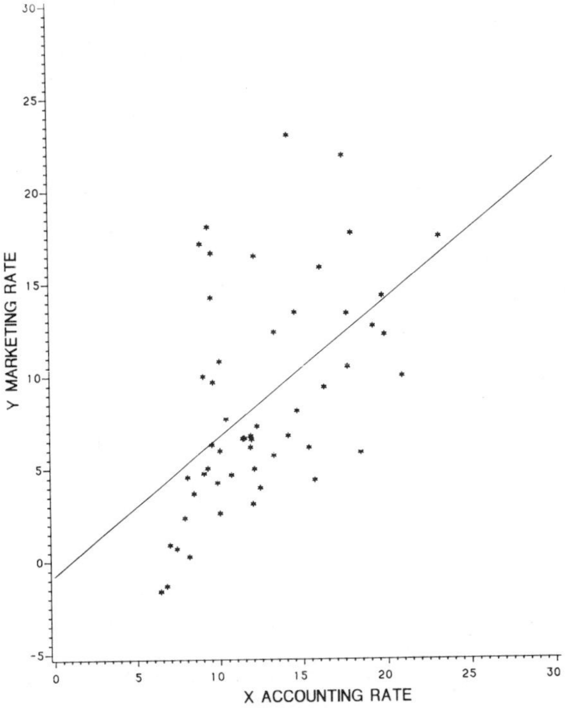

ESTIMATION OF ERROR VARIANCE (RESIDUAL DEGREES OF FREEDOM)

In practical situations, an estimate of the error variance, σ^2, is required. The estimate is used in the calculation of estimated standard errors of coefficients for hypothesis testing and, in many instances, plays a major role in assessing quality of fit and prediction capabilities of the regression model $\hat{y} = b_0 + b_1 x$. (See Chapter 4). It is not difficult to visualize intuitively what the estimator is. The residuals,

$y_i - \hat{y}_i$, which are the observed *errors of fit* are clearly the *empirical counterparts* to the ε_i, the model errors which are not observed. Thus it is reasonable that the sample variance of the residuals should provide an estimator of σ^2. If we divide the residual sum of squares, $\sum_{i=1}^{n} (y_i - \hat{y}_i)^2$ by the appropriate denominator, $n-2$, an unbiased estimator is produced. (See Exercise 2.3.) Thus we define

$$s^2 = \frac{\sum_{i=1}^{n} (y_i - \hat{y}_i)^2}{n-2}$$

as the estimator of σ^2.

The quantity s^2 is often called the *error mean square*. It should be emphasized that s^2 is unbiased *under the important assumption that the model is correct*. Details regarding bias in s^2 if the model is inadequate are discussed in Chapters 3 and 4. The denominator is often referred to as the *error or residual degrees of freedom*. In the text we often refer to degrees of freedom as df. The reason for the use of this term will become more apparent in the next section when we discuss the distributional properties of this statistic. The reader, at this point, can visualize residual degrees of freedom in a regression context as the number of data points, n, or *pieces of information*, reduced by the number of parameters estimated (two parameters: slope and intercept). In the accounting return data in Example 2.1, the estimate, s^2, is given by

$$s^2 = \frac{\sum_{i=1}^{n} (y_i - \hat{y}_i)^2}{n-2} = \frac{1344.949}{52} = 25.8644$$

We can now use this estimate of the model error variance in place of σ^2 in estimating variances and hence in estimating standard errors of the slope and intercept. Making use of Eq. (2.9) and (2.10), we have

Estimated standard error of $b_0 = 1.9765$

Estimated standard error of $b_1 = 0.14316$

Use of these standard errors will be illustrated in subsequent sections. (Also see Exercise 2.7.)

For students studying data analysis, the concept of degrees of freedom often remains mysterious and a source of frustration. We have used the explanation here that the residual degrees of freedom is $n-2$, since there are 2 degrees of freedom lost due to the requirement of estimation of two parameters. A similar, but perhaps more insightful, explanation emphasizes s^2 as a sample variance of the residuals. Residual degrees of freedom can always be viewed as the sample size reduced by the *number of restrictions* on the residuals. These restrictions are induced through the requirement of parameter estimation. The student of elementary statistics knows that when one is sampling i.i.d. from $N(0, \sigma)$ with random variables denoted by $y_1, y_2, \ldots, y_n$, the degrees of freedom for estimation of σ^2 is $n-1$. The loss of

1 degree of freedom is derived from the restriction $\sum_{i=1}^{n}(y_i - \bar{y}) = 0$ on the residuals, which in this case are the $y_i - \bar{y}$. Now, in the simple linear regression case, there are two restrictions on the residuals, the $y_i - \hat{y}_i$. These restrictions are given by

$$\sum_{i=1}^{n} (y_i - \hat{y}_i) = 0$$

$$\sum_{i=1}^{n} (y_i - \hat{y}_i)x_i = 0$$

and are easily verified by the use of Eq. (2.4) and (2.5). (See Exercise 2.2.) One can easily connect these restrictions to the requirement of parameter estimation since (2.4) and (2.5) emerge out of the need to estimate β_0 and β_1, respectively.

2.4 MAXIMUM LIKELIHOOD ESTIMATION

In Section 2.3 the method of least squares was discussed. Bias and variance properties were given. We pointed out how these properties make use of the assumptions that the ε_i are uncorrelated with mean zero and common variance σ^2. However, there was no assumption made concerning the *distribution type* on the ε_i. In much of what follows in this chapter, the normal theory assumption is made. This distributional assumption is made so that inference can be derived from tests of hypotheses and confidence interval estimation. Those students of statistics should recall that when one is willing to assume an underlying distribution of the basic random variable (in this case the ε_i) the method of *maximum likelihood* results in an intuitively appealing form of estimation with outstanding asymptotic properties. See, for example, Roussas (1973).

ESTIMATION OF REGRESSION COEFFICIENTS

Suppose we assume the ε_i are i.i.d. $N(0, \sigma^2)$. The joint density function, or likelihood function is given by

$$L = \prod_{i=1}^{n} f(\varepsilon_i)$$

where $f(\varepsilon_i)$ is the density function of ε_i, the ith disturbance or random error in the sample. As a result, we can write the likelihood as

$$L = \frac{1}{(2\pi)^{n/2}\sigma^n} e^{(-1/2\sigma^2)\sum_{i=1}^{n}\varepsilon_i^2}$$

$$= \frac{1}{(2\pi)^{n/2}\sigma^n} e^{(-1/2\sigma^2)\sum_{i=1}^{n}(y_i - \beta_0 - \beta_1 x_i)^2}$$

Finding estimators for β_0 and β_1 that maximize L involves minimization of the exponential part of L. That is, we seek to minimize $[1/(\sigma^2)]\sum_{i=1}^{n}(y_i - \beta_0 - \beta_1 x_i)^2$. This, of course, is identical to the *least squares* criterion, namely minimization of the residual sum of squares. As a result, if *one is willing to make the normal theory assumption, the maximum likelihood estimators of the regression coefficients β_0 and β_1 are the least squares estimators* given in Eq. (2.7) and (2.8). It should be made clear that if we cannot make the normal theory assumption, the least squares estimator is not the maximum likelihood estimator. As we indicated previously, Chapter 3, which deals with multiple regression, contains a formal discussion regarding properties of least squares estimators when the normal theory assumption is appropriate. We contrast it with the case when the analyst is unable to make the normal theory assumption.

ESTIMATION OF ERROR VARIANCE

In Section 2.3 the least squares estimators of regression coefficients were developed and the important estimator, s^2, of σ^2 was discussed. This unbiased estimator of σ^2 is that most often used by data analysts. But is it the maximum likelihood estimator in the normal theory case? Consider the likelihood function with b_0 and b_1 substituted for β_0 and β_1

$$L = \frac{1}{(2\pi)^{n/2}(\sigma^2)^{n/2}} e^{(-1/2\sigma^2)\sum_{i=1}^{n}(y_i - \hat{y}_i)^2}$$

If we differentiate the log likelihood with respect to σ^2, we obtain

$$\frac{\partial \log L}{\partial \sigma^2} = \frac{-n}{2\sigma^2} + \frac{1}{2\sigma^4}\sum_{i=1}^{n}(y_i - \hat{y}_i)^2$$

The maximum likelihood estimator $\hat{\sigma}^2$ is found by setting the derivative to zero. Thus we find

$$\hat{\sigma}^2 = \sum_{i=1}^{n}\frac{(y_i - \hat{y}_i)^2}{n}$$

As a result, the maximum likelihood estimator of σ^2 is the biased estimator which can be written as

$$\boxed{\hat{\sigma}^2 = s^2\left(\frac{n-2}{n}\right)}$$

2.5

PARTITIONING TOTAL VARIABILITY

In any regression problem, the analyst will observe variation in the response. One can view the fitted model as an "explanation" of this variation. Of course, it is important that the $\hat{y}_i$, the fitted values, follow the y_i closely. If this is accomplished, then the variation in the $\hat{y}_i$ around $\bar{y}$ will be close to the variation in the y_i around $\bar{y}$. The reader should note that (using the centered model)

$$\begin{aligned}\bar{\hat{y}} &= \sum \frac{\hat{y}_i}{n} \\ &= \frac{1}{n} \sum_{i=1}^{n} [b_0^* + b_1(x_i - \bar{x})] \\ &= \frac{1}{n} \sum_{i=1}^{n} [\bar{y} + b_1(x_i - \bar{x})] \\ &= \bar{y}\end{aligned}$$

As a result, it is natural to consider the sources of variation, the total sum of squares, $\sum_{i=1}^{n} (y_i - \bar{y})^2$ and the *regression sum of squares*, $\sum_{i=1}^{n} (\hat{y}_i - \bar{y})^2$. The former measures variation in observed response and the latter expresses variation "explained" by the regression.

Many of the procedures that produce statistical inference depend heavily on the relationship between the total and regression sums of squares. The relationship is as follows

$$\boxed{\sum_{i=1}^{n} (y_i - \bar{y})^2 = \sum_{i=1}^{n} (\hat{y}_i - \bar{y})^2 + \sum_{i=1}^{n} (y_i - \hat{y}_i)^2} \tag{2.11}$$

i.e.,

$$SS_{\text{Total}} = SS_{\text{Reg}} + SS_{\text{Res}}$$

Here of course, SS_{Res} is the familiar residual sum of squares. Equation (2.11) respresents the following conceptual identity

$$\begin{pmatrix}\text{Total variability} \\ \text{in response}\end{pmatrix} = \begin{pmatrix}\text{Variability explained} \\ \text{by model}\end{pmatrix} + \begin{pmatrix}\text{Variability} \\ \text{unexplained}\end{pmatrix}$$

With the model that includes the intercept (model (2.2)), $\sum_{i=1}^{n} (y_i - \hat{y}_i) = 0$. (See Exercise 2.2.) As a result, these two important components of variability, regression sum of squares and residual sum of squares, represent respectively, *variation due to the regression line* and *variation around the regression line.*

The reader should gain a clear intuitive feel for the components of variation that we call regression sum of squares and residual sum of squares. Clearly, a desirable situation for the analyst is to achieve a large SS_{Reg} in comparison to SS_{Res}. The former depicts y-variation that one can classify as *produced by changes in the regressor* x. But SS_{Res} can be viewed as mere *chance variation* or variation due to the ε_i in the model. The development of the important identity in Eq. (2.11) is derived from

$$y_i - \bar{y} = \hat{y}_i - \bar{y} + y_i - \hat{y}_i$$

Now if we square both sides and sum, we have

$$\sum_{i=1}^{n} (y_i - \bar{y})^2 = \sum_{i=1}^{n} (\hat{y}_i - \bar{y})^2 + \sum_{i=1}^{n} (y_i - \hat{y}_i)^2 + 2 \sum_{i=1}^{n} (\hat{y}_i - \bar{y})(y_i - \hat{y}_i)$$

The last term on the right-hand side is zero. This is easily seen from the following

$$\begin{aligned}
\sum_{i=1}^{n} (\hat{y}_i - \bar{y})(y_i - \hat{y}_i) &= \sum_{i=1}^{n} \hat{y}_i(y_i - \hat{y}_i) \\
&= \sum_{i=1}^{n} [b_0 + b_1(x_i - \bar{x})][y_i - \hat{y}_i] \\
&= b_1 \sum_{i=1}^{n} x_i(y_i - \hat{y}_i) \\
&= b_1 \sum_{i=1}^{n} x_i[(y_i - \bar{y}) - b_1(x_i - \bar{x})] \\
&= b_1[S_{xy} - b_1 S_{xx}] \\
&= 0
\end{aligned}$$

In order to build a foundation for inferences regarding the slope of the regression, we follow with a discussion of the distributional properties of the sums of squares under the normal theory assumption.

DISTRIBUTIONAL PROPERTIES OF SUMS OF SQUARES

One of the advantages of the least squares procedure is that it allows us to form simple criteria such as SS_{Reg} and SS_{Res}, whose properties are very easily studied and understood. For purposes of hypothesis testing, it is vital to know the distributional properties of these two components of the total variability. Let us again make the assumption (in addition to those discussed in Section 2.2) that the ε_i are normal. Simply stated, we say that the *ε_i are independent* $N(0, \sigma^2)$.

The reader should now easily understand the role of the normality assumption in the context of the model of Eq. (2.2) and other assumptions that we have given. A straight line connects the *means* of the y with x, and the scatter around the line is produced by a Gaussian distribution around $E(y \mid x)$. Figure 2.4 depicts the model. Note that the variances are equal for each distribution, reflecting the homogeneous variance assumption. The dots on Figure 2.4 represent data points. The line connecting the means represents the "true" regression line.

FIGURE 2.4 **Simple linear regression model**

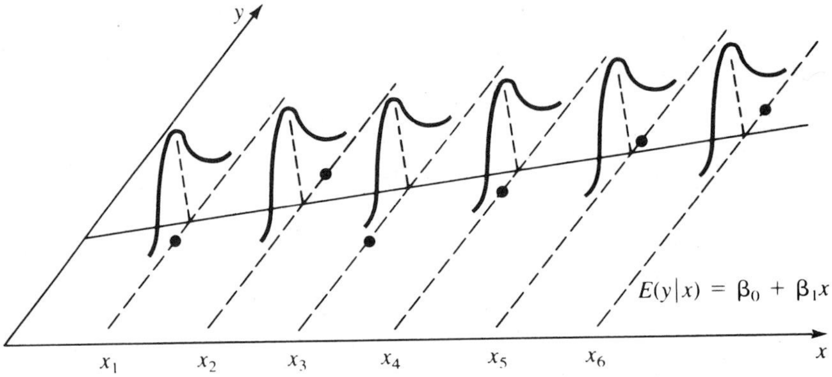

The utility of the partitioning of SS_{Total} in Eq. (2.11) for the case of simple linear regression is, quite simply put, to aid in the determination of whether the variation explained by the regression model is real, or is no more than chance variation. The term *real* here implies that there is indeed a linear trend produced by the relationship between y and x; that is, there is a regression of y on x. From the standpoint of the model of Eq. (2.2), we say that there is a regression of y on x when the regression slope $\beta_1 \neq 0$. Figure 2.5 shows two hypothetical data sets, one for which the regression of y on x appears to be statistically significant, and the other for which the variation is mere chance variation and the true regression is likely a horizontal line. In Figure 2.5(a) we observe pictorially that the SS_{Reg} explains an appreciable portion of SS_{Total}. This, of course, leads us to conclude that, despite some random variation around the fitted line, there is a positive slope in the regression, and $E(y|x)$ increases with an increase in x. On the other hand, Figure 2.5(b) shows a horizontal fitted regression line. Clearly $SS_{\text{Reg}} = 0$ and the *entire variation* in y is error variation around the fitted line $\hat{y} = \bar{y}$.

FIGURE 2.5

(a) **Data set depicting a positive regression trend**

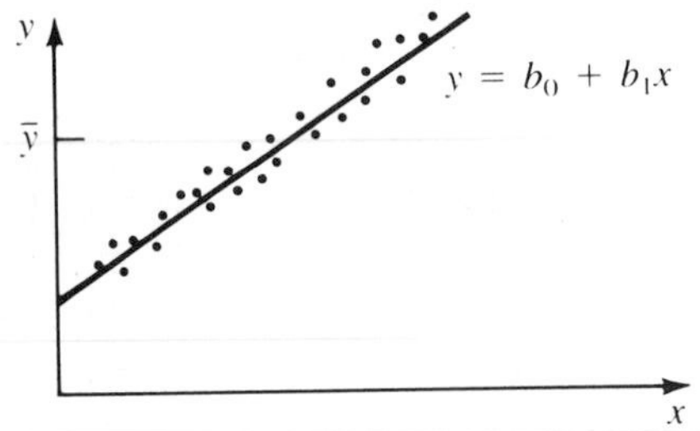

(b) **Data set depicting no significant regression**

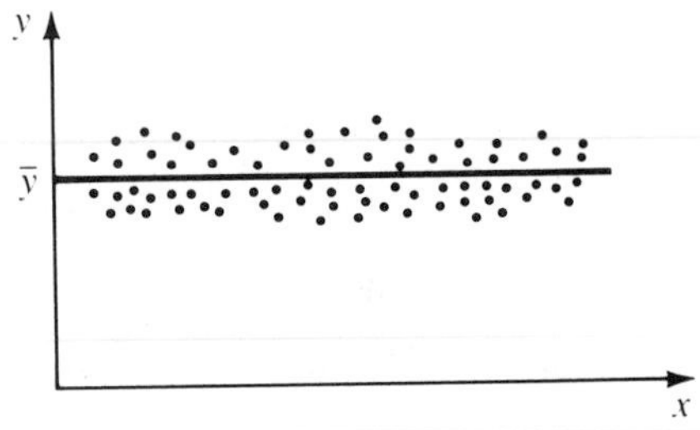

Obviously, in Figure 2.5, it is clear that a significant regression is present in one case and not in the other. But, in many cases, a statistical test of hypothesis is required to determine whether the regression is significant. It should be obvious that the distributional properties of these sums of squares will prove beneficial.

PARTITIONING OF DEGREES OF FREEDOM

Under our normal theory assumptions, SS_{Res} has a $\sigma^2\chi^2_{n-2}$ distribution. For theoretical justification, the reader should refer to Graybill (1976) and Searle (1971). Under the condition that there is no regression; i.e., when $\beta_1 = 0$, SS_{Reg}/σ^2 has χ^2_1. As a result, the notion of *residual degrees of freedom* and *regression degrees of freedom* should take on a meaning to the reader. In fact, we now can partition degrees of freedom much like we partition sums of squares

$$SS_{\text{Total}} = SS_{\text{Reg}} + SS_{\text{Res}} \qquad \text{(sums of squares)}$$
$$n - 1 = 1 + (n - 2) \qquad \text{(degrees of freedom)}$$

Also, if the reader views the fitted model with the regressor in centered form, $\hat{y} = \bar{y} + b_1(x - \bar{x})$, the partitioning of degrees of freedom is intuitively very appealing. If we view n, the number of data points, as the total number of available degrees of freedom, we have the partitioning as follows:

Regression df	1	(Degree of freedom used for estimation of slope)
Residual df	$n - 2$	(Number of data points, with two df eliminated for estimation of two parameters in the model)
Total df	$n - 1$	(Number of data points, with one eliminated for estimation of mean $\bar{y}$)

The purpose of using $n - 1$ for total evolves from the fact that the total sum of squares represents variation *around the mean*, $\bar{y}$.

As in the case of all data analysis tools involving degrees of freedom, the source of the degrees of freedom must be the χ^2 distribution. As a result, we may view the purpose of this partitioning as that of providing a foundation for the test of the hypothesis

$$H_0\colon \beta_1 = 0$$
$$H_1\colon \beta_1 \neq 0$$

In this regard we have the following additional rationale for partitioning of degrees of freedom:

SS	df	Distribution
SS_{Reg}	1	$\sigma^2\chi^2_1$ (under H_0)
SS_{Res}	$n - 2$	$\sigma^2\chi^2_{n-2}$
SS_{Total}	$n - 1$	$\sigma^2\chi^2_{n-1}$ (under H_0)

In addition SS_{Reg} and SS_{Res} are independent. For theoretical justification, the reader is again referred to Graybill (1976) or Seber (1977). These results allow for ease in developing tests of hypotheses on the slope and a better understanding of their use. Perhaps more importantly, they lay a foundation for more extensive inferences in the multiple regression model discussed in Chapter 3.

2.6

TESTS OF HYPOTHESES ON SLOPE AND INTERCEPT

At this point the reader would be advised to give serious thought to what information, in his or her own field, must be extracted from the fitted regression line. The possibilities can be enumerated as follows:

1. Does x, the regression variable, truly influence y, the response?
2. Is there an adequate fit of the data to the model?
3. Will the model adequately predict responses (either through interpolation or extrapolation)?

In the first case, success can quite often be achieved in answering the question through hypothesis testing on the slope β_1. As we have indicated, a hypothesis that is often of interest is given by

$$H_0: \beta_1 = 0$$
$$H_1: \beta_1 \neq 0 \tag{2.12}$$

Of course if H_0 is true, the implication is that the model reduces to $E(y) = \beta_0$, suggesting that, indeed, the regressor variable does not influence the response (at least not through the linear relationship implied by (2.1)). Rejection of H_0 in favor of H_1 leads one to the conclusion that x significantly influences the response in a linear fashion.

There is an ever-present danger of reading too much into the rejection of H_0 in (2.12). For this reason the results of the test do not possess the importance often attributed to them. It is entirely possible that x is found to significantly affect the response, but yet the model does not provide a successful solution to the problem that motivated the regression analysis. The rejection of H_0 merely means that a trend is detected. Nothing is implied concerning the quality of fit of the regression line with regard to one's preconceived standards. In addition, and perhaps more importantly, nothing is implied regarding the capability of the model to *predict* according to some standard set down by the scientist.

As one might expect, the test of H_0 in (2.12) makes use of the distributional properties of SS_{Reg} and SS_{Res}, and, as one would expect, the conclusion that is drawn depends on the relative magnitudes. The details of the test appear in the following material.

THE ANALYSIS OF VARIANCE

The normal theory distributional properties to which we alluded in Section 2.5 allow for considerable ease in testing the hypothesis of Eq. (2.12). A simple F-test, produced through the computation outlined in the *analysis of variance* in Table 2.2 can be used. Since the statistic

$$\frac{SS_{\text{Reg}}/1}{SS_{\text{Res}}/(n-2)} = \frac{MS_{\text{Reg}}}{s^2}$$

is a $(\chi_1^2/1)/[\chi_{n-2}^2/(n-2)]$ under H_0, then MS_{Reg}/s^2 follows $F_{1,n-2}$ under H_0 and is thus a candidate for a test statistic for the hypothesis in (2.12).

Table 2.2 represents a standard analysis of variance table for simple linear regression. The "bottom line" information, of course, comes from the F-statistic. Some insight can be gained by observing that the F-value is a ratio of mean squares (recall that s^2 is referred to as the *error mean square*). Here, the F-statistic may be viewed as a ratio that expresses *variance explained by the model* divided by *variance due to model error* or experimental error. As a result, large values of $F = MS_{\text{Reg}}/s^2$ are evidence in favor of H_1. An alternate view can be explored by observing the expected values of the two mean squares. We know that $E(s^2) = \sigma^2$, and it turns out that (see Exercise 2.5)

$$E(MS_{\text{Reg}}) = \sigma^2 + \beta_1^2 S_{xx}$$

As a result, detection of a significant slope through the analysis of variance is essentially a detection of a statistically significant value of $\beta_1^2 S_{xx}$ *over and above mere experimental error variance.* The positive nature of $\beta_1^2 S_{xx}$ provides the user reinforcement in the use of the upper tail, one-tailed F-test.

TABLE 2.2 **Analysis of variance**

Source	Sum of Squares (SS)	df	Mean Square (MS)	F
Regression	SS_{Reg}	1	$SS_{\text{Reg}}/1$	$F = \frac{MS_{\text{Reg}}}{s^2}$
Residual	SS_{Res}	$n-2$	s^2	
Total	SS_{Total}	$n-1$		

Example 2.2 ***Wood Density Data***

In the manufacture of commercial wood products, it becomes important to estimate the relationship between the density of a wood product and its stiffness. A relatively new type of particleboard is being considered, which can be formed with considerably more ease than the accepted commercial product. It is necessary to know at what density the stiffness of the product compares to the well-known, well-documented commercial product. Thirty particleboards were produced at densities ranging from roughly 8 to 26 pounds per cubic foot, and the stiffness was measured in pounds per square inch. Table 2.3 shows the data.

The fitted *regression line* relating the two mechanical properties of the particleboard was computed using Eq. (2.7) and (2.8) and is given by

$$\hat{y} = -25{,}433.739 + 3{,}884.976x$$

The residual estimate of variance is calculated as follows:

$$s^2 = \frac{3.7822 \times 10^9}{28} = 1.35081 \times 10^8 \frac{\text{lb}^2}{(\text{in}^2)^2}$$

We use the analysis of variance as a mechanism for gaining information in favor of the hypothesis that there is a linear relationship between density and stiffness in the new material. That is, we test

$$H_0\colon \beta_1 = 0$$
$$H_1\colon \beta_1 \neq 0$$

The analysis of variance table appears in Table 2.4. From Table 2.4, $F = MS_{\text{Reg}}/s^2 = 110.243$, which is significant at a level less than 0.0001. Thus H_0 is rejected in favor of H_1, with the conclusion that there is, indeed, a *statistically significant linear trend between density and stiffness.*

TABLE 2.3 **Density and stiffness for 30 particleboards**

Density (x) lb/ft^3	Stiffness (y) lb/in.^{2}
9.50	14814.00
8.40	17502.00
9.80	14007.00
11.00	19443.00
8.30	7573.00
9.90	14191.00
8.60	9714.00
6.40	8076.00
7.00	5304.00
8.20	10728.00
17.40	43243.00
15.00	25319.00
15.20	28028.00
16.40	41792.00
16.70	49499.00
15.40	25312.00
15.00	26222.00
14.50	22148.00
14.80	26751.00

TABLE 2.3 Continued

Density and stiffness for 30 particleboards

Density (x) lb/ft^3	Stiffness (y) lb/in.^{2}
13.60	18036.00
25.60	96305.00
23.40	104170.00
24.40	72594.00
23.30	49512.00
19.50	32207.00
21.20	48218.00
22.80	70453.00
21.70	47661.00
19.80	38138.00
21.30	53045.00

Terrance E. Conners, "Investigation of Certain Mechanical Properties of a Wood-Foam Composite" (M.S. Thesis, Department of Forestry and Wildlife Management, University of Massachusetts, Amherst, Massachusetts, 1979).

TABLE 2.4

Analysis of variance for wood density data of Table 2.3

Source	SS	df	MS	F
Regression	1.48917×10^{10}	1	1.48917×10^{10}	110.243
Residual	3.78227×10^{9}	28	1.35081×10^{8}	
Total	1.86740×10^{10}	29		

THE t-TESTS AND CONFIDENCE INTERVAL ESTIMATION

Separate tests of hypotheses on both slope and intercept can be accomplished using t-tests. In addition to the two-sided hypothesis of Eq. (2.12), a test involving a *one-sided* hypothesis can be tested with a general formulation that will be provided in this section. We observed earlier that in order to use hypothesis testing we must be willing to make the *normal theory assumption.* Under this condition, since the slope is a linear function of the y_i, the latter being normal, we can write

$$b_1 \sim N\left(\beta_1, \frac{\sigma^2}{S_{xx}}\right)$$

Some very straightforward applications of standard relationships between distributions will allow us to write (see Graybill (1976))

$$\frac{(b_1 - \beta_1)}{s}\sqrt{S_{xx}} \sim t_{n-2}$$

where t_{n-2} is student's t-distribution with $n-2$ degrees of freedom. As a result, if one is interested in testing

$$H_0: \beta_1 = \beta_{1,0}$$
$$H_1: \beta_1 \neq \beta_{1,0}$$

where $\beta_{1,0}$ is a specified constant, the test is a *two-tailed t-test*, making use of the statistic

$$t = \frac{(b_1 - \beta_{1,0})}{s}\sqrt{S_{xx}} \tag{2.13}$$

For the interesting special case in which $\beta_{1,0} = 0$,

$$t = \frac{b_1\sqrt{S_{xx}}}{s} \tag{2.14}$$

and thus

$$t^2 = \frac{b_1^2 S_{xx}}{s^2} = \sum_{i=1}^{n} \frac{(\hat{y}_i - \bar{y})^2}{s^2} = \frac{SS_{\text{Reg}}}{s^2}$$

which is our F-statistic developed through the analysis of variance approach.

If one is interested in a hypothesis involving a one-sided alternative, for example,

$$H_0: \beta_1 = \beta_{1,0}$$
$$H_1: \beta_1 > \beta_{1,0}$$

the t-statistic given by (2.13) still applies with the use of the upper tail critical region of the t_{n-2} distribution. Of course, if the hypothesis is written

$$H_0: \beta_1 = \beta_{1,0}$$
$$H_1: \beta_1 < \beta_{1,0}$$

the lower tail critical region is used.

For the intercept, we can, again, make use of the fact that b_0 is a linear combination of normal random variables and thus observe that

$$b_0 \sim N\left[\beta_0, \sigma^2\left(\frac{1}{n} + \frac{\bar{x}^2}{S_{xx}}\right)\right]$$

As a result, if one is interested in testing

$$H_0: \beta_0 = \beta_{0,0}$$
$$H_1: \beta_0 \neq \beta_{0,0}$$

the appropriate test statistic is given by

$$t = \frac{b_0 - \beta_{0,0}}{s\sqrt{(1/n) + (\bar{x}^2/S_{xx})}}$$

which follows t_{n-2} under $H_0\colon \beta_0 = \beta_{0,0}$. As in the case of the slope, if the alternative hypothesis were one-sided, the corresponding one-tailed t-test is appropriate.

Two items are worth emphasis at this point. Notice that for testing the hypothesis of a zero slope, the square of the t-statistic of Eq. (2.14) gives the F-statistic produced in the analysis of variance approach. This implies that the two methods arrive at the same conclusion when the test is a two-tailed test. Also notice that the t-statistic in (2.14) is of the form

$$t = \frac{\text{coefficient}}{\text{standard error of coefficient}}$$

where the coefficient is, of course, b_1 and the standard error is $s/\sqrt{S_{xx}}$. The latter is, of course, the estimate of the standard deviation of b_1. Example 2.3 provides an illustration of the use of the t-test.

Example 2.3 ***Wood Density Data***

Consider again the data of Table 2.3. Suppose we wish to use the t-statistic given by Equation (2.14) to test the hypothesis

$$H_0\colon \beta_1 = 0$$
$$H_1\colon \beta_1 \neq 0$$

The slope of the least squares line of stiffness on density is 3,884.976 with a standard error ($s/\sqrt{S_{xx}}$) of 370.009. Thus the t-statistic with 28 degrees of freedom is given by

$$t = \frac{3884.976}{370.009} = 10.500$$

which is, of course, statistically significant at a level below 0.0001. Note that this t-value is the square root of $F = 110.243$, the latter obtained in the analysis of variance for the same data. The t-test of the hypothesis in (2.12) is equivalent to the analysis of variance procedure outlined previously in this section.

There are often circumstances under which the data analyst may want to establish a confidence interval estimate on either the slope, β_1, or the intercept, β_0. The machinery for doing so has been established. *Under normal theory assumptions* on the ε_i,

$$\frac{(b_1 - \beta_1)\sqrt{S_{xx}}}{s} \sim t_{n-2}$$

As a result, we can write the inequality

$$\Pr\left\{-t_{\alpha/2,n-2} < \frac{(b_1-\beta_1)\sqrt{S_{xx}}}{s} < t_{\alpha/2,n-2}\right\} = 1-\alpha$$

where $t_{\alpha/2,n-2}$ is the upper $\alpha/2$ percent point of the t-distribution with $n-2$ degrees of freedom. We can now say with probability $1-\alpha$ that β_1 is covered by the interval described by

$$b_1 \pm t_{\alpha/2,n-2}\sqrt{\frac{s^2}{S_{xx}}}$$

In a similar fashion, the sampling distribution of the statistic

$$\frac{b_0-\beta_0}{s\sqrt{\frac{1}{n}+\frac{\bar{x}^2}{S_{xx}}}}$$

allows us to write a $100(1-\alpha)\%$ confidence interval on β_0 as

$$b_0 \pm t_{\alpha/2,n-2}\,s\sqrt{\frac{1}{n}+\frac{\bar{x}^2}{S_{xx}}}$$

As one often experiences with confidence interval estimation, the confidence interval is of the form

Point estimate $\pm$ (Constant)(Standard error of point estimate)

The following example illustrates the computation of the confidence intervals for the case of the wood density data.

Example 2.4 ***Wood Density Data; Confidence Interval Estimates of Coefficients***

Consider again the data of Table 2.3. A 95% confidence interval estimate on the slope, β_1, of the regression is given by

$$3884.976 \pm 2.048(370.009) = 3884.976 \pm 757.778$$

In addition, the confidence interval on the intercept is given by

$$-25{,}433.739 \pm 2.048(6{,}104.704) = -25{,}433.739 \pm 12{,}502.434$$

2.7

SIMPLE REGRESSION THROUGH THE ORIGIN (FIXED INTERCEPT)

The model in Eq. (2.2) presumes that the data analyst requires the estimation, from the data, of both slope and intercept. There are certainly many real phenomena in which it is known that the intercept is a fixed constant. In fact, many problems arise in which it is known that $\beta_0 = 0$; thus what is required is a least squares procedure that produces a regression line through the origin.

It is interesting that many regression analysis users are reluctant to make use of the fixed intercept notion, fearing that it is of dubious value. Fixing the intercept allows for estimating the slope with retention of the virtues of least squares. One is merely performing least squares with extra information, namely the known value of β_0, introduced into the model.

Suppose that data (x_i, y_i), $i = 1, 2, \ldots, n$ are collected, and one needs to deal with the general case, i.e., where there is a known intercept β_0. Then b_1, the estimator for the slope, must satisfy

$$\frac{\partial}{\partial b_1}\left[\sum_{i=1}^{n} (y_i - \beta_0 - b_1 x_i)^2\right] = 0 \tag{2.15}$$

which reduces to

$$\sum_{i=1}^{n} (y_i - \beta_0 - b_1 x_i)x_i = 0$$

Thus, the estimator b_1 becomes

$$b_1 = \frac{\sum_{i=1}^{n} y_i x_i - \beta_0 \sum_{i=1}^{n} x_i}{\sum x_i^2} \tag{2.16}$$

Of course, for the special case of regression through the origin, the estimator, b_1, reduces to

$$b_1 = \frac{\sum y_i x_i}{\sum x_i^2} \tag{2.17}$$

Developing techniques for inference on the slope for regression through the origin follows closely that given in Sections 2.2–2.6. For example, under the assumptions of uncorrelated errors and homogeneous variance, $\text{Var}(b_1) = (\sigma^2/\sum_{i=1}^{n} x_i^2)$. It follows easily, then, that the t-statistic for hypothesis testing on β_1 is given by

$$t = \frac{(b_1 - \beta_{1,0})}{s}\sqrt{\sum_{i=1}^{n} x_i^2} \tag{2.18}$$

where, in this case, $s^2 = \sum_{i=1}^{n} [(y_i - \hat{y}_i)^2/(n-1)]$, and of course $\hat{y}_i$, the fitted response at the ith data point, is given by $\hat{y}_i = b_1 x_i$. If the errors are $N(0, \sigma^2)$ and independent, the t-statistic in (2.18) follows t_{n-1} under $H_0\colon \beta_1 = \beta_{1,0}$. The residual degrees of freedom, $n-1$, follow from the fact that $\sum_{i=1}^{n} [(y_i - \hat{y}_i)^2/\sigma^2]$ is a χ^2_{n-1} variate. The notion of $n-1$ degrees of freedom rather than n continues to have intuitive appeal since there is only one parameter to estimate, namely β_1. Thus only 1 degree of freedom is used in estimation.

If inference on the slope is required in the case where β_0 is known but nonzero, $\hat{y}_i = \beta_0 + b_1 x_i$ is, of course, the fitted value and (2.18) still applies.

THE ANALYSIS OF VARIANCE APPROACH FOR THE FIXED INTERCEPT CASE

It is quite simple to use the foregoing methodology to develop a t-statistic for a test on the slope for the fixed intercept case. Thus for hypothesis testing, the analysis of variance approach is interesting but not necessary. However, it is important that we observe and understand what, in the fixed intercept case, is the analogy to the analysis of variance approach in the general case where we need to estimate β_0. First, let us emphasize that the partitioning of the sum of squares $\sum_{i=1}^{n} (y_i - \bar{y})^2$ as in (2.11) no longer holds. However, consider, with β_0 known, the identity

$$\sum_{i=1}^{n} (y_i - \beta_0)^2 = \sum_{i=1}^{n} (\hat{y}_i - \beta_0)^2 + \sum_{i=1}^{n} (y_i - \hat{y}_i)^2 \tag{2.19}$$

Equation (2.19) is clearly an identity since

$$\sum_{i=1}^{n} (\hat{y}_i - \beta_0)(y_i - \hat{y}_i) = b_1 \sum_{i=1}^{n} x_i(y_i - \hat{y}_i) = 0$$

from the result of (2.15). Here, we are dealing with a partition of the variation around β_0 rather than around $\bar{y}$. The component,

$$\sum_{i=1}^{n} (\hat{y}_i - \beta_0)^2 = b_1^2 \sum_{i=1}^{n} x_i^2$$

corresponds to the regression sum of squares, and, with the normal theory assumption, follows $\sigma^2\chi_1^2$ under the condition that $\beta_1 = 0$. In much the same fashion as before, $\sum_{i=1}^{n} (\hat{y}_i - \beta_0)^2$ and SS_{Res} are independent, and we have

$$\frac{b_1^2 \sum_{i=1}^{n} x_i^2}{s^2} \sim F_{1,n-1}$$

under $H_0\colon \beta_1 = 0$. As a result, the partition of variability approach remains valid, albeit reconstructed, and the concept allows an upper-tailed F-test for testing the hypothesis of *no linear relationship* between the response and regressor variable. The denominator degrees of freedom change from $n-2$ to $n-1$ for obvious reasons, and the reader will again observe that the square root of the F-statistic above is the t-statistic in (2.18) when the hypothesized value, $\beta_{1,0}$, is zero.

Example 2.5 **Navy Manpower Data**

Regression analysis is an important tool in industrial and government manpower applications, e.g., in situations where the response variable represents man-hours (per month, say) in various similar work installations and the regressor variable (in most cases there are more than one) represents a measure of workload. In this example, 22 specific Naval installations were sampled and, in this particular functional area, which is clerical in nature, "items processed" represents the single regressor variable. The data are shown in Table 2.5, and Figure 2.6 shows a plot of the data with the least squares regression.

TABLE 2.5 **Monthly man-hours expended as a function of items processed**

Installation	Items Processed, x	Monthly Man-hours, y
1	15	85
2	25	125
3	57	203
4	67	293
5	197	763
6	166	639
7	162	673
8	131	499
9	158	657
10	241	939
11	399	1546
12	527	2158
13	533	2182
14	563	2302
15	563	2202
16	932	3678
17	986	3894
18	1021	4034
19	1643	6622
20	1985	7890
21	1640	6610
22	2143	8522

Procedures and Analyses for Staffing Standards Development: Data/Regression Analysis Handbook (San Diego, California: Navy Manpower and Material Analysis Center, 1979).

Theory suggests that as the workload regressor approaches zero, the personnel required, as measured by monthly man-hours should approach zero. Thus, the model fitted is the *no intercept* model

$$y_i = \beta_1 x_i + \varepsilon_i$$

with b_1, the estimate of β_1 (in (2.17)) given by

$$b_1 = 3.9910$$

The analysis of variance is shown in Table 2.6. Refer to (2.19) with $\beta_0 = 0$. Clearly the F-statistic gives information strongly in favor of a nonzero slope.

FIGURE 2.6 **Plot of y, monthly man-hours, against x, items processed and regression line (through the origin): Navy manpower data of Table 2.5**

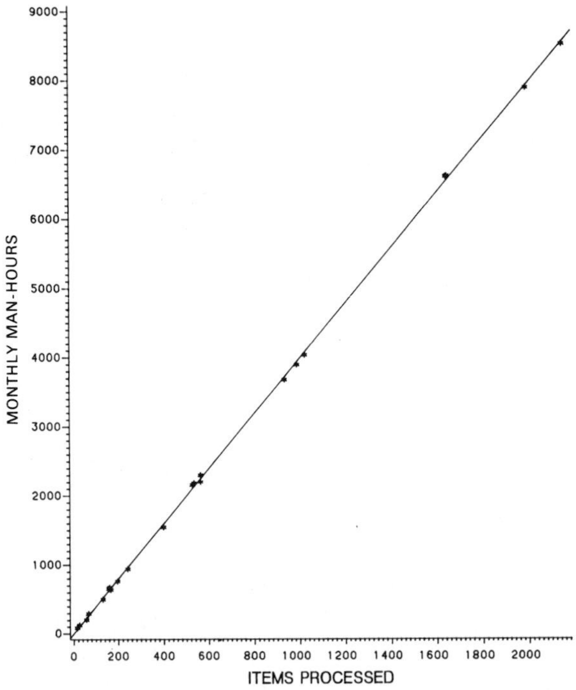

TABLE 2.6 **Analysis of variance for data of Table 2.5**

Source	*SS*	df	*MS*	*F*
Regression	292,460,792	1	292,460,792	176,676.264
Residual	34,762.319	21	1,655.349	
Total	292,495,554	22		

2.8

QUALITY OF FITTED MODEL

The reader should recall that in Section 2.6 the notion of hypothesis testing on the slope was motivated by the obvious need to determine if there is a significant linear relationship, or to ascertain the reasonableness of a particular slope $\beta_{1,0}$. However, we also alluded to other questions that are answered by the fitted regression. It is true that, in many regression problems, hypothesis testing is not the form of inference that will solve the analyst's problems. Often the success of the analytical exercise depends on the proper choice of quantitative criteria, which determine the quality of the fitted model. Two questions quite naturally arise.

1. Does the data fit the model adequately?
2. Will the model predict the response well enough?

In this important section, we deal with these problems specifically for the simple linear regression case, though the concepts will extend to the multiple regression case, and the reader will be reminded throughout the text of the distinction between *fit* and *prediction.*

COEFFICIENT OF DETERMINATION (R^2)

The coefficient of determination, often referred to, symbolically, as R^2 is a much-used, sometimes misunderstood, measure of the fit of the regression line. If we consider the model given by Eq. (2.2), the definition is, simply,

$$R^2 = \frac{SS_{\text{Reg}}}{SS_{\text{Total}}} = \frac{\sum_{i=1}^{n} (\hat{y}_i - \bar{y})^2}{\sum_{i=1}^{n} (y_i - \bar{y})^2} \tag{2.20}$$

From the fundamental partition identity given in (2.11), R^2 can alternatively be written

$$R^2 = 1 - \frac{SS_{\text{Res}}}{SS_{\text{Total}}}$$

Model fitters quote values of R^2 quite often, perhaps due to its ease of interpretation. From (2.20) one can easily notice that it represents the *proportion of variation in the response data that is explained by the model.* Clearly $0 \leqslant R^2 \leqslant 1$, and the upper bound is achieved when the fit of the model to the data is perfect, i.e., all residuals are zero. What is an acceptable value for R^2? This is a difficult question to answer, and, in truth, what is acceptable depends on the scientific field from which the data were taken. A chemist, charged with doing a linear calibration of a high precision piece of equipment, certainly expects to experience a very high R^2 value (perhaps exceeding 0.99), while a behavioral scientist, dealing in data reflecting human behavior, may feel fortunate to observe an R^2 as high as 0.70. An experienced model fitter senses when the value of R^2 is large enough, given the situation confronted. Clearly, some scientific phenomena lend themselves to modeling with considerably more accuracy than others.

Although the coefficient of determination is easy to interpret and can be understood by most experimenters regardless of their training in statistics, there are some pitfalls in its use that are worth noting. For example, it is a dangerous criterion for comparison of candidate models simply because any additional model terms (e.g., a quadratic term in x_i added to (2.2)) will decrease SS_{Res} (at least not increase it) and thus increase R^2 (at least not decrease it). (See Exercise 2.6.) This implies that R^2 can be made artificially high by a rather unwise practice of *overfitting*, i.e., the inclusion of too many model terms. An increase in R^2 does not imply that the additional model term is needed. Indeed, we will see later that if prediction is the purpose of the model, a more complicated structure with R^2 higher than that of a simpler model does not necessarily represent the superior prediction equation. Thus one should not subscribe to a model selection process that solely involves the consideration of R^2. The notion of overfitting and its role in assessing prediction capabilities will be discussed extensively in Chapters 3 and 4.

The coefficient of determination also can appear to be artificially high either because the slope of the regression is large or because the spread of the regressor data $x_1, x_2, \ldots, x_n$ is great. We can illustrate this without much difficulty by observing structures of the numerator and denominator separately. We have already seen that

$$E(SS_{\text{Reg}}) = \sigma^2 + \beta_1^2 S_{xx}$$

and that

$$E(SS_{\text{Res}}) = \sigma^2(n-2)$$

As a result,

$$\begin{aligned} E(SS_{\text{Total}}) &= E(SS_{\text{Reg}} + SS_{\text{Res}}) \\ &= \sigma^2(n-1) + \beta_1^2 S_{xx} \end{aligned}$$

Now, we *cannot* state that $E(R^2)$ is simply the ratio of the expected values. However, we can define a *parametric R^2-type quantity* as

$$\mathscr{R}^2 = \frac{(1/n) + (\beta_1^2 S_{xx}/n\sigma^2)}{[(n-1)/n] + (\beta_1^2 S_{xx}/n\sigma^2)} = \frac{n^{-1} + \omega}{1 - n^{-1} + \omega}$$

where $\omega = \beta_1^2 S_{xx}/n\sigma^2$ can be visualized as a *signal-to-noise ratio*. When ω is large, it describes a condition in which the ratio of slope to the standard error of the estimated slope is large in magnitude. This is clearly a desirable situation, one that should certainly produce a high quality regression. A simplification of $\mathscr{R}^2$, for moderate or large n, can be written

$$\mathscr{R}^2 \cong \frac{\omega}{\omega + 1}$$

It is clear that ω can be made large, for a fixed n, if the variation in the measured regressors is unusually large, i.e., if S_{xx} is large. While a large S_{xx} is often desirable (see (2.9)), it can produce an artificially high coefficient of determination; it is artificial in the sense that the R^2-statistic may be large even though values of the residuals and s^2 may not reflect the same optimism. As a result, it is not merely the quality of fit which influences R^2. The general nature of the model and the data range also play important roles.

R^2 IN THE ZERO INTERCEPT CASE

The analyst faces a dilemma in the computation of R^2 for the case of a zero or fixed intercept model. The fundamental partition identity of (2.11) allows for an appealing R^2 interpretation (in (2.20)), which is given by

$$R^2 = \frac{\text{Variation explained by regression}}{\text{Variation observed}}$$

where the word *variation* is on the response variable. In the case of regression through the origin, we replace the fundamental identity by

$$\sum_{i=1}^{n} y_i^2 = \sum_{i=1}^{n} \hat{y}_i^2 + \sum_{i=1}^{n} (y_i - \hat{y}_i)^2$$

which is a special case of Eq. (2.19). Thus the R^2-analog for the no intercept model may be taken to be

$$R_{(0)}^2 = \frac{\sum_{i=1}^{n} \hat{y}_i^2}{\sum_{i=1}^{n} y_i^2} = \frac{SS_{\text{Reg}}}{SS_{\text{Total}}}$$

Some computer software packages do, indeed, compute $R_{(0)}^2$ as shown above. However, there is a marked difference between the expression for $R_{(0)}^2$ and that of R^2 in the intercept model (2.20). In the zero intercept model, the variation described by numerator and denominator represents *dispersion around zero.* In the intercept model, the R^2-statistic is a ratio of dispersion in the $\hat{y}$'s to that of the y's, the dispersion being around the mean $\bar{y}$. The redefinition of the total and regression sums of squares in the zero intercept case necessitates consideration of an alternative to $R_{(0)}^2$. The major difficulty with this statistic stems from the fact that we cannot use it for performance comparison with the intercept model. There will be a strong tendency for $R^2_{(0)}$ to be larger than R^2 in Eq. (2.20), even though the quality of fit (or prediction) may not be superior. This results from the fact that *uncorrected sums of squares* are used. Thus, for the case of nearly equivalent performance in terms of the residual sum of squares, $R^2_{(0)}$ may be considerably larger than R^2. The upshot is a severely misleading comparison with the zero intercept model quite possibly being ranked higher erroneously.

There are alternate ways of computing R^2 for the zero intercept case so that we can make reasonable comparisons between competing models, though R^2 is certainly not the ideal statistic for that purpose. One possible choice is given by

$$R^2_{(0)*} = 1 - \frac{\sum_{i=1}^{n} (y_i - \hat{y}_i)^2}{\sum_{i=1}^{n} (y_i - \bar{y})^2}$$

Here, of course,

$$\sum_{i=1}^{n} (y_i - \bar{y}_i)^2 = \sum_{i=1}^{n} y_i^2 - \frac{\left(\sum_{i=1}^{n} y_i x_i\right)^2}{\sum_{i=1}^{n} x_i^2}$$

Though it serves as a more reasonable basis for comparison than $R^2_{(0)}$, the statistic $R^2_{(0)*}$ can be negative in situations where $\sum_{i=1}^{n} (y_i - \hat{y}_i)^2$ is relatively large. Further discussion of the zero intercept model is presented in Casella (1983) and Hahn (1977). For an illustration of the comparison between an intercept and a zero intercept model, see Exercise 2.12.

COEFFICIENT OF VARIATION

The coefficient of variation (CV) is a reasonable criterion for representing quality of fit and measuring spread of *noise* around the regression line. The CV is defined as

$$CV = (s/\bar{y}) \times 100 \qquad \textbf{(2.21)}$$

So the interpretation is that CV is the residual estimate of error standard deviation, measured as a percent of the average response value. Its use is motivated by the fact that the estimate of error standard deviation, s, is often not satisfactory as a measure of quality of fit due to the fact that it is not scale free. For example, an inexperienced analyst would not be likely to identify with a value of $s = 14$ parts per million as an error standard deviation. However, if this value of s leads to a CV of, say, 5%, the analyst now knows that the *natural dispersion* around the line, as measured by s, is only 5% of the average response measurement.

Example 2.6 ***Wood Density Data; Standard Regression Analysis***

Consider again the wood density data of Table 2.3. Table 2.7 provides the reader an annotated SAS (1985) printout of a PROC REG run on these data. The reader should note the location of such items as the analysis of variance, R^2, estimated regression coefficients with standard errors, and t-statistics accompanied by p-values (indicating level of significance). Also note the root mean squared error, s, and the coefficient of variation, CV.

TABLE 2.7 **Wood density data**

DEP VARIABLE: Y

ANALYSIS OF VARIANCE

SOURCE	DF	SUM OF SQUARES	MEAN SQUARE	F VALUE	PROB>F
MODEL	1	14891739363	14891739363	110.243	0.0001
ERROR	28	3782270481	135081089		
C TOTAL	29	18674009844			

ROOT MSE	11622.44	R-SQUARE	0.7975
DEP MEAN	34666.83	ADJ R-SQ	0.7902
C.V.	33.52611		

PARAMETER ESTIMATES

VARIABLE	DF	PARAMETER ESTIMATE	STANDARD ERROR	T FOR H0: PARAMETER=0	PROB > |T|
INTERCEP	1	-25433.73872	6104.70422	-4.166	0.0003
X	1	3884.97557	370.00948	10.500	0.0001

2.9

CONFIDENCE INTERVALS ON MEAN RESPONSE AND PREDICTION INTERVALS

Perhaps more often than not, the scientist or engineer is quite interested in whether or not the fitted regression can be used as a mechanism for prediction of response values or estimation of mean response. Prediction is so vitally important in data analysis that we dedicate much of Chapters 3 and 4 to it. In many regression situations, quality (or lack of quality) of prediction is not so easy to identify as quality of fit. The concepts that must be addressed are perhaps more complex than what inexperienced analysts are accustomed to assimilating. At this point we merely wish to illustrate criteria that enable the user to ascertain the quality of prediction or estimation of mean response with the regression equation in the present setting, namely the simple linear regression model.

Before we begin, we should make clear the nature of the statistic

$$\hat{y}(x_0) = b_0 + b_1 x_0$$

This may be viewed as the *estimated response*. It is the estimator of the mean y at $x = x_0$. This becomes very clear when we consider that $E(y|x_0) = \beta_0 + \beta_1 x_0$. Thus we interpret a standard error of $\hat{y}(x_0)$ as the standard error of the estimator of *mean response conditional on* x_0. The notion *standard error*, of course, evokes the image of precision or variation. In this case, it reflects the variation of $\hat{y}$ at x_0, if repeated regressions were conducted, based on the same x-levels and new observations on y each time. We focus on Var $\hat{y}(x_0)$ and notice that

$$\begin{aligned}\text{Var } \hat{y}(x_0) &= \text{Var}[b_0 + b_1 x_0] \\ &= \text{Var}[\bar{y} + b_1(x_0 - \bar{x})]\end{aligned}$$

Since $\bar{y}$ and b_1 are independent (see Section 2.6),

$$\text{Var } \hat{y}(x_0) = \sigma^2\left[\frac{1}{n} + \frac{(x_0 - \bar{x})^2}{S_{xx}}\right] \tag{2.22}$$

If we substitute the estimator s^2 for σ^2, we can define *standard error of prediction* as

$$s_{\hat{y}(x_0)} = s\sqrt{\frac{1}{n} + \frac{(x_0 - \bar{x})^2}{S_{xx}}} \tag{2.23}$$

Actually, (2.23) defines an *estimated* standard error of prediction since σ is replaced by s. As a result, under the condition of normal errors, $\hat{y}(x_0)$ is normal, and a $100(1-\alpha)\%$ confidence interval for $E(y\,|\,x_0)$ can be written

$$\boxed{\hat{y}(x_0) \pm t_{\alpha/2,n-2}\, s\sqrt{\frac{1}{n} + \frac{(x_0 - \bar{x})^2}{S_{xx}}}} \tag{2.24}$$

The expression in (2.24) is, indeed, that of a confidence interval and is not to be confused with the prediction interval on a *new response observation* at $x = x_0$. The latter reflects bounds in which the analyst can realistically expect an observation y_0 at $x = x_0$ to fall. These bounds will be developed and illustrated after the following example.

Consider the accounting return data in Example 2.1. Table 2.8 shows columns

TABLE 2.8 **Further output for accounting and stock return data**

Company	y_i	$\hat{y}_i$	$y_i - \hat{y}_i$	$s_{\hat{y}_i}$	UCL	LCL
1	17.730	11.810	5.920	0.999	13.813	9.806
2	4.540	5.798	−1.258	0.978	7.759	3.836
3	3.960	8.453	−4.493	0.695	9.848	7.057
4	8.120	9.820	−1.700	0.737	11.299	8.341
5	6.780	8.111	−1.331	0.708	9.531	6.691
6	9.690	6.750	2.940	0.835	8.425	5.075
7	12.370	8.996	3.374	0.695	10.390	7.602
8	15.880	10.680	5.200	0.828	12.342	9.019
9	−1.340	4.986	−6.326	1.120	7.234	2.738
10	18.090	6.591	11.499	0.856	8.309	4.873

TABLE 2.8
Continued

Further output for accounting and stock return data

Company	y_i	$\hat{y}_i$	$y_i - \hat{y}_i$	$s_{\hat{y}_i}$	UCL	LCL
11	17.170	6.317	10.853	0.896	8.114	4.519
12	6.780	9.496	−2.716	0.714	10.930	8.063
13	4.740	6.414	−1.674	0.881	8.183	4.646
14	23.020	9.533	13.487	0.717	10.971	8.095
15	7.680	7.214	0.466	0.779	8.777	5.650
16	14.320	12.896	1.424	1.195	15.294	10.497
17	−1.630	4.766	−6.396	1.161	7.096	2.436
18	16.510	8.270	8.240	0.701	9.676	6.863
19	17.530	15.002	2.528	1.623	18.259	11.744
20	12.690	12.566	0.124	1.133	14.840	10.292
21	4.660	7.415	−2.755	0.759	8.938	5.893
22	3.670	6.030	−2.360	0.940	7.916	4.144
23	10.490	11.651	−1.161	0.972	13.601	9.700
24	10.000	6.402	3.598	0.883	8.174	4.630
25	21.900	11.510	10.390	0.949	13.415	9.606
26	5.860	12.109	−6.249	1.050	14.216	10.001
27	10.810	6.988	3.822	0.805	8.603	5.373
28	5.710	8.965	−3.255	0.694	10.358	7.573
29	13.380	11.626	1.754	0.968	13.569	9.684
30	13.430	9.753	3.677	0.732	11.221	8.285
31	10.000	13.628	−3.628	1.339	16.315	10.941
32	16.660	6.719	9.941	0.839	8.403	5.036
33	9.400	10.809	−1.409	0.845	12.504	9.113
34	0.240	5.847	−5.607	0.969	7.792	3.901
35	4.370	10.455	−6.085	0.800	12.061	8.849
36	3.110	8.184	−5.074	0.704	9.597	6.771
37	6.630	7.830	−1.200	0.724	9.284	6.377
38	14.730	20.733	−6.003	2.897	26.545	14.920
39	6.150	8.105	−1.955	0.708	9.525	6.684
40	5.960	6.988	−1.028	0.805	8.603	5.373
41	6.300	6.707	−0.407	0.841	8.394	5.020
42	0.680	5.371	−4.691	1.051	7.479	3.262
43	12.220	12.981	−0.761	1.212	15.413	10.550
44	0.900	5.102	−4.202	1.099	7.307	2.897
45	2.350	5.670	−3.320	0.999	7.674	3.665
46	5.030	6.548	−1.518	0.862	8.279	4.818
47	6.130	10.247	−4.117	0.777	11.806	8.688
48	6.580	8.141	−1.561	0.706	9.559	6.724
49	14.260	6.683	7.577	0.844	8.376	4.990
50	2.600	6.982	−4.382	0.806	8.599	5.365
51	4.970	8.239	−3.269	0.702	9.648	6.830
52	6.650	7.885	−1.235	0.721	9.331	6.439
53	4.250	6.902	−2.652	0.815	8.539	5.266
54	7.300	8.337	−1.037	0.699	9.738	6.935

indicating y_i (market returns), $\hat{y}_i$ (fitted market returns), $y_i - \hat{y}_i$ (the residuals), $s_{\hat{y}_i}$ (standard error of prediction), and the upper and lower 95% confidence limits on mean y for each of the 54 companies.

An example of an interpretation of the results in Table 2.8 is as follows. Company 16, R.J. Reynolds, has a predicted or rather *fitted* market return of 12.896% with an *actual* return of 14.32%. The standard error of prediction is truly a standard error of the fitted value here since the regressor values being considered are those in the data set that built the regression. For R.J. Reynolds this standard error is 1.195%. For the fixed value of accounting return in the case of R.J. Reynolds, we are 95% confident that the *mean* market return lies between 10.497% and 15.294%.

Several items are worth noting here. We make a distinction between fitted values, say fitted market returns, and prediction. *Prediction* and the corresponding standard error and confidence limits apply to regressor values where interpolation or extrapolation was necessary, i.e., where the value of the regressor variable was not contained in the data set. The standard errors of the fitted values are certainly not unimportant in the simple linear regression case. They do give some indication of the capability of the regression model for interpolation. However, they do not reveal the model's performance in the area of extrapolation. The standard error of prediction (or fit) has become a standard part of regression computer printout with most computer packages. It can be valuable information to the analyst. However, unless values other than the x_i are used in Eq. (2.23), prediction, certainly extrapolation, is not being properly assessed.

Note that the standard error of prediction is not constant for all x_0. Thus it should not be confused with s, the error standard deviation. The quantity $s_{\hat{y}(x_0)}$ reflects the fact that the quality of the regression line is very much a function of the *location* at which one is predicting. For example, $s_{\hat{y}}$ for R.J. Reynolds is 1.195% while for Ford Motors (Company 7) this standard error is 0.695%. This non-uniformity in prediction performance is discussed further in Section 2.12 and illustrated in Figure 2.7, which reflects the confidence intervals on mean response for the accounting return data.

There is often confusion regarding the implication of the word *prediction*. Obviously the statistic $\hat{y}(x_0)$, the point on the regression line at $x = x_0$, serves the dual purpose as the estimate of mean response *and the predicted value*. The standard error of prediction, given by (2.23), is used in constructing a confidence interval on the mean response. However, it is not appropriate for establishing any form of inference on a future single observation. Suppose the *mean response* at a fixed $x = x_0$ is not of interest. Rather, one is interested in some type of bound on a single response observation at x_0. Consider a single observation at $x = x_0$ denoted symbolically by y_0, independent of $\hat{y}(x_0)$. We can construct a prediction interval on y_0 by beginning with $y_0 - \hat{y}(x_0)$. We can standardize by considering

$$\begin{aligned}\operatorname{Var}(y_0 - \hat{y}(x_0)) &= \sigma^2 + \sigma^2\left(\frac{1}{n} + \frac{(x_0 - \bar{x})^2}{S_{xx}}\right)\\ &= \sigma^2\left(1 + \frac{1}{n} + \frac{(x_0 - \bar{x})^2}{S_{xx}}\right)\end{aligned}$$

As a result, since $E[y_0 - \hat{y}(x_0)] = 0$,

$$\frac{y_0 - \hat{y}(x_0)}{\sigma\sqrt{1 + \frac{1}{n} + \frac{(x_0 - \bar{x})^2}{S_{xx}}}} \sim N(0, 1)$$

under the normal theory assumptions. We can now replace σ by s (see Graybill (1976)) and

$$\frac{y_0 - \hat{y}(x_0)}{s\sqrt{1 + \frac{1}{n} + \frac{(x_0 - \bar{x})^2}{S_{xx}}}} \sim t_{n-2} \qquad \textbf{(2.25)}$$

From (2.25) a probability bound or prediction interval can be placed on y_0, i.e., *an interval in which y_0 is contained with a fixed probability* $(1 - \alpha)$. This prediction interval is given by

$$\boxed{\hat{y}(x_0) \pm t_{\alpha/2,n-2}\, s\sqrt{1 + \frac{1}{n} + \frac{(x_0 - \bar{x})^2}{S_{xx}}}} \qquad \textbf{(2.26)}$$

FIGURE 2.7 **Confidence intervals on mean response; accounting return data of Table 2.1**

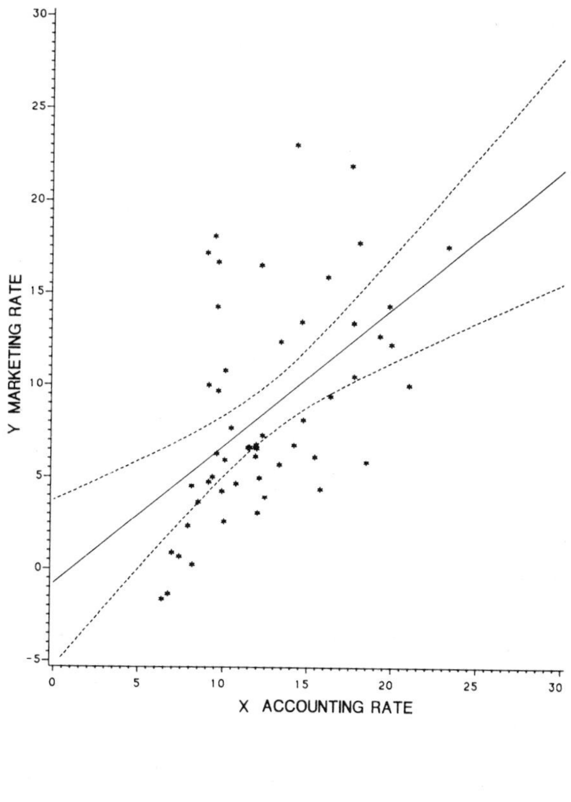

Figure 2.8 illustrates the prediction intervals for the accounting return data on the same plot with the confidence intervals. Of course the prediction intervals are wider, reflecting the larger quantity inside the square root in (2.26). The quantity

$$s\sqrt{1+\frac{1}{n}+\frac{(x_0-\bar{x})^2}{S_{xx}}}$$

is used in a *standard error type role* in developing the prediction interval. However, it is *not* the standard error of the predicted value $\hat{y}(x_0)$. Rather, it is the standard error of the difference $y_0 - \hat{y}(x_0)$.

FIGURE 2.8 **Confidence and prediction intervals; accounting return data of Table 2.1**

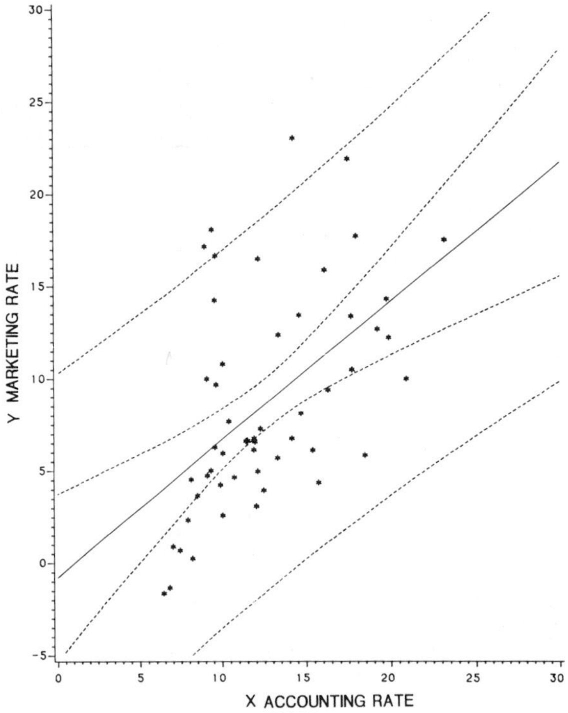

Often the inexperienced analyst is confused about the interpretation associated with a data point falling outside the bounds revealed in Figure 2.8. There may be a temptation to interpret a point outside one of the bands as a signal of a suspect observation. The inner bands reflect confidence intervals on the population *mean* response; hence if an observation falls outside these bands, there is no special interpretation attached to it. Clearly it is more rare for an observation to fall outside the outer bands (i.e., those constructed for an individual observation), and they would certainly seem more appropriate as guidelines for gauging the adequacy of fit of an individual observation to a fitted model. However, in general, the presence of an observation outside the prediction intervals does not give the analyst license for eliminating the observation from the data set. A detailed presentation of *detection of outliers* through analysis of residuals is given in Chapter 5.

FACTORS ALTERING PREDICTION CAPABILITIES OF THE MODEL

Intuition certainly suggests which factors influence the prediction capabilities of a regression model. For example, an increase in the sample size obviously decreases the prediction variance in (2.22), assuming all other things are equal. In addition, prediction is improved by an increase in the quantity $S_{xx} = \sum_{i=1}^{n} (x_i - \bar{x})^2$, implying that the larger the spread in the data range of the regressor variable, the better the prediction capabilities of the model. This could be an important consideration in situations where the experimenter has control on the x-values. One note of caution follows: Prediction variance is an appropriate measure of quality of prediction *assuming that the model is specified properly*. If the model is underspecified, i.e., if too few model terms are included, *bias* in the predicted response becomes a factor and must somehow be taken into account. Clearly if the data analyst uses an excessive spread in x-levels in order to achieve a large S_{xx}, there may be a risk of badly straining the model assumption. For example, a model linear in x in a narrow region may be palatable while a large range in x may require a model that contains curvature. Chapter 4 gives much attention to bias when discussing consideration of underspecification and overspecification of the model.

Finally, another result that is consistent with intuition concerns the location of x_0, the point at which one wishes to predict. Equations (2.22) and (2.23) clarify that the prediction will be at its best when x_0 is close to $\bar{x}$, the average of the regressor values. We can then see that poor results could occur via prediction when we extrapolate. Figures 2.7 and 2.8 illustrate this.

2.10 SIMULTANEOUS INFERENCE IN SIMPLE LINEAR REGRESSION

Often a data analyst makes the mistake of building say a 95% confidence interval estimate on several parameters with the presumption that 95% is the confidence coefficient, or the degree of confidence, on all the parameter statements simultaneously. Students of statistics normally learn about *joint confidence regions* quite early in their academic careers. Too often users ignore this important concept in their attempts at drawing conclusions about parameters. In a regression setting, for example, it may be of interest to develop a *joint confidence region* on β_0 and β_1, i.e., a region in which we can say that there is 95% confidence that β_0 *and* β_1 fall inside the region.

In this same context, a similar argument can be made that plots of confidence intervals on mean response given, for example, in Figure 2.7 may be misleading. The plots represent a depiction of the continuous relationship that exists among the upper 95% confidence intervals on mean marketing rate for the 54 companies involved. The same plot is given for the lower confidence bounds. However, the user cannot infer that, with probability 0.95, the intervals for all 54 data locations cover the mean marketing rate *simultaneously*. Often data analysts use plots such as that given in Figure 2.7 to conclude that the true regression line $E(y \mid x) = \beta_0 + \beta_1 x$ lies inside the plotted confidence intervals with 95% confidence. This is just simply not true! In this section we discuss simultaneous confidence regions in the context of simple linear regression, and where appropriate, we discuss simultaneous tests of hypotheses. It should be pointed out that the material in this section will be extended to multiple regression in Chapter 3. As a result, the reader will view it again couched in the somewhat more complicated setting of multiple linear regression.

JOINT CONFIDENCE REGION ON β_0 and β_1

Joint estimation of the slope and intercept is often very important in applications. The confidence region approach allows the analyst to gain some insight into the precision of the point estimators b_0 and b_1. When the $100(1-\alpha)\%$ confidence region is found, the implication is that *with repeated experiments*, and thus repeated computation of the confidence region, $100(1-\alpha)\%$ of such computed regions will contain both β_0 and β_1.

As in the case of the confidence regions discussed in Sections 2.6 and 2.9, we must make the normal theory assumptions. From the theory of linear models (see Searle (1971) or Graybill (1976)),

$$[b_0 - \beta_0, b_1 - \beta_1]\begin{bmatrix} n & \sum_{i=1}^{n} x_i \\ \sum_{i=1}^{n} x_i & \sum_{i=1}^{n} x_i^2 \end{bmatrix}\begin{bmatrix} b_0 - \beta_0 \\ b_1 - \beta_1 \end{bmatrix} \sim 2s^2 F_{2,n-2}$$

This expression can be used for developing a confidence region. The reader should be aware of the fact that apart from σ^2 the matrix in the above expression is the inverse of the variance-covariance matrix of $[b_0, b_1]$. We can now use this to develop the confidence region since the inequality

$$[b_0 - \beta_0, b_1 - \beta_1]\begin{bmatrix} n & \sum_{i=1}^{n} x_i \\ \sum_{i=1}^{n} x_i & \sum_{i=1}^{n} x_i^2 \end{bmatrix}\begin{bmatrix} b_0 - \beta_0 \\ b_1 - \beta_1 \end{bmatrix} \leqslant 2s^2 F_{\alpha,2,n-2}$$

holds with probability $1-\alpha$. As a result, all β_0 and β_1 combinations that satisfy the inequality fall inside a $100(1-\alpha)\%$ confidence region. Here, of course, $F_{\alpha,2,n-2}$ is an upper tail αth percentage point of the F-distribution.

The joint confidence region on β_0 and β_1 forms an elliptical region that is easy to use in a graphical way. The ellipse will always be centered at the point (b_0, b_1). Clearly, the equation that generates the ellipse is derived by merely *equating* the left-hand side of the inequality to the right-hand side. The following example provides an illustration.

Example 2.7 Consider again the wood density data of Table 2.3. The values of b_0 and b_1 are, respectively, $-25{,}433.739$ and $3{,}884.976$. The expression for the joint confidence region was used to compute the ellipse given in Figure 2.9. Any (β_0, β_1) combination inside the ellipse is inside the joint 95% confidence region.

FIGURE 2.9 **Joint confidence region for slope and intercept for the wood density data**

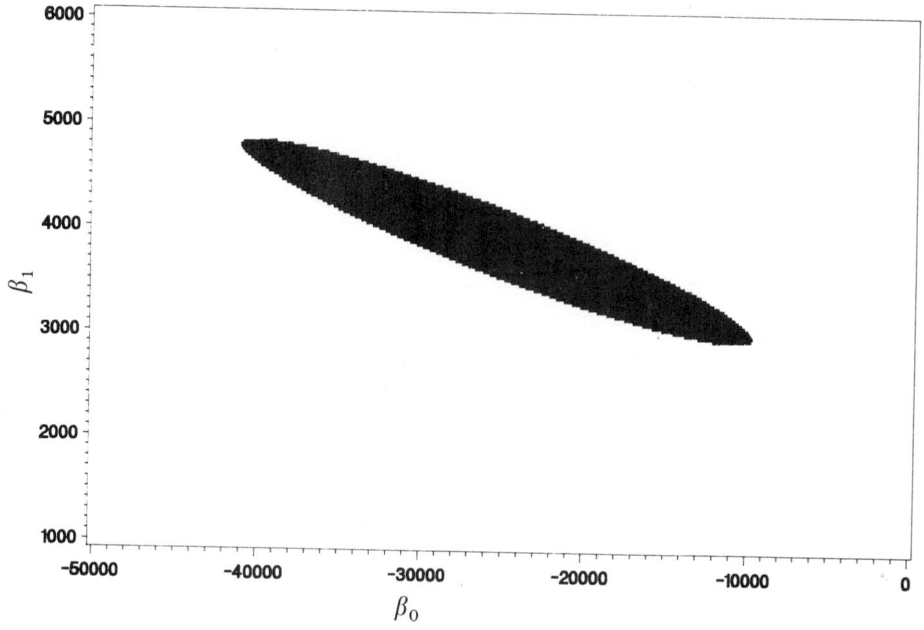

CONFIDENCE BAND FOR THE LINEAR REGRESSION LINE (WORKING-HOTELLING PROCEDURE)

In many applications of simple linear regression, it is of interest to have a confidence region on the entire regression line. It should be obvious to the reader that this is developed by consideration of all possible regressions produced by the β_0, β_1 combinations in the elliptical plot of the joint confidence region on β_0 and β_1. A hyperbolic confidence region *around the fitted* regression line is easily computed using an expression similar to that in Eq. (2.24), namely

$$\hat{y}(x_0) \pm \text{constant} \cdot s\sqrt{\frac{1}{n} + \frac{(x_0 - \bar{x})^2}{S_{xx}}}$$

In other words, the width of the band at any location x_0 still involves the standard error of prediction. However, the appropriate constant is not a percentage point of the t-distribution. Rather, the boundary of the $100(1-\alpha)\%$ confidence band is given by

$$\hat{y}(x_0) \pm \sqrt{2 \cdot F_{\alpha,2,n-2}}\, s \sqrt{\frac{1}{n} + \frac{(x_0 - \bar{x})^2}{S_{xx}}}$$

As in the case of the joint confidence region on β_0 and β_1, the confidence band on the regression line may be best used by the experimenter when he or she can view it graphically. The following example using the wood density data may be helpful.

Example 2.8

Consider the wood density data of Table 2.3. Figure 2.10 shows the fitted regression line and the 95% confidence band on the "true" regression line $E(y|x) = \beta_0 + \beta_1 x$. Also shown is the standard plot of the confidence interval on $E(y|x)$ discussed in Section 2.9. Note how the Working-Hotelling approach, which represents a band on the regression line, is more conservative than the plot of *conditional* confidence intervals.

FIGURE 2.10 **Fitted regression and 95% confidence band on the regression line for the wood density data**

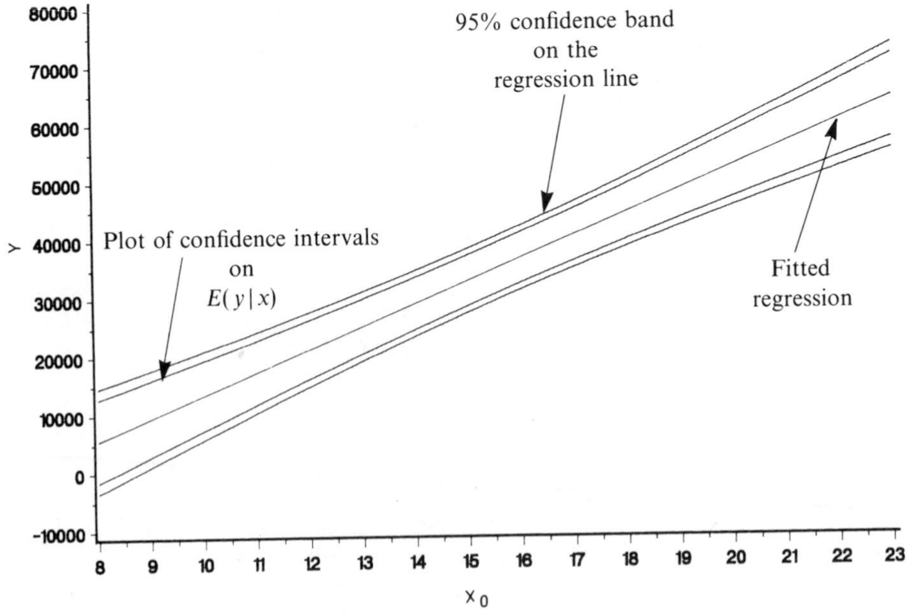

BONFERRONI APPROACH TO SIMULTANEOUS INFERENCE IN SIMPLE LINEAR REGRESSION

At this point the reader has been exposed to a simultaneous confidence region on β_0 and β_1 as well as a confidence band on the regression line. In both cases the tools can be used in an expeditious fashion *through the use of plots* as we have illustrated in the examples with the wood density data. There are practical situations, however, in which plots may not be available. It is often convenient to use confidence regions that involve *separate* intervals on each parameter but, of course, intervals for which the joint coverage is at least $100(1-\alpha)\%$. In addition, *in the multiple regression* case, a joint confidence region on regression coefficients, or a confidence region on the regression relationship (similar to the confidence band on the regression) might be unruly if one were to rely on plots. Thus, it is of interest to have, say, for regression coefficients, confidence intervals of the type

$$b_0 \pm \text{constant} \cdot (\text{standard error of } b_0)$$
$$b_1 \pm \text{constant} \cdot (\text{standard error of } b_1)$$

and have the constant determined so that we have confidence at least $100(1-\alpha)\%$ that the coefficients β_0 and β_1 are simultaneously in the interval. As one would expect, the implication is that if repeated samples were taken and repeated regressions were computed with different values of b_0 and b_1, at least 95% of such combinations of b_0 and b_1 would have both intervals covering the appropriate parameter.

The Bonferroni procedure for finding confidence intervals on two parameters is based on the fact that if a $100(1-\alpha)\%$ confidence interval is applied to each parameter

$$\boxed{\Pr[\bar{B}_0 \cap \bar{B}_1] \geqslant 1 - 2\alpha}$$

This equation is an application of the *Bonferroni inequality* to simple linear regression. Here $\Pr[B_0]$ is the probability that the confidence interval on β_0 does not cover the true parameter, and $\Pr[B_1]$ is the probability that the confidence interval on β_1 does not cover the true parameter. Thus $\Pr[\bar{B}_0 \cap \bar{B}_1]$ is the probability that the two confidence intervals simultaneously cover their respective parameters. This form of the Bonferroni inequality can be developed by using some fairly elementary notions of probability. We can write

$$\Pr[B_0 \cup B_1] = \Pr[B_0] + \Pr[B_1] - \Pr[B_0 \cap B_1]$$

Since

$$\Pr[\bar{B}_0 \cap \bar{B}_1] = 1 - \Pr[B_0 \cup B_1]$$

we have that

$$\Pr[\bar{B}_0 \cap \bar{B}_1] = 1 - (\Pr[B_0] + \Pr[B_1]) + \Pr[B_0 \cap B_1]$$

Since $\Pr[B_0 \cap B_1] \geqslant 0$,

$$\boxed{\Pr[\bar{B}_0 \cap \bar{B}_1] \geqslant 1 - \Pr[B_0] - \Pr[B_1]}$$

Now, $\Pr[B_0] = \Pr[B_1] = \alpha$, and thus we have

$$\Pr[\bar{B}_0 \cap \bar{B}_1] \geqslant 1 - 2\alpha$$

Now, if the *confidence coefficients* for the confidence intervals on β_0 and β_1 are both .95, i.e., $1 - 0.05$, then the confidence coefficient for joint estimation of β_0 and β_1 will be at least 0.90. Thus at least 90% of such pairs of intervals will simultaneously cover the two parameters.

It should now be clear to the reader how we can use the Bonferroni inequality to produce intervals for which the *joint confidence coefficient* is controlled to be at least $1 - \alpha$. This can be accomplished by computing individual confidence intervals on β_0 and β_1, each with confidence coefficient $1 - \alpha/2$. A $100(1 - \alpha)\%$ joint confidence interval on β_0 and β_1 is given by

$$\boxed{\begin{array}{l} b_0 + t_{\alpha/4,n-2}\,(\text{standard error of } b_0) \\ b_1 + t_{\alpha/4,n-2}\,(\text{standard error of } b_1) \end{array}}$$

Example 2.9 **Bonferroni Confidence Bounds for Wood Density Data**

Consider the wood density data in Table 2.3. A pair of 95% joint confidence intervals are given by

$$b_0 \pm t_{0.0125,28}(6104.704)$$
$$b_1 \pm t_{0.0125,28}(370.009)$$

The value

$$t_{0.0125,25} = 2.3685,\ b_0 = -25{,}433.739, \text{ and } b_1 = 3{,}884.976$$

As a result, the numerical intervals are indicated by

$$-25{,}433.739 \pm 14{,}458.991$$
$$3{,}884.976 \pm 876.366$$

In order to illustrate the contrast between the elliptical joint confidence region on β_0 and β_1 given in Example 2.7 and the Bonferroni confidence intervals for the wood density data, consider the plot given in Figure 2.11. Notice that the area covered by the Bonferroni confidence region is considerably larger. The rectangular region produces a much less efficient confidence region than the elliptical one. The reader should keep in mind that the Bonferroni approach results in a region for which the confidence is *at least* 95%. In general, it will produce a more conservative joint confidence region.

FIGURE 2.11 Bonferroni confidence region and elliptical confidence region on β_0 and β_1 for the wood density data

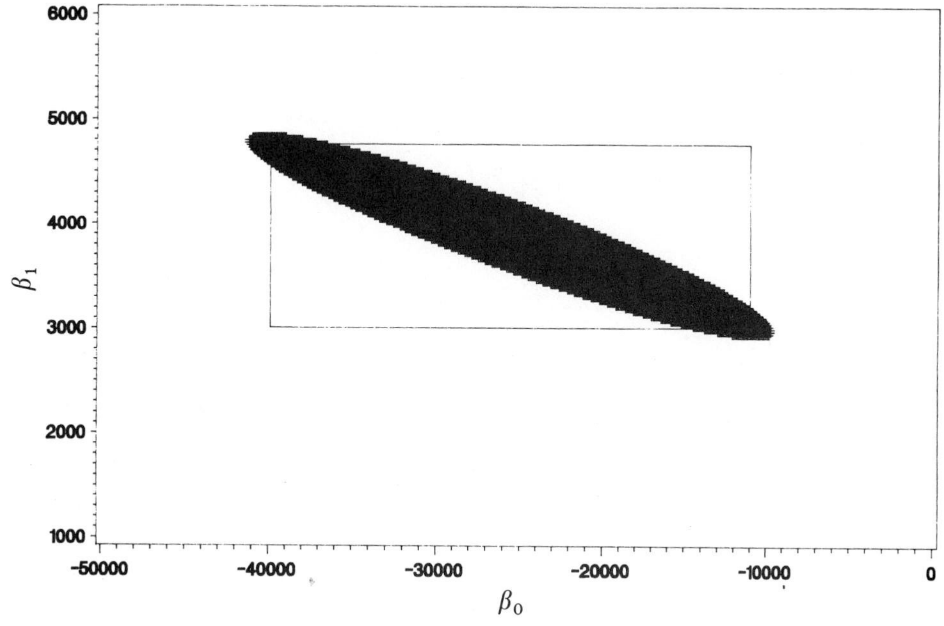

Example 2.10 ***Bonferroni Confidence Intervals for Mean Response for Wood Density Data***

Suppose we consider the wood density data in Table 2.3. Suppose it is of interest to simultaneously estimate mean stiffness for four different values of wood density. These values are

$$x_1 = 16.0 \text{ lb/ft}^3$$
$$x_2 = 18.0 \text{ lb/ft}^3$$
$$x_3 = 21.0 \text{ lb/ft}^3$$
$$x_4 = 23.0 \text{ lb/ft}^3$$

The point estimates of mean stiffness and joint 95% confidence intervals are given by

$$\hat{y}_i \pm t_{.05/8,28}\, s\sqrt{\frac{1}{30} + \frac{(x_i - \bar{x})^2}{S_{xx}}} \qquad i = 1, 2, 3, 4$$

The factor 8 is used because four parameters are being estimated, i.e., there are four mean responses to be estimated. The value $t_{.05/8,28}$ is given by 2.6695. As a result the confidence intervals are

$$[31{,}0372.2, \quad 42{,}414.5]$$
$$[38{,}304.6, \quad 50{,}687.1]$$
$$[48{,}281.7, \quad 64{,}019.8]$$
$$[54{,}571.6, \quad 73{,}269.8]$$

USE OF THREE CONFIDENCE INTERVAL METHODS FOR MEAN RESPONSE

Example 2.10 provides another illustration of Bonferroni confidence intervals. Obviously if the data analyst desires to use the fitted regression to estimate mean response and to attach confidence intervals, three methods are available: the standard confidence intervals on $E(y|x)$ discussed in Section 2.9, the confidence band on the regression line, and the Bonferroni approach. For joint confidence intervals, the first method is not appropriate. In general, if only a few predictions are made, the Bonferroni method will be more efficient. In both cases the "controlled" confidence coefficient is a lower bound. (See Exercise 2.14.)

EXTENSION TO SIMULTANEOUS TESTS IN SIMPLE LINEAR REGRESSION

We have developed and illustrated simultaneous confidence regions in simple linear regression. We have focused on joint *estimation* of the slope and intercept and joint estimation of mean response. As we indicated earlier, the purpose of a joint confidence region is to have control over the confidence (say 95%) that all parameters which are estimated simultaneously fall inside the computed region. The same concept can be employed for *hypothesis testing* on multiple parameters. The notion of *joint* in hypothesis testing implies that the probability of Type I error or α error refers to the joint statement on multiple parameters. For example in the case of the slope and intercept, we may wish to test

$$H_0: \begin{cases} \beta_0 = \beta_{0,0} \\ \beta_1 = \beta_{1,0} \end{cases}$$

H_1: at least one statement of H_0 is not true

To test this hypothesis at the 0.05 level implies a probability 0.05 of concluding that *at least* one of the statements in H_0 is false when in fact the entire hypothesis is true. In the case of one parameter, it is easy to verify that rejection of

$$H_0\colon \beta_1 = \beta_{1,0}$$

at level α is identical to computing a $100(1-\alpha)\%$ confidence interval estimate on β_1 and finding that $\beta_{1,0}$ is outside the confidence interval

$$b_1 \pm t_{\alpha/2,n-2} \cdot (\text{standard error of } b_1)$$

which, of course, can be written

$$b_1 \pm t_{\alpha/2,n-2} \frac{s}{\sqrt{S_{xx}}}$$

The same can be noted of the test on the intercept. Thus, if one tests both individual hypotheses, the probability of Type I error for the joint statement is the probability that at least one of the null values ($\beta_{0,0}$ or $\beta_{1,0}$) falls outside the confidence interval on the parameter. This probability is given by

$$1 - \Pr[\bar{B}_0 \cap \bar{B}_1] \leqslant 1 - (1 - 2\alpha) \leqslant 2\alpha$$

Thus if an $\alpha = 0.05$ level test is conducted on the two parameters, the joint probability of *Type I error* has an upper bound of 0.10.

The reader should surmise by now that the same type of relationship exists between simultaneous tests and simultaneous confidence intervals as exists between tests and confidence interval estimation for individual parameters.

SIMULTANEOUS TESTS ON β_0 AND β_1

Rejection of the joint hypothesis on β_0 and β_1 should occur with probability α when $\beta_0 = \beta_{0,0}$ and $\beta_1 = \beta_{1,0}$. We know from the development of the elliptical confidence region on β_0 and β_1 that under the null hypothesis, i.e., when $\beta_0 = \beta_{0,0}$ and $\beta_1 = \beta_{1,0}$,

$$[b_0 - \beta_{0,0}, b_1 - \beta_{1,0}] \begin{bmatrix} n & \sum_{i=1}^{n} x_i \\ \sum_{i=1}^{n} x_i & \sum_{i=1}^{n} x_i^2 \end{bmatrix} \begin{bmatrix} b_0 - \beta_{0,0} \\ b_1 - \beta_{1,0} \end{bmatrix} \sim 2s^2 F_{2,n-2}$$

A proper test procedure (see Graybill (1976)) for the test is to reject H_0 if

$$[b_0 - \beta_{0,0}, b_1 - \beta_{1,0}] \begin{bmatrix} n & \sum_{i=1}^{n} x_i \\ \sum_{i=1}^{n} x_i & \sum_{i=1}^{n} x_i^2 \end{bmatrix} \begin{bmatrix} b_0 - \beta_{0,0} \\ b_1 - \beta_{1,0} \end{bmatrix} \geqslant 2s^2 F_{\alpha,2,n-2}$$

This is seen to be identical to the procedure that *rejects the joint hypothesis*, namely,

$$H_0: \begin{cases} \beta_0 = \beta_{0,0} \\ \beta_1 = \beta_{1,0} \end{cases}$$

at a level α *when the combination* $(\beta_{0,0}, \beta_{1,0})$ falls outside a $100(1-\alpha)\%$ confidence region on β_0 and β_1. Again, this is easily accomplished by observing the plot of the elliptical confidence region.

The Bonferroni confidence regions can also be used for testing the joint hypothesis. One merely computes the individual $100(1-\alpha/2)\%$ confidence intervals for β_0 and β_1 and rejects H_0 at a level α if at least one of the null values (either $\beta_{0,0}$ or $\beta_{1,0}$) falls outside the confidence interval on that parameter. As in the case of confidence intervals, one pays the price of conservation with the Bonferroni approach. In the case of hypothesis testing, since the rectangular area covered by the Bonferroni confidence regions is much larger than that of the elliptic confidence region (see Figure 2.11), more combinations of β_0 and β_1 are included in the acceptance region; this results in lower probability of rejection or lower power of the test. Please keep in mind that the chosen α level that governs the size of the confidence interval is an upper bound for the true probability of Type I error. For an illustration of simultaneous testing of β_0 and β_1, the reader should do Exercise 2.15.

2.11

A COMPLETE ANNOTATED COMPUTER PRINTOUT

Example 2.11 **Plastic Hardness Data**

In this example an annotated computer printout using PROC REG from SAS is given. Included is the analysis of variance, *t*-statistics, significance probabilities, predicted values (fitted values), residuals, confidence limits, and prediction limits on a new observation. The data involve the use of 10 plastic specimen, which are prepared and allowed to age for selected periods of time. Measurements of Brinell hardness are made on each specimen after aging. The data are as follows.

Specimen	1	2	3	4	5	6	7	8	9	10
Time (hr)	10	15	25	40	45	60	70	85	90	95
Hardness	216	222	237	262	271	308	324	369	375	379

The model $E(y|x) = \beta_0 + \beta_1 x$ was fit to the data. The response y is the Brinell hardness, and x is time. Table 2.9 shows the PROC REG run. This example will be used again as an illustration of analysis of residuals. See Exercise 2.17.

TABLE 2.9 PROC REG using plastic hardness data

ANALYSIS OF VARIANCE

SOURCE	DF	SUM OF SQUARES	MEAN SQUARE	F VALUE	PROB>F
MODEL	1	36194.79128	36194.79128	879.291	0.0001
ERROR	8	329.30872	41.16358989		
C TOTAL	9	36524.10000			

ROOT MSE	6.415886	R-SQUARE	0.9910
DEP MEAN	296.3	ADJ R-SQ	0.9899
C.V.	2.165334		

PARAMETER ESTIMATES

VARIABLE	DF	PARAMETER ESTIMATE	STANDARD ERROR	T FOR H0: PARAMETER=0	PROB > |T|
INTERCEP	1	187.81397	4.18344929	44.895	0.0001
TIME	1	2.02777620	0.06838386	29.653	0.0001

OBS	ACTUAL	PREDICT VALUE	STD ERR PREDICT	LOWER95% MEAN	UPPER95% MEAN	LOWER95% PREDICT	UPPER95% PREDICT
1	216.0	208.1	3.6007	199.8	216.4	191.1	225.1
2	222.0	218.2	3.3238	210.6	225.9	201.6	234.9
3	237.0	238.5	2.8133	232.0	245.0	222.4	254.7
4	262.0	268.9	2.2290	263.8	274.1	253.3	284.6
5	271.0	279.1	2.1105	274.2	283.9	263.5	294.6
6	308.0	309.5	2.0770	304.7	314.3	293.9	325.0
7	324.0	329.8	2.3215	324.4	335.1	314.0	345.5
8	369.0	360.2	2.9591	353.4	367.0	343.9	376.5
9	375.0	370.3	3.2166	362.9	377.7	353.8	386.9
10	379.0	380.5	3.4886	372.4	388.5	363.6	397.3

OBS	RESIDUAL
1	7.9083
2	3.7694
3	-1.5084
4	-6.9250
5	-8.0639
6	-1.4805
7	-5.7583
8	8.8250
9	4.6862
10	-1.4527

2.12 A LOOK AT RESIDUALS

In this section we provide a preliminary presentation of the type of information that can be gained from the ordinary residuals, i.e., $e_i = y_i - \hat{y}_i$, called the *errors of fit*. Many details will not be presented here; they must be relegated to a point where the notions in multiple regression will provide the machinery to study the statistical properties of residuals. In fact, Chapter 5 is completely dedicated to the study of residuals. However, the student of simple linear regression can, at this point, gain some insight into information that can be recovered by analysis and plotting of residuals.

The reader should view residuals as quantities through which we determine whether or not conditions are ideal. By *ideal* we refer to the assumptions made on the ε_i in Section 2.2. Severe violations of the assumptions $E(\varepsilon_i) = 0$ and $E(\varepsilon_i)^2 = \sigma^2$ for $i = 1, 2, \ldots, n$ can often be ascertained through analysis of residuals. The reader uses residuals as the quantities which most closely emulate the model errors, the ε_i. The ideal appearance of a plot of residuals against x, the regressor variable, depicts random values centered around zero as in Figure 2.12. By the nature of the least squares procedure, $\sum_{i=1}^{n} (y_i - \hat{y}_i) = 0$.

FIGURE 2.12 **Ideal plot of $(y_i - \hat{y}_i)$ against x_i for simple linear regression**

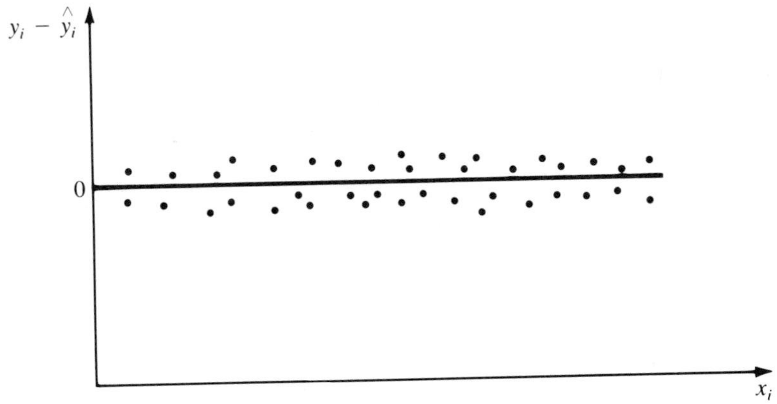

WHAT HAPPENS IF MODEL IS MISSPECIFIED?

In order to better understand how we detect nonideal conditions, it can be instructive to study how the residuals behave when one of our assumptions is violated. Let us presume that the data $(x_1, y_1), (x_2, y_2), \ldots, (x_n, y_n)$ are generated from the model

$$y_i = \beta_0^* + \beta_1(x_i - \bar{x}) + \beta_2 x_i^2 + \varepsilon_i \qquad (i = 1, 2, \ldots, n) \tag{2.27}$$

and $E(\varepsilon_i) = 0$ and $E(\varepsilon_i^2) = \sigma^2$ for $i = 1, 2, \ldots, n$. In other words, the conditions associated with the *quadratic regression* model of (2.27) are ideal. However, the analyst does not fit the model of (2.27) but, rather, assumes a simple linear regression model

$$y_i = \beta_0^* + \beta_1(x_i - \bar{x}) + \varepsilon_i^* \qquad (i = 1, 2, \ldots, n) \tag{2.28}$$

Obviously unknown to the analyst, $\varepsilon_i^* = \varepsilon_i + \beta_2 x_i^2$, and hence the error for the analyst's model is not truly random but has a systematic component, namely, $E(\varepsilon_i^*) = \beta_2 x_i^2$. Now, what impact does this have on the results and how is it transmitted into the residuals? Would the residuals of the *incorrect* simple linear model of (2.28) still look random around a mean of zero?

Consider the fitted simple linear model

$$\hat{y}_i = b_0^* + b_1(x_i - \bar{x})$$

The least squares estimates are given by (see Eq. (2.7) and (2.8))

$$b_0^* = \bar{y} = \frac{\sum_{i=1}^{n} [\beta_0^* + \beta_1(x_i - \bar{x}) + \beta_2 x_i^2 + \varepsilon_i]}{n} = \beta_0^* + \beta_2 \overline{x^2} + \bar{\varepsilon}$$

and

$$b_1 = \frac{\sum_{i=1}^{n} (x_i - \bar{x})[\beta_0^* + \beta_1(x_i - \bar{x}) + \beta_2 x_i^2 + \varepsilon_i]}{S_{xx}}$$

$$= \beta_1 + \frac{n\beta_2(\overline{x^3} - \bar{x}\overline{x^2})}{S_{xx}} + \frac{\sum_{i=1}^{n} (x_i - \bar{x})\varepsilon_i}{S_{xx}}$$

Now it is obvious that both b_0^* and b_1 are *biased estimators.* In fact

$$E(b_0^*) = \beta_0^* + \beta_2 \overline{x^2} \tag{2.29}$$

and

$$E(b_1) = \beta_1 + \frac{n\beta_2(\overline{x^3} - \bar{x}\overline{x^2})}{S_{xx}} \tag{2.30}$$

Now, what about the residuals? We need to deal with the properties of the $y_i - \hat{y}_i$, where $\hat{y}_i = b_0^* + b_1(x_i - \bar{x})$. Using (2.27), (2.29) and (2.30), we have

$$E(y_i - \hat{y}_i) = \beta_0^* + \beta_1(x_i - \bar{x})$$
$$+ \beta_2 x_i^2 - \left[\beta_0^* + \beta_2 \overline{x^2} + \beta_1(x_i - \bar{x}) + \frac{(x_i - \bar{x})n\beta_2(\overline{x^3} - \bar{x}\overline{x^2})}{S_{xx}}\right]$$
$$= \beta_2\left[(x_i^2 - \overline{x^2}) + n(x_i - \bar{x})\frac{(\overline{x^3} - \bar{x}\overline{x^2})}{S_{xx}}\right]$$

and thus

$$\boxed{E(y_i - \hat{y}_i) = \beta_2(x_i^2 + rx_i + s)} \tag{2.31}$$

where

$$r = -n(\overline{x^3} - \bar{x}\overline{x^2})/S_{xx}$$

$$s = -\left[\overline{x^2} + \frac{n\bar{x}(\overline{x^3} - \bar{x}\overline{x^2})}{S_{xx}}\right]$$

For the special case in which $\beta_2 = 0$, i.e., the fitted model is correct, the residuals have mean zero which we certainly expect. However, if $\beta_2 \neq 0$, the expected value of the ith residual is a quadratic function of x_i. As a result, a plot of the residuals against x should produce a curvilinear type of appearance as we notice from the relationship in (2.31). The reader should also observe from (2.29) and (2.30) that the ignored model term $\beta_2 x_i^2$ transmits bias to model coefficients b_0 and b_1. A pictorial representation of what one might observe in a residual plot appears in Figure 2.13. The vertical deviations in Figure 2.13(a) are transmitted to the residual plot in Figure 2.13(b). Notice that the residuals still fluctuate around zero, but in a nonrandom fashion. Additional study of residual plots to detect model misspecification with focus on different forms of the residuals will be presented in Chapter 5.

FIGURE 2.13 ***(a)*** A curvilinear set of data ***(b)*** A residual plot for the data of part (a)

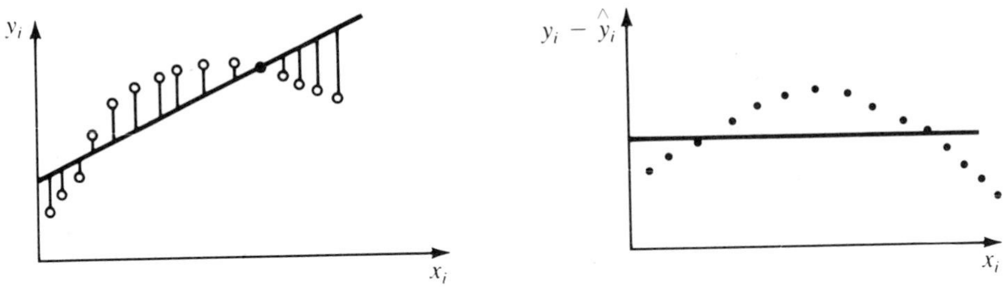

AID IN DETECTION OF HETEROGENEOUS VARIANCE

Treatment of residuals can play an important role in diagnosing the problem of heterogeneous variance. While plotting of ordinary residuals may be useful, the use of *studentized residuals* discussed in detail in Chapter 5 are preferable for this purpose. As a result, the reader should refer to Chapter 5 where the material is presented in the context of *multiple regression*.

DETECTION OF NORMALITY (NORMAL PROBABILITY PLOTS)

As we have indicated in previous sections, the assumption of normality of the ε_i is made in the case of some of the techniques in simple linear regression. The t-tests, F-tests, and confidence limits on parameters require the assumption of normality.

Further motivation for detection of violations of the normality assumption lies in the fact that the *method of least squares possesses more desirable properties when the errors are normal.* The details regarding the properties of least squares under normal and nonnormal error distributions are given in Chapter 3 as we generalize to multiple linear regression.

Before we embark on methods for detection of normality, we should indicate that minor departures from normality have little impact on the regression results. In addition, false detection of a departure from normality could arise from a *model misspecification.* A naive analyst, who has not thoroughly searched for the proper model, may detect a problem and diagnose it as a departure from normality on the errors when the anomaly was produced from a misspecified model. As we shall show in the development that follows, the detection of nonnormal errors depends on the assumption that $E(\varepsilon_i) = 0$, a condition that will not hold if the model is grossly unspecified.

The detailed properties of residuals in multiple linear regression will reveal that residuals are not uncorrelated nor do they have common variance, even if the standard assumptions on the ε_i hold. (See Chapter 5.) When ideal conditions on the ε_i hold, the correlation among residuals is not severe. However in some regression data sets, the residuals may experience large differences in their variances. Many computer plotting routines do not take this into account. For the simple linear model case consider

$$\begin{aligned}\operatorname{Var}(y_i - \hat{y}_i) &= \operatorname{Var} y_i + \operatorname{Var} \hat{y}_i - 2\operatorname{Cov}(y_i, \hat{y}_i) \\ &= \sigma^2 + \sigma^2\left(\frac{1}{n} + \frac{(x_i - \bar{x})^2}{S_{xx}}\right) - 2\operatorname{Cov}(y_i, \bar{y} + b_1(x_i - x))\end{aligned}$$

The covariance term in this expression is written as

$$\operatorname{Cov}\left[y_i, \bar{y} + \frac{S_{xy}}{S_{xx}}(x_i - \bar{x})\right] = \frac{\sigma^2}{n} + \frac{\sigma^2(x_i - \bar{x})^2}{S_{xx}}$$

Thus we can write the variance of the ith residual as

$$\boxed{\operatorname{Var}(y_i - \hat{y}_i) = \sigma^2\left[1 - \left(\frac{1}{n} + \frac{(x_i - \bar{x})^2}{S_{xx}}\right)\right]} \tag{2.32}$$

The reader should recognize the portion

$$\frac{1}{n} + \frac{(x_i - \bar{x})^2}{S_{xx}}$$

as $\operatorname{Var} \hat{y}_i/\sigma^2$, which played an important role in development of confidence intervals on mean response discussed in Section 2.9. Notice that residuals at x-locations close to the data center have larger variances than those located at more extreme locations. In conceptualizing the notion of a variance of the ith residual, the reader should visualize the variability that occurs in the residual at $x = x$, if one resamples, i.e., observes, new y observations over and over at the same levels of x.

The presence of unequal variances among the residuals produces some difficulty in the use of residuals as a mechanism for detection of nonideal conditions in the ε_i, for example, nonnormality. In what follows we discuss the types of plots that can be used for detection of the violation of the normality assumption. We eventually make use of what we learned in Eq. (2.32) to standardize the residuals so that they may better emulate the true model errors, i.e., the ε_i.

WHAT SHOULD BE PLOTTED FOR DETECTION OF NONNORMALITY?

In order to detect severe departures from normality, the analyst may view graphically a *normal probability plot* of the residuals. The plot is based on a statistical theory of order statistics that indicates that for an independent sample from a normal distribution, with mean μ and variance σ^2, the expected value of the ith *smallest is given by*

$$\mu + \sigma\mu_{(i)} \qquad \text{where} \quad \mu_{(i)} \cong z\left(\frac{i - 0.375}{n + 0.25}\right)$$

Here $z(w)$ denotes the $w(100)$ percentile of the standard normal distribution; in other words, if z is a standard normal variate,

$$\Pr(z \leqslant z(w)) = w$$

Thus the *true model errors*, i.e., the ε_i, when plotted against their expected value, $\mu_{(i)}$ should produce a plot that is reasonably close to a *straight line* through the origin since $\mu = 0$ for the ε_i. Now, of course, the residuals are the quantities that the analyst has available. Thus the *normal probability plot* is a plot of residuals against their expected values. If the residuals are suitably standardized, the straight line should go through the origin with a slope of 1.0. Two possible approaches are as follows:

(i) Plot $\frac{(y_i - \hat{y}_i)}{s}$ against $z\left(\frac{i - 0.375}{n + 0.25}\right)$. Here $(y_i - \hat{y}_i)$ is the ith smallest residual.

why so? to get 45° but won't affect test using CORR

(ii) Form a standardized form of the residual, i.e., form

$$\boxed{r_i = \frac{(y_i - \hat{y}_i)}{s_{(y_i - \hat{y}_i)}}} \qquad (2.33)$$

where, from Eq. (2.32), the standard error of the residual is given by

$$s_{(y_i - \hat{y}_i)} = s\sqrt{1 - \left[\frac{1}{n} + \frac{(x_i - \bar{x})^2}{S_{xx}}\right]} \tag{2.34}$$

Again, $y_i - \hat{y}_i$ is the ith smallest residual.

In both cases plot the standardized form of the residual against

$$z\left(\frac{i - 0.375}{n + 0.25}\right)$$

A departure from a straight line appearance is an indication of departures from normality. The utility of method (ii), of course, lies in the fact that since, in general, the residuals do not have equal variance, one is *standardizing each properly*.

The residual given in (2.33) is often called a *studentized residual*. The studentized residuals will resurface in Chapter 5 as we provide further development on the use of residuals. In Example 2.12 we use both methods to produce normal probability plots on the accounting return data of Section 2.3. The value

$$z\left(\frac{i - 0.375}{n + 0.25}\right)$$

provides an excellent approximation of the expected value of the ith order statistic. The true values appear in Table C.5 of Appendix C. They are often called rankits. They are tabled for n up to 20. Many commercial software packages have procedures for normal probability plots.

In the ideal situation the plot of the ranked studentized residuals against their expected values will display a straight line through the origin with a slope of 1.0. Deviations from this ideal picture theoretically reflect departures from normality. Experience is necessary for the development of sound judgment in the use of these plots. Certainly there will be random deviations from the straight line. As in the case of all types of diagnostic plots, sample size is a factor. Very little can be learned in cases where the sample size is small.

It is often very difficult to pinpoint exactly the outcome of the apparent departure from normality on the basis of the normal plots. In real data sets, failure of the homogeneous variance assumption may, indeed, produce the appearance of a deviation from normality. In addition, isolated cases of outliers with unusually large residuals may show similar departure from the straight line. A thorough discussion of outliers appears in Chapter 5. However, it is informative to have access to specific appearances in the deviation from the straight line. They are shown in Figure 2.14. In the plots the ranked studentized residuals are on the vertical axis. The heavy-tailed error distributions reflect a higher frequency of extreme observations than the normal distribution. In the case of the light-tailed appearance, fewer extreme observations appear than expected. Heterogeneous variance will often produce a light-tailed error distribution. Model functional form misspecification will often result in an apparently skewed error distribution.

FIGURE 2.14 **Typical residual plots of ranked residuals**

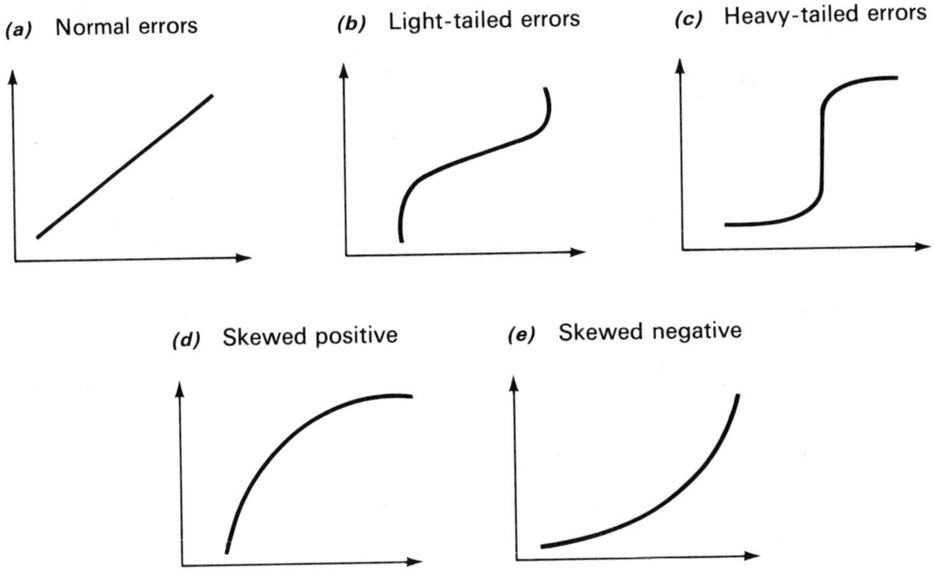

Example 2.12 ***Accounting Return Data—Normal Probability Plot***

Consider the data of Table 2.1. In Figures 2.15 and 2.16 normal probability plots on the residuals are shown. In Figure 2.15 the plot involves method (i), i.e., a plot of the ranked residuals standardized by the root error mean square, s. The abscissa is

$$z\left(\frac{i-0.375}{n+0.25}\right)$$

Figure 2.16 uses method (ii) with studentized residuals.

In this case there does seem to be a departure from the straight line, thus suggesting some type of departure from the ideal. The figures depict a deviation from symmetry in the distribution of residuals. This condition is apparent in the original plot of the data in Figure 2.3. Notice how the residuals that are larger in magnitude are above the regression line. The dispersion created by a few companies with relatively high marketing rates has created the nonsymmetry.

FIGURE 2.15 **Normal probability plot for accounting return data**

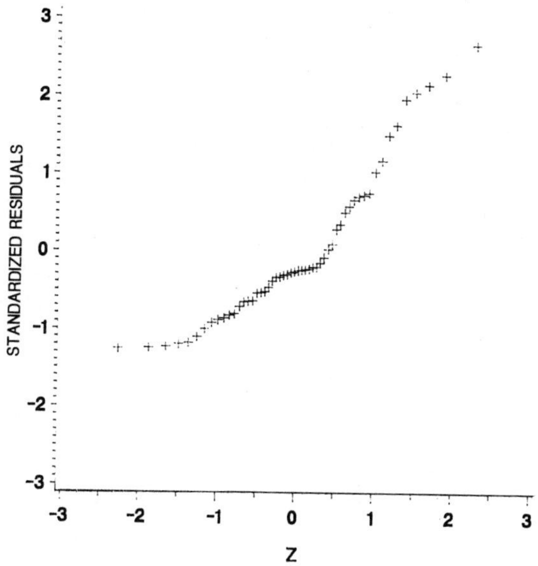

FIGURE 2.16 **Normal probability plot for accounting return data**

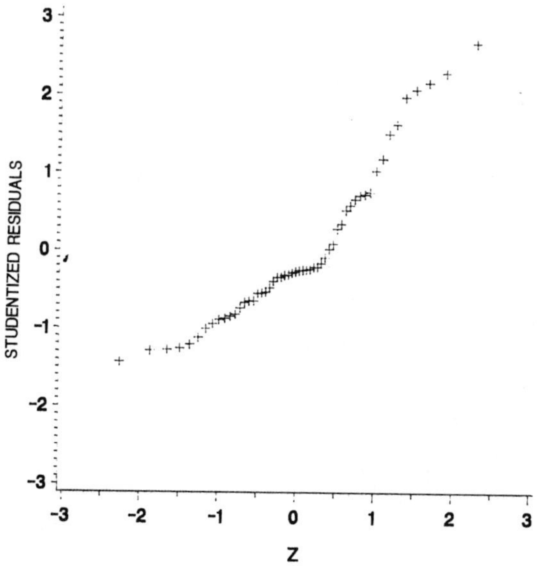

ADDITIONAL COMMENTS

The serious student or analyst interested in residual plots and residual analysis should read the material in Chapter 5. More needs to be developed regarding residual analysis in the context of multiple regression. The task of uncovering violation of assumptions is often more difficult in the case of multiple regressor variables. It should be clearly understood that detection of nonnormality is often a difficult condition to interpret. Again, paraphrasing the point to which we alluded earlier in this section, the use of normal probability plots implies that the $y_i - \hat{y}_i$ are indeed emulating the ε_i and, of course, we assume the latter are independent with mean zero and common variance. In many cases, a condition that appears to suggest nonnormal errors has merely uncovered a violation of another assumption.

In Chapters 5 and 6, we will have access to more technical machinery that allows a better understanding of residuals and residual plots. Residuals are used extensively to construct other types of diagnostics.

Aid in Isolating Outliers or Erroneous Data Points

Clearly, when data points exist where residuals are unusually high, the temptation is to suspect that something went wrong and that the data point is an outlying observation (or what is often referred to as an outlier). The obvious question is "How large is too large?" The issue is important and rather complex. At this point it is sufficient to declare that the analyst should not provide any arbitrary criterion. Details are given in Chapter 5 with illustrations.

Aid in Detection of High Influence Points

In Chapter 6, much attention is focused on the implication of high influence data points, i.e., those points that have a large amount of impact on the fitted regression. The analyst needs to know where these specific points are and how significant the influence is. Information that quantifies this influence can be garnered from residuals.

2.13 BOTH x AND y RANDOM

In all of the preceding development, one of the basic assumptions made revolved around the nature of x, the *regressor* variable. We assumed throughout that the x_i in the simple linear regression model are nonrandom. All standard errors of coefficients, tests of hypotheses, and confidence and prediction intervals depend on this assumption. Two other situations occur quite frequently in practice, however:

1. The variables x and y are both random and are observations from a joint density function.

2. The variable x is measured with nonnegligible error.

Some treatment will be given to case (2) in Chapter 7. In the first case, generally less interest is associated with prediction than in the case where x is assumed not random. When x and y are jointly distributed, there is a natural requirement to learn, from the data, something about the structure of the relationship and the degree of association. This leads to the topic of *correlation analysis.*

THE CORRELATION COEFFICIENT—CORRELATION ANALYSIS

As an example of a practical situation in which the structure of the dependency between two random variables is important, consider an anthropological study to determine the *strength* of the linear association between x and y, the length and circumference respectively of a particular bone in the adult body. Thus n experimental units (adults) were chosen randomly and the measurements (x_i, y_i) $i = 1, 2, \ldots, n$, become realizations of jointly distributed random variables having joint density function $f(x, y)$.

It is convenient to assume that the conditional distribution $f(y \mid x)$ is normal with mean

$$E(y \mid x) = \beta_0 + \beta_1 x \tag{2.35}$$

and variance

$$\sigma^2_{y|x} = \sigma^2$$

At this point there appears to be a strong resemblance between the present model and that assumed by Eq. (2.2) with the normal theory assumption. In a *conditional sense* they are alike, but the assumed scientific structure that produced the data are different. In the use of (2.2), the two scientific variables x and y relate mathematically in a manner that we postulate is linear in x. The randomness, brought about by the ε_i, appears principally from our inability to model precisely; thus we have the term *model error*. In the present case, the linear relationship of Eq. (2.35) only appears as a parameter in the conditional distribution $f(y \mid x)$.

The conditional distribution $f(y \mid x)$, producing a *linear model* for the conditional expectation in (2.35), is a product of the assumption of a *bivariate normal* distribution $f(y, x)$, given by

$$f(y, x) = \frac{1}{(2\pi)\sigma_x\sigma_y\sqrt{1-\rho^2}} \exp\left\{-\frac{1}{2(1-\rho^2)}\left[\frac{y-\mu_y}{\sigma_y}\right]^2 + \left[\frac{x-\mu_x}{\sigma_x}\right]^2 - 2\rho\left[\frac{x-\mu_x}{\sigma_x}\right]\left[\frac{y-\mu_y}{\sigma_y}\right]\right\}$$

for $-\infty < x < \infty$ and $-\infty < y < \infty$. Here μ_x, μ_y, σ_x, and σ_y are means and standard deviations of x and y. The parameter ρ is the *correlation coefficient* defined as the ratio of the covariance of x and y to the product of the standard deviations; namely

$$\rho = \frac{E(y-\mu_y)(x-\mu_x)}{\sigma_x \sigma_y} = \frac{\sigma_{xy}}{\sigma_x \sigma_y} \tag{2.36}$$

The interpretation of coefficient ρ is easily seen when we investigate how it relates to the other parameters. The linear structure in Eq. (2.35) is termed the *regression line*. Using straightforward methods of finding the conditional distribution $f(y \mid x)$ (see Graybill (1976)), we obtain the following normal density.

$$f(y \mid x) = \frac{1}{\sqrt{2\pi}\,\sigma} \exp\left[-\frac{1}{2}\left(\frac{y-\beta_0-\beta_1 x}{\sigma}\right)^2\right]$$

where

$$\beta_1 = \frac{\sigma_y}{\sigma_x} \cdot \rho \tag{2.37}$$

$$\beta_0 = \mu_y - \mu_x \rho \frac{\sigma_y}{\sigma_x} \tag{2.38}$$

$$\sigma^2 = \sigma_y^2 (1-\rho^2) \tag{2.39}$$

Equations (2.37), (2.38), and (2.39) help to illustrate the role of ρ and to draw a parallel between the model described here and that given by the fixed x case in (2.2). From (2.37) and (2.39),

$$\rho^2 = 1 - \frac{\sigma^2}{\sigma_y^2} \tag{2.40}$$

$$\rho^2 = \beta_1^2 \frac{\sigma_x^2}{\sigma_y^2} \tag{2.41}$$

Thus the value of $\rho = 0$ when $\beta_1 = 0$; i.e., there is no linear regression. Since σ^2 is the variance in a conditional distribution, $\sigma^2 \leqslant \sigma_y^2$, and hence $\rho^2 \leqslant 1$. As a result, we have

$$-1 \leqslant \rho \leqslant 1$$

From (2.37), the sign carried by ρ is that of β_1, the slope of the regression line. A value of $\rho = \pm 1$ can occur if $\sigma^2 = 0$, and of course, this implies a perfect linear relationship between y and x. Thus ρ describes the *degree of linear association* between the variables.

PARAMETER ESTIMATION AND HYPOTHESIS TESTING

It is clear that a correlation analysis should involve estimation of ρ. The interpretation of the strength of association would depend on the closeness of the estimate to 1.0 (or -1.0). Tests of hypotheses on ρ can also be part of the analyst's arsenal of tools here.

The estimators of ρ, β_0, β_1, and σ^2 generate more parallels between the present case and the fixed x situation that prevailed in this chapter prior to Section 2.13. We shall consider estimation of ρ first. The usual estimator is the *sample correlation coefficient*, r, given by

$$r = \frac{S_{xy}}{\sqrt{S_{xx}S_{yy}}} \tag{2.42}$$

Further intuition is gained from the fact that, using the least squares estimator

$$b_1 = \frac{S_{xy}}{S_{xx}} \tag{2.43}$$

as an estimator for β_1, we have

$$r^2 = \frac{S_{xy}^2}{S_{xx}S_{yy}} = \frac{b_1 S_{xy}}{S_{yy}} = \frac{SS_{\text{Reg}}}{S_{yy}} = 1 - \frac{SS_{\text{Res}}}{S_{yy}} \tag{2.44}$$

The result in (2.44) is the R^2 or coefficient of determination discussed in Section 2.8. Thus the *squared sample correlation coefficient* is equivalent to the *proportion of variation explained.*

Equations (2.42) and (2.43) represent expressions for the estimators of the two important parameters ρ and β_1, the slope of the regression. The intercept β_0 is estimated by

$$b_0 = \bar{y} - b_1 \bar{x} \tag{2.45}$$

It should come as no surprise that the estimators for slope and intercept for the random x case and for the fixed x case are the same. In addition to the estimates of ρ, β_0, β_1, the data analyst may also be keenly interested in a hypothesis testing mechanism for assessing the evidence, either for or against a significant linear association between the variables x and y. Thus interest centers around the hypothesis

$$\begin{aligned} H_0&: \rho = 0 \\ H_1&: \rho \neq 0 \end{aligned} \tag{2.46}$$

Since $\rho = 0$ when $\beta_1 = 0$, one might expect the test of H_0 in (2.46) to resemble the test on β_1 discussed in Section 2.6. It turns out that

$$\frac{r\sqrt{n-2}}{\sqrt{1-r^2}} \sim t_{n-2}$$

under $H_0: \rho = 0$ (see Graybill (1976)). Thus

$$t = \frac{r\sqrt{n-2}}{\sqrt{1-r^2}} \tag{2.47}$$

is an appropriate statistic for testing H_0 of (2.46). A close inspection reveals that

$$\frac{r\sqrt{n-2}}{\sqrt{1-r^2}} = \frac{\dfrac{S_{xy}}{\sqrt{S_{yy}S_{xx}}}\sqrt{n-2}}{\sqrt{1-\dfrac{S_{xy}^2}{S_{yy}S_{xx}}}} = \frac{b_1\sqrt{S_{xx}}}{\sqrt{\dfrac{S_{yy}-S_{xy}^2/S_{xx}}{n-2}}}$$

which is the t-statistic in (2.14) for testing $H_0\colon \beta_1 = 0$ in the fixed x case. A two-tailed t-test of H_0 in (2.46) is made using the test statistic in (2.47). The degrees of freedom for the t-distribution are $n-2$. If the hypothesis is one sided, then the corresponding one-tailed test should be employed.

In cases where it is of interest to test a more general hypothesis

$$H_0\colon \rho = \rho_0$$

the analyst may take advantage of the approximate normality of the statistic

$$\boxed{\tfrac{1}{2}\ln\frac{1+r}{1-r}} \qquad \textbf{(2.48)}$$

For large samples ($n \geqslant 25$), the statistic in (2.48) is approximately normal with mean $\frac{1}{2}\ln[(1+\rho)/(1-\rho)]$ and variance $1/(n-3)$. Thus the test would involve standardizing the statistic in (2.48), namely forming the statistic

$$z = \frac{\sqrt{n-3}}{2}\left[\ln\left(\frac{1+r}{1-r}\right) - \ln\left(\frac{1+\rho_0}{1-\rho_0}\right)\right] \qquad \textbf{(2.49)}$$

and comparing to the critical points of the standard normal distribution.

Example 2.13 ***Loblolly Pine Data***

Consider the data of Table 2.10 used to study the correlation between certain important variables in loblolly pine trees. A set of 29 trees were selected randomly with the following data taken on specific gravity and a crucial mechanical property called modulus of rupture in kilopascals (KPa). The correlation between the two variables is important to assess. In addition, the analyst desires to test

$$H_0\colon \rho = 0$$
$$H_1\colon \rho > 0$$

From the data, we compute the following statistics.

$$S_{xx} = 0.11273$$
$$S_{yy} = 11{,}807{,}324{,}786$$
$$S_{xy} = 34{,}422.75972$$

The sample correlation coefficient is computed using Eq. (2.42). The result is given by

$$r = 0.9435$$

The value of r would suggest a reasonably strong linear association between modulus of rupture and specific gravity. The t-statistic given by (2.47) can be used to test the hypothesis. We can write the equivalent form

$$t = \frac{r\sqrt{n-2}}{\sqrt{1-r^2}} = \frac{0.9435\sqrt{27}}{\sqrt{1-0.9435^2}} = 14.79$$

The t-statistic is significant at well below the 0.0001 level, giving strong statistical evidence of a linear association between the two variables.

TABLE 2.10 **Modulus of rupture and specific gravity for loblolly pine trees**

Specific Gravity x (gm/cm^3)	Modulus of Rupture y (KPa)	Specific Gravity x (gm/cm^3)	Modulus of Rupture y (KPa)
0.414	29,186	0.548	67,095
0.383	29,266	0.581	85,156
0.399	26,215	0.557	69,571
0.402	30,162	0.550	84,160
0.442	38,867	0.531	73,466
0.422	37,831	0.550	78,610
0.466	44,576	0.556	67,657
0.500	46,097	0.523	74,017
0.514	59,698	0.602	87,291
0.530	67,705	0.569	86,836
0.569	66,088	0.544	82,540
0.558	78,486	0.557	81,699
0.577	89,869	0.530	82,096
0.572	77,369	0.547	75,657
		0.585	80,490

"Quantitative Anatomical Characteristics of Plantation Grown Loblolly Pine (*Pinus taeda* L.) and Cottonwood (*Populus deltoides* Bart. ex Marsh.) and Their Relationships to Mechanical Properties," Department of Forestry and Forest Products, Virginia Polytechnic Institute and State University, Blacksburg, Virginia, 1983.

EXERCISES FOR CHAPTER 2

2.1 Consider the simple linear regression model given by Eq. (2.2). Consider the least squares *residuals*, given by $y_i - \hat{y}_i$ $(i = 1, 2, \ldots, n)$. Show that

$$\sum_{i=1}^{n} \frac{\hat{y}_i}{n} = \bar{y}$$

2.2 Show that, for the simple linear regression model,

(a) $\sum_{i=1}^{n} (y_i - \hat{y}_i) = 0$ (b) $\sum_{i=1}^{n} (y_i - \hat{y}_i)x_i = 0$

2.3 The estimator of the error variance, σ^2, discussed in Section 2.3 is given by

$$s^2 = \sum_{i=1}^{n} \frac{(y_i - \hat{y}_i)^2}{(n-2)}$$

Show that the estimator is unbiased; i.e., prove $E(s^2) = \sigma^2$.

2.4 We learned in Section 2.12 that if the simple linear regression model is misspecified and the true model is quadratic, then the residuals show a quadratic trend. In the development we also learned that the estimators b_0^* and b_1 are not unbiased. Is s^2 a biased estimator of σ^2? If so, show the bias. (*Comment*: Model underspecification is considered quite generally in Section 4.3.)

2.5 The regression sum of squares for the simple linear regression model of Eq. (2.2) is given by

$$SS_{\text{Reg}} = \sum_{i=1}^{n} (\hat{y}_i - \bar{y})^2$$

Show that $E(SS_{\text{Reg}}) = E(MS_{\text{Reg}}) = \sigma^2 + \beta_1^2 S_{xx}$. Make use of Eq. (B.3) in Appendix B.1.

2.6 Consider the set of data (y_i, x_i), $i = 1, 2, \ldots, n$, and the following two candidate models

$$y_i = \beta_0 + \beta_1 x_i + \varepsilon_i \qquad (i = 1, 2, \ldots, n) \qquad \text{(Model A)}$$
$$y_i = \gamma_0 + \gamma_1 x_i + \gamma_2 x_i^2 + \varepsilon_i \qquad (i = 1, 2, \ldots, n) \qquad \text{(Model B)}$$

Suppose both are fit to the same data. Show $SS_{\text{Res},A} \geqslant SS_{\text{Res},B}$. (*Hint*: Consider model A as a rival to model B. Which of the following will result in the smallest residual *SS*? Model B with $\gamma_0 = b_0$, $\gamma_1 = b_1$, $\gamma_2 = 0$ (b_0 and b_1 are estimated from model A); or Model B with γ_0, γ_1 and γ_2 replaced by the least squares estimators.)

2.7 For the accounting return data of Table 2.1, compute a t-statistic for testing

$$H_0: \beta_1 = 0$$
$$H_1: \beta_1 \neq 0$$

where β_1 is the slope of the regression model given by Eq. (2.2).

2.8 Consider the zero intercept model given by

$$y_i = \beta_1 x_i + \varepsilon_i \qquad (i = 1, 2, \ldots, n)$$

with the ε_i normal, independent, with variance σ^2. Let x_0 denote an arbitrary value of x. Show that the $100(1-\alpha)\%$ confidence interval on $E(y \mid x_0)$ is given by

$$b_1 x_0 \pm t_{\alpha/2, n-1} s \sqrt{\frac{x_0^2}{\sum_{i=1}^n x_i^2}}$$

where $s = \sqrt{\sum_{i=1}^n (y_i - b_1 x_i)^2/(n-1)}$ and $b_1 = \dfrac{\sum_{i=1}^n y_i x_i}{\sum_{i=1}^n x_i^2}$.

2.9 An experiment was conducted to study the mass of a tracer material exchanged between the main flow of an open channel and the "dead zone" caused by a sudden open channel expansion. Researchers need this information to improve the water quality modeling capability of a river. It is important to determine the exchange constant K for varying flow conditions. The value of K describes the exchange process when a dead zone appears. In this study, values of the Froude Numbers (N_F) were used to predict K. Froude Numbers are functions of upstream channel velocity and water depth. The data are as follows:

Observation	N_F	K
1	0.012500	−0.12562
2	0.023750	−0.12062
3	0.025625	−0.09625
4	0.030000	−0.08062
5	0.033125	−0.07937
6	0.038125	−0.07312
7	0.038125	−0.07250
8	0.038125	−0.07187
9	0.041250	−0.09375
10	0.043125	−0.05312
11	0.045000	−0.05750
12	0.046875	−0.05750
13	0.047500	−0.02000
14	0.050000	−0.04375
15	0.051250	−0.02000
16	0.056250	−0.05125
17	0.062500	−0.03125
18	0.068750	−0.04875
19	0.077500	−0.01687
20	0.044000	−0.04625

Linda Sharon Weiss, "Laboratory Study of Tracer Trapping and Release in Dead Zones of an Open Channel Expansion" (Masters Thesis, Department of Civil Engineering, Virginia Polytechnic Institute and State University, Blacksburg, Virginia, 1981.)

The negative sign of the K values indicate "flushing," the direction of mass transfer out of the dead zone.

(a) Use the least squares procedure to fit the model

$$K_i = \beta_0 + \beta_1 (N_F)_i + \varepsilon_i$$

(b) Compute R^2, s^2 and confidence limits on the mean response at the data locations.

(c) Compute the residuals at each data point.

2.10 Physical fitness testing is an important aspect of athletic training. A common measure of the magnitude of cardiovascular fitness is the maximum volume of oxygen uptake during a strenuous exercise. A study was conducted on 24 middle-aged men to study the influence of the time that it takes to complete a 2-mile run. The oxygen uptake measure was accomplished with standard laboratory methods as the subjects performed on a motor driven treadmill. The data are as follows:

Subject	Maximum Volume of O_2 (y)	Time in Seconds (x)
1	42.33	918
2	53.10	805
3	42.08	892
4	50.06	962
5	42.45	968
6	42.46	907
7	47.82	770
8	49.92	743
9	36.23	1045
10	49.66	810
11	41.49	927
12	46.17	813
13	48.18	858
14	43.21	860
15	51.81	760
16	53.28	747
17	53.29	743
18	47.18	803
19	56.91	683
20	47.80	844
21	48.65	755
22	53.69	700
23	60.62	748
24	56.73	775

P.M. Ribisl and Kachadorian, "Maximal Oxygen Intake Prediction in Young and Middle Aged Males," *Journal of Sports Medicine*, 9 (1969): 17–22.

(a) Estimate the parameters of a simple linear regression

$$y_i = \beta_0 + \beta_1 x_i + \varepsilon_i \qquad (i = 1, 2, \ldots, 24)$$

(b) Does the time it takes to run a distance of two miles have a significant influence on maximum oxygen uptake? Use

$$H_0: \beta_1 = 0$$
$$H_1: \beta_1 \neq 0$$

to answer the question.

2.11 In an experiment to determine the influence of certain physical measures on the performance of punters in American football, 13 punters were used as subjects in an experiment in which the average distance on 10 punts was measured. In addition, measures of left leg and right leg strength (lb lifted) were taken via a weight lifting test. The following data were taken. All subjects use their right legs for punting.

Subject	Left Leg (lb)	Right Leg (lb)	Average Punting Distance
1	170	170	162 ft 6 in.
2	130	140	144 ft 0 in.
3	170	180	147 ft 6 in.
4	160	160	163 ft 6 in.
5	150	170	192 ft 0 in.
6	150	150	171 ft 9 in.
7	180	170	162 ft 0 in.
8	110	110	104 ft 10 in.
9	110	120	105 ft 8 in.
10	120	130	117 ft 7 in.
11	140	120	140 ft 3 in.
12	130	140	150 ft 2 in.
13	150	160	165 ft 2 in.

Data analyzed for the Department of Health, Physical Education and Recreation by the Statistical Consulting Center, Virginia Polytechnic Institute and State University, Blacksburg, Virginia, 1983.

(a) Fit a simple linear regression with punting distance as the response and right leg strength as the independent or regressor variable.
(b) Fit a simple linear regression with left leg strength as the regressor variable.

2.12 Consider the Naval manpower data of Example 2.5.
(a) Fit a regression model containing an intercept.
(b) Compute R^2 (intercept model), $R^2_{(0)}$ and $R^2_{(0)*}$ for the zero intercept model.
(c) Compute s^2 for both the intercept and the zero intercept model.
(d) Compute the confidence limits on $E(y \mid x_i)$ at the 22 sample installations. Perform the computations for both the intercept and zero intercept models.
(e) Use the information in (a)–(d) to choose between the intercept model and zero intercept model.

2.13 Discuss the computation of a $100(1-\alpha)\%$ confidence interval on the slope for the case of the zero intercept model.

2.14 (a) Consider the data of Exercise 2.10. Attach 95% simultaneous confidence intervals on the mean max volume of O_2 for the following values of x, the time to run 2 miles.
(i) $x = 750$ (ii) $x = 775$ (iii) $x = 800$ (iv) $x = 825$ (v) $x = 850$
(b) Attach joint 95% confidence intervals on the slope and intercept. Use a plot to illustrate your results.

2.15 The study "Development of LIFTEST, A Dynamic Technique to Assess Individual Capability to Lift Material" was conducted at Virginia Polytechnic Institute and State University in 1982 to determine if certain static arm strength measures have influence on the "dynamic lift" characteristics of an individual. Twenty-five individuals were subjected to strength tests and then were asked to perform a weight-lifting test in which weight was dynamically lifted overhead. The data are as follows:

Individual	Arm Strength, x	Dynamic Lift, y
1	17.3	71.4
2	19.3	48.3
3	19.5	88.3
4	19.7	75.0
5	22.9	91.7
6	23.1	100.0
7	26.4	73.3
8	26.8	65.0
9	27.6	75.0
10	28.1	88.3
11	28.2	68.3
12	28.7	96.7
13	29.0	76.7
14	29.6	78.3
15	29.9	60.0
16	29.9	71.7
17	30.3	85.0
18	31.3	85.0
19	36.0	88.3
20	39.5	100.0
21	40.4	100.0
22	44.3	100.0
23	44.6	91.7
24	50.4	100.0
25	55.9	71.7

(a) Estimate, with least squares, the linear regression line.
(b) It is hypothesized that $\beta_0 = 0$, $\beta_1 = 2.2$. Test the appropriate joint hypothesis. Use a 95% joint confidence region on β_0 and β_1 to illustrate your conclusions.

(c) Studentized residuals were discussed in Section 2.12. Plot the studentized residuals against x. (*Comment*: Refer to Chapter 5 for further discussion regarding studentized residuals and how they may be used for further analysis of these data.)

2.16 Observations on the yield of a chemical reaction taken at various temperatures were recorded as follows:

x (°C)	y (%)
150	77.4
150	76.7
150	78.2
200	84.1
200	84.5
200	83.7
250	88.9
250	89.2
250	89.7
300	94.8
300	94.7
300	95.9

(a) Fit a simple linear regression, estimating β_0 and β_1.
(b) Compute 95% confidence intervals on $E(y\,|\,x)$ at the 4 levels of temperature in the data. Plot the upper and lower confidence intervals around the regression line.
(c) Plot a 95% confidence region on the regression line. Plot on the same graph for part (b). Comment.

2.17 Consider the data of Example 2.11 in Section 2.11. Plot the residuals on a graph against x. Comment on the appropriateness of the simple linear regression model.

2.18 Consider the wood density data of Example 2.2 in Section 2.6. Construct a normal probability plot, and comment regarding the normal error assumption.

2.19 An entomological experiment was conducted at Virginia Polytechnic Institute and State University to study the survivability of stalk borer larvae. Nine chambers were used to simulate different conditions. It was of interest to develop a model relating the mean size of larvae as a function of the stalk head diameter.

Chamber	Head Diameter (cm), x	Size of Larvae (cm), y
1	333	2.274
	624	2.765
	1102	2.912
	1352	2.831
	1643	2.859

TABLE Continued

TABLE Continued

Chamber	Head Diameter (cm), x	Size of Larvae (cm), y
2	333	2.081
	624	2.726
	1102	2.869
	1352	2.881
	1643	2.814
3	397	2.129
	744	2.725
	1314	2.823
	1609	2.875
	1950	2.921
4	397	2.533
	744	2.823
	1314	2.921
	1609	2.873
	1950	2.903
5	461	2.281
	864	2.669
	1526	2.730
	1871	2.818
	2274	2.767
6	461	2.133
	864	2.577
	1526	2.692
	1871	2.832
	2274	2.769
7	295	2.056
	558	2.617
	831	2.780
	1070	2.793
	1362	2.871
8	237	1.407
	444	2.148
	784	2.570
	962	2.661
	1169	2.757
9	109	1.352
	204	1.603
	360	1.929
	442	2.103
	537	2.180

(a) Estimate (and plot) the linear regression line that relates larvae size to stalk head diameter.

(b) Plot the residuals against x, the head diameter. Comment on appropriateness of the simple linear regression model. Can you comment on the effect of the different chamber conditions? These data will be the subject of another exercise in Chapter 3.

2.20 Consider the fixed zero intercept model of Section 2.7. The appropriate estimator of σ^2 is given by

$$s^2 = \sum_{i=1}^{n} \frac{(y_i - \hat{y}_i)^2}{(n-1)}$$

Show that s^2 is unbiased for σ^2.

2.21 Suppose the experimenter postulates a model

$$y_i = \beta_0 + \beta_1 x_i + \varepsilon_i \qquad (i = 1, 2, \ldots, n)$$

where β_0 is known.

(a) What is the appropriate least squares estimator of β_1? Justify your answer.
(b) What is the variance of the estimator of the slope in (a)?
(c) Write out an expression for the confidence interval on $E(y \mid x)$ for this model.

2.22 Consider the studentized residuals discussed in Section 2.12:

$$\frac{y_i - \hat{y}_i}{s\sqrt{1 - (1/n) - [(x_i - \bar{x})^2/S_{xx}]}}$$

The denominator is found by merely constructing the variance of $y_i - \hat{y}_i$, namely

$$\text{Var}(y_i - \hat{y}_i) = \sigma^2\left[1 - \frac{1}{n} - \frac{(x_i - \bar{x})^2}{S_{xx}}\right]$$

and then standardizing $y_i - \hat{y}_i$.

(a) Show that

$$\sum_{i=1}^{n} \frac{\text{Var}(y_i - \hat{y}_i)}{\sigma^2} = n - 2$$

(b) Under the conditions that the ε_i are i.i.d. $N(0, \sigma)$, does the studentized residual have a t-distribution with $n-2$ degrees of freedom? If not, why not?

2.23 Consider a situation in which the regression data set is divided into two parts as follows.

x	y	
x_1	y_1	Portion 1
x_2	y_2	
$\vdots$	$\vdots$	
x_{n_1}	y_{n_1}	
x_{n_1+1}	y_{n_1+1}	Portion 2
$\vdots$	$\vdots$	
$x_{n_1+n_2}$	$y_{n_1+n_2}$	

The model is given by

$$\begin{aligned} y_i &= \beta_0^{(1)} + \beta_1 x_i + \varepsilon_i \qquad i = 1, 2, \ldots, n_1 \\ &= \beta_0^{(2)} + \beta_1 x_i + \varepsilon_i \qquad i = n_1 + 1, \ldots, n_1 + n_2 \end{aligned}$$

In other words there are two regression lines with common slope. Using the centered model,

$$\begin{aligned} y_i &= \beta_0^{(1)*} + \beta_1(x_i - \bar{x}_1) + \varepsilon_i \qquad i = 1, 2, \ldots, n_1 \\ &= \beta_0^{(2)*} + \beta_1(x_i - \bar{x}_2) + \varepsilon_i \qquad i = n_1 + 1, \ldots, n_1 + n_2 \end{aligned}$$

where

$$\bar{x}_1 = \sum_{i=1}^{n_1} \frac{x_i}{n_1} \qquad \bar{x}_2 = \sum_{i=n_1+1}^{n_1+n_2} \frac{x_i}{n_2}$$

Show that the least squares estimate of β_1 is given by

$$\hat{\beta}_1 = \frac{\sum_{i=1}^{n} (x_i - \bar{x}_1) y_i + \sum_{i=n_1+1}^{n_1+n_2} (x_i - \bar{x}_2) y_i}{\sum_{i=1}^{n_1} (x_i - \bar{x}_1)^2 + \sum_{i=n_1+1}^{n_1+n_2} (x_i - \bar{x}_2)^2}$$

(*Hint*: Set up $\sum_{i=1}^{n} (y_i - \hat{y}_i)^2$ according to this centered model. Then find estimators for $\beta_0^{(1)*}$, $\beta_0^{(2)*}$, and β_1 that minimize $\sum_{i=1}^{n} (y_i - \hat{y}_i)^2$. *Comment*: For further development concerning this type of problem, see Section 3.10.)

2.24 From Section 2.12 it is apparent that an ideal set of residuals from a regression analysis take on a random scatter around a value of zero. Note from Exercise 2.2 that the residuals sum to zero. Do the residuals sum to zero for the model that is forced through the origin? Explain.

2.25 Consider the simple linear regression model of Eq. (2.3). Show that the least squares slope of Eq. (2.7) is given by

$$b_1 = \beta_1 + \sum_{i=1}^{n} d_i \varepsilon_i \qquad \text{where} \quad d_i = \frac{(x_i - \bar{x})}{S_{xx}}$$

Also show that

$$b_0 = \beta_0 + \bar{\varepsilon} - \bar{x} \sum_{i=1}^{n} d_i \varepsilon_i$$

REFERENCES FOR CHAPTER 2

1. Acton, F.S. 1959. *Analysis of Straight-Line Data*. New York: Dover.
2. Barnett, V.D. 1970. Fitting straight lines—the linear functional relationship with replicated observations. *Applied Statistics* 19: 135–144.
3. Bartlett, M.S. 1949. Fitting a straight line when both variables are subject to error. *Biometrics* 5: 207–212.
4. Berkson, J. 1950. Are there two regressions? *Journal of American Statistical Association* 45: 164–180.
5. Box, G.E.P. 1966. Use and abuse of regression. *Technometrics* 8: 625–629.

6. Casella, G. 1983. Leverage and regression through the origin. *American Statistician*, vol. 37, no. 2 (May): 147.

7. Draper, N.R., and H. Smith. 1981. *Applied Regression Analysis*, 2d ed. New York: John Wiley & Sons.

8. Ezekiel, M. and K.A. Fox. 1959. *Methods of Correlation and Regression Analysis*. New York: John Wiley & Sons.

9. Folks, J.L. 1967. Straight line confidence regions for linear models. *Journal of the American Statistical Association* 62: 1365–1374.

10. Graybill, F.A. 1976. *Theory and Application of the Linear Model*. Boston, Massachusetts: Duxbury Press.

11. Gunst, R.F., and R.L. Mason. 1980. *Regression Analysis and Its Applications: A Data Oriented Approach*. New York: Marcel Dekker.

12. Hahn, G.J. 1977. Fitting regression models with no intercept term. *Journal of Quality Technology* 9: 56.

13. Hahn, G.J. 1973. The coefficient of determination exposed. *Chemical Technology*, vol. 9, no. 2: 56–61.

14. Hahn, G.J. 1972. Simultaneous prediction intervals for a regression model. *Technometrics* 14: 203–214.

15. Halperin, M. 1961. Fitting of straight lines and prediction when both variables are subject to error. *Journal of the American Statistical Association* 56: 657–669.

16. Madansky, A. 1959. The fitting of straight lines when both variables are subject to error. *Journal of the American Statistical Association* 54: 173–205.

17. Miller, R.G. Jr. 1981. *Simultaneous Inference*, 2d ed. New York: McGraw-Hill Springer-Verlag.

18. Montgomery, D.C., and E.A. Peck. 1982. *Introduction to Linear Regression Analysis*. New York: John Wiley & Sons.

19. Roussas, George, G. 1973. *A First Course in Mathematical Statistics*. Reading, Massachusetts: Addison-Wesley.

20. SAS Institute Inc. 1985. *SAS User's Guide: Statistics, Version 5*. Cary, North Carolina: SAS Institute Inc.

21. Searle, S.R. 1971. *Linear Models*. New York: Wiley.

22. Seber, G.A.F. 1977. *Linear Regression Analysis*. New York: John Wiley & Sons.

23. Stigler, S.M. 1981. Gauss and the invention of least squares. *Annals of Statistics* 99: 465–474.

24. Working, H. and H. Hotelling. 1929. Applications of the theory of error to the interpretation of trends. *Journal of the American Statistical Association Supplement (Proceedings)* 24: 73–85.

CHAPTER 3

THE MULTIPLE LINEAR REGRESSION MODEL

3.1 MODEL DESCRIPTION AND ASSUMPTIONS

More often than not, the scientist engaged in model building must consider more than one regressor variable. As a result, Eq. (2.1) and the resulting model of (2.2) must be extended to the *multiple linear regression* situation. Consider an experiment in which data are generated of the type:

$\mathbf{y}$	$\mathbf{x}_1$	$\mathbf{x}_2$	$\cdots$	$\mathbf{x}_k$
y_1	x_{11}	x_{21}	$\cdots$	x_{k1}
y_2	x_{12}	x_{22}	$\cdots$	x_{k2}
$\vdots$	$\vdots$	$\vdots$		$\vdots$
y_n	x_{1n}	x_{2n}	$\cdots$	x_{kn}

Each row of the above array represents a *data point*. If the experimenter is willing to assume that in the region of the x's defined by the data, y_i is related approximately linearly to the regressor variables, then the model formulation

$$y_i = \beta_0 + \beta_1 x_{1i} + \beta_2 x_{2i} + \cdots + \beta_k x_{ki} + \varepsilon_i \qquad (i = 1, 2, \ldots, n; n \geqslant k + 1) \qquad \textbf{(3.1)}$$

may be a reasonable one. As in the case of simple regression, ε_i is a model error, assumed uncorrelated from observation to observation, with mean zero and constant variance σ^2. In addition, the x_{ji} are not random and are measured with negligible error.

The multiple linear regression model of Eq. (3.1) represents a significant departure from the simple linear regression model discussed at length in Chapter 2. We can no longer easily display data in a simple picture or plot as we did in Chapter 2. The meaning of the β_j in (3.1) requires more discussion, and a pictorial display of the workings of the least squares procedure is more complicated and requires more development. There is often confusion among subject matter scientists doing statistical modeling concerning what is meant by the term *linear*. The following subsection is designed to deal with this universal problem in terminology.

WHAT IS A LINEAR MODEL?

The reader needs to understand the difference between a linear model and a model that is nonlinear.

> *A linear model is defined as a model that is linear in the parameters, i.e., linear in the coefficients, the β's in Eq. (3.1).*

In this text we deal with linear models in Chapters 1–8. In Chapter 9 we present material and illustrations associated with nonlinear models. Examples of both types of models will add to the reader's understanding. Consider the situation in which the analyst wants to create a model of the relationship between y and x, but the relationship is, say, polynomial in nature. This is an example of a *linear model*. For example, a model quadratic in x (but of course linear in the β's) is given by

$$y = \beta_0 + \beta_1 x + \beta_2 x^2 + \varepsilon$$

Another example of a linear model is one which contains interaction among a pair of regressor variables. We have

$$y = \beta_0 + \beta_1 x_1 + \beta_2 x_2 + \beta_{12} x_1 x_2 + \varepsilon$$

In some applications there is a need to perform transformations on the regressor variables. For example, consider the case of three regressor variables x_1, x_2, and x_3. The following is a linear model in which the natural log transformation is made on each variable:

$$y = \beta_0 + \beta_1 \ln x_1 + \beta_2 \ln x_2 + \beta_3 \ln x_3 + \varepsilon$$

In each of the three illustrations provided here, *transformations* are made on the regressor variables but the model remains linear in the parameters.

The response variable, y, is often involved in transformations. For example, the analyst may be interested in using a log transformation on y and, say, reciprocal transformations on x_1 and x_2. As a result, the *linear model* is written

$$\ln y = \beta_0 + \beta_1\left(\frac{1}{x_1}\right) + \beta_2\left(\frac{1}{x_2}\right) + \varepsilon$$

All of the models described here are linear models by our definition. It is important to note that data analysts fit linear models on transformed data. There is no intention here of discussing the utility and hazards involved with transformations. Rather, we hope to make it clear that the multiple linear regression procedures that we cover in this chapter apply even where the variables involved are not in their natural form. A thorough discussion of transformations appears in Chapter 7.

At this point the reader may be curious about what kinds of examples fall into the category of nonlinear models, i.e., models not linear in the parameters. Given a response y and two regressors, x_1 and x_2, the following represent two examples of nonlinear models:

$$y = \beta_0 + \beta_1 x_1^{\gamma_1} + \beta_2 x_2^{\gamma_2} + \varepsilon$$

$$y = \frac{\beta_0}{1 + e^{-(\beta_1 x_1 + \beta_2 x_2)}} + \varepsilon$$

The first model contains 5 parameters and the second model contains 3 parameters. In both cases, parameters enter the model in a nonlinear way.

INTERPRETATION OF MODEL AND MODEL PARAMETERS

One important goal of the regression analysis is, of course, to estimate the β's in the model Eq. (3.1). The β's are often called *partial regression* coefficients. By this we mean that in a multiple regression with, say, parameters β_1, β_2, and β_3, the parameter β_1 is interpreted as the expected change in response (positive or negative) per unit change in x_1 *with the other x's held constant.* Similar interpretations can be made for the other β's. It is quite simple to deal with the *straight line* interpretation of the simple linear regression model of Eq. (2.1). In the multiple linear regression case, the interpretation of (3.1) is an interesting extension. Consider Figure 3.1 depicting data and the *true regression,* which is a *regression plane.* The ε_i's are vertical distances to the plane and the points on the plane represent the expected response. The regression coefficients β_1 and β_2 are the slopes of the regression plane in the direction of x_1 and x_2 respectively. The intercept, β_0, is also displayed in the figure. Obviously the purpose of the regression analysis is to estimate the three parameters and thus estimate the *population regression plane.*

FIGURE 3.1 **Multiple linear regression in two variables. (The notation * represents the data and · represents points on the regression plane.)**

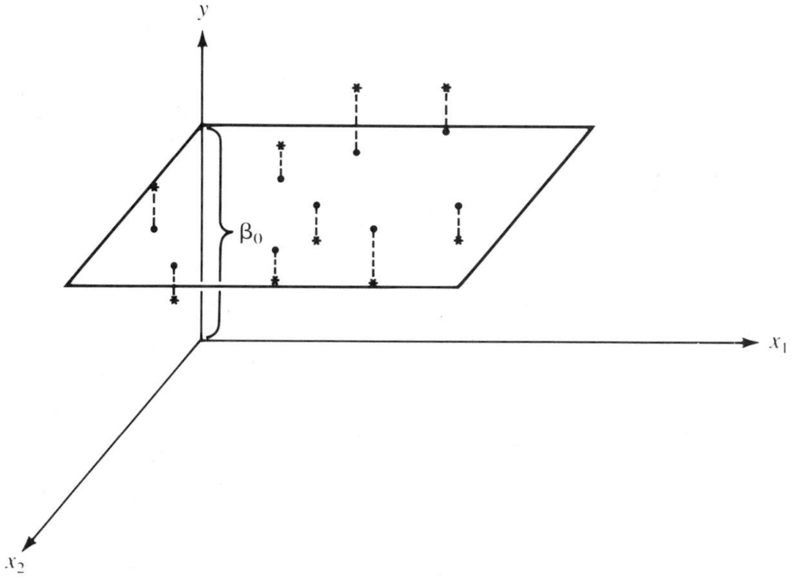

3.2

THE GENERAL LINEAR MODEL AND THE LEAST SQUARES PROCEDURE

We shall, once again, consider least squares as the method for estimation of the parameters. It is convenient at this point to reintroduce the model of Eq. (3.1) in matrix notation. The model can be written

$$\boxed{\mathbf{y} = \mathbf{X}\boldsymbol{\beta} + \boldsymbol{\varepsilon}} \tag{3.2}$$

where

$$\mathbf{y} = \begin{bmatrix} y_1 \\ y_2 \\ \vdots \\ y_n \end{bmatrix} \qquad \mathbf{X} = \begin{bmatrix} 1 & x_{11} & x_{21} & \cdots & x_{k1} \\ 1 & x_{12} & x_{22} & \cdots & x_{k2} \\ \vdots & \vdots & \vdots & & \vdots \\ 1 & x_{1n} & x_{2n} & \cdots & x_{kn} \end{bmatrix} \qquad \boldsymbol{\beta} = \begin{bmatrix} \beta_0 \\ \beta_1 \\ \beta_2 \\ \vdots \\ \beta_k \end{bmatrix}$$

and the vector $\boldsymbol{\varepsilon}$, the column of model errors, is given by

$$\boldsymbol{\varepsilon} = \begin{bmatrix} \varepsilon_1 \\ \varepsilon_2 \\ \vdots \\ \varepsilon_n \end{bmatrix}$$

The reader should become quite comfortable with the model (3.2), the *general linear model.* Many of the concepts discussed in this text can be understood and the illustrative examples can be followed without becoming overly preoccupied with matrices; however, it is not possible to achieve *maximum appreciation* for developments in regression analysis without following the matrix manipulations. Important and very timely topics, such as *multicollinearity, biased estimation, robust regression,* and *analysis of residuals,* will become merely a set of recipes to regression users who do not equip themselves to handle matrices properly.

Based on the assumptions mentioned in Section 3.1, the error vector, $\boldsymbol{\varepsilon}$, and the response vector, $\mathbf{y}$, are random vectors. The $\mathbf{X}$ matrix is of dimension $n \times p$, where $p = k + 1$ is the total number of model parameters. The jth column of $\mathbf{X}$ contains readings on the regressor x_j measured with negligible error. This matrix is the *data matrix* or *model matrix*. At times it is even called the *design matrix*. Actually, it is a combination of all three of these concepts. The $\mathbf{X}$ matrix displays the regressor data and thus the *experimental design* in cases in which the levels of the regressor variables are planned. However, $\mathbf{X}$ is obviously a function of the model. For the reader who is unfamiliar with the use of matrices in conjunction with linear regression, a few illustrations of the use of the $\mathbf{X}$ matrix may be helpful.

Consider a situation in which the yield of a chemical reaction is to be modeled as a function of temperature. The data are given by

y (yield, %)	x (temperature, °F)
77	160
79	165
82	170
83	175
85	180
86	185
87	190

A simple linear regression model

$$y_i = \beta_0^* + \beta_1(x_i - \bar{x}) + \varepsilon_i \qquad (i = 1, 2, \ldots, 7)$$

is postulated. Notice we are illustrating with the centered model. In the form of

Eq. (3.2), we have

$$\begin{bmatrix} 77 \\ 79 \\ 82 \\ 83 \\ 85 \\ 86 \\ 87 \end{bmatrix} = \begin{bmatrix} 1 & -15 \\ 1 & -10 \\ 1 & -5 \\ 1 & 0 \\ 1 & 5 \\ 1 & 10 \\ 1 & 15 \end{bmatrix} \begin{bmatrix} \beta_0^* \\ \beta_1 \end{bmatrix} + \begin{bmatrix} \varepsilon_1 \\ \varepsilon_2 \\ \varepsilon_3 \\ \varepsilon_4 \\ \varepsilon_5 \\ \varepsilon_6 \\ \varepsilon_7 \end{bmatrix}$$

Suppose the experimental plan is to vary the reaction time as well as the temperature, with the experiment as follows:

y (yield, %)	x_1 (temperature, °F)	x_2 (time, hr)
77	160	1
79	160	2
82	165	1
83	165	2
85	170	1
88	170	2
90	175	1
93	175	2

The multiple linear regression model (both variables centered) is written

$$y_i = \beta_0^* + \beta_1(x_{1i} - \bar{x}_1) + \beta_2(x_{2i} - \bar{x}_2) + \varepsilon_i$$

and in the matrix form of Eq. (3.2), we have

$$\begin{bmatrix} 77 \\ 79 \\ 82 \\ 83 \\ 85 \\ 88 \\ 90 \\ 93 \end{bmatrix} = \begin{bmatrix} 1 & -7.5 & -0.5 \\ 1 & -7.5 & 0.5 \\ 1 & -2.5 & -0.5 \\ 1 & -2.5 & 0.5 \\ 1 & 2.5 & -0.5 \\ 1 & 2.5 & 0.5 \\ 1 & 7.5 & -0.5 \\ 1 & 7.5 & 0.5 \end{bmatrix} \begin{bmatrix} \beta_0^* \\ \beta_1 \\ \beta_2 \end{bmatrix} + \begin{bmatrix} \varepsilon_1 \\ \varepsilon_2 \\ \varepsilon_3 \\ \varepsilon_4 \\ \varepsilon_5 \\ \varepsilon_6 \\ \varepsilon_7 \\ \varepsilon_8 \end{bmatrix} \tag{3.3}$$

We will use the same examples for illustration in future sections. In what follows we outline the development of the least squares procedure for estimation of parameters in multiple linear regression.

LEAST SQUARES DEVELOPMENT

The least squares estimator **b** for the regression coefficients in $\boldsymbol{\beta}$ is the vector that satisfies

$$\frac{\partial}{\partial \mathbf{b}}[(\mathbf{y}-\mathbf{X}\mathbf{b})'(\mathbf{y}-\mathbf{X}\mathbf{b})]=\mathbf{0}$$

Here, the expression $(\mathbf{y}-\mathbf{X}\mathbf{b})'(\mathbf{y}-\mathbf{X}\mathbf{b})$ represents the residual sum of squares. Performing the indicated differentiation produces the expression

$$-2\mathbf{X}'\mathbf{y}+2(\mathbf{X}'\mathbf{X})\mathbf{b}=\mathbf{0}$$

and thus the least squares normal equations, for this general case are given by

$$\boxed{(\mathbf{X}'\mathbf{X})\mathbf{b}=\mathbf{X}'\mathbf{y}}$$

As a result, assuming $\mathbf{X}$ is of full column rank,

$$\boxed{\mathbf{b}=(\mathbf{X}'\mathbf{X})^{-1}\mathbf{X}'\mathbf{y}} \tag{3.4}$$

The $\mathbf{X}'\mathbf{X}$ matrix of (3.4) is a $(k+1)\times(k+1)$ symmetric matrix whose diagonal elements are the sums of squares of the elements in columns of the $\mathbf{X}$ matrix, and whose off diagonal elements are sums of cross products of elements in the same columns. The nature of $\mathbf{X}'\mathbf{X}$ plays an important role in the properties of the estimators in **b** and will often be a large factor in the success (or failure) of ordinary least squares as an estimation procedure. More motivation and detail will be presented in Sections 3.3, 3.8 (discussion of multicollinearity), and in Chapter 8.

Consider the yield, temperature, and time data in Eq. (3.3). The least squares normal equations for estimation of β_0^*, β_1, and β_2 are given by

$$(\mathbf{X}'\mathbf{X})\mathbf{b}=\mathbf{X}'\mathbf{y}$$

$$\begin{bmatrix} 8 & 0 & 0 \\ 0 & 250 & 0 \\ 0 & 0 & 2 \end{bmatrix}\begin{bmatrix} b_0^* \\ b_1 \\ b_2 \end{bmatrix}=\begin{bmatrix} 677 \\ 222.5 \\ 4.5 \end{bmatrix}$$

with the solution given by $b_0^*=84.625$, $b_1=0.89$, and $b_2=2.25$.

ESTIMATION OF σ^2

It is necessary to obtain a good estimate of σ^2 in multiple regression. We use the estimate in variable screening via hypothesis testing, or for assessing model quality.

Before we discuss the estimator, we should reinvestigate the identity of Eq. (2.11) in light of the partition of total degrees of freedom for multiple linear regression. In the case of multiple regression, identity (2.11) still holds. However, the regression sum of squares, SS_{Reg}, explains variation that accounts for *k model terms*. Thus the total degrees of freedom partition as follows

$$n-1=k+(n-k-1)$$

The unbiased estimator, s^2, once again expresses variation in the residuals, i.e., variation about the regression $\hat{\mathbf{y}}=\mathbf{Xb}$, with the denominator now becoming $n-p$, *where p is the number of parameters estimated*. In the notation of the model in (3.2), $p=k+1$. As a result,

$$s^2=\frac{(\mathbf{y}-\mathbf{Xb})'(\mathbf{y}-\mathbf{Xb})}{n-p}=\sum_{i=1}^{n}\frac{(y_i-\hat{y}_i)^2}{n-p} \tag{3.5}$$

where, as in the case of simple regression, $\hat{y}_i$ is the predicted or fitted response at the ith data point. As in the simple linear regression case of Chapter 2, this estimator, *the residual mean square*, expresses natural variation or experimental error variance and is an unbiased estimator, assuming that the model postulated, and thus fitted, is correct. (See Appendix B.1.) The reader's intuition should suggest that if the model is underspecified, then s^2 is, *on the average*, an overestimate of σ^2. More on the effect of model underspecification will be presented in Section 3.6 and in Chapter 4. We note that a change in the model via elimination of model terms will increase SS_{Res} and also increase $n-p$, the residual degrees of freedom. Thus the sample estimate, s^2, may or may not increase. Adding regressor variables can also either reduce or increase s^2. We emphasize this because often the residual mean square serves as more than merely an estimator of error variance. Many data analysts use it (in a comparative sense) as a quick model selection criterion for discriminating among competing models; i.e., one favors the model with the smallest residual mean square. In Chapter 4 more insight will be provided regarding the role of s^2 in model discrimination. Of course, there are several criteria (discussed in Chapter 4), and the reader should not take this as a definitive decision-making procedure for model selection.

GEOMETRY OF LEAST SQUARES

For those that have a background in vector geometry, it is fairly simple to understand the geometry of the least squares procedure. In addition, the sum of squares identity of (2.11) for multiple linear regression is illustrated. Consider the picture given in Figure 3.2, which depicts a regression situation with $n=3$ observations and $p=2$ parameters. The three-dimensional axis system shown is in the *y-observation space*. The $\mathbf{y}$ vector represents the observation vector in the observation space. The two-dimensional plane of the figure is the *estimation space*. By the estimation space, we mean the space that contains points of the form $\mathbf{X}\hat{\boldsymbol{\beta}}$ where $\hat{\boldsymbol{\beta}}$, of course, is an estimate of $\boldsymbol{\beta}$. One should consider the point $\mathbf{y}^*$ as an

FIGURE 3.2 **The geometry of the least squares procedures**

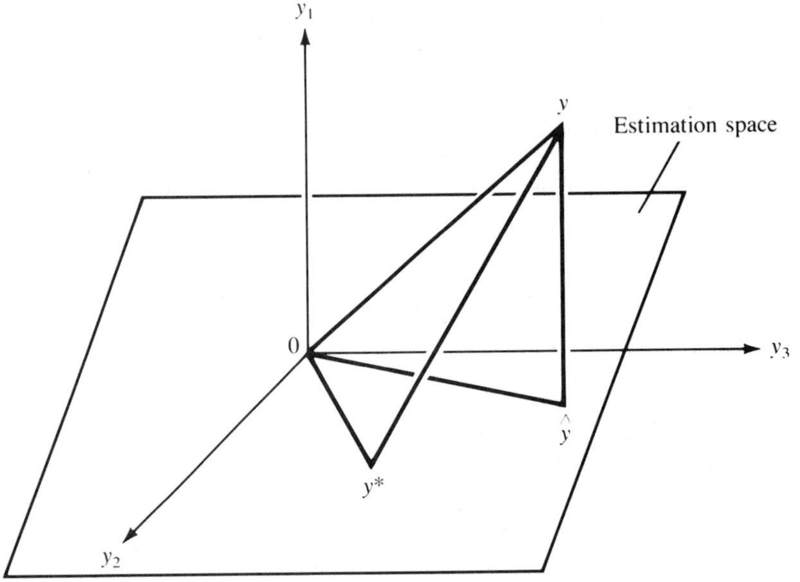

arbitrary candidate point in the estimation space. Now, what point in the estimation space produces a $\hat{\mathbf{y}}$ for which the residual sum of squares is minimized? The squared distance from $\mathbf{y}^*$ to $\mathbf{y}$ is indeed $(\mathbf{y}-\mathbf{y}^*)'(\mathbf{y}-\mathbf{y}^*)$, or letting $\mathbf{y}^* = \mathbf{X}\hat{\boldsymbol{\beta}}$, the squared distance becomes $(\mathbf{y}-\mathbf{X}\hat{\boldsymbol{\beta}})'(\mathbf{y}-\mathbf{X}\hat{\boldsymbol{\beta}})$. So, the least squares procedure applies when we choose the point in the estimation space that minimizes this squared distance. It is obvious that this is accomplished at the point $\hat{\mathbf{y}}$ when we drop a *perpendicular* from $\mathbf{y}$ to the estimation space. Now, we know that when this is done, this shortest vector $\mathbf{y}-\hat{\mathbf{y}}$ must be such that

$$\mathbf{X}'(\mathbf{y}-\hat{\mathbf{y}}) = 0$$

Using the usual notation $\hat{\mathbf{y}} = \mathbf{X}\mathbf{b}$, for this point in the estimation space,

$$\mathbf{X}'(\mathbf{y}-\mathbf{X}\mathbf{b}) = 0$$

Now, it is simple to see that the above implies that $\mathbf{b}$ is given by

$$(\mathbf{X}'\mathbf{X})\mathbf{b} = \mathbf{X}'\mathbf{y}$$

which represents the normal equations with the solutions being the least squares estimators given by Eq. (3.4).

3.3

PROPERTIES OF LEAST SQUARES ESTIMATORS UNDER IDEAL CONDITIONS

At this point the reader should recall that the ideal conditions of the model in (3.1) are:

1. the ε_i have mean zero (model functional form is correct),
2. the ε_i are uncorrelated, and have common variance σ^2 (homogeneous variance).

It should be anticipated that to carry out hypothesis testing on the β's, the assumption of normality on the ε_i must be made. We review these assumptions in order to consider them in the discussion of properties of the **b** vector. Also, the reader better appreciates the reason why there is often a need to deviate from ordinary least squares if we focus on what is desirable (or undesirable) about the method, and under what conditions its performance is less than ideal.

BIAS AND VARIANCE PROPERTIES OF THE PARAMETER ESTIMATES

Under the condition that $E(\boldsymbol{\varepsilon}) = \mathbf{0}$, **b** in (3.4) is an unbiased estimator for $\boldsymbol{\beta}$. This can be easily verified. Since $E(\mathbf{y}) = \mathbf{X}\boldsymbol{\beta}$ and **X** is not random

$$\begin{aligned} E(\mathbf{X}'\mathbf{X})^{-1}\mathbf{X}'\mathbf{y} &= (\mathbf{X}'\mathbf{X})^{-1}\mathbf{X}'E(\mathbf{y}) \\ &= (\mathbf{X}'\mathbf{X})^{-1}\mathbf{X}'\mathbf{X}\boldsymbol{\beta} \\ &= \boldsymbol{\beta} \end{aligned}$$

It is convenient and insightful to express the vector of estimates, **b**, in terms of the true coefficients, i.e. the β's. We can write (see Exercise 3.23)

$$\mathbf{b} = \boldsymbol{\beta} + \mathbf{R}\boldsymbol{\varepsilon}$$

where $\mathbf{R} = (\mathbf{X}'\mathbf{X})^{-1}\mathbf{X}$. The **R** matrix plays a relatively important role in the influence diagnostics in Chapter 6. Notice, from the above expression, that unbiasedness of **b** follows immediately.

For the variance, or the dispersion properties of **b**, we seek the variance-covariance matrix $E(\mathbf{b} - \boldsymbol{\beta})(\mathbf{b} - \boldsymbol{\beta})'$. We must make assumption 2, given at the beginning of Section 3.3. The variance-covariance matrix of **b** is given by (see Exercise 3.1)

$$\boxed{\text{Var}(\mathbf{b}) = \sigma^2(\mathbf{X}'\mathbf{X})^{-1}} \tag{3.6}$$

Notice that Eq. (3.6) implies that the variances of the coefficients appear on the diagonal elements of $(\mathbf{X}'\mathbf{X})^{-1}$, apart from the error variance σ^2. Similarly, covariances among the b's appear as the off-diagonal elements of the same matrix.

For the cases of conditions 1 and 2, but without assuming normality on the ε_i, we state a very important optimality property for the least squares estimator of $\boldsymbol{\beta}$. The result is the *Gauss–Markoff Theorem.*

In the case of the model in (3.2), if $E(\boldsymbol{\varepsilon}) = \mathbf{0}$ *and* $\text{Var}(\boldsymbol{\varepsilon}) = \sigma^2\mathbf{I}$ *(variance-covariance matrix), then the estimators achieve minimum variance of the class of linear unbiased estimators.*

For a proof of this theorem, the reader is referred to Graybill (1976).

It is often stated that the least squares estimators are BLUE (Best Linear Unbiased Estimators). The term "best" here is used in the sense of minimum variance. It turns out that the properties of the least squares estimators in the case of the linear model are stronger if the normality assumption holds. The properties of $\mathbf{b}$ in (3.4), under the normality assumption, are outlined in what follows.

EFFECT OF NORMAL ERRORS ON PROPERTIES OF ESTIMATORS

Neither the unbiasedness nor the dispersion properties given by Eq. (3.6) rely on a normality assumption. However, the practicing data analyst, who attempts to choose between the least squares estimators and an alternative form of estimation, should understand the properties of the least squares estimators when the errors are normal. These properties can be illustrated with the following result.

If the errors are normal and independent with mean zero and common variance, then the least squares estimators for the elements of $\boldsymbol{\beta}$ *in model (3.2) achieve uniformly minimum variance in the class of all unbiased estimators (UMVU).*

For the proof of this result, the reader should consult Graybill (1976).

In the case of the BLUE property, we say that the minimum variance property is confined to *linear* unbiased estimators, while in the UMVU case, the minimum variance property is confined to only unbiased estimators. By a linear estimator, we imply that the coefficients are linear functions of the y observations. In other words, a linear estimator is one of the form $\mathbf{Ry}$ where, in this case, $\mathbf{R}$ is a $(k+1) \times n$ matrix. Clearly the least squares estimator in (3.4) is a linear estimator with

$$\mathbf{R} = (\mathbf{X}'\mathbf{X})^{-1}\mathbf{X}'$$

In addition to attaining the UMVU property under normality, the least squares estimators are also *maximum likelihood estimators.* We developed the maximum likelihood estimators in Chapter 2 for the simple linear regression case. For multiple regression, the maximum likelihood development is given in Appendix B.3.

Careful study of these properties suggests that when errors are not normal, there is more room for improvement over the performance of the least squares estimators. Indeed, when errors deviate considerably from Gaussian, other estimators of the regression model coefficients, namely robust regression estimators, do perform better than the ordinary least squares estimator. Robust regression will be discussed in Chapter 7.

The following example illustrates the computation of the regression coefficients and s^2, the residual mean square, for a multiple regression example.

Example 3.1 **Breadwrapper Data**

Consider an experiment in which the researcher wants to determine an empirical relationship between the seal strength (y) in grams per inch of a breadwrapper stock and regressor variables: sealing temperature (x_1), cooling bar temperature (x_2), and percent polyethylene in the stock (x_3). The combinations of the regressor variables were chosen in advance by the scientist. The data are given in Table 3.1.

The researcher felt that a model that is quadratic in the regressor variables would be appropriate. Thus, terms that are of order two were used. The multiple regression model is as follows:

$$y_i = \beta_0 + \beta_1 x_{1i} + \beta_2 x_{2i} + \beta_3 x_{3i} + \beta_{11} x_{1i}^2 + \beta_{22} x_{2i}^2 + \beta_{33} x_{3i}^2 + \beta_{12} x_{1i} x_{2i} + \beta_{13} x_{1i} x_{3i} + \beta_{23} x_{2i} x_{3i} + \varepsilon_i$$

In this case, since the model involves squares and second-order cross products, there are 9 regressor variables and hence 10 *parameters* to be estimated. The **X** matrix, then, is given by

$\mathbf{X} =$

	x_1	x_2	x_3	x_1^2	x_2^2	x_3^2	x_1x_2	x_1x_3	x_2x_3
1	225	46	0.5	50625	2116	.25	10350	112.5	23
1	285	46	0.5	81225	2116	.25	13110	142.5	23
1	225	64	0.5	50625	4096	.25	14400	112.5	32
1	285	64	0.5	81225	4096	.25	18240	142.5	32
1	225	46	1.7	50625	2116	2.89	10350	382.5	78.2
1	285	46	1.7	81225	2116	2.89	13110	484.5	78.2
1	225	64	1.7	50625	4096	2.89	14400	382.5	108.8
1	285	64	1.7	81225	4096	2.89	18240	484.5	108.8
1	204.5	55	1.1	41820.25	3025	1.21	11247.5	224.95	60.5
1	305.5	55	1.1	93330.25	3025	1.21	16802.5	336.05	60.5
1	255	39.9	1.1	65025	1592.01	1.21	10174.5	280.5	43.89
1	255	70.1	1.1	65025	4914.01	1.21	17875.5	280.5	77.11
1	255	55	0.09	65025	3025	.0081	14025	22.95	4.95
1	255	55	2.11	65025	3025	4.4521	14025	538.05	116.05
1	255	55	1.1	65025	3025	1.21	14025	280.5	60.5
1	255	55	1.1	65025	3025	1.21	14025	280.5	60.5
1	255	55	1.1	65025	3025	1.21	14025	280.5	60.5
1	255	55	1.1	65025	3025	1.21	14025	280.5	60.5
1	255	55	1.1	65025	3025	1.21	14025	280.5	60.5
1	255	55	1.1	65025	3025	1.21	14025	280.5	60.5

The **X** matrix can then be used to compute **X′X** and $(\mathbf{X}'\mathbf{X})^{-1}$. The coefficients are then computed from Eq. (3.4). The fitted least squares regression equation is as follows:

$$\begin{aligned}\hat{y} = &-104.8568 + 0.4948x_1 + 1.7303x_2 + 14.2620x_3 \\ &-0.0008425x_1^2 - 0.01291x_2^2 - 3.1846x_3^2 \\ &-0.0013x_1x_2 - 0.02778x_1x_3 + 0.02778x_2x_3\end{aligned}$$

with the residual sum of squares

$$\sum_{i=1}^{n} (y_i - \hat{y}_i)^2 = 11.8678$$

and the residual estimate of variance, s^2 (10 residual degrees of freedom)

$$s^2 = \frac{11.8678}{10} = 1.18678$$

TABLE 3.1 **Breadwrapper stock data**

y (g/in.)	x_1 (°F)	x_2 (°F)	x_3 (weight in %)
6.6	225	46	0.5
6.9	285	46	0.5
7.9	225	64	0.5
6.1	285	64	0.5
9.2	225	46	1.7
6.8	285	46	1.7
10.4	225	64	1.7
7.3	285	64	1.7
9.8	204.5	55	1.1
5.0	305.5	55	1.1
6.9	255	39.9	1.1
6.3	255	70.1	1.1
4.0	255	55	0.09
8.6	255	55	2.11
10.1	255	55	1.1
9.9	255	55	1.1
12.2	255	55	1.1
9.7	255	55	1.1
9.7	255	55	1.1
9.6	255	55	1.1

Data originally in *Tappi*, vol. 41, no. 6 (June 1958): 295–300. Reprinted and analyzed in *Response Surface Methodology* by Raymond H. Myers (Department of Statistics, Virginia Polytechnic Institute and State University, Blacksburg, Virginia, 1976), p. 79.

3.4

HYPOTHESIS TESTING IN MULTIPLE LINEAR REGRESSION

In many traditional scientific areas, the primary function of the model-building exercise is to determine which regressor variables truly influence the response y. Quite often a new system, which has not been thoroughly studied, requires a preliminary investigation in order to determine what factors are truly relevant. While our total statistical arsenal is quite plentiful for handling such a situation, regression analysis is often the chosen technique. It becomes important, then, to determine which regressor variables are responsible for a significant variation in the response y. The reader should acquire a thorough understanding of how the standard procedures work for hypothesis testing on the individual β_i. Questions relating to the drawbacks of using hypothesis testing as a form of variable screening are equally important.

PARTITIONING OF VARIABILITY

As in the case of simple linear regression, we have the following fundamental identity relating total variability to residual and regression sums of squares:

$$\sum_{i=1}^{n} (y_i - \bar{y})^2 = \sum_{i=1}^{n} (\hat{y}_i - \bar{y})^2 + \sum_{i=1}^{n} (y_i - \hat{y}_i)^2 \tag{3.7}$$

where $\hat{y}_i$ is the fitted regression given by $\hat{y}_i = b_0 + b_1 x_{1i} + b_2 x_{2i} + \cdots + b_k x_{ki}$. The coefficient of determination has the same interpretation as in simple linear regression and is, again, given y

$$R^2 = \frac{\sum_{i=1}^{n} (\hat{y}_i - \bar{y})^2}{\sum_{i=1}^{n} (y_i - \bar{y})^2} = \frac{SS_{\text{Reg}}}{SS_{\text{Total}}} \tag{3.8}$$

In (3.7), the regression sum of squares can be partitioned into meaningful portions that help assess the worth of a *single* regression variable in the traditional sense. SS_{Reg} can be partitioned into *sequential* regression sums of squares as follows:

$$\begin{aligned} R(\beta_1, \beta_2, \ldots, \beta_k \mid \beta_0) = {} & R(\beta_1 \mid \beta_0) + R(\beta_2 \mid \beta_1, \beta_0) + R(\beta_3 \mid \beta_2, \beta_1, \beta_0) \\ & + \cdots + R(\beta_k \mid \beta_{k-1}, \beta_{k-2}, \ldots, \beta_2, \beta_1, \beta_0) \end{aligned} \tag{3.9}$$

The notation $R(\cdot \mid \cdot)$ implies "regression explained by ..." with the vertical line denoting "in the presence of." For example, $R(\beta_2 \mid \beta_1, \beta_0)$ is the increase in the regression sum of squares when the regressor x_2 is added to a model involving only

x_1 and the constant term. As a result, for the components in (3.9), each represents the *incremental increase* in regression explained by a particular regressor variable in the model containing a subset. Equation (3.9) represents a meaningful partitioning of SS_{Reg} into "single degree of freedom" contributions in which the reader visualizes a model beginning with only a constant term, and each regressor variable is added sequentially. As we place each regressor into the model, the regression sum of squares increases. From (3.7) it should be clear that the residual sum of squares decreases by the same amount as SS_{Reg} increases. Thus the term in (3.9) can also be viewed as the individual *reduction in the residual sum of squares* produced by the sequential introduction of regressors. Thus, for example, $R(\beta_3 \mid \beta_0, \beta_1, \beta_2)$ is both the increase in SS_{Reg} or the reduction in SS_{Res}, due to entrance of $\beta_3 x_3$ into the model containing β_0, $\beta_1 x_1$, and $\beta_2 x_2$.

Equation (3.9) can help us assess the worth of individual regressors. Other types of partitioning are very useful if we require information regarding the worth of a *subset of regressors.* For example, suppose, for $k = 4$ ($p = 5$), the worth of the regressors x_3 and x_4 is questionable. We can write

$$R(\beta_1, \beta_2, \beta_3, \beta_4 \mid \beta_0) = R(\beta_1, \beta_2 \mid \beta_0) + R(\beta_3, \beta_4 \mid \beta_0, \beta_1, \beta_2) \tag{3.10}$$

where both terms on the right-hand side represent two degrees of freedom partitions. $R(\beta_1, \beta_2 \mid \beta_0)$ would not be useful in general for inference on β_1 and β_2 since there has been no adjustment for x_3 and x_4. However, $R(\beta_3, \beta_4 \mid \beta_0, \beta_1, \beta_2)$ would be vital in explaining the importance of x_3 and x_4 collectively.

As in the simple linear case, the regression sum of squares, $R(\beta_1, \ldots, \beta_k \mid \beta_0) = \sum_{i=1}^{n} (\hat{y}_i - \bar{y})^2$ quantifies the amount of the total sum of squares in the response that is explained by the model. Isolating increments of SS_{Reg}, as in the cases of the preceding illustrations, can be used to determine which variable or subsets are responsible for the quality of fit quantified by SS_{Reg} or R^2. Hopefully the reader can gain some sense of generality from Eq. (3.9) and (3.10). In the illustration of (3.10), if one is interested in inference on β_3 and β_4, $R(\beta_3, \beta_4 \mid \beta_0, \beta_1, \beta_2)$ may be computed as

$$R(\beta_3, \beta_4 \mid \beta_0, \beta_1, \beta_2) = R(\beta_3 \mid \beta_0, \beta_1, \beta_2) + R(\beta_4 \mid \beta_0, \beta_1, \beta_2, \beta_3)$$

In the following we attempt to shed light on tests of hypotheses regarding subsets of the parameters.

TESTS ON SUBSETS OF THE REGRESSION PARAMETERS

Partitions of the regression sum of squares, as shown in (3.9) and (3.10), generate rather traditional tests of hypotheses on individual regression coefficients, and, generally, on subsets of regression coefficients. These tests can be developed quite succinctly and generally by using matrix notation. Suppose $\boldsymbol{\beta}$ and $\mathbf{X}$ are subdivided such that

$$\mathbf{X} = [\mathbf{X}_1 \;\vdots\; \mathbf{X}_2] \qquad \boldsymbol{\beta} = \begin{bmatrix} \boldsymbol{\beta}_1 \\ \cdots \\ \boldsymbol{\beta}_2 \end{bmatrix}$$

where $\mathbf{X}_1$ is $n \times p_1$ $(p_1 + p_2 = p)$ and $\boldsymbol{\beta}_1$ contains the p_1 scalar elements associated with the columns of $\mathbf{X}_1$. Thus the general linear regression model in (3.2) can now be written

$$y = \mathbf{X}_1 \boldsymbol{\beta}_1 + \mathbf{X}_2 \boldsymbol{\beta}_2 + \boldsymbol{\varepsilon} \tag{3.11}$$

Suppose $\boldsymbol{\beta}_1$ contains the *questionable* parameters (or parameter). Thus we wish to test

$$\boxed{\begin{aligned} H_0&: \boldsymbol{\beta}_1 = \mathbf{0} \\ H_1&: \boldsymbol{\beta}_1 \neq \mathbf{0} \end{aligned}} \tag{3.12}$$

In the model in (3.11), the constant term, β_0, may appear in either $\boldsymbol{\beta}_1$ or $\boldsymbol{\beta}_2$. For example, if

$$\boldsymbol{\beta} = \begin{bmatrix} \beta_0 \\ \beta_1 \\ \beta_2 \\ \beta_3 \end{bmatrix}$$

and interest is in the test of

$$H_0: \beta_1 = \beta_2 = 0$$

then our partitioning would be

$$\boldsymbol{\beta}_1 = \begin{bmatrix} \beta_1 \\ \beta_2 \end{bmatrix} \qquad \boldsymbol{\beta}_2 = \begin{bmatrix} \beta_0 \\ \beta_3 \end{bmatrix}$$

and the columns of $\mathbf{X}_1$ and $\mathbf{X}_2$ would be arranged accordingly.

Thus for the general hypothesis on a subset of the parameters, namely that of (3.12), the appropriate test must involve the regression sum of squares explained by $\boldsymbol{\beta}_1$ or, equivalently, the reduction in residual sum of squares enjoyed by the introduction of $\mathbf{X}_1\boldsymbol{\beta}_1$ into a model containing $\mathbf{X}_2\boldsymbol{\beta}_2$. From the earlier development, it should be clear that the test involves

$$R(\boldsymbol{\beta}_1 \mid \boldsymbol{\beta}_2) = R(\boldsymbol{\beta}_1, \boldsymbol{\beta}_2) - R(\boldsymbol{\beta}_2)$$

The term $R(\boldsymbol{\beta}_1, \boldsymbol{\beta}_2)$ is the regression explained by *all* model terms, including the constant. The notation $R(\boldsymbol{\beta}_2)$ represents the regression sum of squares explained by a model involving only $\mathbf{X}_2\boldsymbol{\beta}_2$. The *extra sum of squares* regression, $R(\boldsymbol{\beta}_1 \mid \boldsymbol{\beta}_2)$, is written (see Appendix A.2)

$$\begin{aligned} R(\boldsymbol{\beta}_1 \mid \boldsymbol{\beta}_2) &= \mathbf{y}'\mathbf{X}(\mathbf{X}'\mathbf{X})^{-1}\mathbf{X}'\mathbf{y} - \mathbf{y}'\mathbf{X}_2(\mathbf{X}_2'\mathbf{X}_2)^{-1}\mathbf{X}_2'\mathbf{y} \\ &= \mathbf{y}'[\mathbf{X}(\mathbf{X}'\mathbf{X})^{-1}\mathbf{X}' - \mathbf{X}_2(\mathbf{X}_2'\mathbf{X}_2)^{-1}\mathbf{X}_2']\mathbf{y} \end{aligned} \tag{3.13}$$

With the standard assumptions on the vector $\boldsymbol{\varepsilon}$ (including normality), $R(\boldsymbol{\beta}_1 \mid \boldsymbol{\beta}_2)/\sigma^2$ is $\chi^2_{p_1}$ under H_0 (see Graybill (1976)) and is independent of the residual sum of squares. Thus, under H_0, $[R(\boldsymbol{\beta}_1 \mid \boldsymbol{\beta}_2)/p_1]/s^2$ is $F_{p_1, n-p}$. So, the test statistic

$$F = \frac{R(\boldsymbol{\beta}_1 \mid \boldsymbol{\beta}_2)/p_1}{s^2}$$

provides an upper tail one-tailed test criterion using the F-distribution, thereby determining if the *extra sum of squares regression* or the *residual sum of squares reduced* is sufficient to warrant inclusion of $\boldsymbol{\beta}_1$ in the model.

TESTS ON SINGLE REGRESSION COEFFICIENTS (PARTIAL F AND t-TESTS)

The above development provides a basis for testing any subset of the regression parameters. In the case where $\boldsymbol{\beta}_1$ is a scalar, i.e., a single regression coefficient is being tested, $p_1 = 1$ and the F-test with $(1, n-p)$ degrees of freedom is called the *partial F-test*. A two-tailed t-test can provide the same information. As we indicated in Section 3.3, the variance of the jth regression coefficient is $\sigma^2 \cdot c_{jj}$ where c_{jj} is the jth diagonal element of $(\mathbf{X}'\mathbf{X})^{-1}$. Thus a t-test of

$$\left.\begin{aligned} H_0&: \beta_j = 0 \\ H_1&: \beta_j \neq 0 \end{aligned}\right\} \quad (j = 0, 1, 2, \ldots, k)$$

can be accomplished by using

$$t = \frac{b_j}{s\sqrt{c_{jj}}} \tag{3.14}$$

The quantity $s\sqrt{c_{jj}}$, the estimated standard deviation of b_j, is called the *standard error* of b_j.

The test described above is called a partial t-test and $t^2 = F$ where F is the partial F-statistic for the same hypothesis. The distinction between the t- and partial F-test on a regression coefficient comes solely from the fact that the t-test gives an indication of *direction*. From (3.14) we see that the sign of the t-value comes from the sign of the regression coefficient. However, a significant partial F gives no indication of direction on the coefficient.

Thus the extra sum of squares principle provides the user with one of the main attractions of the least squares estimation procedure, namely a mechanism for testing hypotheses on a single regression coefficient or a subset. We exploit what is known about the statistical properties of SS_{Reg}, the variation explained. Simply put, evidence in favor of rejection of H_0 of Eq. (3.12) tells the analyst that the extra variation explained when $\mathbf{X}_1\boldsymbol{\beta}_1$ is placed in the model in the presence of $\mathbf{X}_2\boldsymbol{\beta}_2$ is more than we attribute to chance.

SEQUENTIAL F-TESTS ON INDIVIDUAL PARAMETERS

The partial F-tests just discussed, which we will illustrate in Example 3.2, are not formed from sums of squares contributions that add to SS_{Reg}. That is, in general,

$$\begin{aligned} R(\beta_1 \mid \beta_0, \beta_2, \beta_3, \ldots, \beta_k) + R(\beta_2 \mid \beta_0, \beta_1, \beta_3, \ldots, \beta_k) + \cdots \\ + R(\beta_k \mid \beta_0, \beta_1, \beta_2, \ldots, \beta_{k-1}) \neq R(\beta_1, \beta_2, \ldots, \beta_k \mid \beta_0) \end{aligned}$$

Thus the extra sums of squares used in the partial F-tests do not form a complete partition of SS_{Reg}. As a result, in general, the sums of squares and, indeed, the tests themselves *are not independent*. One can, however, generate tests from independent sums of squares by using the components of the right-hand side of Eq. (3.9). These F-tests are called *sequential F-tests*. Both sequential and partial F-tests are illustrated in example 3.2.

Example 3.2 **Squid Data**

An experiment was conducted in order to study the size of squid eaten by sharks and tuna. The regressor variables are characteristics of the beak or mouth of the squid. The regressor variables and response considered for the study are

x_1: Rostral length in inches
x_2: Wing length in inches
x_3: Rostral to notch length
x_4: Notch to wing length
x_5: Width in inches
y: Weight in pounds

The study involved measurements and weight taken on 22 specimen. The data are shown in Table 3.2. The model is given by

$$y_i = \beta_0 + \beta_1 x_{1i} + \beta_2 x_{2i} + \beta_3 x_{3i} + \beta_4 x_{4i} + \beta_5 x_{5i} + \varepsilon_i$$

In this example we begin by illustrating the identity in Eq. (3.7) through the following analysis of variance.

Source	SS	df	MS	F
Regression (model)	208.007	5	$\frac{SS_{\text{Reg}}}{5} = 41.6015$	$F = \frac{MS_{\text{Reg}}}{s^2} = 84.070$
Residual (error)	7.918	16	$s^2 = 0.4948$	
Total	215.925	21		

The five degrees of freedom for regression represent the number of model terms apart from the intercept. The residual degrees of freedom are, of course, $n - p = 22 - 6 = 16$. One immediately uses the above analysis of variance table to test a hypothesis using $R(\beta_1, \beta_2, \beta_3, \beta_4, \beta_5 \mid \beta_0)$. This is merely what we call the regression sum of squares (5 df) in (3.7) with the hypothesis being

$$H_0: \begin{bmatrix} \beta_1 \\ \beta_2 \\ \beta_3 \\ \beta_4 \\ \beta_5 \end{bmatrix} = \mathbf{0}$$

This can be visualized as a very special case of testing on subsets of the parameters. The F-statistic is statistically significant ($F = 84.070$) indicating that at least one of the regressors influences squid weight. As a simple by-product of the above portion of the analysis, note that

$$R^2 = \frac{208.007}{215.925} = 0.9633$$

with the implication that the five-variable regression model explains 96.33% of the variation in squid weight. The sequential and partial sums of squares are as follows.

Sequential	Partial
$R(\beta_1 \mid \beta_0) = 199.145$	$R(\beta_1 \mid \beta_0, \beta_2, \beta_3, \beta_4, \beta_5) = 0.298731$
$R(\beta_2 \mid \beta_1, \beta_0) = 0.126664$	$R(\beta_2 \mid \beta_0, \beta_1, \beta_3, \beta_4, \beta_5) = 0.868761$
$R(\beta_3 \mid \beta_0, \beta_1, \beta_2) = 4.119539$	$R(\beta_3 \mid \beta_0, \beta_1, \beta_2, \beta_4, \beta_5) = 0.078273$
$R(\beta_4 \mid \beta_0, \beta_1, \beta_2, \beta_3) = 0.263496$	$R(\beta_4 \mid \beta_0, \beta_1, \beta_2, \beta_3, \beta_5) = 0.982690$
$R(\beta_5 \mid \beta_0, \beta_1, \beta_2, \beta_3, \beta_4) = 4.352193$	$R(\beta_5 \mid \beta_0, \beta_1, \beta_2, \beta_3, \beta_4) = 4.352193$

Note that the sequential sums of squares add to 208.007, the regression sum of squares explained by all of the regressors; the partial sums of squares do not add to anything meaningful. The partial sums of squares, used individually, can test hypotheses on single regression coefficients. The sequential sums of squares can be used to illustrate tests on subsets. For example, suppose we wish to test

$$H_0\colon \begin{bmatrix} \beta_4 \\ \beta_5 \end{bmatrix} = \mathbf{0}$$

the implication being that if we fail to reject H_0, the coefficients on width and notch to wing length are simultaneously not found to differ significantly from zero. The proper sum of squares is found by computing $R(\beta_4, \beta_5 \mid \beta_0, \beta_1, \beta_2, \beta_3)$, given by

$$\begin{aligned} &R(\beta_1, \beta_2, \beta_3, \beta_4, \beta_5 \mid \beta_0) - R(\beta_1, \beta_2, \beta_3 \mid \beta_0) \\ &\quad = R(\beta_1, \beta_2, \beta_3, \beta_4, \beta_5 \mid \beta_0) \\ &\qquad - [R(\beta_1 \mid \beta_0) + R(\beta_2 \mid \beta_1, \beta_0) + R(\beta_3 \mid \beta_0, \beta_1, \beta_2)] \\ &\quad = 208.007 - (199.145 + 0.126664 + 4.119539) \\ &\quad = 4.615689 \end{aligned}$$

Alternatively, we can write

$$\begin{aligned} R(\beta_4, \beta_5 \mid \beta_0, \beta_1, \beta_2, \beta_3) &= R(\beta_4 \mid \beta_0, \beta_1, \beta_2, \beta_3) \\ &\quad + R(\beta_5 \mid \beta_0, \beta_1, \beta_2, \beta_3, \beta_4) \\ &= 0.263496 + 4.352193 \\ &= 4.615689 \end{aligned}$$

The test statistic is then

$$F = \frac{4.615689/2}{s^2} = \frac{2.307845}{0.494845} = 4.663683$$

The numerator and denominator degrees of freedom are 2 and 16 respectively. The hypothesis is rejected at the 0.025 level.

The partial sums of squares can be used to test hypotheses on individual parameters using F-statistics. Equivalent tests, of course, come from t-statistics given by Eq. (3.14). A standard part of most regression computer packages is a listing of the regression coefficients and their *standard errors*, the latter given by the denominators of the t-statistics in (3.14). For this example, the listing is as follows.

	Coefficient	Standard Error	t	Prob $> \|t\|$
b_0	−6.512215	0.933561	−6.976	0.0001
b_1	1.999413	2.573338	0.777	0.4485
b_2	−3.675096	2.773660	−1.325	0.2038
b_3	2.524486	6.347495	0.398	0.6961
b_4	5.158082	3.660283	1.409	0.1779
b_5	14.401162	4.855994	2.966	0.0091

The latter column above gives the p-value, or level at which the t-statistic is significant. Individual inferences drawn here indicate that b_1, b_2, b_3, and b_4 are not significant. Caution should be suggested here however. The t-tests are certainly valid and, indeed, informative when one wants to draw conclusions concerning an individual regressor *in the presence of the others*. However, it can be dangerous to use sets of them simultaneously to screen variables and hence build models. Thus it would be presumptuous of us to conclude at this time that the only regressor to be retained in the model is x_5, the beak width. Chapter 4 deals solely with criteria for model building and thus will contain additional discussion along these lines.

TABLE 3.2 **Squid weight and beak measurements**

x_1	x_2	x_3	x_4	x_5	y
1.31	1.07	0.44	0.75	0.35	1.95
1.55	1.49	0.53	0.90	0.47	2.90
0.99	0.84	0.34	0.57	0.32	0.72
0.99	0.83	0.34	0.54	0.27	0.81
1.05	0.90	0.36	0.64	0.30	1.09
1.09	0.93	0.42	0.61	0.31	1.22
1.08	0.90	0.40	0.51	0.31	1.02
1.27	1.08	0.44	0.77	0.34	1.93
0.99	0.85	0.36	0.56	0.29	0.64

TABLE 3.2 Continued

TABLE 3.2 Continued

Squid weight and beak measurements

x_1	x_2	x_3	x_4	x_5	y
1.34	1.13	0.45	0.77	0.37	2.08
1.30	1.10	0.45	0.76	0.38	1.98
1.33	1.10	0.48	0.77	0.38	1.90
1.86	1.47	0.60	1.01	0.65	8.56
1.58	1.34	0.52	0.95	0.50	4.49
1.97	1.59	0.67	1.20	0.59	8.49
1.80	1.56	0.66	1.02	0.59	6.17
1.75	1.58	0.63	1.09	0.59	7.54
1.72	1.43	0.64	1.02	0.63	6.36
1.68	1.57	0.72	0.96	0.68	7.63
1.75	1.59	0.68	1.08	0.62	7.78
2.19	1.86	0.75	1.24	0.72	10.15
1.73	1.67	0.64	1.14	0.55	6.88

Rudolf J. Freund, SAS Tutorial, "Regression with SAS with Emphasis on PROC REG" (Paper presented at Eighth Annual SAS Users Group International Conference, New Orleans, Louisiana, January 16–19, 1983).

FURTHER COMMENTS ON SEQUENTIAL AND PARTIAL TESTS

Many regression software packages contain both sequential and partial F-statistics. Both types of tests have very definite drawbacks if one is interested in model building through the process of variable screening. Perhaps the most effective way to begin an illustration of these drawbacks is to remind the reader what each is designed to do.

Sequential F-tests

These tests give information regarding the contribution of a variable in a model containing the preceding regressor variables. The order of entry, then, can have a profound effect on the results. If regressor variable 4 is adjusted for 1, 2, and 3, it is quite possible that its contribution to SS_{Reg} will be quite different than if it were adjusted for, say, only variable 1. In other words, the *appropriateness of a regressor variable often depends on what regressor variables are in the model with it*. Thus a full scale variable screening cannot be accomplished effectively using sequential F-tests unless they are used in harmony with an imaginative selection of order and, then, only when many stages are implemented into the procedure. This, in fact, provides the operative logic of the Stagewise Regression Procedures (Forward Selection, Backward Elimination, Stepwise Regression) discussed in Chapter 4.

Partial F-tests

The partial F-test gives information regarding the importance of a single regressor in a model involving all other regressors. Certainly this can be informative if one is interested in the role of a particular regressor variable. However, in a large scale variable screening procedure, one can often encounter difficulties with decision making by using only information from k partial F-tests on k regressor variables. Certainly, if a variable is statistically significant in the presence of all regressor variables, this does not imply that it would be a very important contributor in a subset. On the other hand, an insignificant regressor could be rendered important in a particular subset. Clearly the interrelationships and multiple association among the regressor variables (*multicollinearity*) make it very difficult to draw conclusions regarding the issue of *best model* by using only partial F-tests.

In the case of the squid data, for example, variable x_1, rostral length, is clearly unimportant in the presence of all of the other variables ($t = 0.78$, $p = 0.4485$). However, as a single regressor, x_1 would appear to be important, with

$$F = \frac{199.145}{16.780/20} = 237.360$$

The numerator of the above test statistic is the sequential regression sum of squares with x_1 representing the initial entrance into the model. The s^2 is computed for a regression with x_1 being the only regressor. Clearly this is evidence that x_1 explains a significant portion of the variation in the response. But the t-test (or partial F) on x_1 would imply that the variation explained by x_1 as a lone regressor is accounted for by the other regressors.

The role of partial and sequential F-tests in large-scale screening problems is discussed in Chapter 4.

TEST OF THE GENERAL LINEAR HYPOTHESIS

Tests on individual regression parameters and on subsets of parameters can be put into a more general framework that allows much more flexibility. For example, consider a biologist who builds two growth models with two independent data sets, or consider a sociologist who constructs two regression equations, one for males and one for females. In each case it may be very important for the user to gain some insight into whether or not the parameters of the regressions differ significantly across the two data sets. These and other types hypotheses can be tested through the use of the *general linear hypothesis*

$$\begin{aligned} H_0&: \mathbf{C}\boldsymbol{\beta} = \mathbf{d} \\ H_1&: \mathbf{C}\boldsymbol{\beta} \neq \mathbf{d} \end{aligned} \tag{3.15}$$

This hypothesis accommodates the important hypothesis in Eq. (3.12) as a special case. In (3.15), the matrix $\mathbf{C}$ is an $r \times p$ matrix that is of rank r and $r \leqslant p$. In other words, the hypothesis allows us to postulate $r \leqslant p$ nonredundant and non-contradictory statements about the parameters.

Illustrations of the General Linear Hypothesis (Test for Parallelism and Equality of Regressions Across Data Sets)

We alluded to the importance of comparing independent regression equations or individual parameters across independent data sets. Suppose we have two independent data sets

$$
\left.\begin{array}{cc}
\mathbf{y} & \mathbf{x} \\
y_1 & x_1 \\
y_2 & x_2 \\
\vdots & \vdots \\
y_{n_1} & x_{n_1}
\end{array}\right\} \text{Data set 1}
$$

$$
\left.\begin{array}{cc}
y_{n_1+1} & x_{n_1+1} \\
\vdots & \vdots \\
y_{n_1+n_2} & x_{n_1+n_2}
\end{array}\right\} \text{Data set 2}
$$

and two models are postulated

$$
\begin{aligned}
y_i &= \beta_0 + \beta_1 x_i + \varepsilon_i \qquad i = 1, 2, \ldots, n_1 \\
y_i &= \gamma_0 + \gamma_1 x_i + \varepsilon_i \qquad i = n_1 + 1, \ldots, n_1 + n_2
\end{aligned}
$$

Suppose it is of interest to test equality of the slope of the two regression lines. We may write these two separate models as a *single model.* Namely,

$$
\begin{bmatrix} y_1 \\ y_2 \\ \vdots \\ y_{n_1} \\ \hline y_{n_1+1} \\ \vdots \\ y_{n_1+n_2} \end{bmatrix}
=
\begin{bmatrix}
1 & 0 & x_1 & 0 \\
1 & \cdot & x_2 & \cdot \\
\vdots & \vdots & \vdots & \vdots \\
1 & 0 & x_{n_1} & 0 \\ \hline
0 & 1 & 0 & x_{n_1+1} \\
\vdots & \vdots & \vdots & \vdots \\
0 & 1 & 0 & x_{n_1+n_2}
\end{bmatrix}
\begin{bmatrix} \beta_0 \\ \gamma_0 \\ \beta_1 \\ \gamma_1 \end{bmatrix}
+ \boldsymbol{\varepsilon}
$$

As a result, the hypothesis for equality of slopes is easily written

$$
H_0\colon [0 \quad 0 \quad 1 \quad -1] \begin{bmatrix} \beta_0 \\ \gamma_0 \\ \beta_1 \\ \gamma_1 \end{bmatrix} = 0
$$

If we require testing the equality of the entire regression line, the hypothesis would

become
$$H_0: \begin{bmatrix} 1 & -1 & 0 & 0 \\ 0 & 0 & 1 & -1 \end{bmatrix} \begin{bmatrix} \beta_0 \\ \gamma_0 \\ \beta_1 \\ \gamma_1 \end{bmatrix} = \begin{bmatrix} 0 \\ 0 \end{bmatrix}$$

Suppose we wish to test the hypothesis that the slopes in the two regressions are both equal to a specific value, say 2.0. We then simply test

$$H_0: \begin{bmatrix} 0 & 0 & 1 & 0 \\ 0 & 0 & 0 & 1 \end{bmatrix} \begin{bmatrix} \beta_0 \\ \gamma_0 \\ \beta_1 \\ \gamma_1 \end{bmatrix} = \begin{bmatrix} 2 \\ 2 \end{bmatrix}$$

Development of the Test of the General Linear Hypothesis

As we indicated earlier, the general linear hypothesis is designed to test hypotheses involving linear statements on the parameters of the multiple linear regression model. To test the hypothesis we must, of course, make the usual normal theory assumption on the ε_i. Namely, we assume $\boldsymbol{\varepsilon} \sim N(\mathbf{0}, \sigma^2 I_n)$. Recall from Section 3.3 that, under these conditions, $\mathbf{b} \sim N(\boldsymbol{\beta}, \sigma^2(\mathbf{X}'\mathbf{X})^{-1})$. Here, $\sigma^2(\mathbf{X}'\mathbf{X})^{-1}$ is the variance-covariance matrix of the coefficient vector $\mathbf{b}$. The parameter vector under test is $\mathbf{C}\boldsymbol{\beta}$, which is estimated by $\mathbf{Cb}$. The elements of $\mathbf{Cb}$ are linear combinations of normally distributed random variables. The vector $\mathbf{Cb}$ is unbiased for $\mathbf{C}\boldsymbol{\beta}$ and the variance-covariance matrix of $\mathbf{Cb}$ is $\sigma^2\mathbf{C}(\mathbf{X}'\mathbf{X})^{-1}\mathbf{C}'$. (Note the rule given in Exercise 3.1.) Thus we have

$$\mathbf{Cb} \sim N(\mathbf{C}\boldsymbol{\beta}, \sigma^2\mathbf{C}(\mathbf{X}'\mathbf{X})^{-1}\mathbf{C}')$$

Theory of linear models indicates that, under H_0: $\mathbf{C}\boldsymbol{\beta} = \mathbf{d}$,

$$\frac{[\mathbf{Cb} - \mathbf{d}]'[\mathbf{C}(\mathbf{X}'\mathbf{X})^{-1}\mathbf{C}']^{-1}[\mathbf{Cb} - \mathbf{d}]}{\sigma^2} \sim \chi_r^2$$

and
$$\frac{[\mathbf{Cb} - \mathbf{d}]'[\mathbf{C}(\mathbf{X}'\mathbf{X})^{-1}\mathbf{C}']^{-1}[\mathbf{Cb} - \mathbf{d}]}{rs^2} \sim F_{r,n-p}$$

See Graybill (1976). This statistic can be viewed as a ratio of mean squares as in the case of the other F-statistics in this section. Notice that the numerator represents a sense of "squared distance" between $\mathbf{Cb}$ and $\mathbf{d}$, standardized by the variance-covariance matrix of $\mathbf{Cb}$. The quantity r is the numerator degrees of freedom; it represents the number of parameters or linear combinations of parameters under test. The test procedure for the hypothesis of (3.15) is, then, to reject H_0 on the basis of the statistic

$$F = \frac{[\mathbf{Cb} - \mathbf{d}]'[\mathbf{C}(\mathbf{X}'\mathbf{X})^{-1}\mathbf{C}']^{-1}[\mathbf{Cb} - \mathbf{d}]}{rs^2} \qquad \textbf{(3.16)}$$

Rejection of H_0 is made on the basis of *an upper-tailed one-tailed F-test*; the concepts are similar to those in the other analysis of variance F-tests discussed in this section. In fact the test statistic on a subset of the $\boldsymbol{\beta}$ vector (notice Eq. (3.13)) is a special case of that given in Eq. (3.16).

Example 3.3 ***Exercise and Cholesterol Data***

Consider the data in Table 3.3. Twenty-six subjects were chosen to take part in a study to determine the effect of exercise activity and weight on HDL cholesterol. Eight subjects were placed in a control group, eight in a group that took part in a relatively rigorous running program, and ten subjects were placed in a program involving rigorous running and weightlifting. The weights and HDL cholesterol of the subjects were recorded after ten weeks of the program. Simple linear regression models were postulated for the three exercise groups. It was important for the scientist to test for *parallelism*, i.e., determine if the three regression lines fit to the three groups of subjects have a common slope. As a result, the three following models are postulated.

$$\begin{aligned} y_i &= \beta_0 + \beta_1 x_i + \varepsilon_i \qquad i = 1, 2, \ldots, 8 \qquad \text{(Group 1)} \\ y_i &= \gamma_0 + \gamma_1 x_i + \varepsilon_i \qquad i = 9, 10, \ldots, 16 \qquad \text{(Group 2)} \\ y_i &= \delta_0 + \delta_1 x_i + \varepsilon_i \qquad i = 17, 18, \ldots, 26 \qquad \text{(Group 3)} \end{aligned}$$

where y_i is the HDL cholesterol and x_i is the weight of the ith subject.

The hypothesis of interest is given by $H_0\colon \beta_1 = \gamma_1 = \delta_1$, with the alternative being that the three slopes are not equal. The $\mathbf{X}$ matrix and the $\boldsymbol{\beta}$ vector are given by

$$\mathbf{X} = \begin{bmatrix} 1 & 0 & 0 & x_1 & 0 & 0 \\ \vdots & \vdots & \vdots & x_2 & \vdots & \vdots \\ & & & \vdots & & \\ 1 & 0 & 0 & x_8 & 0 & 0 \\ \hline 0 & 1 & 0 & 0 & x_9 & 0 \\ \vdots & \vdots & \vdots & \vdots & x_{10} & \vdots \\ & & & & \vdots & \\ 0 & 1 & 0 & 0 & x_{16} & 0 \\ \hline 0 & 0 & 1 & 0 & 0 & x_{17} \\ \vdots & \vdots & \vdots & \vdots & \vdots & x_{18} \\ & & & & & \vdots \\ 0 & 0 & 1 & 0 & 0 & x_{26} \end{bmatrix} \qquad \boldsymbol{\beta} = \begin{bmatrix} \beta_0 \\ \gamma_0 \\ \delta_0 \\ \beta_1 \\ \gamma_1 \\ \delta_1 \end{bmatrix} \tag{3.17}$$

The $\mathbf{C}$ matrix is designed to test equality of the three slopes. As a result, we have

$$\mathbf{C} = \begin{bmatrix} 0 & 0 & 0 & 1 & -1 & 0 \\ 0 & 0 & 0 & 1 & 0 & -1 \end{bmatrix} \qquad \mathbf{d} = \begin{bmatrix} 0 \\ 0 \end{bmatrix}$$

The least squares fits to the three simple linear regression models for the data of Table 3.3 give

$$\mathbf{b} = \begin{bmatrix} \hat{\beta}_0 \\ \hat{\gamma}_0 \\ \hat{\delta}_0 \\ \hat{\beta}_1 \\ \hat{\gamma}_1 \\ \hat{\delta}_1 \end{bmatrix} = \begin{bmatrix} 23.054 \\ 14.255 \\ 76.880 \\ 0.2495 \\ 0.250 \\ -0.082 \end{bmatrix}$$

The $(\mathbf{X}'\mathbf{X})^{-1}$ matrix is given by

$$(\mathbf{X}'\mathbf{X})^{-1} = \begin{bmatrix} 12.166 & 0 & 0 & -0.0752 & 0 & 0 \\ 0 & 2.8334 & 0 & 0 & -0.0187 & 0 \\ 0 & 0 & 3.0721 & 0 & 0 & -0.0187 \\ -0.0752 & 0 & 0 & 0.00047 & 0 & 0 \\ 0 & -0.0187 & 0 & 0 & 0.00013 & 0 \\ 0 & 0 & -0.0187 & 0 & 0 & 0.000118 \end{bmatrix}$$

and the residual mean square, $s^2 = 85.6179$. We can now use Eq. (3.16) to test the hypothesis of equality of slope.

$$F = \frac{(\mathbf{Cb} - \mathbf{d})'[\mathbf{C}(\mathbf{X}'\mathbf{X})^{-1}\mathbf{C}')^{-1}(\mathbf{Cb} - \mathbf{d})}{2s^2} = 2.95$$

The F-value for 2 and 20 degrees of freedom is significant at level 0.075. As a result, there is certainly some evidence that suggests that the slopes are not equal. Incidentally, a quick observation of the point estimates makes it clear that significance (though perhaps marginally) of the F-statistic is because of the relative "flatness" of the cholesterol vs. weight relationship for the third treatment group, i.e., the running *and* weightlifting group. The slope estimates are strikingly similar for the first two groups. Obviously, more analysis is required of these data. It will be treated further in illustrations later in the chapter.

TABLE 3.3 **Data for Example 3.3 involving three sets of subjects**

	Group	Weight (lb)	HDL Cholesterol (mg/decaliter)
Control	1	163.5	75
	1	180	72.5
	1	178.5	62
	1	161.5	60
	1	127	53
	1	161	53
	1	165	65
	1	144	63.5
Running	2	141	49
	2	162	53.5
	2	134	30
	2	121	40.5

TABLE 3.3 Continued

TABLE 3.3 Continued

Data for Example 3.3 involving three sets of subjects

	Group	Weight (lb)	HDL Cholesterol (mg/decaliter)
Running	2	145	51.5
	2	106	57.5
	2	134	49
	2	216.5	74
Running and Weightlifting	3	136.5	54.5
	3	142.5	79.5
	3	145	64
	3	165	69
	3	226	50.5
	3	122	58
	3	193	63.5
	3	163.5	76
	3	154	55.5
	3	139	68

Data analyzed by VPI and SU Consulting Center, Blacksburg, Virginia, in collaboration with the Department of Health and Physical Education at VPI and SU, 1986.

The General Linear Hypothesis and the Extra Sum of Squares Principle

In the hypothesis testing development prior to the general linear hypothesis, we spent a considerable amount of effort on the *extra sum of squares* principle. The fairly simple idea centers around the notion that the usefulness of a regressor variable (or subset) in a model is dependent on the proportion of variation explained (regression sum of squares) by the variable (or subset) in the presence of the other regressors in the model. Obviously the determination of what is a *significant contribution* is made on the basis of the F-test. Alternatively this *variation explained* can just as easily be viewed as *residual sum of squares reduced.* As an illustration, consider Example 3.2 with the squid data. The test of

$$H_0: \begin{bmatrix} \beta_4 \\ \beta_5 \end{bmatrix} = \mathbf{0}$$

was made by using

$$R(\beta_4, \beta_5 \mid \beta_0, \beta_1, \beta_2, \beta_3) = R(\beta_1, \beta_2, \beta_3, \beta_4, \beta_5 \mid \beta_0) - R(\beta_1, \beta_2, \beta_3 \mid \beta_0) \quad \textbf{(3.18)}$$

This equation represents the difference in SS_{Reg} between two models, the *full model* containing all five regressor variables, and the *reduced model* containing all regressor variables except those under test. The reader should think of the reduced model as being the model that exists under the constraint of the null hypothesis H_0. It should be clear that Eq. (3.18) can also be viewed as the *difference in the residual sum of squares* between the full and reduced model since SS_{Reg} and SS_{Res} add to SS_{Total}. In other words, for the squid data, the amount that the presence of β_4 and β_5 increase SS_{Reg} is equal to the amount that SS_{Res} is reduced when β_4 and β_5 are brought into the model. Thus

$$R(\beta_4, \beta_5 \mid \beta_0, \beta_1, \beta_2, \beta_3) = SS_{\text{Res,reduced}} - SS_{\text{Res,full}}$$

which is the reduction in residual sum of squares. We point out this equivalence because, in the use of the general linear hypothesis, the use of residual sum of squares allows for more flexibility. At this point we should note that Eq. (3.13), which provides the extra sum of squares for testing that a parameter subset is zero, could easily have been written

$$\begin{aligned} R(\boldsymbol{\beta}_1 \mid \boldsymbol{\beta}_2) &= (\mathbf{y}'\mathbf{y} - \mathbf{y}'\mathbf{X}_2(\mathbf{X}_2'\mathbf{X}_2)^{-1}\mathbf{X}_2'\mathbf{y}) - (\mathbf{y}'\mathbf{y} - \mathbf{y}'\mathbf{X}(\mathbf{X}'\mathbf{X})^{-1}\mathbf{X}'\mathbf{y}) \\ &= SS_{\text{Res,reduced}} - SS_{\text{Res,full}} \end{aligned}$$

This notion of the comparison of the full and reduced model extends nicely to the general linear hypothesis. We know that the test statistic for the general linear hypothesis of Eq. (3.15) is given in (3.16). The theory of linear models (see Graybrill (1976) or Searle (1971) has privided theoretical results for the hypothesis of (3.15) that allows us to compute the numerator of (3.16) via the extra sum of squares principle. That is, it turns out that

$$\boxed{(\mathbf{Cb} - \mathbf{d})'[\mathbf{C}(\mathbf{X}'\mathbf{X})^{-1}\mathbf{C}']^{-1}(\mathbf{Cb} - \mathbf{d}) = SS_{\text{Res,reduced}} - SS_{\text{Res,full}}} \quad \textbf{(3.19)}$$

We will not supply the details that produced Eq. (3.19). The full model, i.e., that which has no constraints on the parameters allows for the smallest residual sum of squares. The null hypothesis induces constraints on the parameters, and thus the residual sum of squares increases. The *null hypothesis is supported* when the increase in SS_{Res} under H_0 is not statistically significant. On the other hand, the alternative hypothesis is supported when inducing H_0 on the model results in a significant increase in SS_{Res}. The rejection of H_0 results from large values of the test statistic in (3.16). Example 3.4 outlines the use of the extra sum of squares principle for the situation of Example 3.3.

Example 3.4 Consider again the data of Table 3.3. Recall that the scientist was interested in testing equality of the slopes for the three exercise groups. First consider the full model described by three separate regressions with separate slopes and intercepts. The **X** matrix is indicated in Eq. (3.17). The analysis, including analysis of variance information, is shown in Table 3.4. The reduced model essentially involves invoking H_0: $\beta_1 = \gamma_1 = \delta_1$. Thus we can write the reduced model as

$$\begin{aligned} y_i &= \beta_0 + \beta_1 x_i + \varepsilon_i \qquad & i &= 1, 2, \ldots, 8 \\ y_i &= \gamma_0 + \beta_1 x_i + \varepsilon_i \qquad & i &= 9, 10, \ldots, 16 \\ y_i &= \delta_0 + \beta_1 x_i + \varepsilon_i \qquad & i &= 17, 18, \ldots, 26 \end{aligned}$$

The **X** matrix and parameter vector for the reduced model is then written

$$\mathbf{X} = \begin{bmatrix} 1 & 0 & 0 & x_1 \\ \vdots & \vdots & \vdots & x_2 \\ & & & \vdots \\ 1 & 0 & 0 & x_8 \\ \hdashline 0 & 1 & 0 & x_9 \\ \vdots & \vdots & \vdots & x_{10} \\ & & & \vdots \\ 0 & 1 & 0 & x_{16} \\ \hdashline 0 & 0 & 1 & x_{17} \\ \vdots & \vdots & \vdots & x_{18} \\ & & & \vdots \\ 0 & 0 & 1 & x_{26} \end{bmatrix} \qquad \boldsymbol{\beta} = \begin{bmatrix} \beta_0 \\ \gamma_0 \\ \delta_0 \\ \beta_1 \end{bmatrix}$$

Here, of course, there are four parameters to estimate. The full model contains six parameters. Table 3.5 contains the analysis for the reduced model.

The analyses of Tables 3.4 and 3.5 allow the analyst a pragmatic approach to testing procedures using the general linear hypothesis. For application with modern computer software, one merely conducts separate regression analyses using the full and reduced model as in Tables 3.4 and 3.5. The test statistic is then

$$F = \frac{(SS_{\text{Res,reduced}} - SS_{\text{Res,full}})}{\Delta df \cdot s^2}$$

The quantity s^2 is the error mean square for the full model, and Δdf is the difference in the error degrees of freedom for the full and reduced model. This is equivalent to r, the number of statements in the null hypothesis. (See Eq. (3.16).)

For the illustration involving the exercise data, and the test for parallelism for the three regression lines, notice that the residual sums of squares (error sums of squares) appear in the analysis of variance. Thus we have

$$F = \frac{2217.41 - 1712.36}{2(85.618)} = 2.95$$

which, of course, is equivalent to the value obtained in Example 3.3. Again, the F-value is significant at the 0.075 level.

TABLE 3.4 **Regression analysis for the full model using the separate regressions for the exercise and cholesterol data of Table 3.3**

Source	df	*SS*
Regression	6	93,701.39
Error	20	1,712.36
Total	26	95,413.75

TABLE 3.5 **Regression analysis for the reduced model using common slope for the data of Table 3.3**

Source	df	*SS*
Regression	4	93,196.34
Error	22	2,217.41
Total	26	95,413.75

In Example 3.4 we formed the test statistic by using the difference in the residual sum of squares. It is easy to determine that the same result would have been produced by using the difference in the regression sum of squares. This, of course, is because the total sum of squares is the same for the full and reduced model. However, this will not always be the same for all applications of the general linear hypothesis. For example, consider the simple linear regression model

$$y_i = \beta_0 + \beta_1 x_1 + \varepsilon_i \qquad (i = 1, 2, \ldots, n)$$

which, for our illustration, can be viewed as the full model. Consider the hypothesis

$$H_0\colon \beta_1 = 2$$
$$H_1\colon \beta_1 \neq 2$$

Now, the reduced model is given by

$$z_i = \beta_0 + \varepsilon_i \qquad (i = 1, 2, \ldots, n)$$

where $z_i = y_i - 2x_i$. Here, of course, the total sum of squares using the z_i is not equivalent to the total sum of squares of the y_i.

3.5

CONFIDENCE INTERVALS AND PREDICTION INTERVALS IN MULTIPLE REGRESSION

Confidence intervals on the mean response, given specific levels $x_{1,0}, x_{2,0}, \ldots, x_{k,0}$ of the regressor variables, play the same important role as do the corresponding confidence intervals in simple linear regression discussed in Chapter 2. Of equal importance is the prediction interval on a new observation. This interval gives probabilistic bounds on a new observation at fixed conditions on the regressor variables.

In order to develop the bounds, the parameter $\text{Var}(\hat{y}|\mathbf{x} = \mathbf{x}_0)$ must be determined. Here

$$\mathbf{x}' = (1, x_1, x_2, \ldots, x_k)$$

and the vector

$$\mathbf{x}_0' = (1, x_{1,0}, x_{2,0}, \ldots, x_{k,0})$$

give specific values of the regressor variables. In the multiple regressor case, $\text{Var}(\hat{y}\,|\,\mathbf{x} = \mathbf{x}_0)$ is given by (see Appendix A.2)

$$\text{Var}(\hat{y}\,|\,\mathbf{x}_0) = \sigma^2 \cdot \mathbf{x}_0'(\mathbf{X}'\mathbf{X})^{-1}\mathbf{x}_0 \tag{3.20}$$

Thus, at $\mathbf{x} = \mathbf{x}_0$, assuming normal errors, $100(1-\alpha)\%$ confidence bounds on $E(y\,|\,\mathbf{x} = \mathbf{x}_0)$ are given by

$$\boxed{\hat{y}(\mathbf{x}_0) \pm t_{\alpha/2,n-p}\, s\sqrt{\mathbf{x}_0'(\mathbf{X}'\mathbf{X})^{-1}\mathbf{x}_0}} \tag{3.21}$$

Here, the quantity $s\sqrt{\mathbf{x}_0'(\mathbf{X}'\mathbf{X})^{-1}\mathbf{x}_0} = s_{\hat{y}(\mathbf{x}_0)}$ is the *standard error of prediction* for the general regression model. One can easily see that $\mathbf{x}_0'(\mathbf{X}'\mathbf{X})^{-1}\mathbf{x}_0$ involves variances and covariances of all regression coefficients (apart from σ^2). Indeed, the quadratic form, $\mathbf{x}_0'(\mathbf{X}'\mathbf{X})^{-1}\mathbf{x}_0$ plays an important role in much of the material in succeeding portions of the text. At this point the reader should become familiar with it as the prediction variance (apart from σ^2) at $\mathbf{x} = \mathbf{x}_0$.

Using an argument similar to that used in simple regression in Chapter 2, the prediction interval on a new observation at $\mathbf{x} = \mathbf{x}_0$ can be obtained from

$$\boxed{\hat{y}(\mathbf{x}_0) \pm t_{\alpha/2,n-p}\, s\sqrt{1 + \mathbf{x}_0'(\mathbf{X}'\mathbf{X})^{-1}\mathbf{x}_0}} \tag{3.22}$$

Example 3.5 ***Forestry Data***

The principal objective of many data collection exercises in forestry is developing models to use in predicting volume and general value of trees in a forested tract. The data in Table 3.6 give values of characteristics of a particular stand of trees, including: AGE, the age of a particular pine stand; HD, the average height of dominant trees in feet; N, the number of pine trees per acre at age, AGE; and MDBH, the average diameter at breast height (measured at 4.5 feet above ground) at age, AGE.

The data were used to build a model to predict MDBH. Theory suggests that a reasonable definition of regressor variables are $x_1 = \text{HD}$, $x_2 = \text{AGE} \cdot \text{N}$, $x_3 = \text{HD/N}$ with the response variable $y = \text{MDBH}$. Thus the following model was postulated.

$$y_i = \beta_0 + \beta_1 x_{1i} + \beta_2 x_{2i} + \beta_3 x_{3i} + \varepsilon_i \qquad (i = 1, 2, \ldots, 20)$$

Table 3.7 reveals the results from a computer printout giving many of the statistics discussed in this chapter. Note that R^2 is 0.8672, and the residual estimate of variance, s^2, is 0.0862 in^2. Also note that if one were to test

$$H_0: \beta_3 = 0$$
$$H_1: \beta_3 \neq 0$$

TABLE 3.6 **Stand characteristics for pine trees**

AGE	HD	N	MDBH
19	51.5	500	7.0
14	41.3	900	5.0
11	36.7	650	6.2
13	32.2	480	5.2
13	39.0	520	6.2
12	29.8	610	5.2
18	51.2	700	6.2
14	46.8	760	6.4
20	61.8	930	6.4
17	55.8	690	6.4
13	37.3	800	5.4
21	54.2	650	6.4
11	32.5	530	5.4
19	56.3	680	6.7
17	52.8	620	6.7
15	47.0	900	5.9
16	53.0	620	6.9
16	50.3	730	6.9
14	50.5	680	6.9
22	57.7	480	7.9

Harold E. Burkhart, Robert C. Parker, Mike R. Strub, and Richard Oderwald, "Yields of Old-field Loblolly Pine Plantations," Division of Forestry and Wildlife Resources Pub. FWS-3-72, Virginia Polytechnic Institute and State University, Blacksburg, Virginia, 1972.

TABLE 3.7 **Regression model statistics for data of Table 3.6 (Dependent variable: MDBH)**

Source	df	*SS*	*MS*	*F*-Value	Pr > *F*	*R*-Square
Regression	3	9.0064	3.0021	34.83	0.0001	0.8672
Error	16	1.3791	0.0862			
Correcting Total	19	10.3855				

Source	df	Sequential *SS*	*F*-Value	Pr > *F*	df	Partial *SS*	*F*-Value	Pr > *F*
x_1	1	6.2070	72.01	0.0001	1	1.2678	14.71	0.0015
x_2	1	2.7846	32.31	0.0001	1	0.6710	7.78	0.0131
x_3	1	0.0148	0.17	0.6844	1	0.0148	0.17	0.6844

Parameter	Estimate	*T* for H_0: Parameter = 0	Pr > |*T*|	Std. Error of Estimate	Residual Std. Dev.	MDBH Mean
Intercept	3.2357	9.33	0.0001	0.3466	0.2936	6.265
b_1	0.09740	3.84	0.0015	0.02540		
b_2	−0.0001688	−2.79	0.0131	0.00006052		
b_3	3.4668	0.41	0.6844	8.3738		

Observation	Actual	Fitted Value	Residual	Std. Error of Prediction	LCL on Mean	UCL on Mean	Lower Prediction Limit	Upper Prediction Limit
1	7.000	7.005	−0.005	0.144	6.700	7.310	6.312	7.698
2	5.000	5.290	−0.290	0.137	4.999	5.581	4.603	5.977
3	6.200	5.799	0.401	0.124	5.537	6.061	5.124	6.474
4	5.200	5.551	−0.351	0.148	5.238	5.864	4.854	6.248
5	6.200	6.153	0.047	0.108	5.924	6.382	5.490	6.816
6	5.200	5.072	0.128	0.146	4.763	5.381	4.377	5.767
7	6.200	6.349	−0.149	0.080	6.179	6.519	5.704	6.994
8	6.400	6.211	0.189	0.105	5.989	6.434	5.550	6.872
9	6.400	6.345	0.055	0.192	5.937	6.753	5.601	7.089
10	6.400	6.971	−0.571	0.121	6.713	7.228	6.297	7.644
11	5.400	5.275	0.125	0.119	5.022	5.527	4.603	5.946
12	6.400	6.499	−0.099	0.125	6.234	6.765	5.823	7.176
13	5.400	5.630	−0.230	0.128	5.358	5.902	4.950	6.309
14	6.700	6.825	−0.125	0.094	6.626	7.025	6.172	7.479
15	6.700	6.894	−0.194	0.096	6.690	7.099	6.239	7.549
16	5.900	5.715	0.185	0.118	5.464	5.966	5.044	6.386
17	6.900	7.020	−0.120	0.123	6.759	7.280	6.345	7.694
18	6.900	6.402	0.498	0.086	6.221	6.583	5.754	7.050
19	6.900	6.804	0.095	0.151	6.484	7.126	6.104	7.505
20	7.900	7.490	0.410	0.205	7.055	7.925	6.730	8.249

the partial F-statistic and, of course, the t-statistic indicate that the null hypothesis is not rejected. This implies that x_3 is not needed in the presence of x_1 and x_2.

Consider, for example, the conditions at observation 5, the fifth data point. The actual MDBH is 6.2 in. and the *fitted value* y is 6.153 in. with a computed residual of 0.047 in. The *standard error of prediction* of MDBH at this location (see Eq. (3.21)) is 0.108 in., resulting in a 95% confidence interval on the mean MDBH of

$$[5.924, 6.382]$$

at the conditions of observation 5. The 95% prediction interval on a new observed MDBH at this condition is given by

$$[5.490, 6.816]$$

Further Comments

To this point in the text, all examples have been used to illustrate concepts or regression statistics discussed in that section. Now it is realistic to ask some questions not necessarily answered by the computer printout in the case of the present data set:

1. Since x_3 appears, on the basis of t or partial F, to explain a statistically insignificant portion of variation in the response, should the term $\beta_3 x_3$ be eliminated from the model? Would the model *excluding* $\beta_3 x_3$ more effectively predict MDBH?

We attempt to tackle this issue in Chapter 4.

2. What is the real utility of the standard error of prediction?

Obviously we can use it to compute the confidence limits on the mean response (MDBH in this illustration). These results, including the prediction interval on a new observed MDBH, are standard in most regression computer packages, and they are certainly helpful. But the practitioner should surely be reminded that the conditions *at the data points do not represent locations where there is great likelihood that one needs to predict*. In fact, one already has response data at these locations. More interest is centered on confidence limits and standard errors of prediction at other combinations of the regressor variables, those that represent locations where one might truly wish to use the equation to predict. For example, in the forestry illustration, consider the following information, indicating combinations of the x's not in the data. Standard errors of prediction as well as confidence limits are shown.

x_1	x_2	x_3	$s_{\hat{y}(\mathbf{x}_0)}$	LCL	UCL
10	2,500	0.02	0.2725	3.279	4.435
80	6,000	0.1333	0.6839	9.027	11.927
75	25,000	0.075	0.3450	5.848	7.311

Some computer packages allow easy access to this type of information as well as prediction-oriented information at the data points that constructed the regression. In this illustration the quality of $\hat{y}$ as a *fitted value* would appear to be superior to its value as a true predictor. In fact, at all three selected locations, where the regressor values represent a slight degree of extrapolation, the capability of the prediction equation drops off substantially. It is this kind of information that can contribute heavily to our efforts to quantify prediction or extrapolation capabilities of a regression model and to choose the best.

As the material on basic multiple linear regression nears completion, the reader should notice the emergence of several often repeated points. If model selection from a set of candidate models is important, several criteria need to be observed. Certainly the partial F-statistics can provide some evidence but cannot be the sole device used. Some criteria evolve from the standard error of prediction mentioned in this section while others are based on the intuitive concept that we should force validation of the candidate models through the exercise of predicting independent data. The good analyst understands that models are put forth because they are scientifically sensible. Ideally, the one that makes sense structurally survives the tournament that ensues as the statistician treks from one criterion to the next. Often the model selected is not the one that makes maximum sense. It is quite possible that the model that is theoretically justifiable is one that the data involved cannot support, though we certainly cannot adopt a prediction model that is totally devoid of theoretical underpinnings. Much of Chapter 4 contains information and examples designed to produce criteria for choice of best model for prediction.

3.6 DATA WITH REPEATED OBSERVATIONS

In areas such as the biological and physical sciences, the experimenter is often able to *design the experiment* on the x-levels; thus the levels can be controlled and repeated observations on the response can be taken at each x-combination. For example, the chemist, who is interested in a regression function relating reaction rate y to temperature x_1 and reaction time x_2, benefits from obtaining multiple observations or making multiple runs at each experimental point. From this type of experiment, we obtain an estimate of σ^2 from a source that represents the *reproducibility* of experiment, or experimental error variance. This source of variation can then be separated from the usual residual variation, producing a *model-independent* portion of variation that can be useful in ascertaining validity of the model.

CONCEPT OF LACK OF FIT

The terminology *lack of fit* has come to imply an exercise of *checking for model adequacy* with the lack of fit mean square representing a component of variation that can be used in comparison to the *pure* error mean square (developed from the repeated response observations).

Let y_{ij} represent the jth response at the ith experimental combination. Here, $i = 1, 2, \ldots, m$; $j = 1, 2, \ldots, n_i$ and $\sum_{i=1}^{m} n_i = n$. In other words, there are $m \geqslant p$ distinct combinations of the regressor variables in the experiment, and n_i experimental runs at the ith combination. Consider now the following partition of the residual sum of squares.

$$\sum_{i=1}^{m} \sum_{j=1}^{n_i} (y_{ij} - \hat{y}_i)^2 = \sum_{i=1}^{m} \sum_{j=1}^{n_i} (y_{ij} - \bar{y}_i)^2 + \sum_{i=1}^{m} n_i(\bar{y}_i - \hat{y}_i)^2 \qquad \textbf{(3.23)}$$

where $\hat{y}_i$ is the fitted value of the response at the ith combination of the regressor variables and $\bar{y}_i$ is the average response at the ith combination. The quantity $\sum_{i=1}^{m} \sum_{j=1}^{n_i} (y_{ij} - \bar{y}_i)^2$ measures variation due to repeated observations as previously described and, of course,

$$\sum_{i=1}^{m} \sum_{j=1}^{n_i} (y_{ij} - \bar{y}_i)^2 \bigg/ \sum_{i=1}^{m} (n_i - 1)$$

the pure error mean square, unbiasedly estimates σ^2 whether or not the fitted model is the correct one. The quantity $\sum_{i=1}^{m} n_i(\bar{y}_i - \hat{y}_i)^2$ is called the *lack of fit sum of squares* and has $m - p$ associated degrees of freedom. If the model is correct, one would assume that the $\bar{y}_i$ do not deviate appreciably from the fitted value $\hat{y}_i$; and in fact,

$$E(MS_{\text{LOF}}) = E\left[\sum_{i=1}^{m} n_i(\bar{y}_i - \hat{y}_i)^2/(m - p)\right] = \sigma^2$$

If the model is underspecified, i.e., if there are terms in $E(y_i)$ involving powers and products in $x_{1i}, x_{2i}, \ldots$, then (MS_{LOF}) will overestimate σ^2. Indeed, the quantity by which σ^2 is overestimated measures the difference between $E(y_i)$, the true mean response, and the quantity that the analyst assumes is the true response. As an example, if $k = 1$ and the postulated model is

$$y_i = \beta_0 + \beta_1 x_i + \varepsilon_i$$

while, truly,

$$E(y_i) = \beta_0 + \beta_1 x_i + \beta_{11} x_i^2 \qquad \textbf{(3.24)}$$

then

$$E(MS_{\text{LOF}}) = \sigma^2 + \frac{\sum_{i=1}^{m} n_i[\beta_{11} x_i^2]^2}{m - 2} \qquad \textbf{(3.25)}$$

The details regarding the general development of $E(MS_{\text{LOF}})$ are presented in the next subsection.

The test for model adequacy via the lack of fit approach is then given by

$$F = \frac{SS_{\text{LOF}}/(m-p)}{SS_{\text{Pure error}}/(n-m)} = \frac{MS_{\text{LOF}}}{MS_{\text{Pure error}}}$$

and is an upper-tailed one-tailed F-test with $m-p$ and $n-m$ numerator and denominator degrees of freedom respectively. A significant value of the F-statistic implies that there is a detectable contribution due to terms of order above those included in the model.

Example 3.6 ***Breadwrapper Data***

Consider the breadwrapper illustration of Example 3.1 with the data contained in Table 3.1. In this illustration, one location in the regressor variables contains 6 repeated experimental runs. This is not uncommon in experimental design situations. The location in question is the point $x_1 = 255°\text{F}$, $x_2 = 55°\text{F}$, and $x_3 = 1.1\%$. With the experiment employed here, various additional terms involving powers and cross products *could have been introduced into the model.* We should begin with a "count" on the degrees of freedom. In this case, $n = 20$, $m = 15$, and the value of p, the number of model terms, is 10. Using (3.23), we have

$$SS_{\text{Pure error}} = 4.96$$

with 5 degrees of freedom. This represents the corrected sum of squares of the seal strength values at the single location where the repeated observations are taken. The residual sum of squares with 10 degrees of freedom had already been illustrated and is given by $SS_{\text{Res}} = 11.8678$. Thus, the partition of variability, degrees of freedom, and F-test are shown in the following table.

Partition of variation, test for lack of fit

Source	SS	df	MS	F
Regression	70.3022	9	7.8113	—
Lack of fit	6.9078	5	1.3816	$F = 1.39$
Pure error	4.9600	5	0.9920	—

The F-statistic is not significant, and thus we infer that *lack of fit is insignificant.* The implication here is that the model with 9 regression variables and a constant term is adequate. Put another way, we conclude that the additional 5 terms (5 lack of fit degrees of freedom) are, as a group, not statistically significant.

ADDITIONAL DEVELOPMENT AND COMMENTS

In this section we investigate the lack of fit notion using a matrix development. The approach and results will provide much needed machinery in Chapter 4 where we investigate the impact of model misspecification. The result of erroneously under- and overspecification of a model must be clearly understood if the reader expects to make serious judgments regarding proper model selection.

In a very large sense, the concept of lack of fit deals with underspecification and, in fact, provides the analyst with a formal test of hypothesis, which is informative regarding the adequacy of the model that was fit. To illuminate this further, suppose the experimenter postulates a model

$$\mathbf{y} = \mathbf{X}_1 \boldsymbol{\beta}_1 + \boldsymbol{\varepsilon}^* \tag{3.26}$$

where $\mathbf{X}_1$ is $n \times p$ and $\boldsymbol{\beta}_1$ contains p elements. The "true" model however is given by

$$\mathbf{y} = \mathbf{X}_1 \boldsymbol{\beta}_1 + \mathbf{X}_2 \boldsymbol{\beta}_2 + \boldsymbol{\varepsilon}$$

where $\mathbf{X}_2$ is $n \times (m - p)$ and contains terms in powers and products that could be estimated, given the sample size and the data at hand. As a result, $E(\boldsymbol{\varepsilon}^*) \neq \mathbf{0}$. If one ignores $\mathbf{X}_2 \boldsymbol{\beta}_2$, the least squares estimate of $\boldsymbol{\beta}_1$ is $\mathbf{b}_1 = (\mathbf{X}_1' \mathbf{X}_1)^{-1} \mathbf{X}_1' \mathbf{y}$. Suppose we consider the residual sum of squares, $SS_{\text{Res}} = (\mathbf{y} - \mathbf{X}_1 \mathbf{b}_1)'(\mathbf{y} - \mathbf{X}_1 \mathbf{b}_1)$. It turns out that under the assumption of uncorrelated errors and common error variance (see Appendix B.2),

$$E(SS_{\text{Res}}) = \sigma^2(n - p) + \boldsymbol{\beta}_2' [\mathbf{X}_2' \mathbf{X}_2 - \mathbf{X}_2' \mathbf{X}_1 (\mathbf{X}_1' \mathbf{X}_1)^{-1} \mathbf{X}_1' \mathbf{X}_2] \boldsymbol{\beta}_2 \tag{3.27}$$

The term $\boldsymbol{\beta}_2' [\mathbf{X}_2' \mathbf{X}_2 - \mathbf{X}_2' \mathbf{X}_1 (\mathbf{X}_1' \mathbf{X}_1)^{-1} \mathbf{X}_1' \mathbf{X}_2] \boldsymbol{\beta}_2$ represents a quadratic form in $\boldsymbol{\beta}_2$, the ignored parameters. We know that the pure error sum of squares has expectation

$$E(SS_{\text{Pure error}}) = \sigma^2 \left[\sum_{i=1}^{m} (n_i - 1) \right]$$

where $\sum n_i = n$, the total sample size. Thus $E(MS_{\text{LOF}})$ is given by

$$\begin{aligned} E(MS_{\text{LOF}}) &= \frac{E(SS_{\text{Res}}) - E(SS_{\text{Pure error}})}{m - p} \\ &= \frac{\sigma^2(m - p) + \boldsymbol{\beta}_2' [\mathbf{X}_2' \mathbf{X}_2 - \mathbf{X}_2' \mathbf{X}_1 (\mathbf{X}_1' \mathbf{X}_1)^{-1} \mathbf{X}_1' \mathbf{X}_2] \boldsymbol{\beta}_2}{m - p} \\ &= \sigma^2 + \frac{\boldsymbol{\beta}_2' [\mathbf{X}_2' \mathbf{X}_2 - \mathbf{X}_2' \mathbf{X}_1 (\mathbf{X}_1' \mathbf{X}_1)^{-1} \mathbf{X}_1' \mathbf{X}_2] \boldsymbol{\beta}_2}{m - p} \end{aligned} \tag{3.28}$$

Thus the F-test for lack of fit is designed to detect the contribution of the positive definite quadratic form (see Appendix A.2)

$$\boldsymbol{\beta}_2' [\mathbf{X}_2' \mathbf{X}_2 - \mathbf{X}_2' \mathbf{X}_1 (\mathbf{X}_1' \mathbf{X}_1)^{-1} \mathbf{X}_1' \mathbf{X}_2] \boldsymbol{\beta}_2$$

where $\boldsymbol{\beta}_2$ represents the vector of regression parameters beyond the fitted model that can be estimated. As a result, the sensitivity of the lack of fit test depends directly on the value of this quadratic form; and thus on the size of the coefficients in $\boldsymbol{\beta}_2$ and the elements in $\mathbf{X}_1$ and $\mathbf{X}_2$.

3.7 SIMULTANEOUS INFERENCE IN MULTIPLE REGRESSION

In Section 2.10 we motivated the need to consider simultaneous inference in simple linear regression. Simultaneous tests on slope and intercept are developed as well as simultaneous confidence regions on these two parameters. In addition, simultaneous confidence intervals on the mean response were considered at the data locations. The extension of much of this material to multiple regression is fairly straightforward given the motivation provided in Chapter 2 and the machinery produced in this chapter. We shall proceed by dealing initially with inference on individual parameters.

SIMULTANEOUS TESTS ON COEFFICIENTS

The reader should realize that testing the hypothesis

$$\begin{aligned} H_0\colon \beta_0 &= \beta_{0,0} \\ \beta_1 &= \beta_{1,0} \\ &\vdots \\ \beta_k &= \beta_{k,0} \end{aligned}$$

in the framework of the multiple linear regression model of Eq. (3.1) involves nothing more than applying the general linear hypothesis of Section 3.4. The hypothesis becomes (in matrix notation)

$$\begin{aligned} H_0\colon \boldsymbol{\beta} &= \boldsymbol{\beta}_0 \\ H_1\colon \boldsymbol{\beta} &\neq \boldsymbol{\beta}_0 \end{aligned}$$

which is a special case of the hypothesis of equation (3.15) with $\mathbf{C} = I_p$ and $\mathbf{d} = \mathbf{0}$. Thus the use of the test statistic of Eq. (3.16) gives

$$F = \frac{(\mathbf{b} - \boldsymbol{\beta}_0)'(\mathbf{X}'\mathbf{X})(\mathbf{b} - \boldsymbol{\beta}_0)}{ps^2} \tag{3.29}$$

which is $F_{p,n-p}$ *under* H_0 when we make the usual normal theory assumption. Thus the usual uppertail F-test is used. Tests on subsets of the coefficients using the hypothesis of Eq. (3.12) can also be made as a special case of the general linear hypothesis. In fact, the test statistic in (3.13) is a special case of (3.16) with $C = [I_p \vdots 0]$.

SIMULTANEOUS CONFIDENCE REGION ON COEFFICIENTS

The methodology that is used for simultaneous testing on elements in $\boldsymbol{\beta}$ allows for computing a simultaneous confidence region on the parameters. This is an extension of the elliptical confidence region for β_0 and β_1 illustrated in Example 2.7. From the result of (3.29)

$$\Pr\left[\frac{(\mathbf{b}-\boldsymbol{\beta})'(\mathbf{X}'\mathbf{X})(\mathbf{b}-\boldsymbol{\beta})}{ps^2} \leqslant F_{\alpha,p,n-p}\right] = 1-\alpha$$

As a result, any set of parameters $\boldsymbol{\beta}$ that satisfy the inequality

$$\boxed{\frac{(\mathbf{b}-\boldsymbol{\beta})'(\mathbf{X}'\mathbf{X})(\mathbf{b}-\boldsymbol{\beta})}{ps^2} \leqslant F_{\alpha,p,n-p}}$$

fall inside the $100(1-\alpha)\%$ confidence region for $\boldsymbol{\beta}$.

BONFERRONI CONFIDENCE REGION FOR REGRESSION COEFFICIENTS

In Section 2.10 we discussed the Bonferroni approach to confidence limits for simple linear regression. Bonferroni confidence limits allow for a rectangular confidence region in such a way that a bound on the *joint confidence coefficient* is controlled. As in the simple linear regression case, the Bonferroni confidence intervals are of the form

$$b_j \pm t\cdot(\text{standard error of } b_j)$$

where the t constant represents an appropriate percentile point of the t-distribution. In the simple linear case the bound on the joint confidence region is controlled by using the $t_{\alpha/4}$ percentile point (with the appropriate degrees of freedom). Simply put, *the generalization is to use $t_{\alpha/2p}$, where p is the number of parameters involved in the joint confidence region.* Thus a $100(1-\alpha)\%$ confidence region on $\beta_0, \beta_1, \beta_2, \ldots, \beta_k$ is given by

$$\begin{aligned}
&b_0 \pm t_{\alpha/2p,n-p}\cdot(\text{standard error of } b_0)\\
&b_1 \pm t_{\alpha/2p,n-p}\cdot(\text{standard error of } b_1)\\
&\vdots\\
&b_k \pm t_{\alpha/2p,n-p}\cdot(\text{standard error of } b_k)
\end{aligned}$$

where $p = k + 1$. Here, of course, the exact confidence coefficient is *at least* $1-\alpha$. The standard errors of the coefficients are described in Eq. (3.14). It should be noted that the Bonferroni confidence bounds may be more helpful in some instances than the exact confidence region merely because it is difficult to use the exact method graphically as one can in simple linear regression. As a result, the Bonferroni bounds, though more conservative, are easier to use.

SIMULTANEOUS CONFIDENCE BOUNDS ON MEAN RESPONSE

In Section 2.10 we touched on the determination of confidence intervals on mean response in the simple linear regression case when multiple estimations, or predictions, are being made simultaneously. The idea put forth dealt with confidence limits that are conservative enough to allow the analyst to put a bound on the joint confidence coefficient. We used the Bonferroni approach and the Working-Hotelling approach. As one might expect, the Bonferroni approach extends nicely to multiple linear regression for the reader that has been exposed to the material in Section 3.5. Suppose it is of interest to estimate mean response at r locations, i.e., at r combinations of the regressor variables. Denote these combinations by $\mathbf{x}'_1, \mathbf{x}'_2, \ldots, \mathbf{x}'_r$. In the case of an intercept in the model, the initial element in $\mathbf{x}_j$ is 1. Of course, these need not be locations that represent data points. As in the case of simple linear regression (see Example 2.10), the t percentile point is increased to take into account the simultaneous nature of the coverage of the confidence interval. Thus, if r mean responses are to be estimated, the appropriate form of the confidence interval is

$$\hat{y}(\mathbf{x}_j) \pm t_{\alpha/2r,n-p} s_{\hat{y}(\mathbf{x}_j)}$$

The $\hat{y}(\mathbf{x}_j)$ represents the estimated response at the location $\mathbf{x}_j$ while $s_{\hat{y}(\mathbf{x}_j)}$ is the standard error of prediction at the location $\mathbf{x}_j$. Now, if we make use of Eq. (3.20), the Bonferroni approach allows us to be at least $100(1-\alpha)\%$ confident that the following intervals cover the mean response values

$$\hat{y}(\mathbf{x}_j) \pm t_{\alpha/2r,n-p} s\sqrt{\mathbf{x}'_j(\mathbf{X}'\mathbf{X})^{-1}\mathbf{x}_j} \qquad (j = 1, 2, \ldots, r)$$

The Working-Hotelling procedure can also be used for confidence limits around simultaneous predicted values, i.e., joint confidence limits on mean response. Of course, if the number of model parameters is large and the value of r is relatively small, the Bonferroni approach will be considerably more efficient. Joint confidence intervals at r locations using the Working-Hotelling approach are given by

$$\hat{y}(\mathbf{x}_j) \pm \sqrt{pF_{a,p,n-p}}\, s\sqrt{\mathbf{x}'_j(\mathbf{X}'\mathbf{X})^{-1}\mathbf{x}_j} \qquad (j = 1, 2, \ldots, r)$$

Example 3.7 ***Simultaneous Inference Using Forestry Data***

Consider the forestry data of Table 3.6. In this example we use the same model that was used for Example 3.5 (three regressors, x_1, x_2, and x_3). It is of interest to illustrate simultaneous inference for estimation of mean response and compare a set of joint confidence limits with the standard confidence limits on mean response given in Table 3.7. Suppose the biologist is interested in predicting MDBH simultaneously at five locations, namely those at observations 16, 17, 18, 19, and 20 in Table 3.7. The confidence limits that appear on mean MDBH in Table 3.7 are, of course, not joint confidence limits. Rather they are 95% confidence limits on mean MDBH with the 95% coverage being *conditional* on a given location. Notice that the standard errors of prediction at locations 16–20, as well as fitted values, appear in Table 3.7. The appropriate t percentile point for $r = 5$ computed confidence limits and joint coverage of at least 95% is (see Table C.2)

$$t_{.025/5,16} = t_{.005,16} = 2.921$$

Thus the 95% joint confidence limits (at least 95% confidence) are given by the bracketed values below

Location		
16	$5.715 \pm 2.921(0.118)$:	[5.370; 6.060]
17	$7.020 \pm 2.921(0.123)$:	[6.661; 7.379]
18	$6.402 \pm 2.921(0.086)$:	[6.151; 6.653]
19	$6.804 \pm 2.921(0.151)$:	[6.363; 7.245]
20	$7.490 \pm 2.921(0.205)$:	[6.891; 8.089]

Notice how the joint confidence limits are more conservative than the standard 95% intervals in Table 3.7.

3.8 MULTICOLLINEARITY IN MULTIPLE REGRESSION DATA

Regression analysis is employed as an analytic tool for any number of reasons. We alluded earlier to the need for developing an estimate of a functional relationship, which we can use for prediction. On the other hand, the motivation of the regression may be to estimate rates of change of response with respect to particular regressor variables, i.e., estimates of regression coefficients. However, regardless of the purpose for which the regression is computed, problems can arise because the data at hand simply cannot or will not support the scientist's notion of what the model should be. Large experimental error or "noise" in the data simply prohibits a good fit and thus disallows any chance of a quality statistical inference. We categorize this difficulty as a result of an experimental situation in which the model error variance, σ^2, is unduly large.

Another difficulty, which has received considerable attention in the past 20 years, is the problem of multicollinearity. The name *multicollinearity* has been used to such an extent that many potential users of statistical tools are aware that such a problem may exist before they really understand what it is. The term defines itself, *multi* implying many and *collinear* implying linear dependencies. Multicollinearity describes a condition in the *regressor variables.* In this chapter we do *not* presume to deal completely with multicollinearity, but rather to introduce it and ask that the reader attempt to understand its effect on regression results. Methods of detection of serious multicollinearity, formal diagnosis, and procedures for combating it are relegated to Chapter 8. For example, biased forms of estimation, representing alternatives to the usual least squares procedure will be discussed at length in Chapter 8. But, for the reader in the throes of initial exposure to multiple regression, it is important to concentrate on how multicollinearity prohibits precise statistical inference, i.e., how estimation of regression coefficients, or perhaps prediction, is rendered poor due to the condition. The methods of combating the problem, e.g. *ridge regression* or *principal components regression*, are somewhat controversial and, indeed, are often criticized by many statistical researchers. Nevertheless, even if the reader never uses any of the available methods, a clear understanding of the sources and the effects of multicollinearity remain important.

Consider the general linear model in (3.2). The source of multicollinearity results from the nature of the columns of $\mathbf{X}$. Suppose we partition $\mathbf{X}$ such that

$$\mathbf{X} = [\mathbf{1} \quad \mathbf{x}_1 \quad \mathbf{x}_2 \quad \cdots \quad \mathbf{x}_k] \tag{3.30}$$

Each column represents the measurements for a particular regressor variable. View the first column as a column of ones, assuming there is a constant term in the model. Also assume, for ease of understanding, that the regressor variables are centered and scaled; i.e., if x_{ij} is the jth measurement on the regressor variable x_i in the natural units, $\bar{x}_i = \sum_{j=1}^{n} x_{ij}/n$ is subtracted from x_{ij} and $x_{ij} - \bar{x}_i$ is divided by S_i where $S_i = \sqrt{\sum_{j=1}^{n} (x_{ij} - \bar{x}_i)^2}$. This centering and scaling results in $\mathbf{X}'\mathbf{X}$ being a *correlation matrix*. That is, we can write the symmetric matrix

$$\mathbf{X}'\mathbf{X} = \begin{bmatrix} 1 & 0 & 0 & \cdots & 0 \\ 0 & 1 & r_{12} & \cdots & r_{1k} \\ 0 & r_{12} & 1 & \cdots & r_{2k} \\ \vdots & & & & \vdots \\ \vdots & & & & r_{k-1,k} \\ 0 & & & & 1 \end{bmatrix}$$

where r_{ij} is the *simple correlation coefficient* between regressor x_i and x_j. We should state here that the usual column of ones in the first column of $\mathbf{X}$ can be removed resulting in the model form

$$\mathbf{y} = \beta_0 \mathbf{1} + \mathbf{X}^* \boldsymbol{\beta} \tag{3.31}$$

where, in this form, $\boldsymbol{\beta}' = [\beta_1, \beta_2, \ldots, \beta_k]$ is the vector of coefficients, apart from the intercept, and $\mathbf{X}^*$ is the $n \times k$ matrix of centered and scaled regressor variables. The notation $\mathbf{1}$ is used to denote an n-vector of ones.

As a result $\mathbf{X}^{*\prime}\mathbf{X}$ is the $k \times k$ correlation matrix

$$\mathbf{X}^{*\prime}\mathbf{X} = \begin{bmatrix} 1 & r_{12} & r_{13} & \cdots & r_{1k} \\ & 1 & r_{23} & \cdots & r_{2k} \\ & & & & \vdots \\ & & & & r_{k-1,k} \\ & & & & 1 \end{bmatrix}$$

It is interesting that the term *correlation* is a bit of a misnomer here. We are assuming that the regressors are not random variables. Nevertheless, the simple correlation coefficient r_{ij} does measure linear dependency between x_i and x_j *in the data.*

WHAT IS MULTICOLLINEARITY?

A very desirable condition in a set of regression data is to have regressors that are not "moving with each other" in the data set. Near linear dependencies render it more difficult to sort out the impact of each regressor on the response. A simple example should illustrate this.

Consider two somewhat different scenarios both containing two regressor variables. It is our intention to illustrate how near linear dependencies among the regressors can severely inhibit the analyst's ability to estimate regression coefficients. First, suppose for $n = 8$ that we have the following regressor data.

x_1	10	10	10	10	15	15	15	15
x_2	10	10	15	15	10	10	15	15

Next consider the following data set with the same sample size $n = 8$ and same ranges on the regressors.

x_1	10.0	11.0	11.9	12.7	13.3	14.2	14.7	15.0
x_2	10.0	11.4	12.2	12.5	13.2	13.9	14.4	15.0

If it is of interest to fit a $k = 2$ linear regression to the data, the comparison of the variances of the coefficients across data sets is of interest. Computation of correlation coefficients reveals that $r_{12} = 0$ in the case of the first data set and $r_{12} = 0.99215$ for the second data set. Thus we have for data set 1,

$$\mathbf{X}^{*\prime}\mathbf{X}^{*} = \begin{bmatrix} 1 & 0 \\ 0 & 1 \end{bmatrix} \qquad (\mathbf{X}^{*\prime}\mathbf{X}^{*})^{-1} = \begin{bmatrix} 1 & 0 \\ 0 & 1 \end{bmatrix}$$

and for data set 2,

$$(\mathbf{X}^{*\prime}\mathbf{X}^{*}) = \begin{bmatrix} 1 & 0.99215 \\ 0.99215 & 1 \end{bmatrix} \qquad (\mathbf{X}^{*\prime}\mathbf{X}^{*})^{-1} = \begin{bmatrix} 63.94 & -63.44 \\ -63.44 & 63.94 \end{bmatrix}$$

Thus we have the following variances of parameter estimates for the two data sets. (We are using the centered and scaled model of Eq. 3.31.)

$$\text{Data set 1} \qquad\qquad \text{Data set 2}$$

$$\frac{\text{Var } b_1}{\sigma^2} = \frac{\text{Var } b_2}{\sigma^2} = 1.0 \qquad \frac{\text{Var } b_1}{\sigma^2} = \frac{\text{Var } b_2}{\sigma^2} = 63.94$$

Notice how the variances of the coefficients were inflated due to collinearity. This *inflation of variance* is easily accessible information that underscores the damage due to the condition of collinearity. In the first data set the regressors are said to be *orthogonal*; the condition of orthogonality is a very desirable property of an experimental design when one truly does have the capability to control the regressor variables. In data set 2 the regressors are obviously moving together, and this near linear dependency has clearly taken its toll with regard to the quality of estimation. The reader should consider data set 1 and the resulting orthogonal regressors as representing the *ideal case*.

Multicollinearity simply occurs when there are near linear dependencies among the $\mathbf{x}_j^*$, the columns of $\mathbf{X}^*$. That is, there is a set of constants (not all zero) for which

$$\sum_{j=1}^{k} c_j \mathbf{x}_j^* \cong \mathbf{0} \tag{3.32}$$

We write $\cong$ since if the right hand side is identically zero, the linear dependencies are exact and thus $(\mathbf{X}'\mathbf{X})^{-1}$ does not exist. Near dependencies, of course, may exist in real data and produce the effect that we commonly call multicollinearity. One should keep in mind that a regression coefficient is a rate of change or partial derivative of the response with respect to a regressor variable. When the x-data are conditioned in such a way that the regressors are moving with one another, the least squares procedure never is allowed exposure to the data structure that it truly needs to produce a clear estimate of this rate of change.

USE OF EIGENVALUES AND EIGENVECTORS TO EXPLAIN MULTICOLLINEARITY

Suppose we consider the $\mathbf{X}^{*\prime}\mathbf{X}^*$ matrix (correlation form). We know that there exists an orthogonal matrix (see Graybill (1976))

$$\mathbf{V} = [\mathbf{v}_1, \mathbf{v}_2, \ldots, \mathbf{v}_k]$$

such that

$$\mathbf{V}'(\mathbf{X}^{*\prime}\mathbf{X}^*)\mathbf{V} = \text{diag}(\lambda_1, \lambda_2, \ldots, \lambda_k) \tag{3.33}$$

The λ_i are the *eigenvalues* of the correlation matrix. The operation given by Eq. (3.33) is called the *eigenvalue decomposition* of $\mathbf{X}^{*\prime}\mathbf{X}^*$. The columns of $\mathbf{V}$ are normalized eigenvectors associated with the eigenvalues of $(\mathbf{X}^{*\prime}\mathbf{X}^*)$. For our purposes here, we need to denote the ith element of the vector $\mathbf{v}_j$ by v_{ij}. Now if multicollinearity is present, at least one $\lambda_i \cong 0$. (See Appendix A.3.) Thus we can write, for at least one value of j,

$$\mathbf{v}_j'(\mathbf{X}^{*\prime}\mathbf{X}^*)\mathbf{v}_j \cong 0$$

which implies that for at least one eigenvector $\mathbf{v}_j$,

$$\boxed{\sum_{\ell=1}^{k} v_{\ell j}\mathbf{x}_\ell^* \cong \mathbf{0}}$$

Thus the number of small eigenvalues of the correlation matrix relate to the number of multicollinearities according to the definition in (3.32), and the "weights," the c_j in (3.32), are the individual elements in the associated eigenvectors. This development is given at this time for more than mere passing interest. It will become an important aid in diagnosing multicollinearity in the developments in Chapter 8.

TWO EXTREMES: COLLINEARITY AND ORTHOGONALITY

Consider the situation in which the variables are orthogonal to each other; i.e., $\mathbf{X}^{*\prime}\mathbf{X}^*$, the correlation matrix, is the *identity matrix*. In this ideal case, apart from σ^2, $\text{Var}(b_i) = 1.0$ for $i = 1, 2, \ldots, k$. (Here, b_i is taken to be the coefficient of the ith centered and scaled regressor variable.) Now suppose we consider another hypothetical case in which $k = 2$ and

$$\mathbf{X}^{*\prime}\mathbf{X}^* = \begin{bmatrix} 1.0 & 0.975 \\ 0.975 & 1.0 \end{bmatrix}$$

with

$$(\mathbf{X}^{*\prime}\mathbf{X}^*)^{-1} = \begin{bmatrix} 20.2532 & -19.747 \\ -19.747 & 20.2532 \end{bmatrix}$$

Thus, the multicollinearity has produced an inflation from 1.0 to 20.2532 in the variances of the regression coefficients in the scaled and centered model, an increase in *20-fold* over the *ideal case* ($\mathbf{X}^{*\prime}\mathbf{X}^* = I$) when the two regressor variables are orthogonal. We say that the ill-conditioning in the variables (multicollinearity) has resulted in *variance inflation factors* of 20.2532.

The variance inflation factors (VIF) for the ith regression coefficient can be written (see Exercise 3.2)

$$\boxed{\text{VIF} = \frac{1}{1 - R_i^2}} \tag{3.34}$$

where R_i^2 is the coefficient of multiple determination of the regression produced by regressing the variable $\mathbf{x}_i$ against the other regressor variables, the $\mathbf{x}_j (j \neq i)$. Thus the higher the multiple correlation in this artificial regression, the lower the precision in the estimate of the coefficient b_i.

A second approach in illustrating the effect of multicollinearity on the coefficients is the consideration of

$$E(\mathbf{b} - \boldsymbol{\beta})'(\mathbf{b} - \boldsymbol{\beta}) \tag{3.35}$$

Here $\mathbf{b}$ is the vector of least squares estimates of $\boldsymbol{\beta}$ in the model of Eq. (3.31). It should be clear that if the model is correct, the quantity in (3.35) is the sum of the variances of the coefficients. It is also the expected squared distance between the estimate vector $\mathbf{b}$ and the true parameter vector $\boldsymbol{\beta}$. If we, once again, consider centered and scaled regressors (see Appendix A.3),

$$\frac{E(\mathbf{b}-\boldsymbol{\beta})'(\mathbf{b}-\boldsymbol{\beta})}{\sigma^2} = \mathrm{tr}(\mathbf{X}^{*\prime}\mathbf{X}^*)^{-1} = \sum_{i=1}^{k} \frac{1}{\lambda_i} \tag{3.36}$$

where $\lambda_1, \lambda_2, \ldots, \lambda_k$ are the eigenvalues that appear in Eq. (3.33). Thus for an ill-conditioned or near singular $\mathbf{X}^{*\prime}\mathbf{X}^*$, at least one of the eigenvalues will be small, and thus $E(\mathbf{b}-\boldsymbol{\beta})'(\mathbf{b}-\boldsymbol{\beta})$ will be large. For the ideal case, $\sum_{i=1}^{k} (1/\lambda_i) = k$. It becomes clear, then, from Eq. (3.36), that since

$$E(\mathbf{b}-\boldsymbol{\beta})'(\mathbf{b}-\boldsymbol{\beta}) = E(\mathbf{b}'\mathbf{b}) - (\boldsymbol{\beta}'\boldsymbol{\beta})$$

then

$$E(\mathbf{b}'\mathbf{b}) = \boldsymbol{\beta}'\boldsymbol{\beta} + \sigma^2 \sum_{i=1}^{k} \frac{1}{\lambda_i} \tag{3.37}$$

Equation (3.37) underscores the tendency for multicollinearity to produce a vector of regression coefficients that is too long, i.e., coefficients that have the tendency to be *too large* in magnitude. If any of the λ_i are small, obviously $\mathbf{b}'\mathbf{b}$ is heavily biased upward for $\boldsymbol{\beta}'\boldsymbol{\beta}$, and hence one would expect coefficients that are large. This is true in spite of the fact that the b's themselves are unbiased. For example, an eigenvalue taking on a value of 0.0005 is not at all uncommon in highly collinear situations. Clearly, from (3.37), for this situation, $\sum_{i=1}^{k} b_i^2$ is heavily biased, and the result is a tendency for some of the coefficients to be overestimated in magnitude.

As an illustration, suppose in the case of 3 regressors the eigenvalues are

$$\lambda_1 = 2.5 \qquad \lambda_2 = 0.4999 \qquad \lambda_3 = 0.0001$$

Clearly, then, there is one serious dependency. Now, what is the effect on estimation of $\sum_{i=1}^{k} \beta_i^2$? From (3.37),

$$E\left(\sum_{i=1}^{k} b_i^2\right) = \sum_{i=1}^{k} \beta_i^2 + \sigma^2\left(\frac{1}{2.5} + \frac{1}{0.4999} + \frac{1}{0.0001}\right)$$

Thus $\sum_{i=1}^{k} b_i^2$ overestimates $\sum_{i=1}^{k} \beta_i^2$, on the average, by roughly $10{,}000\sigma^2$. In the ideal case, i.e., when there was no collinearity, the magnitude of the bias in $\sum_{i=1}^{k} b_i^2$ would be $3\sigma^2$.

INSTABILITY AND WRONG SIGN

Though it is not our intention to present a thorough discussion of diagnosis of multicollinearity here, we do need to investigate certain quantities to assess the potential severity of the problem. For example, the VIFs, correlation matrix, and eigenvalues of the correlation matrix should be observed. One effect of multicollinearity, not readily illustrated by the foregoing, is the *instability* of the regression coefficients, i.e., coefficients that are very much dependent on the particular data set that generated them. The analyst can detect this through an exercise of artificially altering, or "perturbing" the y-observations and checking for relative stability in the coefficients. In addition, the analyst can remove one of a large set of regressor variables and if multicollinearity is a serious problem, the remaining coefficients may change by large amounts and perhaps change sign. Clearly, this instability is not a satisfactory condition to the experimenter. It would be pleasing if small data perturbations—resulting in data that is no less reasonable than the original data—would generate small changes in the regression coefficients.

EFFECT ON QUALITY OF FIT AND PREDICTION

The undesirable "fallout" from the condition of multicollinearity in the regressor variables does not spread to model's fit. Indeed, the procedure of least squares ensures that the residual sum of squares (and other statistics based on it) will still be as attractive as the model and data will allow. Thus, the reader must visualize a situation where the residuals in the regression analysis may be very small, but yet the coefficients are estimated poorly. Illustration of the deterioration of the coefficients has been given via an argument that traces to the conditioning of $\mathbf{X}'\mathbf{X}$. However, from the standpoint of fit, the residual sum of squares, say, in the case where standard assumptions are made (including normality of the ε_i), has the $\sigma^2\chi^2_{n-p}$ distribution, and thus its distribution is unaffected by the conditioning of $\mathbf{X}'\mathbf{X}$. As a result, the *traditional analysis* of the fit does not signal potential multicollinearity problems. The reader should envision that when regression data are infested with multicollinearity (brought about by *proxies* for one another among the regressor variables) the residual sum of squares surface is extremely *flat*; so there is a large region in the coefficient space in which the residual sum of squares varies very little, and the least squares values for the coefficients would not always be the best choice. Collinearity must be diagnosed. Diagnostic procedures will be discussed in Chapter 8.

If we convince ourselves that the fit is unaffected by severe multicollinearity, then we easily accept the fact that the prediction of response at or near the data combinations of the regressor variables should be relatively unaffected. Prediction may still be good at points in the data range for combinations of the x's that are *not counter* to the multicollinearity described in the data. However, prediction at combinations that are not consistent with relationships in the data, or prediction at points that represent extrapolation outside the range of the data, can be adversely affected by multicollinearity. More depth on this problem is provided in Chapters 4 and 8. At this point, the reader gains a clear insight into why multicollinearity must be diagnosed carefully, and at times, combatted through use of biased estimation techniques. The reader will understand that an ordinary least squares analysis of a highly collinear data set may hide relevant information—not uncover it.

One simple pictorial illustration is the so-called "picket fence" display in Figure 3.3. This illustration shows two regressors with obvious collinearity. The heights of the pickets on the fence represent the y-observations. The task of building a regression model can be likened to balancing a plane on top of the picket fence. Clearly the plane is very unstable and subject to major change if a small change occurs in the height of one of the pickets. This instability produces major changes in the prediction in directions perpendicular to the picket fence. Likewise, major changes can occur in the slopes in the x_1 and x_2 direction. The point of the illustration is that, if collinearity is severe, slight changes in the y-observations can produce major changes in the coefficients while the fitted values remain stable. In addition, if it is necessary to use $\hat{y}$ away from the path described by collinearity, one may expect serious instability in the prediction results.

FIGURE 3.3 **Picket fence illustration of collinearity**

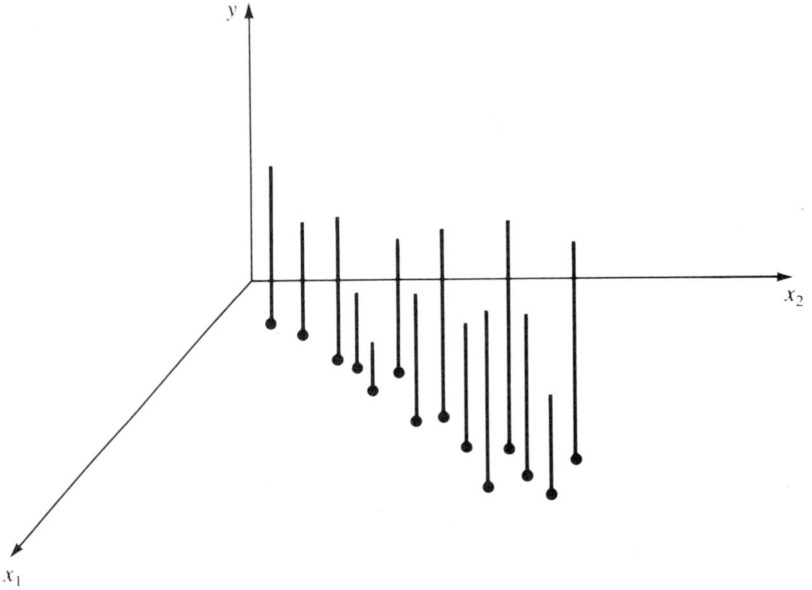

Example 3.8 ***Hospital Data***

Data are given in Table 3.8 that reflect information taken from seventeen U.S. Naval hospitals at various sites around the world. The regressors are workload variables, i.e., items that result in the need for manpower in a hospital installation. A brief description of the variables is as follows:

y: Monthly man-hours

x_1: Average daily patient load

x_2: Monthly X-ray exposures

x_3: Monthly occupied bed days

x_4: Eligible population in the area $\div$ 1000

x_5: Average length of patients' stay in days

The goal here is to produce an empirical equation that will estimate (or predict) manpower needs for Naval hospitals. The following are the least squares regression equation, the estimate of residual standard deviation, and the coefficient of determination.

$$\hat{y} = 1962.948 - 15.8517x_1 + 0.05593x_2 + 1.58962x_3 - 4.21867x_4 - 394.314x_5$$

(In the above fitted regression, the regressor variables are in the natural form, i.e., not centered and scaled.)

$$s = 642.088 \text{ man-hours/month}$$
$$R^2 = 0.99082$$

These results seem to reflect a satisfactory fit between man-hours and the workload variables. However the signs of the coefficients could be classified as alarming if we interpret them literally. The coefficients of the variables x_1 (average daily patient load), x_4 (eligible population), and x_5 (average length of patients' stay) are negative. This implies that, say, in the case of x_1, an increase in patient load, when other x's are held constant, is accompanied by a corresponding decrease in hospital manpower, a conclusion which, of course, is ludicrous. The correlation matrix, showing the empirical linear dependency among these regressor workload variables is as follows:

$$\text{Correlation} = \begin{array}{c} x_1 \\ x_2 \\ x_3 \\ x_4 \\ x_5 \end{array} \overset{\begin{array}{ccccc} x_1 & \quad x_2 & \quad x_3 & \quad x_4 & \quad x_5 \end{array}}{\begin{bmatrix} 1.00000 & 0.90738 & 0.99990 & 0.93569 & 0.67120 \\ & 1.00000 & 0.90715 & 0.91047 & 0.44665 \\ & & 1.00000 & 0.93317 & 0.67111 \\ & & & 1.00000 & 0.46286 \\ & & & & 1.00000 \end{bmatrix}}$$

It would seem that, even though the linear regression model fits the data quite well, the rather curious signs on the regression coefficients may be a result of the effect of multicollinearity. The variance inflation factors, which are the diagonals of the inverse of the correlation matrix, are given by

x_1: VIF = 9,597.57

x_2: VIF = 7.94

x_3: VIF = 8,933.09

x_4: VIF = 23.29

x_5: VIF = 4.28

It is clear that at least two of the regression coefficients, b_1 and b_3, are estimated very poorly in comparison to the ideal, i.e., the condition in which there is no multicollinearity.

In the case of the hospital data set, the correlation between x_1 and x_3 (0.99990) stands out as being noteworthy. If one were to attempt to reduce the multicollinearity, but still confine the estimation procedure to ordinary least squares, the elimination of one of the regressors, either x_1 or x_3, would seem to be a promising or, perhaps necessary, approach. The implication is, perhaps, that a model containing x_1 does not need x_3 or, vice versa. The pair of variables together may prohibit quality estimation of either coefficient. The variable x_1 was eliminated, and the following model and statistics were obtained:

$$\hat{y} = 2032.188 + 0.05608x_2 + 1.0884x_3 - 5.0041x_4 - 410.083x_5$$

$$R^2 = 0.99080$$

$$s = 615.489$$

$$x_2: \quad \text{VIF} = 7.92$$

$$x_3: \quad \text{VIF} = 23.93$$

$$x_4: \quad \text{VIF} = 12.70$$

$$x_5: \quad \text{VIF} = 3.36$$

The regression without x_1 has not only reduced the residual estimate of variance, s^2, while not severely altering R^2, but has also substantially reduced the variance inflation factor on b_3 from 8,933.09 to 23.9268. Thus, from the results shown here, elimination of x_1 would seem to produce reduced multicollinearity and, perhaps, an improved regression.

This example is not intended as a definitive solution to model selection for the hospital data. (The user still cannot be satisfied with the negative coefficients on x_4 and x_5.) Nor is it intended as a final illustration of how to combat multicollinearity. The intention was to illustrate how multicollinearity can influence variances of regression coefficients and to allow the reader to be exposed, relatively early in the text, to a data set that obviously contains multicollinearity. Additional illustrations will be shown in Chapter 8, which contains extensive material dealing with collinearity. In addition, these data are the object of further analysis in Chapters 4 and 8.

TABLE 3.8 **Hospital manpower data**

Site	x_1	x_2	x_3	x_4	x_5	y
1	15.57	2463	472.92	18.0	4.45	566.52
2	44.02	2048	1339.75	9.5	6.92	696.82
3	20.42	3940	620.25	12.8	4.28	1033.15
4	18.74	6505	568.33	36.7	3.90	1603.62
5	49.20	5723	1497.60	35.7	5.50	1611.37
6	44.92	11520	1365.83	24.0	4.60	1613.27
7	55.48	5779	1687.00	43.3	5.62	1854.17

TABLE 3.8 Continued

Hospital manpower data

Site	x_1	x_2	x_3	x_4	x_5	y
8	59.28	5969	1639.92	46.7	5.15	2160.55
9	94.39	8461	2872.33	78.7	6.18	2305.58
10	128.02	20106	3655.08	180.5	6.15	3503.93
11	96.00	13313	2912.00	60.9	5.88	3571.89
12	131.42	10771	3921.00	103.7	4.88	3741.40
13	127.21	15543	3865.67	126.8	5.50	4026.52
14	252.90	36194	7684.10	157.7	7.00	10343.81
15	409.20	34703	12446.33	169.4	10.78	11732.17
16	463.70	39204	14098.40	331.4	7.05	15414.94
17	510.22	86533	15524.00	371.6	6.35	18854.45

Procedures and Analysis for Staffing Standards Development: Data/Regression Analysis Handbook (San Diego, California: Navy Manpower and Material Analysis Center, 1979).

3.9

QUALITY FIT, QUALITY PREDICTION, AND THE HAT MATRIX

Some of the motivation that has preceded this section arises from the researcher's basic need to arrive at a prediction equation, i.e., a function that emulates well the functional relationship existing among the regressor variables. In Chapter 4 details are given regarding the types of quantities that allow for separation or ranking of models from the various candidates available when prediction is a major consideration. We continue to emphasize the important point: *quality fit and quality prediction do not necessarily coincide.* Any researcher, concerned about the prediction capabilities of his working model, should be keenly aware of what quantities should be emphasized in the regression analysis.

STANDARD ERROR OF PREDICTION AND THE HAT MATRIX

The reader should recall that, in Section 3.5, we focused considerable attention on the notion of confidence bounds on mean response. These confidence bounds were based on the standard error of prediction or estimated standard deviation of $\hat{y}$ at a given point

$$\mathbf{x}_0' = [1, x_{1,0}, x_{2,0}, \ldots, x_{k,0}]$$

(the 1, as the initial element, accommodates a model with a constant) given by

$$s_{\hat{y}(\mathbf{x}_0)} = s\sqrt{\mathbf{x}_0'(\mathbf{X}'\mathbf{X})^{-1}\mathbf{x}_0} \tag{3.38}$$

If we consider the development to this point, the experimenter's quest for a conceptual quantity that describes quality of prediction would lead directly to the quantity $h_{00} = \mathbf{x}_0'(\mathbf{X}'\mathbf{X})^{-1}\mathbf{x}_0$, which, apart from σ^2, is the prediction variance. This quantity describes reproducibility of prediction if new y-data are taken and the regression computed with the same sample size and the same combinations of the regressor variables. Now, it is tempting to consider (3.38) as a proper criterion for evaluating prediction; indeed, it would suffice except that it does not readily suggest a *single* norm or provide a single number that an analyst can use. Clearly the quality of $\hat{y}(\mathbf{x}_0)$ as a predictor depends on the location of $\mathbf{x}_0$ in the x-space, and h_{00} may vary considerably over the regressor space. Thus when confronted with the question of whether or not the fitted model predicts well, a data analyst's obvious retort is the question "where?" If the analyst has insight into where prediction is to be accomplished, the standard error of prediction becomes a useful criterion.

In Chapter 4 we carry the issue of evaluating prediction further, and we do suggest single norms that attempt to *summarize* prediction capabilities with one number. The reader should employ, whenever possible, the standard error of prediction as a decision maker to account for the fact that the model does not predict response with uniform quality. We saw this illustrated in the use of the forestry data in Example 3.5.

The quantity $\mathbf{H}$ is defined as the matrix where diagonal elements are the quadratic forms

$$\boxed{h_{ii} = \mathbf{x}_i'(\mathbf{X}'\mathbf{X})^{-1}\mathbf{x}_i}$$

where $\mathbf{x}_i$ reflects the model and location for the ith data point. The off-diagonal elements are $h_{ij} = \mathbf{x}_i'(\mathbf{X}'\mathbf{X})^{-1}\mathbf{x}_j$ for $i \neq j$. Thus the matrix $\mathbf{H}$ can be written

$$\boxed{\mathbf{H} = \mathbf{X}(\mathbf{X}'\mathbf{X})^{-1}\mathbf{X}'} \tag{3.39}$$

The quantity in (3.39) appears in much of the development in Chapters 4, 5, 6, and 7, and is commonly called the "HAT" matrix. It is a symmetric, *idempotent* ($\mathbf{H}^2 = \mathbf{H}$), $n \times n$ matrix that transforms the y's to the $\hat{y}$'s. That is,

$$\begin{aligned}\hat{\mathbf{y}} &= \mathbf{X}\mathbf{b}\\ &= \mathbf{X}(\mathbf{X}'\mathbf{X})^{-1}\mathbf{X}'\mathbf{y}\\ &= \mathbf{H}\mathbf{y}\end{aligned}$$

In addition, as a projection matrix, $\mathbf{H}$ plays an important role in general linear models theory. In fact, by rather straightforward manipulation, we show the vector of residuals can be written

$$\begin{aligned}\mathbf{e} &= \mathbf{y} - \mathbf{X}\mathbf{b}\\ &= \mathbf{y} - \mathbf{X}(\mathbf{X}'\mathbf{X})^{-1}\mathbf{X}'\mathbf{y}\\ &= [\mathbf{I} - \mathbf{H}]\mathbf{y}\end{aligned} \tag{3.40}$$

Also, the residual sum of squares is given by

$$\mathbf{e}'\mathbf{e} = \mathbf{y}'[\mathbf{I} - \mathbf{H}]^2\mathbf{y} \tag{3.41}$$

and since $\mathbf{I} - \mathbf{H}$ is also idempotent (see Exercise 3.12), (3.41) is rewritten

$$\mathbf{e}'\mathbf{e} = \mathbf{y}'[\mathbf{I} - \mathbf{H}]\mathbf{y} \tag{3.42}$$

There are some properties of the HAT matrix that we exploit in future presentations and are worth introducing here.

1. $\text{tr}(\mathbf{H}) = p$, the number of model parameters (see Exercise 3.3).
2. For a model containing a constant term $1/n \leqslant h_{ii} \leqslant 1.0$ (see Exercise 3.4).

Both properties motivate interesting observations. Property 1 implies

$$\sum_{i=1}^{n} \frac{\text{Var } \hat{y}(\mathbf{x}_i)}{\sigma^2} = p \tag{3.43}$$

a result which indicates that *apart from σ^2 the prediction variance, summed over the locations of the data points, equals the number of model parameters.* The implication of this result may or may not be apparent to the reader at this point. However, it would seem to lend some credibility to the choice of *simple models* in the exercise of model building (at least as far as prediction variance is concerned). Property 2 implies that

$$\boxed{\frac{1}{n} \leqslant \frac{\text{Var } \hat{y}(\mathbf{x}_i)}{\sigma^2} \leqslant 1}$$

This result should be intuitively reasonable. It suggests that *the precision in a prediction at the location of a data point is no worse than the error variance in an observation,* i.e., $\text{Var } \hat{y}(\mathbf{x}_i) \leqslant \sigma^2$. In addition, the precision in prediction can be no better than the precision of the average response (σ^2/n) if all of the observations were taken at the same location.

The reader should note that this particular section is an important building block for what follows in subsequent chapters. You can extract a great deal of information from quantities that relate to the elements of the HAT matrix.

3.10

CATEGORICAL OR INDICATOR VARIABLES (REGRESSION MODELS AND ANOVA MODELS)

An extremely important application of regression analysis involves a list of regressor variables that includes *qualitative* variables as well as the usual traditional quantitative variables. For example, the chemical engineer, in modeling yield of reaction y as a function of temperature (x_1) and pressure (x_2), would like to have a model that accounts for the fact that two different catalysts are used, where a *catalyst* represents an example of a categorical variable. Suppose it is of interest to model purchase price, y, of private homes as a function of pertinent regressors, such as x_1 (square feet of living space), x_2 (land acreage), x_3 (number of rooms), etc. However, the data include homes from four distinct geographical locations in the U.S. *Geography* would qualify as a categorical variable. Often the terms *indicator* or even *dummy variable* are used to describe what appear as model terms that may be just as crucial as the usual quantitative variables. They often appear as nuisance variables, possibly unplanned. At times, however, a prudent researcher may plan the data-taking or experimental process, assuming that prechosen levels of a categorical variable are included.

It may be crucial that one determines whether or not the response is influenced by the categorical variable, i.e., through a variable screening exercise. Certainly, if prediction is important, the prediction equation $\hat{y}(\mathbf{x})$ needs to include an *effect* due to the category. The most effective model of illustration of the categorical variable is the special case of one quantitative variable and two categories.

The concept of indicator variables requires the use of an indicator function to describe the role of the variable in the model. This necessitates the use of notation that distinguishes the standard continuous regressors from indicator variables.

SINGLE CATEGORICAL VARIABLE WITH TWO CATEGORIES

Suppose we begin by considering a single quantitative regressor variable x_1, in a situation where there are two categories, for example, the two catalysts in the previous illustration. This produces an additional model term. Let z be defined as follows

$$z = 0 \quad \text{if in first category}$$
$$z = 1 \quad \text{if in second category}$$

Thus for the model

$$y_i = \beta_0 + \beta_1 x_{1i} + \beta_2 z_i + \varepsilon_i \tag{3.44}$$

we have the $\mathbf{X}$ matrix

$$\mathbf{X} = \left[\begin{array}{ccc} 1 & x_{11} & 0 \\ 1 & x_{12} & 0 \\ 1 & x_{13} & 0 \\ \vdots & \vdots & \vdots \\ & & 0 \\ \hline & & 1 \\ \vdots & \vdots & 1 \\ & & \vdots \\ 1 & x_{1n} & 1 \end{array}\right] \begin{array}{l} \left.\begin{array}{l} \\ \\ \\ \\ \\ \end{array}\right\} \text{First category} \\ \left.\begin{array}{l} \\ \\ \\ \\ \end{array}\right\} \text{Second category} \end{array}$$

As a result, for the first category,

$$y_i = \beta_0 + \beta_1 x_{1i} + \varepsilon_i$$

and for the second category,

$$y_i = (\beta_0 + \beta_2) + \beta_1 x_{1i} + \varepsilon_i$$

The single categorical variable results in a mere *shift in intercept* induced by a *constant difference* in response between the categories. Figure 3.4 provides an illustration.

FIGURE 3.4 **Single quantitative variable, one categorical variable with two levels**

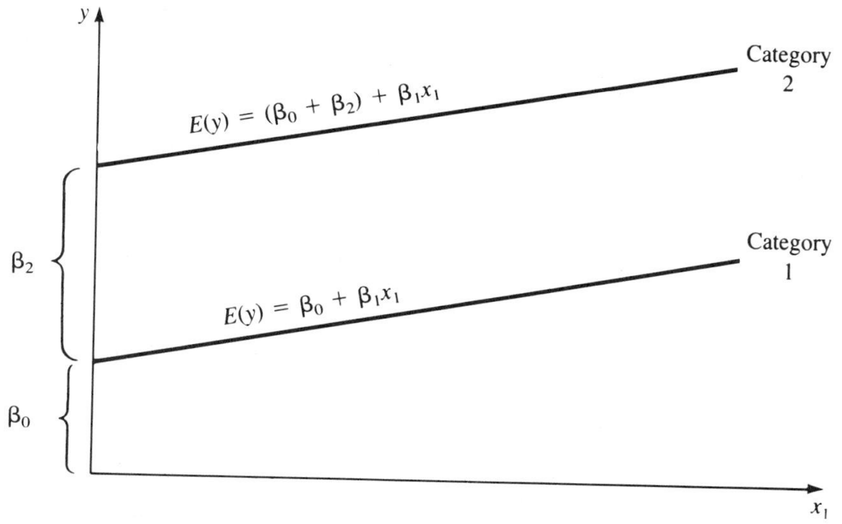

Though the special case considered here is indeed special, an important model assumption made here prevails also in the more general case. *The regression coefficients or rates of change with respect to the quantitative regressor variable or variables are the same across all categories.*

Before extending the discussion to more than one category, we should remind the reader that the 0, 1 values assigned to two categories are purely arbitrary and, in fact, artificial. Clearly if we assign values different than (0, 1), the estimator of β_2 will be different. However, important inferential quantities, such as the t-statistic associated with β_2 and the prediction $\hat{y}(\mathbf{x}_0)$, are independent of the assignment of the levels to the categorical variable.

Example 3.9 ***Teacher Effectiveness Data***

Twenty-three student teachers took part in an evaluation program designed to measure teacher effectiveness and determine what factors are important. Twelve male and eleven female instructors took part. The response measure for each was a quantitative evaluation made on the cooperating teacher. The regressor variables were scores on four standardized tests given to each instructor. The data are given in Table 3.9. Formally, the model fitted is given by

$$y_i = \beta_0 + \sum_{j=1}^{4} \beta_j x_{ji} + \beta_5 z_i + \varepsilon_i \qquad (i = 1, 2, \ldots, 23)$$

where β_5 represents the coefficient of the categorical variable "sex." The assignment of the categorical variables was "1" to females and "0" to males.

A regression analysis was used with a categorical variable providing the distinction between the male and female instructors. Table 3.10 represents a SAS (1985) computer printout showing basic statistics of fit, analysis of variance,

TABLE 3.9 **Results of teacher effectiveness and test scores**

	y	x_1	x_2	x_3	x_4
Males	489	81	151	45.50	43.61
	423	68	156	46.45	44.69
	507	80	165	76.50	54.57
	467	107	149	55.50	43.27
	340	43	134	49.40	49.21
	524	129	163	72.00	49.96
	488	139	159	86.20	53.05
	445	88	135	64.00	49.51
	388	99	141	44.15	39.57
	579	121	145	44.25	51.89
	433	91	129	42.50	53.77
	409	87	115	79.25	56.32
Females	410	69	125	59.00	55.66
	569	57	131	31.75	63.97
	425	77	141	80.50	45.32
	344	81	122	75.00	46.67
	324	0	141	49.00	41.21
	505	53	152	49.35	43.83
	235	77	141	60.75	41.61
	501	76	132	41.25	64.57
	400	65	157	50.75	42.41
	584	97	166	32.25	57.95
	434	76	141	54.50	57.90

TABLE 3.10 Analysis of data of Table 3.9: General linear models procedure

DEPENDENT VARIABLE: Y

SOURCE	DF	SUM OF SQUARES	MEAN SQUARE	F VALUE	PR > F	R-SQUARE	C.V.
MODEL	5	78671.88463065	15734.37692613	3.19	0.0329	0.483695	15.8126
ERROR	17	83975.85449979	4939.75614705		ROOT MSE		Y MEAN
CORRECTED TOTAL	22	162647.73913044			70.28339880		444.47826087

SOURCE	DF	SEQUENTIAL SS	F VALUE	PR > F	DF	PARTIAL SS	F VALUE	PR > F
SEX	1	4364.16337286	0.88	0.3604	1	2176.90083452	0.44	0.5157
X1	1	27717.47643177	5.61	0.0300	1	2045.58014270	0.41	0.5285
X2	1	14215.42495199	2.88	0.1080	1	13483.26174934	2.73	0.1169
X3	1	648.33660071	0.13	0.7216	1	1601.48860776	0.32	0.5765
X4	1	31726.48327332	6.42	0.0214	1	31726.48327332	6.42	0.0214

PARAMETER	ESTIMATE	T FOR H0: PARAMETER=0	PR > |T|	STD ERROR OF ESTIMATE
INTERCEPT	-66.31371596	-0.28	0.7822	236.13316653
SEX	23.30651877	0.66	0.5157	35.10838411
X1	0.43392763	0.64	0.5285	0.67431343
X2	2.04402559	1.65	0.1169	1.23720481
X3	-1.18556859	-0.57	0.5765	2.08217637
X4	83.65495123	2.53	0.0214	33.00908134

OBSERVATION	OBSERVED VALUE	FITTED VALUE	RESIDUAL	LOWER 95% CL FOR MEAN	UPPER 95% CL FOR MEAN
1	489.00000000	435.49594048	53.50405952	377.92253025	493.06935071
2	423.00000000	510.76070237	-87.76070237	428.48056426	593.04084047
3	507.00000000	495.88278667	11.11721333	424.71484208	567.05073125
4	467.00000000	449.31040737	17.68959263	396.90889004	501.71192471
5	340.00000000	368.92334162	-28.92334162	288.30623215	449.54045108
6	524.00000000	508.17640356	15.82359644	436.72305520	579.62975192
7	488.00000000	518.21190011	-30.21190011	444.19946254	592.22433769
8	445.00000000	417.88973667	27.11026333	366.67697211	469.10250123
9	388.00000000	434.21809324	-46.21809324	328.14188088	540.29430560
10	579.00000000	491.90091640	87.09908360	414.51621503	569.28561778
11	433.00000000	444.28958998	-11.28958998	375.72360703	512.85557293
12	409.00000000	416.94018153	-7.94018153	333.48950979	500.39085327
13	410.00000000	465.00896751	-55.00896751	386.75961017	543.25832485
14	569.00000000	464.67555526	104.32444474	391.40869833	537.94241218
15	425.00000000	445.55151474	-20.55151474	397.23366126	493.86936822
16	344.00000000	334.34685151	9.65314849	240.65149073	428.04221230
17	324.00000000	274.17493282	49.82506718	157.07458169	391.27528396
18	505.00000000	462.14781843	42.85218157	395.59810518	528.69753168
19	235.00000000	394.38348314	-159.38348314	327.23239635	461.53456993
20	501.00000000	426.23382636	74.76617364	373.95945711	478.50819561
21	400.00000000	473.60831946	-73.60831946	397.80057597	549.41606294
22	584.00000000	521.42469281	62.57530719	429.08234305	613.76704256
23	434.00000000	469.44403796	-35.44403796	413.22362163	525.66445429

t-statistics on the parameters, sequential and partial F-tests, residuals, and confidence limits on the mean response. (Certain headings were changed to maintain consistency in terminology.)

The role of the categorical variable, sex, can be determined by testing

$$H_0\colon \beta_5 = 0$$
$$H_1\colon \beta_5 \neq 0$$

The F-statistic is given by

$$F = \frac{R(\beta_5|\beta_0, \beta_1, \beta_2, \beta_3, \beta_4)}{s^2} = \frac{2176.9008}{4939.7561} = .44$$

with significance level .5157 and indicates that β_5 does not differ significantly from zero.

Like many data sets, this one still leaves some unanswered questions. The categorical variable explains what appears to be a difference between males and females. But what about model selection? Has the most effective model been adopted here if the purpose of the model is to predict teaching effectiveness? As we indicated earlier in this chapter, model selection is discussed at length in Chapter 4.

ONE CATEGORICAL VARIABLE WITH MULTIPLE LEVELS

In Example 3.9, sex is the categorical variable, and there are two levels; *geography*, as in our original illustration, is a categorical variable and the number of geographical areas is the number of levels. In the case of one categorical variable and multiple levels, say ℓ levels, the model, with k quantitative regressor variables, can be written

$$y = \beta_0 + \beta_1 x_{1i} + \cdots + \beta_k x_{ki} + \beta_{k+1} z_{1i} + \beta_{k+2} z_{2i} + \cdots + \beta_{k+\ell-1} z_{\ell-1,i} + e_i \qquad \textbf{(3.45)}$$

where the $z_1, z_2, \ldots, z_{\ell-1}$ values are either 1 or 0, depending on whether the data point in question is in that category. The $\mathbf{X}$ matrix that accommodates the model of (3.45) is given by

$$
\mathbf{X} = \left[\begin{array}{cccccccc|l}
 & x_1 & x_2 & \cdots & x_k & z_1 & z_2 & \cdots\; z_{\ell-1} & \\
1 & x_{11} & x_{21} & & x_{k1} & 1 & 0 & 0 & \\
1 & x_{12} & x_{22} & & x_{k2} & 1 & 0 & 0 & \text{Category} \\
\vdots & x_{13} & x_{23} & & x_{k3} & \vdots & \vdots & \vdots & \text{level } 1 \\
1 & \vdots & \vdots & & \vdots & 1 & 0 & 0 & \\
\hline
1 & \vdots & \vdots & & \vdots & 0 & 1 & 0 & \\
1 & \vdots & \vdots & & \vdots & 0 & 1 & \vdots & \text{Category} \\
\vdots & \vdots & \vdots & & \vdots & \vdots & \vdots & & \text{level } 2 \\
1 & \vdots & \vdots & & \vdots & 0 & 1 & 0 & \\
\hline
1 & \vdots & \vdots & & \vdots & \vdots & 0 & 1 & \\
\vdots & \vdots & \vdots & & \vdots & \vdots & \vdots & \vdots & \text{Category level } \ell - 1 \\
1 & \vdots & \vdots & & \vdots & \vdots & 0 & 1 & \\
\hline
1 & \vdots & \vdots & & \vdots & 0 & 0 & 0 & \\
\vdots & \vdots & \vdots & & \vdots & \vdots & \vdots & \vdots & \text{Category level } \ell \\
1 & x_{1n} & x_{2n} & & x_{kn} & 0 & 0 & 0 &
\end{array}\right] \qquad \textbf{(3.46)}
$$

Thus there are $(\ell - 1)$ additional columns of the **X**-matrix to account for $\ell - 1$ coefficients of the variables $z_1, \ldots, z_{\ell-1}$, which contribute a *single additive effect* to the ordinary regression model, with the effect depending on which level is relevant. For example, after the coefficients are found through $\mathbf{b} = (\mathbf{X}'\mathbf{X})^{-1}\mathbf{X}'\mathbf{y}$, a predicted response at, say, the *second* category would be given by

$$\hat{y} = b_0 + b_1 x_1 + b_2 x_2 + \cdots + b_k x_k + b_{k+2} \qquad \textbf{(3.47)}$$

as all category levels are set to zero except x_{k+2}, which is set to unity. Thus the categorical coefficients can, as we indicated earlier, be visualized as a shift (positive or negative) in the constant term, depending on which category the prediction is based. For the prediction (3.47), the shift in β_0 is to $(\beta_0 + \beta_{k+2})$. We arbitrarily assigned the final category level ℓ to take on values $z_1 = z_2 \cdots = z_\ell = 0$. In not including a z_ℓ, an arbitrary assignment of $b_{k+\ell} = 0$ is made. The assignment of the zero value can be given to any of the coefficients. Though the $\beta_{k+j} (j = 1, 2, \ldots, \ell)$ are not estimable (uniquely estimated), $(\beta_0 + \beta_{k+j})$ is estimable, and thus the "new intercept" and hence, the predicted response, is independent of the arbitrary assignment.

Example 3.10 Consider the data set in Table 3.11. The response y is the amount of suspended solids in a coal cleansing system. Using a quantitative regressor variable x_1, the pH of the cleansing tank is recorded. The cleansing operation involves the use of a polymer. In this case three different polymers were used in the experiment. All other important conditions were held constant.

The analysis may involve multiple regression that would include a linear model term for pH and the indicator variable "polymer," which contains 3 categories or 3 *classes.* These polymer categories are modeled with two regressor variables containing values of 0 or 1. Thus the model is given by

$$y_i = \beta_0 + \beta_1 x_{1i} + \beta_2 z_{1i} + \beta_3 z_{2i} + \varepsilon_i \qquad (i = 1, 2, \ldots, 18)$$

where x_{1i} represents pH at the ith data point. The values of z_{1i} and z_{2i} are dictated by the polymer number. The $\mathbf{X}$ matrix given by

$$\mathbf{X} = \begin{array}{c} \\ \end{array}\begin{bmatrix} 1 & x_{1,1} & 1 & 0 \\ 1 & x_{1,2} & 1 & 0 \\ 1 & \vdots & 1 & 0 \\ 1 & \vdots & 1 & 0 \\ 1 & \vdots & 1 & 0 \\ 1 & \vdots & 1 & 0 \\ \hdashline 1 & \vdots & 0 & 1 \\ 1 & \vdots & 0 & 1 \\ 1 & \vdots & 0 & 1 \\ 1 & \vdots & 0 & 1 \\ 1 & \vdots & 0 & 1 \\ 1 & \vdots & 0 & 1 \\ \hdashline 1 & \vdots & 0 & 0 \\ 1 & \vdots & 0 & 0 \\ 1 & \vdots & 0 & 0 \\ 1 & \vdots & 0 & 0 \\ 1 & x_{1,17} & 0 & 0 \\ 1 & x_{1,18} & 0 & 0 \end{bmatrix} \begin{array}{l} \left.\begin{array}{c} \\ \\ \\ \\ \\ \\ \end{array}\right\} \text{Polymer 1} \\ \left.\begin{array}{c} \\ \\ \\ \\ \\ \\ \end{array}\right\} \text{Polymer 2} \\ \left.\begin{array}{c} \\ \\ \\ \\ \\ \\ \end{array}\right\} \text{Polymer 3} \end{array}$$

(columns: x_1, z_1, z_2)

The regression analysis was conducted and the linear regression equation is given by

$$\hat{y} = -161.89733 + 54.2940x_{1i} + 89.9981z_{1i} + 27.1657z_{2i}$$

Table 3.12 gives the SAS (analysis of variance) printout for the fitted regression model. The pH clearly has an effect on the cleansing. Obviously the polymer also has an impact. Further investigation of these data will appear in the sections that follow.

TABLE 3.11 **Cleansing data for Example 3.10**

x_1 (pH)	y (amount of suspended solids (mg/L))	Polymer
6.5	292	Polymer 1
6.9	329	Polymer 1
7.8	352	Polymer 1
8.4	378	Polymer 1
8.8	392	Polymer 1
9.2	410	Polymer 1
6.7	198	Polymer 2
6.9	227	Polymer 2
7.5	277	Polymer 2
7.9	297	Polymer 2
8.7	364	Polymer 2
9.2	375	Polymer 2
6.5	167	Polymer 3
7.0	225	Polymer 3
7.2	247	Polymer 3
7.6	268	Polymer 3
8.7	288	Polymer 3
9.2	342	Polymer 3

Data analyzed by the Statistics Consulting Center, Virginia Polytechnic Institute and State University, 1979.

TABLE 3.12 **Analysis of variance printout for coal cleansing data with indicator variables**

DEP VARIABLE: SOLIDS

ANALYSIS OF VARIANCE

SOURCE	DF	SUM OF SQUARES	MEAN SQUARE	F VALUE	PROB>F
MODEL	3	80181.73127	26727.24376	73.676	0.0001
ERROR	14	5078.71318	362.76523		
C TOTAL	17	85260.44444			

ROOT MSE	19.0464	R-SQUARE	0.9404
DEP MEAN	301.5556	ADJ R-SQ	0.9277
C.V.	6.316049		

PARAMETER ESTIMATES

VARIABLE	DF	PARAMETER ESTIMATE	STANDARD ERROR	T FOR H0: PARAMETER=0	PROB > |T|
INTERCEP	1	-161.89733	37.43315576	-4.325	0.0007
Z1	1	89.99806061	11.05228237	8.143	0.0001
Z2	1	27.16569697	11.01042883	2.467	0.0271
X1	1	54.29402597	4.75541126	11.417	0.0001

EQUALITY OF REGRESSOR EFFECTS ACROSS CLASSES

Earlier in this section, we underscored the assumption that the effects of the quantitative regressors remain the same across levels or classes of the categorical variable. This is clear from the parallel lines model description given in Figure 3.4.

To an engineer, biologist, physical scientist, or any scientific investigator this statement is certainly important. In Example 3.9, it is important to emphasize that by using the multiple linear regression model that generated the SAS printout in Table 3.10, the assumption of additivity was made, i.e., the impact of test scores on teacher evaluation is the same for males and females. This additivity assumption is commonly ignored by researchers who blindly submit data and model statements into the computer. Of course, it is often the case that the researcher is not aware of whether or not it is appropriate to make the additivity assumption. Obviously if the assumption does not hold and is not investigated in some sense, the user may be in danger of fitting a grossly incorrect model with resulting interpretations being invalid. In what follows we discuss the methodology for the investigation of the additivity assumption.

USING THE GENERAL LINEAR HYPOTHESIS TO TEST ADDITIVITY

Now consider the simple case illustrated by Figure 3.4, namely one categorical variable with two levels and one quantitative regressor variable denoted by x_1. It is clear from the model of Eq. (3.44) that for each category a common slope coefficient of x_1 is assumed. However, let us investigate a more general model, i.e., that for which two separate slopes on x_1 is assumed. Suppose that n_1 observations are taken from class 1 and n_2 observations are taken from class 2, where $n_1 + n_2 = n$. We may write

$$\begin{aligned} E(y_i) &= \beta_0 + \beta_{11} x_{1i} && (i = 1, 2, \ldots, n_1) \\ &= \beta_0 + \beta_{12} x_{1i} + \beta_2 z_i && (i = n_1 + 1, n_1 + 2, \ldots, n) \end{aligned} \quad \textbf{(3.48)}$$

Here z_i is still defined as 0 in the first category and 1 in the second category as in the model of (3.44). Thus we essentially have two regression models with different intercepts and different slopes. Clearly the test for additivity is one which tests

$$H_0\colon \beta_{11} = \beta_{12}$$
$$H_1\colon \beta_{11} \neq \beta_{12}$$

i.e., a test for equality of slopes for the two regression lines. The test of this hypothesis was discussed in the material dealing with the general linear hypothesis in Section 3.4. In the case of the teacher effectiveness data of Example 3.9 (four quantitative regressors with the indicator variable denoting males and females), the most general model would involve parameters $\beta_{11}, \beta_{12}, \beta_{21}, \beta_{22}, \beta_{31}, \beta_{32}, \beta_{41}$, and β_{42}, where β_{j1} is the coefficient of the jth test score for males and β_{j2} is the coefficient of the jth test score for females. The hypothesis for additivity would be

$$\begin{aligned} H_0\colon \beta_{11} &= \beta_{12} \\ \beta_{21} &= \beta_{22} \\ \beta_{31} &= \beta_{32} \\ \beta_{41} &= \beta_{42} \end{aligned}$$

In any situation, the test is for equality of regression coefficients of each quantitative regressor, or perhaps a subset of the quantitative regressors across classes. Clearly, by inviting separate regression models for each class, the use of the general linear hypothesis certainly applies. This is easily extended to three or more categories as in the case of Example 3.10.

USING INTERACTION TO TEST ADDITIVITY

Though the general linear hypothesis applies nicely for testing the additivity assumption, an alternative approach that produces an equivalent result is worth discussing. Consider first the single quantitative regressor, two category situation in Figure 3.4. If, indeed, the slopes of the regression lines are different, the model is as depicted in Figure 3.5. When this occurs, we say that there is *interaction* between x_1 and the categorical variable z_2. Interaction implies a deviation from additivity. Thus the condition of nonadditivity can be modeled nicely as a model containing interaction. Again let us make the 0 and 1 assignment to the categorical variable z_2

FIGURE 3.5 **Interaction between the quantitative regressor and the categorical variable**

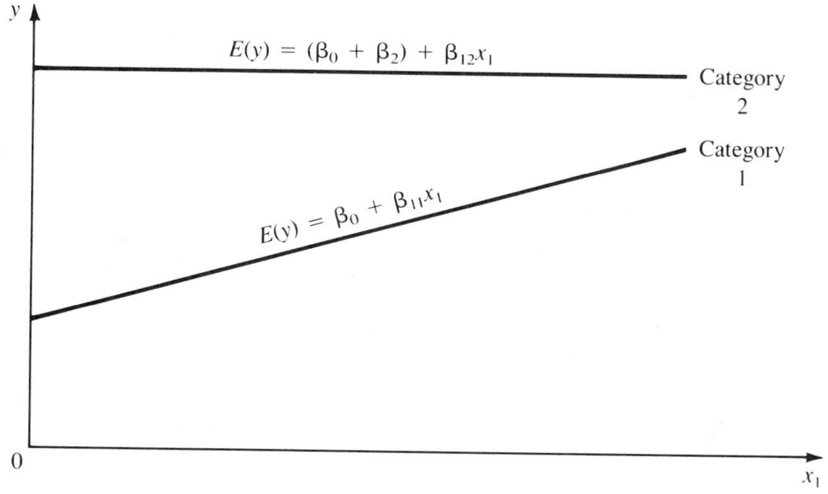

for the two classes. We can consider the model with interaction (and thus inequality of slopes) as

$$y_i = \beta_0 + \beta_{11}x_{1i} + \beta_2 z_i + \beta_{1z}x_{1i}z_i + \varepsilon_i \qquad (i = 1, 2, \ldots, n) \tag{3.49}$$

It should be clear to the reader that this interaction model is equivalent to the model of (3.48) containing separate slopes on the quantitative regressor x_1. In fact, β_{12} in (3.48) is $\beta_{11} + \beta_{1z}$ in (3.49). For the **X** matrix we have

$$\mathbf{X} = \begin{array}{c} \\ \left[\begin{array}{cccc} & x_1 & z_2 & x_1 \cdot z_2 \\ 1 & x_{11} & 0 & 0 \\ 1 & x_{12} & 0 & 0 \\ \vdots & \vdots & \vdots & \vdots \\ 1 & x_{1,n_1} & 0 & 0 \\ \hdashline 1 & x_{1,n_1+1} & 1 & x_{1,n_1+1} \\ 1 & \vdots & 1 & \vdots \\ \vdots & \vdots & \vdots & \vdots \\ 1 & x_{1,n} & 1 & x_{1,n} \end{array}\right] \end{array} \begin{array}{l} \\ \left.\begin{array}{l} \\ \\ \\ \\ \end{array}\right\} \text{First category} \\ \\ \left.\begin{array}{l} \\ \\ \\ \\ \end{array}\right\} \text{Second category} \end{array}$$

This formulation leads to

$$E(y_i) = \beta_0 + \beta_{11}x_{1i} \qquad \text{(first category)}$$
$$E(y_i) = \beta_0 + (\beta_{11} + \beta_{1z})x_{1i} + \beta_2 \qquad \text{(second category)}$$

where z_i is assigned a value 1 in the second category. Obviously if there is no interaction, i.e., $\beta_{1z} = 0$, the model reduces to the additive model or equal slopes model of (3.44).

The upshot of the foregoing is that one may simply model interaction terms and test hypotheses on these terms when there is concern about the additivity assumption. In the scenario surrounding the model of (3.49), a test of

$$H_0\colon \beta_{1z} = 0$$
$$H_1\colon \beta_{1z} \neq 0$$

using a t-test or partial F-test as discussed in Section 3.4 is equivalent to the general linear hypothesis approach to testing equality of the two slopes. Recall that the general linear hypothesis produced an F-test. In the situation of model (3.48), the degrees of freedom for the F will be 1 for the numerator (difference in error df for full and reduced model) and $n_1 + n_2 - 4$ for error (total sample size minus number of parameters estimated). For the case of simply testing $H_0\colon \beta_{1z} = 0$, we have a t-test with $n_1 + n_2 - 4$ degrees of freedom, which produces a test equivalent to the F-test.

We shall now proceed to examine an illustration of the use of interaction for testing the additivity assumption.

Example 3.11 Consider the data of Example 3.10 shown in Table 3.11. The indicator variable "polymer" has three levels or categories, and there is a single quantitative regressor variable, the pH of the system. Consider the plot of the data in Figure 3.6. The plot does not make it clear whether or not the additivity assumption holds. Perhaps the *common slope* assumption for all three polymers is not reasonable. To test this we fit the model

$$y_i = \beta_0 + \beta_1 x_{1i} + \beta_2 z_{1i} + \beta_3 z_{2i} + \beta_{1z_1} x_{1i} z_{1i} + \beta_{1z_2} x_{1i} z_{2i} + \varepsilon_i \qquad (i = 1, 2, \ldots, 18) \tag{3.50}$$

Here, as before, the quantitative regressor x_1 is pH and z_1 and z_2 are used to accommodate the three polymers. If we wish to test the additivity assumption, we merely test

$$H_0: \beta_{1z_1} = \beta_{1z_2} = 0$$

This is easily accomplished by using an F-test as discussed in Section 3.4. If this hypothesis is rejected, we must conclude that the common slope assumption does not hold. Obviously then, the implication would be that *separate regression lines should be fit for each polymer.* Table 3.13 shows a SAS GLM run for the interaction model of Eq. (3.50).

To test the additivity hypothesis, we require

$$R(\beta_1, \beta_2, \beta_3, \beta_{1z_1}, \beta_{1z_2} | \beta_0) - R(\beta_1, \beta_2, \beta_3 | \beta_0) = R(\beta_{1z_1}, \beta_{1z_2} | \beta_0, \beta_1, \beta_2, \beta_3)$$

These two regression sums of squares are those found in Tables 3.12 and 3.13. In addition, it can easily be seen that since the "Type I" sums of squares are sequential sums of squares, $R(\beta_{1z_1}, \beta_{1z_2} | \beta_0, \beta_1, \beta_2, \beta_3)$ is given by $1747.0925 + 778.9528$. Thus

$$F_{2,12} = \frac{2526.0453/2}{212.7223} = 5.94$$

This F statistic is significant at the 0.016 level, indicating that the interaction model terms are significant. One must conclude the following:

(i) The model fit in Example 3.10 is not appropriate.
(ii) The effect of pH is different for the three different polymers.
(iii) For model fitting, one should fit separate regression lines for the three polymers.

FIGURE 3.6 **Plot of the cleansing data of Table 3.11**

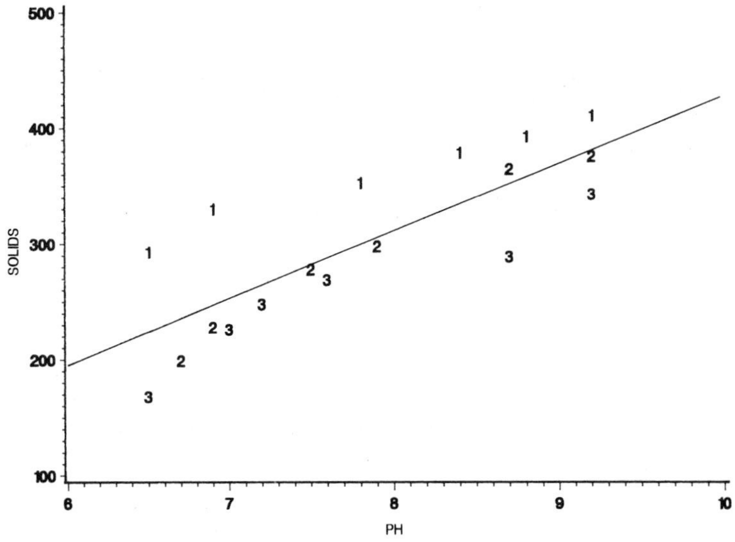

TABLE 3.13 **SAS GLM model for the cleansing data of Table 3.11 and the model of equation (3.50)**

GENERAL LINEAR MODELS PROCEDURE

DEPENDENT VARIABLE: SOLIDS

SOURCE	DF	SUM OF SQUARES	MEAN SQUARE	F VALUE	PR > F	R-SQUARE	C.V.
MODEL	5	82707.77658195	16541.55531639	77.76	0.0001	0.970060	4.8366
ERROR	12	2552.66786250	212.72232187		ROOT MSE		SOLIDS MEAN
CORRECTED TOTAL	17	85260.44444444			14.58500332		301.55555556

SOURCE	DF	TYPE I SS	F VALUE	PR > F	DF	TYPE III SS	F VALUE	PR > F
Z1	1	29526.69444444	138.80	0.0001	1	1756.63797613	8.26	0.0140
Z2	1	3366.75000000	15.83	0.0018	1	498.25669438	2.34	0.1518
X1	1	47288.28682251	222.30	0.0001	1	15759.52941176	74.08	0.0001
X1Z1	1	1747.09252376	8.21	0.0142	1	512.46745932	2.41	0.1466
X1Z2	1	778.95279123	3.66	0.0798	1	778.95279123	3.66	0.0798

PARAMETER	ESTIMATE	T FOR H0: PARAMETER=0	PR > |T|	STD ERROR OF ESTIMATE
INTERCEPT	-158.27450980	-3.26	0.0068	48.51692909
Z1	197.69166266	2.87	0.0140	68.79450541
Z2	-108.73981001	-1.53	0.1518	71.05072781
X1	53.82352941	8.61	0.0001	6.25326858
X1Z1	-13.56098565	-1.55	0.1466	8.73704640
X1Z2	17.39365436	1.91	0.0798	9.08953778

Multiple Categorical Variables

Earlier we discussed the arbitrary assignment of zero to one of the coefficients in the use of categorical variables. One can easily see that if **X** in Eq. (3.46) were augmented with an additional column of zeros and ones, with the latter assigned to the data in the last category level, an interesting condition results. The entries in the last ℓ columns sum to the first, the column of ones. As a result, the **X** matrix is reduced in column rank, and $\mathbf{X}'\mathbf{X}$ becomes singular; so no unique estimator can be found. Thus the arbitrary assignment is a very practical approach.

Multiple categorical variables can be accommodated in a fashion very similar to that discussed for a single variable. Suppose, for example, that a tobacco experiment is being planned in which the effect of nitrogen (x_1), phosphorus (x_2), and potassium (x_3) in the fertilizer are to be modeled with the response being the yield of a tobacco crop (lb/acre). However, the experimental units are such that they must be divided into two types of categories. First, there are three varieties of fertilizer; the tobacco scientist finds that he also must perform the experiment on two farms containing homogeneous soil. Thus there are two categorical variables, one with three levels and the other with two levels. As a result, the portion of the **X** matrix accounting for the categorical variables becomes

$$
\begin{array}{l}
\\ \\ \text{Fertilizer 1} \\ \\ \\ \text{Fertilizer 2} \\ \\ \\ \text{Fertilizer 3} \\ \\
\end{array}
\begin{array}{c}
\underline{\text{Fertilizer}} \quad \underline{\text{Farms}} \\
\begin{array}{ccc} z_1 & z_2 & z_3 \end{array} \\
\left[\begin{array}{ccc}
1 & 0 & 1 \\ \vdots & \vdots & \vdots \\ 1 & 0 & 1 \\ \hdashline
0 & 1 & 1 \\ \vdots & \vdots & \vdots \\ 0 & 1 & 1 \\ \hdashline
0 & 0 & 1 \\ \vdots & \vdots & \vdots \\ 0 & 0 & 1
\end{array}\right]
\end{array}
\left.\begin{array}{c} \\ \\ \\ \\ \\ \\ \\ \\ \\ \end{array}\right\}\text{Farm 1}
\qquad
\begin{array}{l}
\\ \\ \text{Fertilizer 1} \\ \\ \\ \text{Fertilizer 2} \\ \\ \\ \text{Fertilizer 3} \\ \\
\end{array}
\begin{array}{c}
\underline{\text{Fertilizer}} \quad \underline{\text{Farms}} \\
\begin{array}{ccc} z_1 & z_2 & z_3 \end{array} \\
\left[\begin{array}{ccc}
1 & 0 & 0 \\ \vdots & \vdots & \vdots \\ 1 & 0 & 0 \\ \hdashline
0 & 1 & 0 \\ \vdots & \vdots & \vdots \\ 0 & 1 & 0 \\ \hdashline
0 & 0 & 0 \\ \vdots & \vdots & \vdots \\ 0 & 0 & 0
\end{array}\right]
\end{array}
\left.\begin{array}{c} \\ \\ \\ \\ \\ \\ \\ \\ \\ \end{array}\right\}\text{Farm 2}
$$

This assumes, of course, that the two categorical variables are *cross classified*, i.e., each level occurs with every other level (preferably the same number of times). It is difficult to do an in-depth study of multiple categorical variables without including notions of experimental design.

In the subsection that follows, we discuss briefly the idea of indicator variables and their relationships to other types of models, particularly analysis of variance and analysis of covariance models. This section is designed for readers who have had some exposure to these models.

RELATIONSHIP TO ANALYSIS OF VARIANCE AND ANALYSIS OF COVARIANCE MODELS

Suppose we consider the situation of Example 3.10 in which three polymers were used for cleaning coal. However, let us initially assume that the data do not contain pH as a variable. As before, there are six experimental runs taken for each polymer. The data would take the following form:

$$\text{Polymer}\ \begin{matrix}1\\2\\3\end{matrix}\begin{bmatrix} y_{11} & y_{12} & y_{13} & y_{14} & y_{15} & y_{16} \\ y_{21} & y_{22} & y_{23} & y_{24} & y_{25} & y_{26} \\ y_{31} & y_{32} & y_{33} & y_{34} & y_{35} & y_{36} \end{bmatrix}$$

There are various ways of writing the model for this experiment. If we think of it as a regression setting with one indicator variable at three levels (classes) and no quantitative regressors, we may write

$$y_{ij} = \beta_0 + \alpha_i + \varepsilon_{ij} \qquad \begin{matrix} i = 1, 2, 3 \\ j = 1, 2, \ldots, 6 \end{matrix}$$

However, a close observation of the $\mathbf{X}$ matrix and $\boldsymbol{\beta}$ vector reveals

$$\boldsymbol{\beta} = \begin{bmatrix} \beta_0 \\ \alpha_1 \\ \alpha_2 \\ \alpha_3 \end{bmatrix}$$

$$\mathbf{X} = \begin{bmatrix} 1 & 1 & 0 & 0 \\ 1 & \vdots & \vdots & \vdots \\ 1 & & & \\ 1 & & & \\ 1 & \vdots & \vdots & \vdots \\ 1 & 1 & 0 & 0 \\ \hdashline 1 & 0 & 1 & 0 \\ 1 & \vdots & \vdots & \vdots \\ 1 & & & \\ 1 & & & \\ 1 & \vdots & \vdots & \vdots \\ 1 & 0 & 1 & 0 \\ \hdashline 1 & 0 & 0 & 1 \\ 1 & \vdots & \vdots & \vdots \\ 1 & & & \\ 1 & & & \\ 1 & \vdots & \vdots & \vdots \\ 1 & 0 & 0 & 1 \end{bmatrix} \begin{matrix} \left.\begin{matrix} \\ \\ \\ \\ \\ \\ \end{matrix}\right\} \text{Polymer 1} \\ \left.\begin{matrix} \\ \\ \\ \\ \\ \\ \end{matrix}\right\} \text{Polymer 2} \\ \left.\begin{matrix} \\ \\ \\ \\ \\ \\ \end{matrix}\right\} \text{Polymer 3} \end{matrix} \tag{3.51}$$

which is not of full column rank. The rank of $\mathbf{X}$ is 3. Note that the last three columns add to the first column. As a result, $\mathbf{X}'\mathbf{X}$ is singular, and we say that the model is *overparameterized.* Now this implies that the model does not require and, indeed, will not allow four parameters. The reader should recognize this as a standard *one-way analysis of variance* model. One solution to the over-parameterization is to model the categorical variables as we did in the previous subsections. Namely

$$y_i = \beta_0 + \beta_1 z_{1i} + \beta_2 z_{2i} + \varepsilon_i \qquad (i = 1, 2, \ldots, n)$$

where z_{1i} is 1 for observations with the first polymer and 0 otherwise. The variable z_{2i} is 1 for observations with the second polymer and 0 otherwise. Here the **X** matrix is given by

$$\mathbf{X} = \begin{array}{c} \\ \left[\begin{array}{ccc} & z_1 & z_2 \\ 1 & 1 & 0 \\ 1 & \vdots & \vdots \\ 1 & \vdots & \vdots \\ 1 & \vdots & \vdots \\ 1 & \vdots & \vdots \\ 1 & 1 & 0 \\ \hdashline 1 & 0 & 1 \\ 1 & \vdots & \vdots \\ 1 & \vdots & \vdots \\ 1 & \vdots & \vdots \\ 1 & \vdots & \vdots \\ 1 & 0 & 1 \\ \hdashline 1 & 0 & 0 \\ 1 & \vdots & \vdots \\ 1 & \vdots & \vdots \\ 1 & \vdots & \vdots \\ 1 & \vdots & \vdots \\ 1 & 0 & 0 \end{array}\right] \end{array} \begin{array}{l} \left.\begin{array}{c} \\ \\ \\ \\ \\ \end{array}\right\} \text{Polymer 1} \\ \left.\begin{array}{c} \\ \\ \\ \\ \\ \end{array}\right\} \text{Polymer 2} \\ \left.\begin{array}{c} \\ \\ \\ \\ \\ \end{array}\right\} \text{Polymer 3} \end{array} \qquad \textbf{(3.52)}$$

This model is identical to that used in Example 3.10, apart from the quantitative regressor, pH. The **X** matrix in (3.53) has been made *full column* rank (**X′X** is nonsingular) by the arbitrary constraint that $\alpha_3 = 0$. This use of linear constraints on the parameters is common in analysis of variance models. As a result, then, the situation described by (3.51), i.e., the one-way analysis of variance model, can be viewed as a regression model with categorical variables and the assignment of 0's and 1's for the values of z_1 and z_2. In fact, *any analysis of variance model can be viewed as a regression model* with the proper arbitrary constraint (or constraints) to make it full rank and the elements of the **X** matrix being 0's and 1's.

Now, suppose we return to the data of Example 3.10 involving polymers and pH, with the four parameter model. This is, of course, classical use of categorical variable regression. It has been made so because of the arbitrary constraint that the regression coefficient associated with the third polymer is zero. Now, if we do *not* make this constraint, the model is given by

$$y_i = \beta_0 + \beta_1 x_{1i} + \beta_2 z_{1i} + \beta_3 z_{2i} + \beta_4 z_{3i} + \varepsilon_{ij} \tag{3.53}$$

where x_{1i} is pH and z_{1i}, z_{2i} and z_{3i} are zero or one depending on which polymer is involved. In other words,

$$\mathbf{X} = \begin{array}{c} \begin{array}{ccccc} & x_1 & z_1 & z_2 & z_3 \end{array} \\ \left[\begin{array}{ccccc} 1 & x_{11} & 1 & 0 & 0 \\ 1 & x_{12} & 1 & 0 & 0 \\ 1 & \vdots & 1 & 0 & 0 \\ 1 & \vdots & 1 & 0 & 0 \\ 1 & \vdots & 1 & 0 & 0 \\ 1 & \vdots & 1 & 0 & 0 \\ \hdashline 1 & \vdots & 0 & 1 & 0 \\ 1 & \vdots & 0 & 1 & 0 \\ 1 & \vdots & 0 & 1 & 0 \\ 1 & \vdots & 0 & 1 & 0 \\ 1 & \vdots & 0 & 1 & 0 \\ 1 & \vdots & 0 & 1 & 0 \\ \hdashline 1 & \vdots & 0 & 0 & 1 \\ 1 & \vdots & 0 & 0 & 1 \\ 1 & \vdots & 0 & 0 & 1 \\ 1 & \vdots & 0 & 0 & 1 \\ 1 & x_{1,17} & 0 & 0 & 1 \\ 1 & x_{1,18} & 0 & 0 & 1 \end{array}\right] \end{array} \begin{array}{l} \left.\vphantom{\begin{array}{c}1\\1\\1\\1\\1\\1\end{array}}\right\} \text{Polymer 1} \\ \left.\vphantom{\begin{array}{c}1\\1\\1\\1\\1\\1\end{array}}\right\} \text{Polymer 2} \\ \left.\vphantom{\begin{array}{c}1\\1\\1\\1\\1\\1\end{array}}\right\} \text{Polymer 3} \end{array}$$

The $\mathbf{X}$ matrix in this case is *less than full column rank.* The matrix $\mathbf{X}'\mathbf{X}$ is singular and the coefficients β_2, β_3, and β_4 in the model of (3.53) are not estimable. The reader should recognize that this is an *analysis of covariance model* containing a class variable and a continuous regression type variable or covariate. A natural and certainly convenient working model is to set $\beta_4 = 0$ and return to what is recognized as categorical variable regression. This constraint is, however, not unique. Indeed, any one of a large class of linear constraints among the coefficients β_2, β_3 and β_4 would certainly suffice.

The foregoing will allow the student in linear models to have a better idea of where categorical variable regression falls with regard to ANOVA and analysis of covariance model. Strictly speaking, either analysis of variance models or analysis of covariance models can be displayed as categorical regression models.

EXERCISES FOR CHAPTER 3

3.1 In Section 3.3, the bias and variance of the least squares estimators are discussed. The least squares estimators are linear functions of the y-observations. From Eq. (3.4), $\mathbf{b} = (\mathbf{X}'\mathbf{X})^{-1}\mathbf{X}'\mathbf{y}$, illustrating that $\mathbf{b}$ is of the form $\mathbf{b} = \mathbf{A}\mathbf{y}$ where $\mathbf{A}$ is a $p \times n$ matrix. The variance-covariance matrix of $\mathbf{A}\mathbf{y}$ is given by

$$\text{Var}(\mathbf{A}\mathbf{y}) = \mathbf{A}\mathbf{V}\mathbf{A}'$$

where $\mathbf{V}$ is the variance-covariance matrix of $\mathbf{y}$. See Graybill (1976). Use the above expression to show that

$$\text{Var}(\mathbf{b}) = \sigma^2(\mathbf{X}'\mathbf{X})^{-1}$$

3.2 Equation (3.34) gives the relationship between the variance inflation associated with the ith coefficient and the R^2 achieved by regressing $\mathbf{x}_i$ against the other regressors. Prove the result of Eq. (3.34). (*Hint*: Assume the regressors are centered and scaled, and use the result of Equation (A.9) in Appendix A.)

3.3 The HAT matrix appears in Equation (3.39). Show that

$$\text{tr}(\mathbf{H}) = p$$

where p is the number of model parameters. (*Hint*: Make use of the fact that $\text{tr}(\mathbf{AB}) = \text{tr}(\mathbf{BA})$. See Graybill (1976).)

3.4 Let h_{ii} be the ith diagonal of the HAT matrix $\mathbf{H}$.
(a) Prove that, for a multiple regression model with a constant term, $h_{ii} \geqslant 1/n$.
(b) Show that $h_{ii} \leqslant 1$. (*Hint*: Make use of the fact that $\mathbf{H}$ is idempotent.)

3.5 Consider the squid data in Example 3.2.
(a) Generate the residuals for the multiple regression model

$$y_i = \beta_0 + \beta_1 x_{1i} + \beta_2 x_{2i} + \beta_3 x_{3i} + \beta_4 x_{4i} + \beta_5 x_{5i} + \varepsilon_i$$

(b) Compute 95% confidence intervals on the mean response and 95% prediction intervals on a new observation at the conditions of the 22 specimen.
(c) Compute a new multiple regression using regressors x_2, x_4, and x_5. Compute s^2, the standard errors of prediction, and 95% confidence intervals on the mean response at the regressor locations for the 22 specimen.
(d) Use the information from (a), (b), and (c) to make a choice between the full model and the reduced model in part (c).
(e) For the squid data, test

$$H_0\colon \begin{bmatrix} \beta_1 \\ \beta_2 \\ \beta_3 \end{bmatrix} = 0 \qquad H_1\colon \begin{bmatrix} \beta_1 \\ \beta_2 \\ \beta_3 \end{bmatrix} \neq 0$$

Draw conclusions. Comment on the results.

3.6 Consider the data of Example 3.5
(a) Fit a linear regression with the term $\beta_3 x_3$ eliminated.
(b) Compute values of s^2; compute the standard error of prediction at the 20 data locations.
(c) Does the comparison between the results in part (b) and those in Table 3.7 signify a superiority of the reduced model or not? Explain.

3.7 For the reduced model discussed in Exercise 3.6, compute the standard error of prediction at the combinations:

x_1	10	80	75
x_2	2,500	6,000	25,000

Does this reveal anything regarding the relative merit of the full (x_1, x_2, x_3) and reduced (x_1, x_2) models for prediction? Explain.

3.8 In a project[1] to study age and growth characteristics of selected mussel species from Southwestern Virginia, the data below were taken from two distinct locations. Fit a regression with weight as the response, age as the independent variable, and location as a categorical variable. Does it appear that location is significant as a categorical variable?

Location 1

Age	Weight (g)	Age	Weight (g)
3	0.44	11	3.96
3	0.50	11	3.84
3	0.66	12	5.58
3	0.78	12	5.64
4	1.20	12	4.26
4	1.18	13	6.00
4	1.08	13	2.54
6	1.12	13	3.82
6	1.72	14	4.50
7	1.04	14	5.18
7	1.66	14	4.78
7	1.70	14	5.34
8	2.62	14	4.04
9	1.88	15	6.38
10	2.26	15	4.08
11	4.10	16	4.56
11	4.56	22	6.44
11	2.12		

Location 2

Age	Weight (g)	Age	Weight (g)
3	0.76	8	2.52
4	1.38	8	3.90
5	1.20	10	3.94
5	1.76	10	6.22
6	2.60	10	4.96
6	2.16	13	9.02
6	2.64	13	8.20
6	2.52	13	8.26
6	3.08	14	6.40
6	2.12	15	10.06
7	2.72	15	8.60
7	2.96	18	11.06
8	4.54	19	10.78
8	5.26	22	12.04
8	5.60	24	13.92

[1] Data analyzed for the Department of Fisheries by the Statistical Consulting Center, Virginia Polytechnic Institute and State University, Blacksburg, Virginia, 1984.

3.9 Consider the data from Location 1 in Exercise 3.8. Compute a *pure error* mean square and perform an F-test for lack of fit. Draw conclusions regarding the adequacy of the simple linear regression.

3.10 An experiment was designed to study hydrogen embrittlement properties based on electrolytic hydrogen pressure measurements. The solution used was 0.1N NaOH with the material being a certain type of stainless steel. The cathodic charging current density was controlled and varied at four levels. The effective hydrogen pressure was observed as the response. The data follow:

Run	Charging Current Density (x) (ma/cm^2)	Effective Hydrogen Pressure (y) (atm)
1	0.5	86.1
2	0.5	92.1
3	0.5	64.7
4	0.5	74.7
5	1.5	223.6
6	1.5	202.1
7	1.5	132.9
8	2.5	413.5
9	2.5	231.5
10	2.5	466.7
11	2.5	365.3
12	3.5	493.7
13	3.5	382.3
14	3.5	447.2
15	3.5	563.8

Data analyzed for the Department of Materials Engineering by the Statistical Consulting Center, Virginia Polytechnic Institute and State University, Blacksburg, Virginia, 1983.

(a) Run a simple linear regression of y against x.
(b) Compute the *pure error* sum of squares and make a test for lack of fit.
(c) Does the information in part (b) indicate a need for a model in x beyond a *first order regression*? Explain.

3.11 In an effort to model executive compensation for the year 1979, 33 firms were selected, and data were gathered on compensation, sales, profits, and employment. The data on page 156 were gathered for the year 1979.

Firm	Compensation, y (thousands of dollars)	Sales, x_1 (millions of dollars)	Profits, x_2 (millions of dollars)	Employment, x_3
1	450	4600.6	128.1	48000
2	387	9255.4	783.9	55900
3	368	1526.2	136.0	13783
4	277	1683.2	179.0	27765
5	676	2752.8	231.5	34000
6	454	2205.8	329.5	26500
7	507	2384.6	381.8	30800
8	496	2746.0	237.9	41000
9	487	1434.0	222.3	25900
10	383	470.6	63.7	8600
11	311	1508.0	149.5	21075
12	271	464.4	30.0	6874
13	524	9329.3	577.3	39000
14	498	2377.5	250.7	34300
15	343	1174.3	82.6	19405
16	354	409.3	61.5	3586
17	324	724.7	90.8	3905
18	225	578.9	63.3	4139
19	254	966.8	42.8	6255
20	208	591.0	48.5	10605
21	518	4933.1	310.6	65392
22	406	7613.2	491.6	89400
23	332	3457.4	228.0	55200
24	340	545.3	54.6	7800
25	698	22862.8	3011.3	337119
26	306	2361.0	203.0	52000
27	613	2614.1	201.0	50500
28	302	1013.2	121.3	18625
29	540	4560.3	194.6	97937
30	293	855.7	63.4	12300
31	528	4211.6	352.1	71800
32	456	5440.4	655.2	87700
33	417	1229.9	97.5	14600

John B. Guerard, Jr. and Raymond L. Horton, "The Management of Executive Compensation in Large, Dynamic Firms: A Ridge Regression Estimation," Technical Report (Lehigh University, Bethlehem, Pennsylvania, 1984).

Consider the model

$$y_i = \beta_0 + \beta_1 \ln x_{1i} + \beta_2 \ln x_{2i} + \beta_3 \ln x_{3i} + \varepsilon_i \qquad (i = 1, 2, \ldots, 33)$$

(a) Fit the regression with the above model.
(b) Compute the correlation matrix among the regressor variables.
(c) Compute the eigenvalues of the correlation matrix.
(d) Compute the variance inflation factors for the coefficients in the above model.
(e) Make an assessment of the extent of the multicollinearity in this problem.

3.12 The HAT matrix described in Section 3.9 is an idempotent matrix that plays an important role in linear regression analysis. Idempotency of $\mathbf{H}$ implies that $\mathbf{H}^2 = \mathbf{H}$. Show that $\mathbf{I} - \mathbf{H}$ is also idempotent.

3.13 Consider the data of Exercise 3.8. The categorical variable model assumes, of course, that the slopes of the regression lines are equal. Test the hypothesis that the slopes of the regression lines are equal for the two locations. Draw appropriate conclusions. Is one regression slope an appropriate model, or is there evidence that one needs to separate regression lines?

3.14 Consider the situation of a simple linear regression with a single categorical variable with two levels; that is, the model is that given in Exercise 2.23.

(a) Write out the model in $\mathbf{y} = \mathbf{X}\boldsymbol{\beta} + \boldsymbol{\varepsilon}$ form, where each x_i is *centered* with the mean $\bar{x}_j$ for the jth group in question, as given in Exercise 2.23. The vector $\boldsymbol{\beta}$ is a three-parameter vector.

(b) Use $\mathbf{b} = (\mathbf{X}'\mathbf{X})^{-1}\mathbf{X}'\mathbf{y}$; write out an estimator for the common slope, and show its equivalence to that given in Exercise 2.23. Is this estimator of common slope intuitively reasonable?

3.15 Consider the data of Exercise 2.19. The chambers in which the experiments were conducted were held under different conditions of temperature and humidity. There is also some evidence that the model should be curvilinear in head diameter.

(a) Write an appropriate model including linear and quadratic terms in head diameter and an indicator variable to account for the nine chambers.

(b) Fit the model created in part (a) and make appropriate conclusions.

3.16 In an experiment conducted in the civil engineering department at Virginia Polytechnic Institute and State University in 1988, a growth of a certain type of algae in water was observed as a function of time and the dosage of copper added to the water. The following data were collected.

$\mathbf{y}$ (units of algae)	$\mathbf{x}_1$ (mg copper)	$\mathbf{x}_2$ (days)
.3	1	5
.34	1	5
.2	2	5
.24	2	5
.24	2	5
.28	3	5
.2	3	5
.24	3	5
.02	4	5
.02	4	5
.06	4	5
0	5	5
0	5	5
0	5	5
.37	1	12
.36	1	12
.30	2	12

TABLE Continued

TABLE Continued

y (units of algae)	x_1 (mg copper)	x_2 (days)
.31	2	12
.30	2	12
.30	3	12
.30	3	12
.30	3	12
.14	4	12
.14	4	12
.14	4	12
.14	5	12
.15	5	12
.15	5	12
.23	1	18
.23	1	18
.28	2	18
.27	2	18
.25	2	18
.27	3	18
.25	3	18
.25	3	18
.06	4	18
.10	4	18
.10	4	18
.02	5	18
.02	5	18
.02	5	18
.36	1	25
.36	1	25
.24	2	25
.27	2	25
.31	2	25
.26	3	25
.26	3	25
.28	3	25
.14	4	25
.11	4	25
.11	4	25
.04	5	25
.07	5	25
.05	5	25

(a) Consider the following model

$$y_i = \beta_0 + \beta_1 x_{1i} + \beta_2 x_{2i} + \beta_{12} x_{1i} x_{2i} + \varepsilon_i$$

Estimate the coefficients of the model, using multiple linear regression.

(b) Test

$$H_0: \beta_{12} = 0$$
$$H_1: \beta_{12} \neq 0$$

Do you have any reason to change the model from that given in part (a)?

(c) Show a partitioning of total degrees of freedom into those attributed to regression, pure error, and lack of fit.

(d) Using the model you adopted in part (b), make a test for lack of fit and draw conclusions.

(e) Plot residuals of your fitted model against x_1 and x_2, separately, and comment.

3.17 Chemical pruning of fruit trees is a very important activity. A study was done by the horticulture department of Virginia Polytechnic Institute and State University to determine the impact of Terbacil. Three levels of Terbacil, 50 PPM, 75 PPM, and 100 PPM, were used in an experimental apple orchard. The following data set gives the average number of healthy apples per cm^2 of cross sectional area of limb. In another portion of the same orchard, a type of surfactant was added to the Terbacil. This should theoretically reduce the number of healthy apples. The data are as follows:

	Number of Healthy Apples Per cm^2
control (0 PPM)	7.67
50 PPM	6.26
75 PPM	5.73
100 PPM	5.44
50 + surfactant	4.05
75 + surfactant	3.02
100 + surfactant	2.34

Fit an appropriate linear regression model to the data. Use the surfactant as an indicator variable.

3.18 Consider the punting data in Exercise 2.11. Fit three separate models

$$E(y_i) = \beta_0 + \beta_1 x_{1i} \qquad \text{(model 1)}$$
$$E(y_i) = \beta_0 + \beta_1 x_{1i} + \beta_2 x_{2i} \qquad \text{(model 2)}$$
$$E(y_i) = \beta_0 + \beta_2 x_{2i} \qquad \text{(model 3)}$$

Let x_{1i} be right-leg strength; x_{2i} is left-leg strength.

(a) Plot the residuals for these three models. Plot against $\hat{y}$. Comment on which model seems to be most appropriate.

(b) Test

$$H_0: \beta_2 = 0$$
$$H_1: \beta_2 \neq 0$$

Make appropriate conclusion.

3.19 A scientist collects experimental data on the radius of a propellant grain (y) as a function of powder temperature, x_1, extrusion rate, x_2, and die temperature, x_3.

Grain Radius	Powder Temperature (x_1)	Extrusion Rate (x_2)	Die Temperature (x_3)
82	150	12	220
93	190	12	220
114	150	24	220
124	150	12	250
111	190	24	220
129	190	12	250
157	150	24	250
164	190	24	250

(a) Consider the linear regression model with centered regressors

$$y_i = \beta_0^* + \beta_1(x_{1i} - \bar{x}_1) + \beta_2(x_{2i} - \bar{x}_2) + \beta_3(x_{3i} - \bar{x}_3) + \varepsilon_i$$

Write the vector $\mathbf{y}$, the matrix $\mathbf{X}$, and the vector $\boldsymbol{\beta}$ in the model

$$\mathbf{y} = \mathbf{X}\boldsymbol{\beta} + \boldsymbol{\varepsilon}$$

(b) Write out the least squares normal equations

$$(\mathbf{X}'\mathbf{X})\mathbf{b} = \mathbf{X}'\mathbf{y}$$

Comment on what is special about the $\mathbf{X}'\mathbf{X}$ matrix. What characteristic in this experiment do you suppose produces this special form of $\mathbf{X}'\mathbf{X}$?

(c) Estimate the coefficients in the multiple linear regression model.

(d) Test the hypothesis

$$H_0: \beta_1 = 0$$
$$H_0: \beta_2 = 0$$

and make conclusions.

(e) Compute the $100(1-\alpha)\%$ confidence intervals on $E(y \mid x)$ at each of the locations of x_1, x_2, and x_3 described by the data points. Note the relative widths of these confidence intervals.

(f) Compute the HAT diagonals at the eight data points. Comment.

(g) Compute the variance inflation factors of the coefficients b_1, b_2, and b_3. Do you have any explanation as to why these measures of damage due to collinearity give the results that they do?

3.20 The following is a data set generated by the Environmental Protection Agency and analyzed by the Statistical Consulting Center at Virginia Polytechnic Institute and State University. The amount of magnesium uptake is measured at several levels of time. It is anticipated that the two treatments used may result in different regression equations. The data are as follows:

Magnesium (y)	Time (x)	Treatment
2.08	0	1
2.07	0	1
2.09	0	1
2.06	0	1
2.17	1	1
2.15	1	1
2.16	1	1
2.15	1	1
2.08	2	1
2.02	2	1
2.06	2	1
2.04	2	1
1.97	3	1
1.96	3	1
1.98	3	1
1.96	3	1
1.89	4	1
1.87	4	1
1.88	4	1
1.89	4	1
1.81	5	1
1.72	5	1
1.81	5	1
1.75	5	1
1.66	6	1
1.64	6	1
1.61	6	1
1.59	6	1

Magnesium (y)	Time (x)	Treatment
2.08	0	2
2.09	0	2
2.07	0	2
2.04	0	2
2.03	1	2
2.04	1	2
2.01	1	2
2.08	1	2
1.88	2	2
1.86	2	2
1.96	2	2
1.91	2	2
1.77	3	2
1.60	3	2
1.78	3	2
1.80	3	2
1.57	4	2
1.47	4	2
1.59	4	2
1.56	4	2
1.52	5	2
1.43	5	2
1.48	5	2
1.52	5	2
1.53	6	2
1.40	6	2
1.98	6	2
1.45	6	2

(a) A model is postulated in which calcium uptake is regressed against time in a quadratic regression. In other words

$$E(y) = \beta_0 + \beta_1 x + \beta_2 x^2 + \beta_3 z$$

where z is an indicator variable accounting for the treatment. Fit this regression model.

(b) We need to determine if the simple indicator variable is actually appropriate. Suppose we rewrite the model

$$\begin{aligned} E(y_i) &= \beta_0 + \beta_1 x_i + \beta_2 x_i^2 \qquad \text{(treatment 1)} \\ &= \gamma_0 + \gamma_1 x_i + \gamma_2 x_i^2 \qquad \text{(treatment 2)} \end{aligned}$$

Set up the general linear hypothesis for testing

$$H_0\colon \beta_1 = \gamma_1$$
$$H_0\colon \beta_2 = \gamma_2$$

(If this hypothesis is rejected, two separate regressions are necessary.) That is, determine $\mathbf{C}$, $\boldsymbol{\beta}$, and $\mathbf{d}$ for H_0: $\mathbf{C}\boldsymbol{\beta} = \mathbf{d}$.

(c) Test the above hypothesis. The unrestricted model contains six parameters and the restricted model contains four parameters.

3.21 Consider the following data set

y	x_1	x_2
3.9	1.5	2.2
7.5	2.7	4.5
4.4	1.8	2.8
8.7	3.9	4.4
9.6	5.5	4.3
19.5	10.7	8.4
29.3	14.6	14.6
12.2	4.9	8.5

(a) Fit a multiple linear regression model.

(b) Use the general linear hypothesis to test

$$H_0: \beta_1 = \beta_2 = 1$$

and make appropriate conclusions. Use full and restricted model residual sums of squares.

3.22 Consider the following data set. The response involves percent yield.

		Humidity x_2		
		10%	20%	30%
Temperature, x_1	100°	74.5	72.7	71.5
	150°	81.7	80.4	79.5
	200°	95.1	94.2	90.7

(a) Fit a multiple linear regression with the model

$$E(y_i) = \beta_0 + \beta_1 x_1 + \beta_{11} x_1^2 + \beta_2 x_2 + \beta_{22} x_2^2$$

Also compute R^2, $R(\beta_1, \beta_{11} | \beta_0, \beta_2, \beta_{22})$, and test H_0: $\beta_1 = \beta_{11} = 0$.

(b) Rewrite the model using temperature and humidity as indicator variables. Set up the model in $\mathbf{y} = \mathbf{X}\boldsymbol{\beta} + \boldsymbol{\varepsilon}$ form. Fit the model to the set of data.

(c) For the model in part (b) compute R^2 and test the hypothesis that both coefficients of the temperature indicator variables are zero (jointly). Use sums of squares for the full and reduced model.

(d) Are you surprised at the comparison between the results of parts (b) and (c)?

3.23 Consider the general linear regression model $\mathbf{y} = \mathbf{X}\boldsymbol{\beta} + \boldsymbol{\varepsilon}$ and the least squares estimate $\mathbf{b} = (\mathbf{X}'\mathbf{X})^{-1}\mathbf{x}'\mathbf{y}$. Show that

$$\mathbf{b} = \boldsymbol{\beta} + \mathbf{R}\boldsymbol{\varepsilon} \qquad \text{where } \mathbf{R} = (\mathbf{X}'\mathbf{X})^{-1}\mathbf{X}'$$

REFERENCES FOR CHAPTER 3

1. Belsley, D.A., E. Kuh, and R.E. Welsch. 1980. *Regression Diagnostics: Identifying Influential Data and Sources of Collinearity*. New York: John Wiley & Sons.
2. Bowden, D.C. 1970. Simultaneous confidence bands for linear regression models. *Journal of the American Statistical Association* 65: 413–421.
3. Christensen, L.R. 1973. Simultaneous statistical inference in the normal multiple linear regression model. *Journal of the American Statistical Association* 68: 457–561.
4. Cox, D.R. 1970. *The Analysis of Binary Data*. London: Methuen & Co. Ltd.
5. Draper, N.R., and H. Smith. 1981. *Applied Regression Analysis*. 2d ed. New York: John Wiley & Sons.
6. Edwards, A.W.F. 1973. The likelihood treatment of linear regression. *Biometrika* 60: 73–77.
7. Graybill, F.A. 1976. *Theory and Application of the Linear Model*. Boston, Massachusetts: Duxbury Press.
8. Gunst, F.R., and R.L. Mason. 1980. *Regression Analysis and Its Applications: A Data Oriented Approach*. New York: Marcel Dekker.
9. Hoaglin, D.C., and R. Welch. 1978. The hat matrix in regression and ANOVA. *American Statistician* 32: 17–22.
10. Hocking, R.R. 1976. The analysis and selection of variables in linear regression. *Biometrics* 32: 1–51.
11. Kvalseth, T.O. 1985. Cautionary note about R^2. *American Statistician* 39: 279–285.
12. Montgomery, D.C., and E.A. Peck. *Introduction to Linear Regression Analysis*. New York: John Wiley & Sons.
13. Myers, R.H. 1976. *Response Surface Methodology*. Department of Statistics, Virginia Polytechnic Institute and State University, Blacksburg, Virginia.
14. SAS Institute Inc. 1985. *SAS User's Guide: Statistics, Version 5*. Cary, North Carolina: SAS Institute Inc.
15. Schilling, E.G. 1974. The relationship of analysis of variance to regression, Part I. Balanced designs. *Journal of Quality Technology* 6: 74–83.
16. Schilling, E.G. 1974. The relationship of analysis of variance to regression, Part II. Unbalanced designs. *Journal of Quality Technology* 6: 146–153.
17. Searle, S.R. 1971. *Linear Models*. New York: John Wiley & Sons.
18. Silvey, S.D. 1969. Multicollinearity and imprecise estimation. *Journal of the Royal Statistical Society, Series B* 31: 539–552.
19. Sprent, P. 1971. Parallelism and concurrence in linear regression. *Biometrics* 27: 440–444.
20. Suich, R. and G.C. Derringer. 1977. Is the regression equation adequate—one criterion. *Technometrics* 19: 213–216.
21. Younger, M.S. 1985. *Handbook for Linear Regression*. 2d ed. Boston, Massachusetts: Duxbury Press.

CHAPTER 4

CRITERIA FOR CHOICE OF BEST MODEL

The standard model-fitting scenario often involves a scientific investigator armed with data on a set of regressor variables or perhaps *model terms* (transforms of the natural variables). The dilemma evolves from an uncertainty of what terms to include in the model. The decision can be further complicated by the existence of multicollinearity or perhaps by the scientist's prior views and prejudices regarding the importance of individual variables. There may also be difficulties imposed by the quantity or quality of data to be used, and by the fact that the investigator may not have a clear plan regarding what will be done with the adopted model.

Assumptions are made regarding correctness of a postulated model when we are truly trying to find the best approximation that describes the data. The analyst must accept that linear models are merely empirical approximations. The successful model builder will eventually understand that with many data sets, several models can be fit that would be nearly equal in effectiveness. Thus the problem that one deals with is the selection of *one model* from a pool of *candidate models.*

The model builder should learn that the technique used to select from a set of candidate models very well may depend on what will be required of the model. The goal of the experimenter must be considered seriously. Prior to the analysis, the question should always be asked "What will be done with the model?" Perhaps one's model-building assignment will satisfy one or more of the following goals:

1. *Learn something about the system from which the data are taken.* This may be nothing more than the knowledge of a "sign" of a coefficient, say, to an economist, or the slope of a growth curve to a biologist, or optimum operating conditions on regressor variables to, say, a process engineer or a tobacco chemist.
2. *Learn which regressors are important and which are not*; i.e., conduct a variable selection or *variable screening* exercise. Regression is often used for this purpose—independent of any need for prediction. Often, variable screening is a prelude to a more elaborate search for a model. Unfortunately, we often confront variable screening problems in an environment in which the data are fraught with multicollinearity. Clearly 1 and 2 are related.
3. *Prediction*: Selecting from a group of candidate models that which best predicts is often a very difficult task. We touched on this in Chapter 3. Often it is difficult to resist the temptation to make the model overly complicated. One must understand that an adroit job of model selection does not require that one actually locate the correct model. Indeed the correct model may never be found.

We cannot ignore input from experts in the scientific discipline involved. (Statistics is rarely a substitute for sound scientific knowledge and reasoning.) Statistical procedures are vehicles that lead us to conclusions; but scientific logic paves the road along the way. However, a good scientist must remember that to arrive at an adequate prediction equation, balance must be achieved that takes into account what the data can support. There are times when inadequacies in the data and random noise may not allow the true structure to come through. For these reasons, a proper marriage must exist between the experienced statistician and the learned expert in the discipline involved.

To this point in this text, we have focused on fundamentals that can be categorized as traditional least squares model fitting. In this chapter we deal first with relatively modern criteria, used to compare candidate models in terms of their prediction capabilities. Then, in the latter portion of the chapter, we discuss and illustrate sequential variable selection algorithms, which achieved great attention and popularity in the late 1960s and early 1970s.

4.1 STANDARD CRITERIA FOR COMPARING MODELS

In this section we merely borrow on much of what has been developed and comment on some earlier discussed criteria as far as their application to prediction is concerned.

1. *Coefficient of Determination*: R^2. As we indicated earlier, R^2 is surely a measure of the model's capability to fit the present data. The insertion of any new regressor into a model cannot bring about a decrease in R^2. Though there are rules and algorithms (strictly based on R^2) that allow for selection of best model, the statistic itself is not conceptually prediction oriented (i.e., prediction performance based); thus it is not recommended as a sole criteria for choosing the best prediction model from a set of candidate models.
2. *Estimate of Error Variance*: s^2. The statistic, often called *residual mean square*, plays an extremely important role in hypothesis testing in multiple regression, confidence bounds, etc. (See Chapter 3.) It can also provide important information in an exercise from which one hopes to select the best model for prediction. In comparing two models, say a p-term and an m-term model ($m > p$), $s^2(m)$ can exceed $s^2(p)$, implying perhaps that the amount of residual sum of squares reduced by adopting the m-term model did not counteract the loss in residual degrees of freedom. A reasonable, and certainly simple, plan is to choose the candidate model with the smallest value of s^2.
3. *Adjusted* R^2. The use of R^2 in Eq. (3.8) as a statistic for discriminating among competing models can be very hazardous. We know, of course, that any addition of a new model term will result in an increase in R^2 (at least no decrease). In this chapter we will learn about the hazards of overfitting, i.e., the inclusion of an excessive number of model terms. Many software packages compute an R^2-like statistic, which guards against the practice of overfitting. This statistic punishes the user who includes marginally important model terms at the expense of error degrees of freedom.

From Eq. (3.7) and (3.8), we can write

$$R^2 = 1 - \frac{SS_{\text{Res}}}{SS_{\text{Total}}}$$

The adjusted R^2 is a statistic in which SS_{Res} and SS_{Total} are replaced by corresponding mean squares. In other words, adjusted R^2, denoted by $\overline{R^2}$, is given by

$$\overline{R^2} = 1 - \frac{SS_{\text{Res}}/(n-p)}{SS_{\text{Total}}/(n-1)} = 1 - \frac{s^2(n-1)}{SS_{\text{Total}}}$$

where, of course, s^2 is the model residual mean square. It is important to note that in cases in which one is comparing candidate models generated from the same data set, the rank order of models in terms of adjusted R^2 is identical to rank order in terms of s^2. This is obvious since SS_{Total} would be the same for all candidate models.

For further insight and justification surrounding the use of s^2, the reader is reminded of the development in Section 3.6. Recall that if a model is underspecified, i.e., the experimenter fits the model

$$\mathbf{y} = \mathbf{X}_1 \boldsymbol{\beta}_1 + \boldsymbol{\varepsilon}^* \qquad (p \text{ parameters})$$

and the true model is

$$\mathbf{y} = \mathbf{X}_1 \boldsymbol{\beta}_1 + \mathbf{X}_2 \boldsymbol{\beta}_2 + \boldsymbol{\varepsilon} \qquad (m \text{ parameters; } m > p)$$

then the expected value of the residual mean square, s_p^2, for the "short" model is given by (see Appendix B.2)

$$E(s_p^2) = \sigma^2 + \frac{\boldsymbol{\beta}_2'[\mathbf{X}_2'\mathbf{X}_2 - \mathbf{X}_2'\mathbf{X}_1(\mathbf{X}_1'\mathbf{X}_1)^{-1}\mathbf{X}_1'\mathbf{X}_2]\boldsymbol{\beta}_2}{n-p} \tag{4.1}$$

The quantity $n - p$ represents the residual degrees of freedom for the p-term model. One should recall that Eq. (4.1) was used in Chapter 3 to illustrate what the lack of fit test truly detects. The relationship implies that s^2 for an underspecified model is biased upward, with the magnitude of the bias depending in large part on the contribution of the coefficients of the omitted variables, indicated by the vector $\boldsymbol{\beta}_2$. Thus, on the average, if the model is badly underspecified, one expects s^2 to be inflated. Certainly, then, if s_m^2 exceeds s_p^2 where $m > p$, one would assume that the bias inflicted by an underspecification in the p-term model is quite small; or perhaps $\boldsymbol{\beta}_2 = \mathbf{0}$ and there is no underspecification at all. Thus we can view the comparison of residual mean square values in a set of candidate models as a type of empirical investigation of the amount of contribution from the eliminated parameters in $\boldsymbol{\beta}_2$. What adds more, in the way of esthetics, to this argument is that the matrix $\mathbf{X}_2'\mathbf{X}_2 - \mathbf{X}_2'\mathbf{X}_1(\mathbf{X}_1'\mathbf{X}_1)^{-1}\mathbf{X}_1'\mathbf{X}_2$ is the inverse of the variance-covariance matrix (apart from σ^2) of the least squares estimator of $\boldsymbol{\beta}_2$. (See Appendix A.4.) Thus the bias in s^2 due to left out parameters is a *standardized form* of the ignored parameters.

The above two criteria are easy to compute and will be illustrated in examples later in this chapter. In the following section, we deal with a criterion that is less traditional, but actually forces the candidate regression models to perform (predict). We choose to combine this notion of "selection of best model" with *model validation.* The latter suggests a search for a type of model checking against independent data, i.e., evaluation of each candidate by predicting response values that are independent of the data that built the model.

4.2

CROSS VALIDATION FOR MODEL SELECTION AND DETERMINATION OF MODEL PERFORMANCE

Let us assume that it is crucial to select from a set of candidate models the one which will best predict the response. The term "prediction" implies the process of estimating $E[y(\mathbf{x}_0)]$ with $\widehat{y(\mathbf{x}_0)}$ at a future $\mathbf{x} = \mathbf{x}_0$. The *ordinary residuals*, i.e. the $y_i - \hat{y}_i$, are not generally indicative of how the regression model will predict. Indeed, the least squares procedure is designed to produce properties in the regression function that will result in residuals that are smaller than true prediction errors; one must be reminded that $\hat{y}_i$ is not independent of y_i and is, in effect, *drawn to it*. These residuals are measures of quality of fit *and do not assess quality of future prediction.*

As an example, consider the hospital data of Example 3.8. The data appear in Table 3.8. Suppose we consider the fitting of the *full model* to the data, i.e., the model containing variables x_1, x_2, x_3, x_4, and x_5. We repeat the least squares regression equation here, given by

$$\hat{y} = 1962.948 - 15.8517x_1 + 0.05593x_2 + 1.58962x_3 - 4.21867x_4 - 394.314x_5$$

In this case the response is man-hours per month. The residuals from the above fitted model are as follows:

Site	y_i (man-hours)	$\hat{y}_i$	$y_i - \hat{y}_i$
1	566.52	775.025	−208.505
2	696.82	740.670	−43.850
3	1033.15	1103.923	−70.773
4	1603.62	1240.496	363.124
5	1611.37	1564.422	46.948
6	1613.27	2151.272	−538.002
7	1854.17	1689.700	164.470
8	2160.55	1736.236	424.314
9	2305.58	2736.989	−431.409
10	3503.93	3681.853	−177.923
11	3571.89	3239.289	332.601
12	3741.40	4353.333	−611.933
13	4026.52	4257.088	−230.568
14	10343.81	8767.748	1576.061
15	11732.17	12237.027	−504.857
16	15414.94	15038.391	376.549
17	18854.45	19320.697	−466.247

Consider a few of the larger sites in the data set. Specifically we shall focus on sites 15, 16, and 17. Site 17, for example, results in a predicted response of 19320.697 with a residual of −466.247. Now, we illustrate how hopelessly ineffective these three residuals are in measuring quality of prediction, and how much the fitted values, the $\hat{y}_i$, are heavily drawn by the specific y_i values. Consider what would occur if the least squares regression were conducted without the use of say, site 17. Then a comparison can be made between the residual, the fitting error, and the value of $y_{17} - \hat{y}_{17,-17}$, where $\hat{y}_{17,-17}$ denotes the predicted value at site 17, made without the use of site 17 in the regression. The latter error represents the hypothetical prediction error experienced if we used site 17 as a validation of the regression.

These results, along with the same information for sites 15 and 16, are shown as follows:

Site	Ordinary Residual	$y_i - \hat{y}_{i,-i}$
15	−504.857	−2510.842
16	376.549	2242.496
17	−466.247	−3675.121

The results are rather dramatic. At least in the case of these three sites, the fit of a model with the use of the sites may be far superior to the analyst's ability to predict the response at an independent site.

The above illustration suggests a need for a set of residuals that simulate conditions under which the model will perform, i.e., the creation of a set of residuals of the type

$$r_j = \hat{y}(\mathbf{x}_j) - y(\mathbf{x}_j) \qquad (j = 1, 2, \ldots, n^*) \tag{4.2}$$

where $y(\mathbf{x}_j)$ and $\hat{y}(\mathbf{x}_j)$ are *independent*. The error in (4.2) is truly a *prediction error*. The value n^* represents the number of these errors available. Then, of course, a norm on these errors, for example, $\sum_{j=1}^{n^*} |r_j|$ or $\sum_{j=1}^{n^*} (r_j)^2$ could be used for comparative purposes. Now, the obvious question is "How are these residuals developed?"

DATA SPLITTING

It is rather difficult to create the ideal prediction residuals given by Eq. (4.2). In validating a model, one needs to induce as much realism as possible. Clearly, it is not known at the model-building stage the conditions under which the model should perform. These conditions may be forecasting in time, internal prediction, or even extrapolation. One practice, however, that allows for cross validation is *data splitting*, i.e., the partitioning of the data into two subsamples: a *fitting sample* and a *validation sample*. See Montgomery and Peck (1982) and Snee (1977). We denote the data then as follows:

$$\left.\begin{matrix} y_1 & x_{11} & x_{21} & \cdots & x_{k1} \\ y_2 & x_{12} & x_{22} & \cdots & x_{k2} \\ \vdots & \vdots & \vdots & & \vdots \\ y_{n_1} & x_{1,n_1} & x_{2,n_1} & \cdots & x_{k,n_1} \end{matrix}\right\} \text{Fitting sample}$$

$$\left.\begin{matrix} y_{n_1+1} & x_{1,n_1+1} & x_{2,n_1+1} & \cdots & x_{k,n_1+1} \\ \vdots & \vdots & \vdots & & \vdots \\ y_{n_1+n_2} & x_{1,n_1+n_2} & x_{2,n_1+n_2} & \cdots & x_{k,n_1+n_2} \end{matrix}\right\} \text{Validation sample}$$

We can then fit any appropriate candidate model using the fitting sample, estimate coefficients, etc.; then we use the fitted model to estimate the responses in the validation sample. Useful and informative prediction errors can be used to generate norms such as

$$\sum_{j=n_1+1}^{n_1+n_2} (y_j - \hat{y}_j)^2 \qquad \text{and} \qquad \sum_{j=n_1+1}^{n_1+n_2} |y_j - \hat{y}_j|$$

which are criteria to determine the best of the candidate models. The quantity $\hat{y}_j$ is the estimated response at the jth data point, using, of course, the fitted model from the first n_1 observations.

Though there are obvious difficulties with data splitting, it is motivated by a great deal of common sense. Any procedure that determines how the data are to be partitioned should be designed according to the application. For example, if the data are time dependent, it is reasonable to assign the most recent observations in time to the validation sample and thus gain valuable information regarding the forecasting capability of the model. Suppose there is a specific region of the regressor variables where it is anticipated that estimating y is necessary. Some data points in this potentially important region would be reasonable candidates for the validation sample. One note of caution: though there is no universally accepted rule of thumb for the relative size of the two samples, a sufficient number of observations should exist in the fitting sample to provide an adequate number of residual degrees of freedom for the fit. A rather liberal rule here is to make $n \geqslant 2p + 20$, where p is the number of model parameters and n the total sample size, and split the sample roughly equally for validation and fitting. For further elucidation on this matter see Snee (1977).

If data splitting is used as a cross validation mechanism, either for choice of best model or for a study of model stability and general predictive performance, the model that is finally adopted should be fit using the entire data set, making use of all information. The purpose of data splitting should be to study regressor subsets or functional forms (all of which have their own characteristics, or personalities), but the final estimation of the parameters would come from the entire data set.

PRESS *STATISTIC*

In many circumstances involving model building, it is not practical to split data for validation purposes. (Indeed, though, it is surprising how many model builders resist the notion of data splitting even when it *is* practical.) One must be sympathetic with the experimenter's dilemma regarding the expense incurred in gathering data. Certainly, data splitting or validation with newly gathered data cannot be accomplished in all applications.

An interesting and very important criterion, which can be used as a form of validation, very much in the spirit of data splitting, is the PRESS statistic. Again, we are interested in generating prediction errors of the type in (4.2). Consider a set of data in which we withhold or set aside the first observation from the sample, and we use the remaining $n-1$ observations to estimate the coefficients for a particular candidate model. The first observation is then replaced and the second observation withheld with coefficients estimated again. We remove each observation one at a time, and thus the candidate is fit n times. The deleted response is estimated each time, resulting in n prediction errors or *PRESS residuals* $y_i - \hat{y}_{i,-i} = e_{i,-i}$ $(i = 1, 2, \ldots, n)$. These PRESS residuals are true prediction errors with $\hat{y}_{i,-i}$ being independent of y_i. Thus, in this way, the observation y_i *was not simultaneously used for fit and model assessment*, this being the true test of validation. The prediction $\hat{y}_{i,-i}$ is the regression function evaluated at $\mathbf{x} = \mathbf{x}_i$, but y_i was set aside and not used in obtaining the coefficients. Notationally, we have

$$\hat{y}_{i,-i} = \mathbf{x}_i'\mathbf{b}_{-i} \tag{4.3}$$

where $\mathbf{b}_{-i}$ is the set of coefficients computed without the use of the ith observation. Thus each candidate model will have n PRESS residuals associated with it, and PRESS (Prediction Sum of Squares) is defined as

$$\boxed{\begin{aligned}\text{PRESS} &= \sum_{i=1}^{n} (y_i - \hat{y}_{i,-i})^2 \\ &= \sum_{i=1}^{n} (e_{i,-i})^2\end{aligned}} \tag{4.4}$$

So for choice of best model, one might favor the model with the smallest PRESS.

PRESS is important in that one has information in the form of n validations in which the fitting sample for each is of size $n-1$. At first glance, the reader's impression of PRESS would likely bring resistance to the criterion due to complexity of computations involved. However, the computation is not at all complex and will be discussed in a later section. The PRESS residuals can be used to generate another R^2-like statistic that reflects prediction capabilities. This statistic is given by

$$R^2_{\text{Pred}} = 1 - \frac{\text{PRESS}}{\sum_{i=1}^{n} (y_i - \bar{y})^2} \tag{4.5}$$

UTILITY OF THE PRESS *RESIDUALS*

The individual PRESS residuals can be valuable quite apart from their role in computing the PRESS statistic in Eq. (4.4). They give separate measures of the stability of the regression, and they can help the analyst to isolate which data points or observations have a sizable *influence* on the outcome of the regression. For example, suppose an *ordinary residual* of 17.75 gm/in.^{2} is experienced at a particular data point in an experiment in which the strength of a particular breadwrapper is the response. But suppose the PRESS residual is 850.92 gm/in^2. This rather large difference between the PRESS and ordinary residuals would imply that the data point in question is a rather influential observation in the construction of the regression. The implication is that the use of the data point results in a strong attraction of the fitted $\hat{y}_i$ to the observation y_i. We observed this condition in the case of the three isolated sites in the hospital data, illustrated at the beginning of Section 4.2. The sharp distinction between the ordinary residuals and the prediction errors, or PRESS residuals, suggests that these three sites are highly influential.

The PRESS residuals are not the only mechanism for detecting high influence observations, and they are not necessarily the best. The whole of Chapter 6 is dedicated to the diagnostics used in detection of influential data points. At that point in the text and in Chapter 5, where we further develop the study of residuals, there will be a revisitation of the notion of influence and a discussion of other uses of PRESS residuals.

COMPUTATION OF PRESS

As we indicated earlier, the computation of PRESS is quite simple and does not require repeated regressions run by the analyst. We calculate the PRESS residual from the ordinary residual. Namely,

$$e_{i,-i} = \frac{y_i - \hat{y}_i}{1 - \mathbf{x}_i'(\mathbf{X}'\mathbf{X})^{-1}\mathbf{x}_i} = \frac{e_i}{1 - h_{ii}} \tag{4.6}$$

Thus PRESS reduces to

$$\text{PRESS} = \sum_{i=1}^{n} \left(\frac{e_i}{1 - h_{ii}} \right)^2$$

Equation (4.6) is an intriguing, if not remarkable, result. The reader should understand that ease in determining the behavior of a regression, if any data point were set aside, is extremely valuable, not only for validation purposes but also as a diagnostic tool. Other formulae similar to (4.6) for fitted values of response, regression coefficients, variance-covariance matrix, etc. will be exploited in great detail in Chapter 6. The genesis of all of these results is given in Appendix A.5 while, specifically, Eq. (4.6) is derived in Appendix B.4.

One very interesting observation should be made regarding Eq. (4.6). The quantity h_{ii} is, of course, the diagonal of the HAT matrix and, apart from σ^2, represents the prediction variance (see Section 3.9). Thus, the type of result that one would expect does surface. The data points where prediction is poor (h_{ii} close to unity) where, of course, confidence and prediction bounds are relatively loose, is a data point where the PRESS residual is a severe "swelling" of the ordinary residual. Not coincidentally, the point is a potentially high influence point. Another fact that makes this statement perhaps more understandable is that h_{ii} is a *standardized squared distance measure*, the distance being from the point $x_{1i}, x_{2i}, \ldots, x_{ki}$ to the point $\bar{x}_1, \bar{x}_2, \ldots, \bar{x}_k$. For example, in the $k = 1$ case, h_{ii} is written (see Exercise 4.7)

$$h_{ii} = \frac{1}{n} + \frac{(x_i - \bar{x})^2}{\sum_{i=1}^{n} (x_i - \bar{x})^2}$$

A data point that is *extreme* in the direction of any of the x's is one where prediction is relatively poor and where there is a tendency for the PRESS residual to be large (in magnitude).

A simple figure is useful in illustrating ordinary residuals, PRESS residuals, and the notion of influence. Consider Figure 4.1. Here we have a single regressor. Clearly the slope and intercept of the regression change considerably if the single observation, Δ, is set aside. This observation is one that contains a relatively large HAT diagonal. In addition, the PRESS residual is considerably larger in magnitude than the ordinary residual.

FIGURE 4.1 **A $k = 1$ data set with a single influential data point**

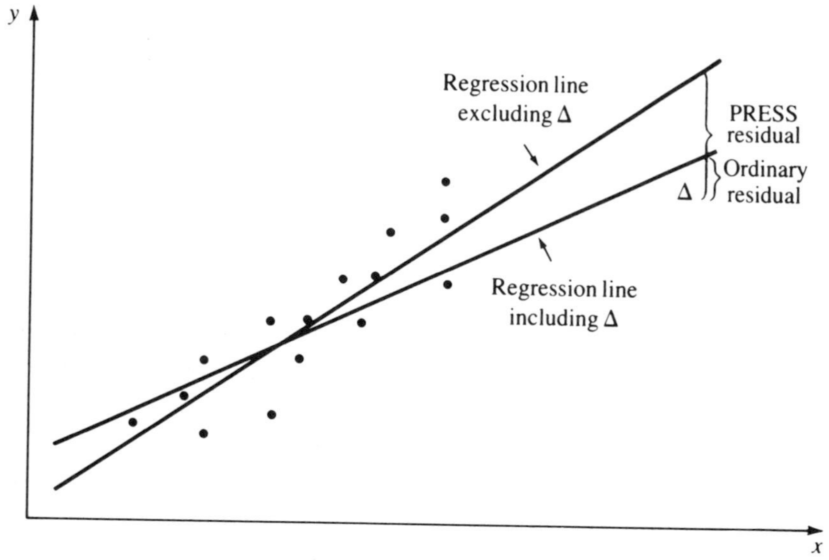

Example 4.1 **Sales Data**

A maker of asphalt roofing shingles is interested in the relationship between sales for a particular year and factors that obviously influence sales: promotional accounts, number of active accounts, number of competing brands, and district potential for the sales districts. Fifteen districts were used, and it is of interest to predict sales from the resulting regression. Table 4.1 supplies the data.

Three models are under serious consideration, based on what is known about the regressor variables. We plan to study the combinations (x_1, x_2, x_3, x_4), (x_1, x_2, x_3), and (x_2, x_3).

The following represents a table of pertinent statistics that can be used for comparing the models.

Model	R^2	R^2_{Pred}	SS_{Res}	s^2	PRESS
x_2, x_3	0.9940	0.9913	534.6621	44.5552	782.1896
x_1, x_2, x_3	0.9970	0.9928	272.7516	24.7956	643.3578
x_1, x_2, x_3, x_4	0.9971	0.9917	262.0728	26.2073	741.7557

Based on this information, the model (x_1, x_2, x_3) appears to be the best for the purposes of predicting sales. The smallest PRESS and smallest s^2 values are achieved with this particular combination of variables.

It is also instructive to compare certain items in the PROC REG output in Tables 4.2, 4.3, and 4.4 across these three models. Notice for the full model,

TABLE 4.1 **Sales data for asphalt shingles**

District	x_1 (Promotional Accounts)	x_2 (Active Accounts)	x_3 (Competing Brands)	x_4 (Potential)	y (Sales in Thousands of Dollars)
1	5.5	31	10	8	79.3
2	2.5	55	8	6	200.1
3	8.0	67	12	9	163.2
4	3.0	50	7	16	200.1
5	3.0	38	8	15	146.0
6	2.9	71	12	17	177.7
7	8.0	30	12	8	30.9
8	9.0	56	5	10	291.9
9	4.0	42	8	4	160.0
10	6.5	73	5	16	339.4
11	5.5	60	11	7	159.6
12	5.0	44	12	12	86.3
13	6.0	50	6	6	237.5
14	5.0	39	10	4	107.2
15	3.5	55	10	4	155.0

John Neter, William Wasserman, and Michael Kutner. *Applied Linear Regression Models* (Homewood, Illinois: Richard D. Irwin, Inc., 1983), p. 439.

TABLE 4.2

Sales data from Example 4.1

DEP VARIABLE: SALES   SALES IN THOUSANDS OF DOLLARS

ANALYSIS OF VARIANCE

SOURCE	DF	SUM OF SQUARES	MEAN SQUARE	F VALUE	PROB>F
MODEL	2	89012.45523	44506.22761	998.901	0.0001
ERROR	12	534.66211	44.55517548		
C TOTAL	14	89547.11733			

ROOT MSE	6.674966	R-SQUARE	0.9940
DEP MEAN	168.9467	ADJ R-SQ	0.9930
C.V.	3.950931		

PARAMETER ESTIMATES

VARIABLE	DF	PARAMETER ESTIMATE	STANDARD ERROR	T FOR H0: PARAMETER=0	PROB > |T|
INTERCEP	1	189.82558	10.12780796	18.743	0.0001
ACT_ACCT	1	3.56916914	0.13328856	26.778	0.0001
COMP_BR	1	-22.27442195	0.70761978	-31.478	0.0001

VARIABLE	DF	VARIABLE LABEL
INTERCEP	1	INTERCEPT
ACT_ACCT	1	ACTIVE ACCOUNTS
COMP_BR	1	COMPETING BRANDS

OBS	ACTUAL	PREDICT VALUE	STD ERR PREDICT	LOWER95% MEAN	UPPER95% MEAN	RESIDUAL
1	79.3000	77.7256	3.1330	70.8994	84.5518	1.5744
2	200.1	207.9	1.9333	203.7	212.1	-7.8345
3	163.2	161.7	3.6467	153.7	169.6	1.5332
4	200.1	212.4	2.2717	207.4	217.3	-12.2631
5	146.0	147.3	2.6069	141.6	152.9	-1.2586
6	177.7	175.9	4.0273	167.2	184.7	1.7565
7	30.9000	29.6076	3.6380	21.6811	37.5341	1.2924
8	291.9	278.3	3.3395	271.1	285.6	13.5731
9	160.0	161.5	2.2699	156.6	166.5	-1.5353
10	339.4	339.0	4.1901	329.9	348.1	0.3972
11	159.6	159.0	2.6197	153.2	164.7	0.6429
12	86.3000	79.5760	2.7456	73.5938	85.5581	6.7240
13	237.5	234.6	2.7841	228.6	240.7	2.8625
14	107.2	106.3	2.3558	101.1	111.4	0.9210
15	155.0	163.4	1.9596	159.1	167.7	-8.3857

TABLE 4.3

Sales data from Example 4.1

DEP VARIABLE: SALES   SALES IN THOUSANDS OF DOLLARS

ANALYSIS OF VARIANCE

SOURCE	DF	SUM OF SQUARES	MEAN SQUARE	F VALUE	PROB>F
MODEL	3	89274.36575	29758.12192	1200.137	0.0001
ERROR	11	272.75158	24.79559845		
C TOTAL	14	89547.11733			

ROOT MSE	4.979518	R-SQUARE	0.9970
DEP MEAN	168.9467	ADJ R-SQ	0.9961
C.V.	2.94739		

PARAMETER ESTIMATES

VARIABLE	DF	PARAMETER ESTIMATE	STANDARD ERROR	T FOR H0: PARAMETER=0	PROB > |T|
INTERCEP	1	178.52062	8.31759314	21.463	0.0001
PROM_EXP	1	2.10554670	0.64785185	3.250	0.0077
ACT_ACCT	1	3.56240254	0.09945492	35.819	0.0001
COMP_BR	1	-22.18799225	0.52855305	-41.979	0.0001

TABLE 4.3 Continued

TABLE 4.3 Continued

Sales data from Example 4.1

VARIABLE	DF	VARIABLE LABEL
INTERCEP	1	INTERCEPT
PROM_EXP	1	PROMOTIONAL EXPENDITURES
ACT_ACCT	1	ACTIVE ACCOUNTS
COMP_BR	1	COMPETING BRANDS

OBS	ACTUAL	PREDICT VALUE	STD ERR PREDICT	LOWER95% MEAN	UPPER95% MEAN	RESIDUAL
1	79.3000	78.6557	2.3547	73.4731	83.8383	0.6443
2	200.1	202.2	2.2759	197.2	207.2	-2.1127
3	163.2	167.8	3.3091	160.5	175.1	-4.5901
4	200.1	207.6	2.2321	202.7	212.6	-7.5414
5	146.0	142.7	2.3970	137.4	148.0	3.2954
6	177.7	171.3	3.3266	164.0	178.6	6.3986
7	30.9000	35.9812	3.3483	28.6115	43.3508	-5.0812
8	291.9	286.0	3.4375	278.5	293.6	5.8749
9	160.0	159.1	1.8568	155.0	163.1	0.9402
10	339.4	341.3	3.2062	334.3	348.4	-1.9221
11	159.6	159.8	1.9705	155.4	164.1	-0.1774
12	86.3000	79.5382	2.0483	75.0300	84.0464	6.7618
13	237.5	236.1	2.1282	231.5	240.8	1.3539
14	107.2	106.1	1.7583	102.2	110.0	1.0979
15	155.0	159.9	1.8054	156.0	163.9	-4.9423

TABLE 4.4

Sales data from Example 4.1

DEP VARIABLE: SALES   SALES IN THOUSANDS OF DOLLARS

ANALYSIS OF VARIANCE

SOURCE	DF	SUM OF SQUARES	MEAN SQUARE	F VALUE	PROB>F
MODEL	4	89285.04450	22321.26113	851.720	0.0001
ERROR	10	262.07283	26.20728326		
C TOTAL	14	89547.11733			

ROOT MSE	5.119305	R-SQUARE	0.9971
DEP MEAN	168.9467	ADJ R-SQ	0.9959
C.V.	3.030131		

PARAMETER ESTIMATES

VARIABLE	DF	PARAMETER ESTIMATE	STANDARD ERROR	T FOR H0: PARAMETER=0	PROB > |T|
INTERCEP	1	177.22856	8.78738255	20.169	0.0001
PROM_EXP	1	2.17021028	0.67369813	3.221	0.0092
ACT_ACCT	1	3.53801385	0.10915205	32.414	0.0001
COMP_BR	1	-22.15833616	0.54537322	-40.630	0.0001
DIST_POT	1	0.20353840	0.31885788	0.638	0.5376

VARIABLE	DF	VARIABLE LABEL
INTERCEP	1	INTERCEPT
PROM_EXP	1	PROMOTIONAL EXPENDITURES
ACT_ACCT	1	ACTIVE ACCOUNTS
COMP_BR	1	COMPETING BRANDS
DIST_POT	1	DISTRICT POTENTIAL

OBS	ACTUAL	PREDICT VALUE	STD ERR PREDICT	LOWER95% MEAN	UPPER95% MEAN	RESIDUAL
1	79.3000	78.8881	2.4480	73.4336	84.3426	0.4119
2	200.1	201.2	2.8274	194.9	207.5	-1.0994
3	163.2	167.6	3.4196	159.9	175.2	-4.3690
4	200.1	208.8	2.9143	202.3	215.3	-8.6881
5	146.0	144.0	3.1627	136.9	151.0	2.0299
6	177.7	172.3	3.7487	163.9	180.6	5.4187
7	30.9000	36.4589	3.5227	28.6097	44.3081	-5.5589
8	291.9	286.1	3.5381	278.2	294.0	5.7671
9	160.0	158.1	2.4757	152.5	163.6	1.9465
10	339.4	342.1	3.5008	334.3	349.9	-2.6749
11	159.6	159.1	2.2665	154.1	164.2	0.4714
12	86.3000	80.2947	2.4163	74.9107	85.6786	6.0053
13	237.5	235.4	2.4647	229.9	240.9	2.0783
14	107.2	105.3	2.2078	100.4	110.2	1.9071
15	155.0	158.6	2.7513	152.5	164.8	-3.6459

(x_1, x_2, x_3, x_4) that the regression coefficient on x_4 is statistically insignificant on the basis of a t-test. Notice also that, for (x_2, x_3) and (x_1, x_2, x_3, x_4), the confidence limits on mean y are not generally as tight as those produced for the model (x_1, x_2, x_3). This is further evidence of the appropriateness of the model (x_1, x_2, x_3).

TABLE 4.5 **Ordinary and PRESS residuals for models (x_2, x_3) and (x_1, x_2, x_3) for sales data**

District	(x_2, x_3) Ordinary Residuals	(x_2, x_3) PRESS Residuals	(x_1, x_2, x_3) Ordinary Residuals	(x_1, x_2, x_3) PRESS Residuals
1	1.574	2.019	0.644	0.830
2	−7.834	−8.552	−2.113	−2.670
3	1.533	2.185	−4.590	−8.220
4	−12.263	−13.870	−7.541	−9.438
5	−1.259	−1.485	3.295	4.289
6	1.756	2.762	6.399	11.556
7	1.292	1.838	−5.081	−9.275
8	13.573	18.105	5.875	11.224
9	−1.535	−1.736	0.940	1.092
10	0.397	0.655	−1.922	−3.283
11	0.643	0.760	−0.177	−0.210
12	6.724	8.093	6.762	8.139
13	2.862	3.465	1.354	1.656
14	0.921	1.052	1.098	1.254
15	−8.386	−9.176	−4.942	−5.690

The PRESS residuals, or *prediction errors* as they may be called (they are, unfortunately, often called *deleted residuals*), are part of the standard computer printout on many regression computer packages. The PRESS residuals and the ordinary residuals for models (x_1, x_2, x_3) and (x_2, x_3) are shown in Table 4.5. The comparisons of the PRESS residuals for the two models illustrated in Table 4.5 point out the difference in performance between the two models. For example, at the second data point, i.e., the second sales district, the model (x_1, x_2, x_3) *overpredicts* sales (without using the second district data in the construction of the regression) by 2.670. Model (x_2, x_3) overpredicts sales in the same district by 8.552.

Note that, at district 3, (x_1, x_2, x_3) overpredicts by 8.220 while (x_2, x_3) has a positive prediction error of 2.185. In fact (x_2, x_3) has a smaller prediction error (in magnitude) in seven of the fifteen districts. The difference in the two PRESS statistics stems from the inability of (x_2, x_3) to predict at districts 2, 4, 8, and 15. While all of the statistics point toward (x_1, x_2, x_3) being a superior model to (x_2, x_3), the three-variable candidate does not *uniformly* predict the district sales better than (x_2, x_3).

ABSOLUTE PRESS *RESIDUALS*

A relatively large PRESS can result because of one or a few large PRESS residuals, which often stand out because the data points in question are extreme in some sense in the space of the x's. Clearly their prediction errors cannot be ignored. However, in order that they not be so influential to the criterion, the analyst may wish to use

$$\sum_{i=1}^{n} |y_i - \hat{y}_{i,-i}| = \sum_{i=1}^{n} |e_{i,-i}|$$

as an alternative to PRESS, the rationale being to avoid squaring the PRESS residuals. In any case, if the analyst uses PRESS to discriminate between models, the PRESS residuals themselves should be observed in order to isolate the location of the large prediction errors.

4.3

CONCEPTUAL PREDICTIVE CRITERIA (THE C_p STATISTIC)

The previous section dealt with model selection criteria, which were based on model validation information. Here we introduce another prediction oriented criterion. It has a sound conceptual base and is easily obtained computationally, although it is not actually based on the formulation of empirical prediction errors. Of course, in any serious search for choice of best model, the investigator should feel committed to consider several criteria, particularly if the choice appears to be a difficult one.

Before we present a new criterion, we will consider what the data analyst faces if he underfits (i.e., ignores or deletes relatively important model terms) or overfits (i.e., includes model terms that have only marginal, or no, contribution). In what follows, the results are based on use of the procedure of least squares with the assumption that the ε_i are uncorrelated with homogeneous variance, σ^2.

IMPACT OF UNDERFITTING

In Chapter 3 we dealt with model underspecification and its impact on the estimator s^2. Equation (4.1) provides an insight into the positive bias of s^2 if the model is underspecified. Proceeding along the same lines, suppose $E(\mathbf{y}) = \mathbf{X}_1\boldsymbol{\beta}_1 + \mathbf{X}_2\boldsymbol{\beta}_2$, where, once again, $\boldsymbol{\beta}_1$ and $\boldsymbol{\beta}_2$ contain p and $m-p$ parameters respectively. As in the case of Eq. (3.26) the model $\mathbf{y} = \mathbf{X}_1\boldsymbol{\beta}_1 + \boldsymbol{\varepsilon}^*$ is fit with $\mathbf{b}_1 = (\mathbf{X}_1'\mathbf{X}_1)^{-1}\mathbf{X}_1'\mathbf{y}$as the least squares estimator for $\boldsymbol{\beta}_1$. Thus

$$\begin{aligned} E(\mathbf{b}_1) &= (\mathbf{X}_1'\mathbf{X}_1)^{-1}\mathbf{X}_1' E(\mathbf{y}) = (\mathbf{X}_1'X_1)^{-1}\mathbf{X}_1'[\mathbf{X}_1\boldsymbol{\beta}_1 + \mathbf{X}_2\boldsymbol{\beta}_2] \\ &= \boldsymbol{\beta}_1 + (\mathbf{X}_1'\mathbf{X}_1)^{-1}\mathbf{X}_1'\mathbf{X}_2\boldsymbol{\beta}_2 = \boldsymbol{\beta}_1 + \mathbf{A}\boldsymbol{\beta}_2 \end{aligned} \quad (4.7)$$

Here $\mathbf{A} = (\mathbf{X}_1'\mathbf{X}_1)^{-1}\mathbf{X}_1'\mathbf{X}_2$ is called the *alias matrix*, and $\mathbf{A}\boldsymbol{\beta}_2$ represents a vector of biases of the coefficients in $\mathbf{b}_1$ due to underspecification.

Consider now, the impact of underspecification on a predicted response $\hat{y}(\mathbf{x}_{1,0})$.

We define
$$\hat{y}(\mathbf{x}_{1,0}) = \mathbf{x}'_{1,0}\mathbf{b}_1 \tag{4.8}$$
where $\mathbf{x}'_{1,0}$ is the vector of terms evaluated at a future point of interest (or at a data point $\mathbf{x}'_{1i}$). Consider now the expectation
$$E[\hat{y}(\mathbf{x}_{1,0})] = \mathbf{x}'_{1,0}[\boldsymbol{\beta}_1 + \mathbf{A}\boldsymbol{\beta}_2]$$
We know that at the point $\mathbf{x}'_0 = [\mathbf{x}'_{1,0}, \mathbf{x}'_{2,0}]$ (involving all the relevant model terms), the mean response, i.e., what one is truly attempting to estimate, is $E[y(\mathbf{x}_0)] = \mathbf{x}'_{1,0}\boldsymbol{\beta}_1 + \mathbf{x}'_{2,0}\boldsymbol{\beta}_2$. Thus the *bias in the prediction* at $\mathbf{x}_0$ is given by
$$E[\hat{y}(\mathbf{x}_{1,0})] - E[y(\mathbf{x}_0)] = [\mathbf{x}'_{1,0}\mathbf{A} - \mathbf{x}'_{2,0}]\boldsymbol{\beta}_2 \tag{4.9}$$
Thus the squared bias can be expressed as the quadratic form
$$\{\text{Bias}[\hat{y}(\mathbf{x}_{1,0})]\}^2 = \boldsymbol{\beta}'_2[\mathbf{x}'_{1,0}\mathbf{A} - \mathbf{x}'_{2,0}]'[\mathbf{x}'_{1,0}\mathbf{A} - \mathbf{x}'_{2,0}]\boldsymbol{\beta}_2 \tag{4.10}$$
From Eq. (4.1) and (4.10), it becomes apparent, after some manipulation, that the bias in s^2 in an underspecified model can be written in terms of the sum of the squared biases in the prediction at the data points. That is,
$$\boxed{E(s_p^2) = \sigma^2 + \frac{1}{n-p}\sum_{i=1}^{n}\{\text{Bias }\hat{y}(\mathbf{x}_i)\}^2} \tag{4.11}$$
Here we again use s_p^2 to indicate the error mean square for the underfitted p-term model.

It is obvious that the impact of an underspecified model is fixed bias in the important quantities, $\hat{y}(\mathbf{x}_{1,0})$, the regression coefficients, and the estimate of error variance. Equation (4.11) is particularly important in developing a model selection criterion, which will be discussed in a later section. Essentially when we underfit or, say, ignore important variables, we deposit the variation accounted for by the ignored variables in the residual sum of squares and hence inflate the residual mean square. This inflation may be viewed as reflecting bias in prediction at the data points.

Example 4.1 and the model (x_2, x_3) nicely illustrates the bias in s^2 for an underfitted model. A comparison of (x_2, x_3) and (x_1, x_2, x_3) in Tables 4.2 and 4.3 certainly suggests that the two-variable model is an underfitted model. This is reflected in the comparison of the values of s^2 for the two models. Next we deal with the impact of overspecification or, more appropriately, *overfitting*.

IMPACT OF OVERFITTING

The reader should realize that, while the underspecified model results in important estimated quantities being biased, the practice of overfitting i.e., fitting model terms that contribute little or nothing, produces results involving *variances* that are larger than those for the simpler model. The important variance results are best summarized by the following statements.

1. For a model containing the regressors $x_1, x_2, \ldots, x_k$ with least squares

estimators $b_0, b_1, b_2, \ldots, b_k$, if the regressor variable x_{k+1} is added to the model producing the new regression coefficients $b_0^*, b_1^*, \ldots, b_k^*, b_{k+1}^*$, then

$$\text{Var } b_i^* \geqslant \text{Var } b_i \qquad (i = 0, 1, 2, \ldots, k) \tag{4.12}$$

2. For the situation just described consider the two regression predictors $\hat{y}_1 = \sum_{i=0}^{k} b_i x_i$ and $\hat{y}_2 = \sum_{i=0}^{k+1} b_i^* x_i$, the latter containing the contribution from the additional regressor. At a point of interest, $\mathbf{x}_0' = (1, x_{1,0}, x_{2,0}, \ldots, x_{k,0}, x_{k+1,0})$,

$$\text{Var } \hat{y}_1(\mathbf{x}_0) \leqslant \text{Var } \hat{y}_2(\mathbf{x}_0) \tag{4.13}$$

The results of items 1 and 2 are very important and allow us to make considerable headway in illustrating the tradeoff that exists between underfitting and overfitting. A model that is too simple may suffer from biased coefficients and biased prediction, while an overly complicated model can result in large variances, both in the coefficients and in the prediction. Therefore, a proper model, in many cases, will be a compromise between a biased model and a model with heavy variance. The magnitude of the variance inflicted by the addition of marginal variables, as one would suspect, depends to a great extent on the multicollinearity induced by the "questionable" regressors.

For proofs of the results in 1 and 2, the reader is referred to Rao (1971).

THE PROPER COMPROMISE AND MALLOWS' C_p

The discussions of underfitting and overfitting make it quite clear what the real task of the model builder or model selector is. We must choose the proper subset of regressors or proper functional form from the candidates so that a suitable balance is reached. For example, for prediction, a reasonable criterion is the mean squared error of prediction. For example at $\mathbf{x}_0$, the mean squared error of $\hat{y}(\mathbf{x}_0)$, the prediction for a particular candidate model, is given by

$$MSE[\hat{y}(\mathbf{x}_0)] = \text{Var } \hat{y}(\mathbf{x}_0) + [E\hat{y}(\mathbf{x}_0) - Ey(\mathbf{x}_0)]^2 \tag{4.14}$$

This, of course, incorporates bias and variance in the same criterion and, indeed, is expressed in the definitive form

$$MSE[\hat{y}(\mathbf{x}_0)] = E[\hat{y}(\mathbf{x}_0) - Ey(\mathbf{x}_0)]^2 \tag{4.15}$$

The expression (4.15) makes sense conceptually, but can it truly be used? Let us consider the formulation previously presented as "Impact of Underfitting." We can use Eq. (4.9), and write (for an underspecified model with ignored parameter vector $\boldsymbol{\beta}_2$)

$$MSE[\hat{y}(\mathbf{x}_{1,0})] = \sigma^2 \mathbf{x}_{1,0}'(\mathbf{X}_1'\mathbf{X}_1)^{-1}\mathbf{x}_{1,0} + [(\mathbf{x}_{1,0}'\mathbf{A} - \mathbf{x}_{2,0}')\boldsymbol{\beta}_2]^2 \tag{4.16}$$

One cannot apply the criterion in Eq. (4.16) directly. The vector $\boldsymbol{\beta}_2$ is unknown, and the criterion suffers, as do several conceptual criteria on prediction, from being a function of the regressor location vectors $\mathbf{x}_{1,0}$ and $\mathbf{x}_{2,0}$.

To hurdle the problem of the location dependency of $MSE\ \hat{y}(\mathbf{x}_0)$, suppose we consider the mean squared error of a *fitted value* $\hat{y}(\mathbf{x}_i)$, summed over the data points. That is, consider the quantity

$$\sum_{i=1}^{n} \frac{[MSE\ \hat{y}(\mathbf{x}_i)]}{\sigma^2} = \sum_{i=1}^{n} \frac{\{[\text{Var}\ \hat{y}(\mathbf{x}_i)] + [\text{Bias}\ \hat{y}(\mathbf{x}_i)]^2\}}{\sigma^2} \tag{4.17}$$

We can view the quantity in (4.17) as a *standardized total error*. Clearly, it does not reflect any extrapolation or interpolation capability of the candidate model. We can, however, produce an estimate of the quantity in (4.17) that achieves a workable balance between bias and variance. Considering the components of (4.17), let us assume, in order to be totally general, that the candidate model in question contains p-parameters and the "true" model contains $m-p$ additional parameters described by the vector $\boldsymbol{\beta}_2$. So, we will denote the prediction at the ith data point as

$$\hat{y}(\mathbf{x}_i) = \mathbf{x}'_{1i}\mathbf{b}_1$$

where $\mathbf{b}_1 = (\mathbf{X}'_1\mathbf{X}_1)^{-1}\mathbf{X}'_1\mathbf{y}$. Thus, we write the variance portion of our criterion as

$$\sum_{i=1}^{n} \frac{\text{Var}\ \hat{y}(\mathbf{x}_{1i})}{\sigma^2} = \sum_{i=1}^{n} \mathbf{x}'_{1i}(\mathbf{X}'_1\mathbf{X}_1)^{-1}\mathbf{x}_{1i} \tag{4.18}$$

We can manipulate (4.18) to obtain a form that is very simple to compute.

$$\begin{aligned}\sum_{i=1}^{n} \mathbf{x}'_{1i}(\mathbf{X}'_1\mathbf{X}_1)^{-1}\mathbf{x}_{1i} &= \sum_{i=1}^{n} \text{tr}\ \mathbf{x}'_{1i}(\mathbf{X}'_1\mathbf{X}_1)^{-1}\mathbf{x}_{1i} \\ &= \sum_{i=1}^{n} \text{tr}\ \mathbf{x}_{1i}\mathbf{x}'_{1i}(\mathbf{X}'_1\mathbf{X}_1)^{-1} \\ &= \text{tr} \sum_{i=1}^{n} \mathbf{x}_{1i}\mathbf{x}'_{1i}(\mathbf{X}'_1\mathbf{X}_1)^{-1} \end{aligned} \tag{4.19}$$

Close inspection of (4.19) reveals that $\sum_{i=1}^{n} \mathbf{x}_{1i}\mathbf{x}'_{1i}$ is, in fact, the matrix $\mathbf{X}'_1\mathbf{X}_1$. Thus the variance portion of the *standardized total mean squared error* becomes

$$\sum_{i=1}^{n} \frac{\text{Var}\ \hat{y}(\mathbf{x}_{1i})}{\sigma^2} = \text{tr}\ I_p = p$$

This may seem to be a rather curious result. *The prediction variance* (apart from σ^2) *summed over the data locations is equal to the number of parameters.* The reader should recognize the development as solving an additional problem, namely the proof of property 1 (in Section 3.9) of the HAT matrix. The quantity $\sum_{i=1}^{n} \mathbf{x}'_{1i}(\mathbf{X}'_1\mathbf{X}_1)^{-1}\mathbf{x}_{1i}$ is the trace of the HAT matrix for the candidate model in question.

Moving on to the bias portion of our criterion in Eq. (4.17), we now need to consider $\sum_{i=1}^{n} [\text{Bias}\ \hat{y}(\mathbf{x}_{1i})]^2$. This quantity can be estimated. Consider Eq. (4.11).

We learned that the estimate, s^2, of residual variance for an underspecified model is biased by the quantity

$$\sum_{i=1}^{n} \frac{[\text{Bias } \hat{y}(\mathbf{x}_i)]^2}{n-p}$$

As a result, *if* σ^2 *were known*, an estimate of the quantity in Eq. (4.17) (the estimate is called the C_p statistic) is given by

$$C_p = p + \frac{(s^2 - \sigma^2)(n-p)}{\sigma^2} \qquad \textbf{(4.20)}$$

There are many different ways of writing Eq. (4.20). Some prefer expressing it in terms of SS_{Res}. At any rate, it expresses *variance + bias* and if an independent estimate of σ^2, say $\hat{\sigma}^2$, can be found, the C_p statistic can be extremely useful as a criterion for discriminating between models. The C_p for a p-parameter regression model would then be written as

$$\boxed{C_p = p + \frac{(s^2 - \hat{\sigma}^2)(n-p)}{\hat{\sigma}^2}} \qquad \textbf{(4.21)}$$

One then favors the candidate model with the *smallest* C_p *value.*

A reasonable norm by which to judge the C_p value of a model is $C_p = p$, a value that suggests that the model contains no estimated bias. That is, all of the error in $\hat{y}$ is variance, and the model is not underspecified. Of course, clear interpretation is often clouded because of the questionable nature of the estimate $\hat{\sigma}^2$. In many practical situations, the residual mean square for the full model or most complete model is used as this estimate. Since the residual mean square for the complete model need not be the smallest estimate of σ^2 among those for the candidate models, it is quite possible that Eq. (4.21) will yield a $C_p < p$ for a few of the candidate models.

Often the C_p values for various candidate models can be displayed on a plot, with the line $C_p = p$ representing the norm. A C_p much larger than p occurs with a heavily biased model. A typical C_p *plot* appears in Figure 4.2. Models A and D appear to be undesirable, having C_p values well above the *variance line*. Model D is clearly the poorest performer. Models B and C appear to be reasonable candidates. For model C, C_p below $p = 3$ implies that the s^2 value is smaller than $\hat{\sigma}^2$.

Example 4.2 **_Sales Data_**

In this section we consider an example in which the C_p statistic is used for model discrimination. This example considers the sales data in Table 4.1. It is always desirable to use an estimate $\hat{\sigma}^2$ that is independent of the s^2 value for the candidate model. Unfortunately such an estimate is not always available. Common practice among data analysts is use of the *residual mean square for the complete* model as $\hat{\sigma}^2$. In the case of the sales data, this is $\hat{\sigma}^2 = 26.2073$. For the model $(x_2, x_3)(p = 3)$

$$C_p = 3 + \frac{(44.5552 - 26.2073)(12)}{26.2073} = 11.4013$$

Clearly, this value is well above 3.0 and thus reflects what would seem to be a biased model. Similar computations reveal that, for the model (x_1, x_2, x_3), $C_p = 3.4075$ with $p = 4$. Of course, for the full model C_p is fixed at 5.0. As a result, the preferable model, from the prediction standpoint, is the model (x_1, x_2, x_3). Using C_p and PRESS statistics lead to the same conclusion regarding the desirability of model (x_1, x_2, x_3).

FIGURE 4.2 **C_p against p plot**

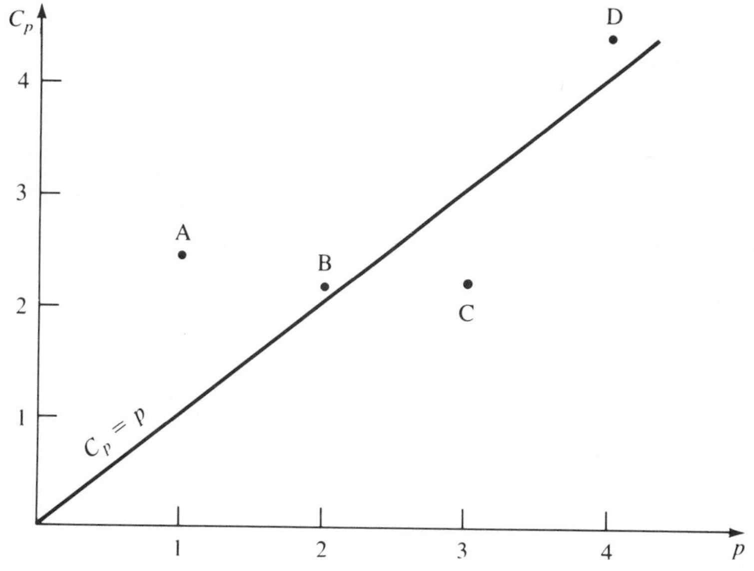

PROPERTIES OF RESIDUAL MEAN SQUARE FOR AN OVERFITTED MODEL

We learned from Eq. (4.11) that the residual mean square is biased upward in cases where the fitted model is biased, i.e., when the analyst is underfitting. In order to gain even more insight into the role of s^2 as a model discriminator, it is instructive to consider its properties when the analyst uses an overfitted model. Suppose one assumes the model

$$\mathbf{y} = \mathbf{X}_1 \boldsymbol{\beta}_1 + \mathbf{X}_2 \boldsymbol{\beta}_2 + \boldsymbol{\varepsilon} \qquad (m \text{ parameters}) \tag{4.22}$$

when in fact $\boldsymbol{\beta}_2 = \mathbf{0}$ and thus the m parameter model is an overfitted model. Letting

$$\mathbf{X} = [\mathbf{X}_1 \,\vdots\, \mathbf{X}_2]$$

the residual mean square for the overfitted model is given by

$$s_m^2 = \frac{\mathbf{y}'[I - \mathbf{X}(\mathbf{X}'\mathbf{X})^{-1}\mathbf{X}']\mathbf{y}}{n-m}$$

The expected value is given by

$$\begin{aligned} E(s_m^2) &= \frac{1}{n-m} E\mathbf{y}'[I - \mathbf{X}(\mathbf{X}'\mathbf{X})^{-1}\mathbf{X}']\mathbf{y} \\ &= \frac{1}{n-m}\{\sigma^2 \operatorname{tr}[I - \mathbf{X}(\mathbf{X}'\mathbf{X})^{-1}\mathbf{X}'] + [E(\mathbf{y})]'[I - \mathbf{X}(\mathbf{X}'\mathbf{X})^{-1}\mathbf{X}'][E(\mathbf{y})]\} \\ &= \frac{1}{n-m}\{\sigma^2(n-m) + \boldsymbol{\beta}_1'\mathbf{X}_1'[I - \mathbf{X}(\mathbf{X}'\mathbf{X})^{-1}\mathbf{X}']\mathbf{X}_1\boldsymbol{\beta}_1\} \end{aligned} \tag{4.23}$$

Details of this development follow the same line as that given in Appendix B.2. From Eq. (4.23), the expected value of the residual mean square for the overfitted model is given by

$$E(s_m^2) = \sigma^2 + \frac{1}{n-m}\boldsymbol{\beta}_1'\mathbf{X}_1'[I - \mathbf{X}(\mathbf{X}'\mathbf{X})^{-1}\mathbf{X}']\mathbf{X}_1\boldsymbol{\beta}_1$$

Since $\mathbf{X}'[I - \mathbf{X}(\mathbf{X}'\mathbf{X})^{-1}\mathbf{X}'] = \mathbf{0}$, $\mathbf{X}_1'[I - \mathbf{X}(\mathbf{X}'\mathbf{X})^{-1}\mathbf{X}'] = \mathbf{0}$, and thus

$$E(s_m^2) = \sigma^2$$

Even if the investigator overfits, i.e., includes model terms that are zero, the residual mean square is unbiased for σ^2. However, the estimator contains fewer degrees of freedom than the error mean square computed from fitting the "correct" model.

In the normal theory case, we know that

$$\frac{s_m^2(n-m)}{\sigma^2} \sim \chi^2_{n-m}$$

See Graybill (1976). As a result,

$$s_m^2 \sim \frac{\sigma^2 \chi^2_{n-m}}{n-m}$$

and

$$\operatorname{Var} s_m^2 = \frac{\sigma^4}{(n-m)^2} \operatorname{Var} \chi^2_{n-m}$$

The variance of a χ^2_v random variable is $2v$, where v is the degrees of freedom. Thus

$$\operatorname{Var}(s_m^2) = \frac{2\sigma^4}{n-m}$$

As a result, a decreasing error degrees of freedom ($n - m$ in the case of the overfitted model) produces a larger variance of s^2. So, *the variance of the estimator of σ^2 for an overfitted model is larger than the variance of the estimator of σ^2 computed from the correct model.* From this we gain some insight into the utility of the residual mean square, s^2, as a discriminator between competing models and as an important component in the C_p statistic.

4.4 SEQUENTIAL VARIABLE SELECTION PROCEDURES

During the early 1960s, the computer field experienced rapid growth, and this era produced several systematic procedures based on sequential F-tests—procedures constructed to arrive efficiently at a reasonable subset of regressor variables in cases where a large number of possible variables exist. The procedures were designed so that a relatively small number of subset regressions are actually run on the computer.

The analyst should bear in mind that these sequential algorithms were developed with efficient computation as the motivation, and in a time when it was often prohibitive to gather information from all 2^k *subset models.* Presently available are many software packages based on computationally swift algorithms for obtaining at least some important results on *all subsets.* (References will be made to some of these developments subsequently.) This fact, coupled with the reality that multicollinearity often clouds the picture, might suggest that the sequential algorithms are not of practical use except in rare cases. Nevertheless, they are an important part of the heritage of least squares regression and are used by a large number of analysts (perhaps because they appear in most standard regression computer software packages). These sequential algorithms are many and varied. The following three subsections describe three general types.

FORWARD SELECTION

Sequentially and systematically, each method adds and/or deletes regressor variables from the existing model on the basis of F-tests. In the case of forward selection, the initial model contains only a constant term. The procedure selects for entry the variable that produces the largest R^2 of any single regressor. Suppose we call this regressor x_1. The second regressor is chosen, which produces the largest increase in R^2 *in the presence of* x_1. Suppose we call this regressor x_2. Note that this is tantamount to choosing the regressor with the largest partial F (partial in the sense of the model which includes that variable teamed with x_1). Thus, in this hypothetical case, at stage 2,

$$F = \frac{R(x_2 \mid x_1)}{s^2(x_1, x_2)}$$

is the largest F for the second stage. The notation $s^2(x_1, x_2)$ is the residual mean square for the model involving regressors x_1 and x_2. We generally compute the residual estimate of variance from the incomplete model, i.e., the existing model at that stage.

The above process continues until, at some stage, the candidate regressor for entry does not exceed a preselected F_{IN} (or, of course, until all regressors have been entered).

STEPWISE REGRESSION

Stepwise regression provides an important modification of forward selection in that, at each stage of selection, all regressors currently in the model are evaluated through the partial F-test. A preselected F_{OUT} critical value is used. Thus, at each stage, a regressor can be entered, and another may be eliminated. The rationale for this should be quite clear. Multicollinearity certainly can render a regressor of little value even though it was an important candidate at an early stage of the procedure. Thus, at all stages following entry, a variable must continue to perform or be eliminated. The procedure terminates when no additional regressors can enter on the basis of F_{IN}, and no regressors in the model can be eliminated on the basis of F_{OUT}.

BACKWARD ELIMINATION

While the last two procedures discussed begin with no regressors in the model, backward elimination begins with all regressors and eliminates one at a time. The first removed is the regressor, which results in the smallest decrease in R^2 (thus the smallest partial F-statistic). The procedure is continued until the candidate regressor for removal experiences a partial F value which exceeds the preselected F_{OUT}.

The following section provides an annotated computer printout illustrating all three sequential algorithms on the same data.

Example 4.3 **Hospital Data**

Consider the hospital data of Example 3.8. The reader should recall that this example illustrated serious multicollinearity. Two models were fit, and variance inflation factors were investigated. Suppose the investigator chooses to pursue the sequential model-building procedures in a search for the "best" model.

Tables 4.6, 4.7, and 4.8 represent annotated computer printouts using the SAS system for forward selection, backward elimination, and stepwise regression, respectively. Notice that the model suggested for forward selection, along with pertinent statistics, is

$$\hat{y} = 2032.1881 + 0.056079x_2 + 1.0884x_3 - 5.0041x_4 - 410.0830x_5$$
$$s^2 = 378{,}826.3203$$
$$R^2 = 0.9908$$
$$C_p = 4.02635$$

The backward elimination procedure and stepwise procedure produce the model

$$\hat{y} = 1523.3892 + 0.052987x_2 + 0.9784x_3 - 320.9508x_5$$
$$s^2 = 378{,}826.3203$$
$$R^2 = 0.9901$$
$$C_p = 2.91769$$

TABLE 4.6 **Hospital manpower data of Table 3.8: Forward selection procedure for dependent variable *y***

STEP 1 VARIABLE X3 ENTERED R SQUARE = 0.97218120 C(P) = 20.38117958

	DF	SUM OF SQUARES	MEAN SQUARE	F	PROB>F
REGRESSION	1	480950231.62604150	480950231.62604150	524.20	0.0001
ERROR	15	13762308.86295839	917487.25753056		
TOTAL	16	494712540.48899990			

	B VALUE	STD ERROR	PARTIAL REG SS	F	PROB>F
INTERCEPT	-28.12861560				
X3	1.11739237	0.04880403	480950231.62604150	524.20	0.0001

STEP 2 VARIABLE X2 ENTERED R SQUARE = 0.98671474 C(P) = 4.94164787

	DF	SUM OF SQUARES	MEAN SQUARE	F	PROB>F
REGRESSION	2	488140157.95096330	244070078.97548168	519.90	0.0001
ERROR	14	6572382.53803656	469455.89557404		
TOTAL	16	494712540.48899990			

	B VALUE	STD ERROR	PARTIAL REG SS	F	PROB>F
INTERCEPT	-68.31395896				
X2	0.07486591	0.01913019	7189926.32492182	15.32	0.0016
X3	0.82287456	0.08295986	46187674.54075647	98.39	0.0001

STEP 3 VARIABLE X5 ENTERED R SQUARE = 0.99006817 C(P) = 2.91769778

	DF	SUM OF SQUARES	MEAN SQUARE	F	PROB>F
REGRESSION	3	489799141.98626880	163266380.66208962	431.97	0.0001
ERROR	13	4913398.50273108	377953.73097931		
TOTAL	16	494712540.48899990			

	B VALUE	STD ERROR	PARTIAL REG SS	F	PROB>F
INTERCEPT	1523.38923568				
X2	0.05298733	0.02009194	2628687.59792946	6.96	0.0205
X3	0.97848162	0.10515362	32726194.93174630	86.59	0.0001
X5	-320.95082518	153.19222065	1658984.03530548	4.39	0.0563

STEP 4 VARIABLE X4 ENTERED R SQUARE = 0.99081100 C(P) = 4.02634991

	DF	SUM OF SQUARES	MEAN SQUARE	F	PROB>F
REGRESSION	4	490166624.64505400	122541656.16126351	323.48	0.0001
ERROR	12	4545915.84394591	378826.32032883		
TOTAL	16	494712540.48899990			

	B VALUE	STD ERROR	PARTIAL REG SS	F	PROB>F
INTERCEPT	2032.18806215				
X2	0.05607934	0.02035863	2874411.76901339	7.59	0.0175
X3	1.08836904	0.15339754	19070221.63728522	50.34	0.0001
X4	-5.00406579	5.08071295	367482.65878517	0.97	0.3441
X5	-410.08295954	178.07810366	2008919.55996109	5.30	0.0400

NO OTHER VARIABLES MET THE 0.5000 SIGNIFICANCE LEVEL FOR ENTRY INTO THE MODEL

TABLE 4.7 **Hospital manpower data of Table 3.8; Backward elimination procedure for dependent variable *y***

STEP 0 ALL VARIABLES ENTERED		R SQUARE = 0.99083295	C(P) = 6.00000000		
	DF	SUM OF SQUARES	MEAN SQUARE	F	PROB>F
REGRESSION	5	490177488.12165090	98035497.62433018	237.79	0.0001
ERROR	11	4535052.36734900	412277.48794082		
TOTAL	16	494712540.48899990			
	B VALUE	STD ERROR	PARTIAL REG SS	F	PROB>F
INTERCEPT	1962.94815647				
X1	-15.85167473	97.65299018	10863.47659691	0.03	0.8740
X2	0.05593038	0.02125828	2853834.33814818	6.92	0.0234
X3	1.58962370	3.09208349	108962.19771764	0.26	0.6174
X4	-4.21866799	7.17655737	142464.93826739	0.35	0.5685
X5	-394.31411702	209.63954082	1458572.00281510	3.54	0.0867
STEP 1 VARIABLE X1 REMOVED		R SQUARE = 0.99081100	C(P) = 4.02634991		
	DF	SUM OF SQUARES	MEAN SQUARE	F	PROB>F
REGRESSION	4	490166624.64505400	122541656.16126351	323.48	0.0001
ERROR	12	4545915.84394591	378826.32032883		
TOTAL	16	494712540.48899990			
	B VALUE	STD ERROR	PARTIAL REG SS	F	PROB>F
INTERCEPT	2032.18806215				
X2	0.05607934	0.02035863	2874411.76901339	7.59	0.0175
X3	1.08836904	0.15339754	19070221.63728595	50.34	0.0001
X4	-5.00406579	5.08071295	367482.65878517	0.97	0.3441
X5	-410.08295954	178.07810366	2008919.55996109	5.30	0.0400
STEP 2 VARIABLE X4 REMOVED		R SQUARE = 0.99006817	C(P) = 2.91769778		
	DF	SUM OF SQUARES	MEAN SQUARE	F	PROB>F
REGRESSION	3	489799141.98626880	163266380.66208962	431.97	0.0001
ERROR	13	4913398.50273108	377953.73097931		
TOTAL	16	494712540.48899990			
	B VALUE	STD ERROR	PARTIAL REG SS	F	PROB>F
INTERCEPT	1523.38923568				
X2	0.05298733	0.02009194	2628687.59792946	6.96	0.0205
X3	0.97848162	0.10515362	32726194.93174914	86.59	0.0001
X5	-320.95082518	153.19222065	1658984.03530548	4.39	0.0563

ALL VARIABLES IN THE MODEL ARE SIGNIFICANT AT THE 0.1000 LEVEL.

The significance levels used in this illustration are 0.500 for entry and 0.100 for removal. In most available computer algorithms, the analyst controls these parameters. Generally, the recommendation is to use a fairly traditional value for removal, i.e., 0.01, 0.05, or 0.10. However, the suggestion is to use a considerably higher value for entry, say 0.25–0.50. We do not choose here to give a detailed rationale for this recommendation. However, the reader should appreciate that in either *forward* procedure, i.e., forward selection or stepwise, the F critical values are not strictly appropriate in any probabilistic sense in the early stages. Recall our discussion in Section 4.3 regarding the impact on s^2 of model underspecification. The models produced in the early stages are often underspecified, with the result that s^2 can be badly overestimated. The tendency, then, is to "stop short" and perhaps not allow important variables to enter due to the deflation of the F-statistics. Since the analyst will learn more about the regressor variables' roles if the algorithm actually generates computations for several models, it is counterproductive to allow the algorithm to terminate prematurely. Obviously this problem is considerably less serious in the case of backward elimination. Mathematical detail concerning the significance level problem can be found in Pope and Webster (1972).

TABLE 4.8

Hospital manpower data of Table 3.8; Stepwise regression procedure for dependent variable *y*

```
STEP 1    VARIABLE X3 ENTERED       R SQUARE = 0.97218120      C(P) =    20.38117958

                         DF          SUM OF SQUARES         MEAN SQUARE           F       PROB>F

        REGRESSION        1      480950231.62604150  480950231.62604150      524.20       0.0001
        ERROR            15       13762308.86295839     917487.25753056
        TOTAL            16      494712540.48899990

                             B VALUE          STD ERROR      PARTIAL REG SS           F       PROB>F

        INTERCEPT       -28.12861560
        X3                1.11739237         0.04880403  480950231.62604150      524.20       0.0001
------------------------------------------------------------------------------------------------------
STEP 2    VARIABLE X2 ENTERED       R SQUARE = 0.98671474      C(P) =     4.94164787

                         DF          SUM OF SQUARES         MEAN SQUARE           F       PROB>F

        REGRESSION        2      488140157.95096330  244070078.97548168      519.90       0.0001
        ERROR            14        6572382.53803656     469455.89557404
        TOTAL            16      494712540.48899990

                             B VALUE          STD ERROR      PARTIAL REG SS           F       PROB>F

        INTERCEPT       -68.31395896
        X2                0.07486591         0.01913019    7189926.32492182       15.32       0.0016
        X3                0.82287456         0.08295986   46187674.54075647       98.39       0.0001
------------------------------------------------------------------------------------------------------
STEP 3    VARIABLE X5 ENTERED       R SQUARE = 0.99006817      C(P) =     2.91769778

                         DF          SUM OF SQUARES         MEAN SQUARE           F       PROB>F

        REGRESSION        3      489799141.98626880  163266380.66208962      431.97       0.0001
        ERROR            13        4913398.50273108     377953.73097931
        TOTAL            16      494712540.48899990

                             B VALUE          STD ERROR      PARTIAL REG SS           F       PROB>F

        INTERCEPT      1523.38923568
        X2                0.05298733         0.02009194    2628687.59792946        6.96       0.0205
        X3                0.97848162         0.10515362   32726194.93174630       86.59       0.0001
        X5             -320.95082518       153.19222065    1658984.03530548        4.39       0.0563
------------------------------------------------------------------------------------------------------
STEP 4    VARIABLE X4 ENTERED       R SQUARE = 0.99081100      C(P) =     4.02634991

                         DF          SUM OF SQUARES         MEAN SQUARE           F       PROB>F

        REGRESSION        4      490166624.64505400  122541656.16126351      323.48       0.0001
        ERROR            12        4545915.84394591     378826.32032883
        TOTAL            16      494712540.48899990

                             B VALUE          STD ERROR      PARTIAL REG SS           F       PROB>F

        INTERCEPT      2032.18806215
        X2                0.05607934         0.02035863    2874411.76901339        7.59       0.0175
        X3                1.08836904         0.15339754   19070221.63728522       50.34       0.0001
        X4               -5.00406579         5.08071295     367482.65878517        0.97       0.3441
        X5             -410.08295954       178.07810366    2008919.55996109        5.30       0.0400
------------------------------------------------------------------------------------------------------
STEP 5    VARIABLE X4 REMOVED       R SQUARE = 0.99006817      C(P) =     2.91769778

                         DF          SUM OF SQUARES         MEAN SQUARE           F       PROB>F

        REGRESSION        3      489799141.98626880  163266380.66208962      431.97       0.0001
        ERROR            13        4913398.50273108     377953.73097931
        TOTAL            16      494712540.48899990

                             B VALUE          STD ERROR      PARTIAL REG SS           F       PROB>F

        INTERCEPT      1523.38923568
        X2                0.05298733         0.02009194    2628687.59792946        6.96       0.0205
        X3                0.97848162         0.10515362   32726194.93174631       86.59       0.0001
        X5             -320.95082518       153.19222065    1658984.03530548        4.39       0.0563
------------------------------------------------------------------------------------------------------

        NO OTHER VARIABLES MET THE 0.5000 SIGNIFICANCE LEVEL FOR ENTRY INTO THE MODEL.
        ALL VARIABLES IN THE MODEL ARE SIGNIFICANT AT THE .1000 LEVEL.
```

The data analyst should view the sequential model-building algorithms not as a *black box*, which produces one final model, but rather as an exercise that allows the user to see several models perform. The sequential procedures can be partially ineffective with data sets involving collinearity among a large number of regressors. Bear in mind that there is no assurance that any specific sequential model-building strategy will result in the "best" variable subset with regard to any standard dealing with theoretical sensibility, prediction performance, etc. The final result is very much dependent on the sequential method chosen and the F_{IN} and F_{OUT} values.

Some commercial software packages allow for the development of several recommended models and, rather than producing a final single model, several models are suggested with final choice left for the analyst. One such method, available in SAS, is called the MAXR procedure.

MAXR *PROCEDURE*

The final product of the MAXR procedure is a set of k models, one each for values of the number of regressors equaling 1, 2, ..., k. This allows the analyst a choice based on any preconceived notion regarding what should be the size of the final model. It also affords a mechanism for forming a manageable number of candidate models from which *one* can be eventually chosen, based on criteria such as PRESS, C_p, a study of residuals, collinearity diagnostics, and other considerations discussed in Chapters 5, 6, and 8.

The methodology of MAXR begins in the same fashion as forward selection or stepwise regression. The initial variable entering the regression is that which produces the largest R^2 as a single regressor. Again, at each stage, the regressor that enters the regression is the one that produces the largest increase in R^2. However, no F-tests are conducted. As a result, no internal decisions are made that are based on probabilities. The procedure does, indeed, proceed until all k regressors are in the regression.

An important feature of MAXR, distinguishing it from other methods, is that, at each stage, it allows for replacement of a model regressor (one that already entered) with a regressor that has not entered *if* the replacement produces a larger R^2. The replaced regressor is relegated to the pool of "waiting" regressors, all of which will eventually enter. For example, once the two-variable model has been found, each model regressor is compared with each not in the model. For a specific model regressor, all switches are considered and that which produces the largest increase in R^2 is made. Then the other regressor is subject to switching in the same fashion. This continues until no *one-on-one* replacements will increase R^2. The MAXR procedure then declares that the surviving model is the "best" two-regressor model. Then another regressor is introduced, and the comparing and replacement process begins again. The output of the MAXR procedure is, then, k models, one for each model size.

The reader should be aware that while the MAXR procedure has several advantages over the other sequential model-building methods, there is still *no assurance* that (i) the "best" model of a particular size is being found in the sense of R^2, and (ii) the best model of a particular size is being found in the sense of PRESS, C_p, or any other criteria. The allowance for replacements requires that more regressions be performed than in the other methods. This is an advantage over forward selection, backward elimination, and stepwise regression.

Example 4.4 ***Teacher Effectiveness Data***

Consider the data of Example 3.9. In particular, let us focus on the data for females. The data are shown in Table 4.9 with three additional test scores (regressors) included. Table 4.10 provides a SAS computer printout that illustrates the use of the MAXR procedure for these data.

From the results of Table 4.10, it is apparent that seven models have been produced by the MAXR procedure. In terms of s^2 and the C_p statistic, the 3, 4, 5, and 6 variable models should be given serious consideration.

TABLE 4.9 **Seven-regressor problem with cooperating female teachers**

y	x_1	x_2	x_3	x_4	x_5	x_6	x_7
410	69	125	93	3.70	59.00	52.5	55.66
569	57	131	95	3.64	31.75	56.0	63.97
425	77	141	99	3.12	80.50	44.0	45.32
344	81	122	98	2.22	75.00	37.3	46.67
324	0	141	106	1.57	49.00	40.8	41.21
505	53	152	110	3.33	49.35	40.2	43.83
235	77	141	97	2.48	60.75	44.0	41.61
501	76	132	98	3.10	41.25	66.3	64.57
400	65	157	95	3.07	50.75	37.3	42.41
584	97	166	120	3.61	32.25	62.4	57.95
434	76	141	106	3.51	54.50	61.9	57.90

TABLE 4.10 **y = quantitative evaluation made by the cooperating female teachers. The regressor variables are scores on standardized tests data of Table 4.9**

```
                         MAXIMUM R-SQUARE IMPROVEMENT FOR DEPENDENT VARIABLE Y

STEP 1    VARIABLE X4 ENTERED              R SQUARE = 0.52849254       C(P) =    20.67324160

                          DF             SUM OF SQUARES         MEAN SQUARE            F       PROB>F

          REGRESSION       1             59033.09686988      59033.09686988        10.09       0.0113
          ERROR            9             52667.81222103       5851.97913567
          TOTAL           10            111700.90909091

                         B VALUE             STD ERROR       PARTIAL REG SS            F       PROB>F

          INTERCEPT     85.98139192                -
          X4           113.49939097        35.73527530       59033.09686988        10.09       0.0113
---------------------------------------------------------------------------------------------------

THE ABOVE MODEL IS THE BEST  1 VARIABLE MODEL FOUND.

STEP 2    VARIABLE X5 ENTERED              R SQUARE = 0.69856452       C(P) =    12.69154819

                          DF             SUM OF SQUARES         MEAN SQUARE            F       PROB>F

          REGRESSION       2             78030.29140126      39015.14570063         9.27       0.0083
          ERROR            8             33670.61768965       4208.82721121
          TOTAL           10            111700.90909091

                         B VALUE             STD ERROR       PARTIAL REG SS            F       PROB>F

          INTERCEPT    316.86421845
          X4            89.90290635        32.27695005       32653.03693881         7.76       0.0227
          X5            -3.00080180         1.41245014       18997.19453138         4.51       0.0664
---------------------------------------------------------------------------------------------------

THE ABOVE MODEL IS THE BEST  2 VARIABLE MODEL FOUND.
```

TABLE 4.10 Continued

TABLE 4.10 Continued

y = quantitative evaluation made by the cooperating female teachers. The regressor variables are scores on standardized tests data of Table 4.9

STEP 3 VARIABLE X3 ENTERED R SQUARE = 0.74385974 C(P) = 12.03312654

	DF	SUM OF SQUARES	MEAN SQUARE	F	PROB>F
REGRESSION	3	83089.80898533	27696.60299511	6.78	0.0177
ERROR	7	28611.10010558	4087.30001508		
TOTAL	10	111700.90909091			

	B VALUE	STD ERROR	PARTIAL REG SS	F	PROB>F
INTERCEPT	-25.30274116				
X3	2.97867091	2.67723336	5059.51758407	1.24	0.3026
X4	92.26016364	31.87803424	34235.91129228	8.38	0.0232
X5	-2.38781323	1.49698428	10399.26746866	2.54	0.1547

MAXIMUM R-SQUARE IMPROVEMENT FOR DEPENDENT VARIABLE Y

STEP 3 X5 REPLACED BY X7 R SQUARE = 0.75549973 C(P) = 11.34996371

	DF	SUM OF SQUARES	MEAN SQUARE	F	PROB>F
REGRESSION	3	84390.00691071	28130.00230357	7.21	0.0151
ERROR	7	27310.90218020	3901.55745431		
TOTAL	10	111700.90909091			

	B VALUE	STD ERROR	PARTIAL REG SS	F	PROB>F
INTERCEPT	-502.56080908				
X3	4.71162788	2.43387743	14621.18472366	3.75	0.0941
X4	67.82806701	37.96244464	12455.13802201	3.19	0.1171
X7	4.87295407	2.81402734	11699.46539403	3.00	0.1269

STEP 3 X4 REPLACED BY X6 R SQUARE = 0.77147027 C(P) = 10.41263724

	DF	SUM OF SQUARES	MEAN SQUARE	F	PROB>F
REGRESSION	3	86173.92999666	28724.64333222	7.88	0.0120
ERROR	7	25526.97909425	3646.71129918		
TOTAL	10	111700.90909091			

	B VALUE	STD ERROR	PARTIAL REG SS	F	PROB>F
INTERCEPT	-841.93605318				
X3	7.95484467	2.76909879	30094.55260327	8.25	0.0239
X6	-9.68825486	4.90292861	14239.06110797	3.90	0.0887
X7	18.47184282	5.66211839	38811.81382343	10.64	0.0138

STEP 3 X3 REPLACED BY X2 R SQUARE = 0.85858744 C(P) = 5.29964363

	DF	SUM OF SQUARES	MEAN SQUARE	F	PROB>F
REGRESSION	3	95904.99769732	31968.33256577	14.17	0.0023
ERROR	7	15795.91139359	2256.55877051		
TOTAL	10	111700.90909091			

	B VALUE	STD ERROR	PARTIAL REG SS	F	PROB>F
INTERCEPT	-928.47178015				
X2	5.59493494	1.33179411	39825.62030393	17.65	0.0040
X6	-10.17114014	3.77780796	16357.09481692	7.25	0.0310
X7	21.02568724	4.62067228	46723.64479966	20.71	0.0026

THE ABOVE MODEL IS THE BEST 3 VARIABLE MODEL FOUND.

STEP 4 VARIABLE X3 ENTERED R SQUARE = 0.89754868 C(P) = 5.01296987

	DF	SUM OF SQUARES	MEAN SQUARE	F	PROB>F
REGRESSION	4	100257.00337158	25064.25084290	13.14	0.0040
ERROR	6	11443.90571933	1907.31761989		
TOTAL	10	111700.90909091			

	B VALUE	STD ERROR	PARTIAL REG SS	F	PROB>F
INTERCEPT	-1097.58998317				
X2	4.18431918	1.53988254	14083.07337492	7.38	0.0348
X3	3.80447559	2.51861384	4352.00567427	2.28	0.1817
X6	-11.74149443	3.62543531	20005.50200232	10.49	0.0177
X7	22.18051897	4.31633188	50365.91962834	26.41	0.0021

THE ABOVE MODEL IS THE BEST 4 VARIABLE MODEL FOUND.

STEP 5 VARIABLE X4 ENTERED R SQUARE = 0.91761675 C(P) = 5.83515459

	DF	SUM OF SQUARES	MEAN SQUARE	F	PROB>F
REGRESSION	5	102498.62561744	20499.72512349	11.14	0.0097
ERROR	5	9202.28347347	1840.45669469		
TOTAL	10	111700.90909091			

	B VALUE	STD ERROR	PARTIAL REG SS	F	PROB>F
INTERCEPT	-1023.63744024				
X2	3.02278623	1.84277644	4952.16571695	2.69	0.1619
X3	4.72304558	2.61032859	6025.31167511	3.27	0.1302
X4	35.08652738	31.79230195	2241.62224585	1.22	0.3200
X6	-10.97292107	3.62877634	16828.66677854	9.14	0.0293
X7	19.27988350	4.98854453	27490.75748319	14.94	0.0118

THE ABOVE MODEL IS THE BEST 5 VARIABLE MODEL FOUND.

TABLE 4.10
Continued

y = quantitative evaluation made by the cooperating female teachers. The regressor variables are scores on standardized tests data of Table 4.9

```
STEP 6    VARIABLE X1 ENTERED          R SQUARE = 0.93063532      C(P) =     7.07108215

                        DF        SUM OF SQUARES        MEAN SQUARE            F      PROB>F

      REGRESSION         6       103952.81101081     17325.46850180         8.94      0.0263
      ERROR              4         7748.09808010      1937.02452003
      TOTAL             10       111700.90909091

                       B VALUE        STD ERROR      PARTIAL REG SS            F      PROB>F

      INTERCEPT    -994.26530554
      X1             -0.59549101      0.68727869     1454.18539337          0.75      0.4351
      X2              2.88252042      1.89742176     4470.45936466          2.31      0.2033
      X3              4.69125568      2.67818575     5943.35860687          3.07      0.1547

                 MAXIMUM R-SQUARE IMPROVEMENT FOR DEPENDENT VARIABLE Y

      X4             48.81896432     36.26263615     3510.69820474          1.81      0.2495
      X6            -10.33612004      3.79461427    14371.92480005          7.42      0.0528
      X7             18.49506142      5.19728494    24529.77803516         12.66      0.0236
--------------------------------------------------------------------------------------------
STEP 6    X4 REPLACED BY X5            R SQUARE = 0.93476955      C(P) =     6.82844017

                        DF        SUM OF SQUARES        MEAN SQUARE            F      PROB>F

      REGRESSION         6       104414.60809504     17402.43468251         9.55      0.0234
      ERROR              4         7286.30099587      1821.57524897
      TOTAL             10       111700.90909091

                       B VALUE        STD ERROR      PARTIAL REG SS            F      PROB>F

      INTERCEPT   -1769.14822710
      X1             -1.13638719      0.87697987     3058.58478560          1.68      0.2648
      X2              6.89576467      2.34303412    15778.10175911          8.66      0.0423
      X3              3.28685500      2.50677029     3131.69264037          1.72      0.2600
      X5              3.00873974      2.03740086     3972.49528898          2.18      0.2138
      X6            -13.01070240      3.67923999    22778.88012975         12.51      0.0241
      X7             28.46100015      5.94666474    41725.40132113         22.91      0.0087
--------------------------------------------------------------------------------------------
THE ABOVE MODEL IS THE BEST  6 VARIABLE MODEL FOUND.

STEP 7    VARIABLE X4 ENTERED          R SQUARE = 0.94888483      C(P) =     8.00000000

                        DF        SUM OF SQUARES        MEAN SQUARE            F      PROB>F

      REGRESSION         7       105991.29837860     15141.61405409         7.96      0.0581
      ERROR              3         5709.61071231      1903.20357077
      TOTAL             10       111700.90909091

                       B VALUE        STD ERROR      PARTIAL REG SS            F      PROB>F

      INTERCEPT   -1537.26966913
      X1             -1.20324215      0.89941840     3406.18145414          1.79      0.2733
      X2              5.27801363      2.98243349     5960.52788299          3.13      0.1749
      X3              4.10332535      2.71480467     4347.90429979          2.28      0.2279
      X4             34.92407344     38.37022264     1576.69028356          0.83      0.4298
      X5              2.30073619      2.22308175     2038.48736779          1.07      0.3768
      X6            -11.75842268      4.00454466    16408.81034829          8.62      0.0607
      X7             24.29261673      7.61059496    19390.83993474         10.19      0.0496
--------------------------------------------------------------------------------------------
THE ABOVE MODEL IS THE BEST  7 VARIABLE MODEL FOUND.
```

4.5

FURTHER COMMENTS AND ALL POSSIBLE REGRESSIONS

The sequential algorithms discussed here have practical use in cases where the analyst does not have access to information on all possible regressions or has an unusually large number of regressor variables, making all possible regressions prohibitive. Nevertheless, one weakness of these algorithms is what one may have considered a strength in the early stages of their development; apart from MAXR, they are *designed to give one answer without displaying results on a large number of subset models*. The reader must remember that there is absolutely no certainty that the truly best model will survive any of the procedures; indeed, there is no certainty that the best model will even be produced computationally. Thus, if at all possible, a user should resort to another mechanism for the final selection.

We assume that any scientist dealing in empirical model building should be armed with as much information as possible for decision-making. Full information is only attained from the use of all possible regressions. For algorithms that produce results from all possible regressions, the reader is referred to Furnival (1971), Furnival and Wilson (1974), or Schatzoff, *et al.* (1968). Certain statistical computer packages, particularly BMDP and SAS, should be explored. Generally, of course, the user should not expect to receive *all* pertinent necessary information on the regressions of all subset models. A reasonable philosophy is to let computer runs on all possible regressions generate sufficient information to reduce the number of candidate models to a relatively small set; then these candidate models can be further investigated with more complete information, the generation of information on residuals, possibly data splitting, outlier and leverage computation, etc. (See Chapters 5, 6 and 7.) An example of a vehicle that produces all possible regression information is a SAS (1987) program written in IML (interactive matrix language). This program generates s^2, R^2, C_p, PRESS, and the sum of the absolute PRESS residuals for all possible models.

Example 4.5 **Hospital Data**

Consider again the hospital data used for illustration in Example 4.3. Recall that the sequential algorithms highlighted variable combinations (x_2, x_3, x_5) and (x_2, x_3, x_4, x_5) for possible further investigation. Here we choose to produce R^2, C_p, s^2, PRESS, and the sum of the absolute PRESS residuals for all possible combinations. Table 4.11 reveals the results. (All combinations have an intercept.) From Table 4.11 we see that several additional models surface as possible candidates. These new models did not appear in the sequential algorithms due to the preoccupation in these procedures with "explained sum of squares." In addition, *the serious multicollinearity created a difficult search for the best prediction model when using sequential methods.* If one considers PRESS, at least five additional models should be given serious consideration. The variable combinations (x_3, x_5) and (x_1, x_5) give small values for the PRESS statistic and the sum of the absolute PRESS residuals. In the case of three variable models, (x_1, x_3, x_5) and (x_3, x_4, x_5) perform relatively well.

This study amplifies the point that one must look at many criteria for a choice of best model. In this case there is no agreement between C_p and the PRESS information in the model selection exercise. The C_p statistic seemed to favor certain three- and four-variable models while PRESS emphasized the need for simplicity, pointing in favor of certain two variable models. The data analyst must know his criteria well and consult with the subject matter scientist for an educated opinion about the relative importance of each variable. Concerning this study, approximately six models should be observed in detail, with studies of residuals, and other diagnostics. In addition, the multicollinearity should be diagnosed in more detail with view toward consideration of biased estimation as an alternative to least squares (see Chapter 8).

Before we abandon these hospital data, let us consider a comparison between the models (x_1, x_2, x_4, x_5) and (x_3, x_5), the latter providing the smallest PRESS and the former enjoying a relatively small C_p of 4.26. Table 4.12 provides a revealing observation of the PRESS and ordinary residuals for the two models.

TABLE 4.11 **Results for all possible regressions**

Number of Variables in Model	$\sum_{i=1}^{n} \lvert e_{i,-i} \rvert$	R^2	C_p	PRESS	s^2	Variables in Model
1	61324.3	0.3348	785.26	455,789,688	21,940,359	x_5
1	22889.9	0.8843	125.87	70,075,727	3,816,879	x_4
1	28031.2	0.8934	114.97	158,243,072	3,517,337	x_2
1	13665.5	0.9715	21.20	22,243,776	939,990	x_1
1	13431.2	0.9722	20.38	21,841,500	917,487	x_3
2	25251.5	0.9104	96.50	71,164,912	3,165,704	x_4, x_5
2	25960.3	0.9239	80.30	128,703,914	2,688,542	x_2, x_5
2	28165.9	0.9306	72.29	140,743,693	2,452,812	x_2, x_4
2	13957.1	0.9725	21.99	22,564,832	971,399	x_1, x_3
2	17327.9	0.9741	20.04	35,786,830	914,181	x_1, x_4
2	16825.2	0.9754	18.57	32,454,824	870,761	x_3, x_4
2	11074.2	0.9840	8.16	12,977,100	564,291	x_1, x_5
2	10914.9	0.9848	7.29	12,628,470	538,720	x_3, x_5
2	12872.9	0.9861	5.66	18,036,834	490,585	x_1, x_2
2	12742.1	0.9867	4.94	17,853,441	469,456	x_2, x_3
3	26241.6	0.9523	48.28	107,102,405	1,816,626	x_2, x_4, x_5
3	18045.5	0.9785	16.80	34,400,018	818,290	x_1, x_3, x_4
3	12296.1	0.9846	9.41	17,828,243	583,845	x_1, x_4, x_5
3	11801.2	0.9850	9.00	16,275,120	570,794	x_3, x_4, x_5
3	11103.7	0.9850	8.97	13,036,635	569,830	x_1, x_3, x_5
3	16227.1	0.9861	7.66	32,814,300	528,257	x_1, x_2, x_4
3	15780.3	0.9868	6.90	30,139,304	504,218	x_2, x_3, x_4
3	14243.3	0.9873	6.21	22,794,229	482,286	x_1, x_2, x_3
3	12163.0	0.9894	3.71	18,051,730	403,215	x_1, x_2, x_5
3	12019.0	0.9901	2.92	17,846,717	377,954	x_2, x_3, x_5
4	12773.4	0.9851	10.92	18,621,398	615,741	x_1, x_3, x_4, x_5
4	17951.9	0.9879	7.54	37,719,883	499,469	x_1, x_2, x_3, x_4
4	13364.2	0.9905	4.34	22,464,723	389,793	x_1, x_2, x_3, x_5
4	14992.3	0.9906	4.26	30,255,902	387,001	x_1, x_2, x_4, x_5
4	14606.1	0.9908	4.03	28,629,419	378,826	x_2, x_3, x_4, x_5
5	16025.0	0.9908	6.00	32,195,222	412,277	x_1, x_2, x_3, x_4, x_5

TABLE 4.12 **Ordinary and PRESS residuals for models (x_1, x_2, x_4, x_5) and (x_3, x_5) for hospital data**

	(x_1, x_2, x_4, x_5) Ordinary Residuals	(x_1, x_2, x_4, x_5) PRESS Residuals	(x_3, x_5) Ordinary Residuals	(x_3, x_5) PRESS Residuals
1	−190.742	−221.344	−239.187	−270.425
2	19.139	24.842	134.214	171.296
3	−83.290	−102.411	−44.389	−50.723
4	389.800	468.426	388.314	461.216
5	95.468	104.522	100.301	109.597
6	−558.012	−685.849	−213.242	−237.293
7	222.787	243.515	173.391	189.307
8	203.618	222.898	288.255	313.497
9	−331.618	−373.978	−538.707	−587.985
10	−287.694	−1028.262	−320.967	−344.822
11	354.867	386.643	519.432	558.552
12	−703.667	−863.476	−1085.508	−1200.408
13	−119.284	−136.412	−403.020	−430.060
14	1582.565	1870.565	2004.738	2210.715
15	−561.584	−2435.280	−469.070	−1448.866
16	424.470	2245.986	−802.734	−1212.467
17	−456.824	−3577.870	508.177	1117.665

For the case of installations 15, 16, and 17, the simpler model gives a prediction error (PRESS residual) considerably smaller than that of model (x_1, x_2, x_4, x_5). In fact, these installations alone nearly account for the difference in the $\sum_{i=1}^{n} |e_{i,-i}|$ for (x_1, x_2, x_4, x_5) and (x_3, x_5), the difference being between 14,992.3 and 10,914.9. Thus it would seem noteworthy that model (x_3, x_5) is more effective at estimating required man-hours for the *large installations*, e.g., those representing data points 15, 16, and 17. Since (x_1, x_2, x_4, x_5) *fits* the data relatively well and produced a smaller residual mean square than (x_3, x_5) ($s^2 = 387{,}001$ as opposed to $s^2 = 538{,}720$ for (x_3, x_5)), this tendency for larger prediction errors was not reflected in the C_p comparison. Potentially large prediction errors in certain areas of the regressor space may have a better chance of being exposed in the PRESS information than in a mere comparison of C_p values. Of course, if extensive extrapolation is the eventual task of the model, neither C_p nor PRESS may be sensitive to poor performance. Example 4.6 will illustrate a more complete search for the most effective prediction model.

The reader should not view this illustration as a true case study. Before any final selection of the "best" model is made, diagnostic information given in Chapters 5, 6, 7, and 8 should be explored.

Example 4.6 **Fitness Data**

In exercise physiology, an objective measure of aerobic fitness is the oxygen consumption in volume per unit body weight per unit time by the individual. To determine if it is feasible to predict this fitness measure, an experiment was conducted in which 31 individuals were tested. The following factors were studied:

x_1: Age in years
x_2: Weight in kilograms
x_3: Time to run $1\frac{1}{2}$ miles
x_4: Resting pulse rate
x_5: Pulse rate at end of run
x_6: Maximum pulse rate during run

The response is oxygen consumption in milliliters (ml) per kilogram (kg) body weight per minute. The data appear in Table 4.13.

All possible regressions were conducted with s^2, R^2, C_p, PRESS, and $\sum_{i=1}^{n} |e_{i,-i}|$ computed for each of the 64 candidate models. We will not display the results for all models. Table 4.14 provides the results for eight models. These are the subsets that are simultaneously superior in terms of PRESS, C_p, $\sum_{i=1}^{n} |e_{i,-i}|$, and s^2. (The ranking of the eight models using the various criteria are not the same.)

Notice that while the subset $(x_1, x_2, x_3, x_5, x_6)$ performs very well with regard to all criteria, there is no obvious clear-cut choice. It was felt that a data splitting exercise might produce additional information. Eleven individuals, selected randomly, were chosen to comprise the validation sample. Each of the 8 models, fit to the data of the remaining 20 individuals, were used to predict the response for the individuals in the validation sample. Table 4.15 contains the 11 *prediction residuals* $(y - \hat{y})$ for each of the 8 models, as well as the sum of squares of the prediction errors and sum of the absolute prediction errors. The data splitting effort reveals some very pleasing results. There is, again, strong support for the model $(x_1, x_2, x_3, x_5, x_6)$, which enjoys the smallest sum of absolute prediction errors and sum of squares of prediction errors. In addition, the ranking of the remaining models is very similar to the ranking according to the PRESS information in Table 4.14.

The information produced by all possible regressions, illustrated by Examples 4.5 and 4.6, can be helpful when there is a need to select the model that performs best as a prediction equation. Statistics such as s^2, C_p, PRESS, and the sum of the absolute PRESS residuals can be invaluable in that regard. However, the final choice of a model should not be made without considering such concepts as analysis of residuals, collinearity diagnostics, transformations, detection of influential observations, and other topics that will be covered in subsequent chapters.

TABLE 4.13 **Oxygen fitness data**

Individual	y	x_1	x_2	x_3	x_4	x_5	x_6
1	44.609	44	89.47	11.37	62	178	182
2	45.313	40	75.07	10.07	62	185	185
3	54.297	44	85.84	8.65	45	156	168
4	59.571	42	68.15	8.17	40	166	172
5	49.874	38	89.02	9.22	55	178	180
6	44.811	47	77.45	11.63	58	176	176
7	45.681	40	75.98	11.95	70	176	180
8	49.091	43	81.19	10.85	64	162	170
9	39.442	44	81.42	13.08	63	174	176
10	60.055	38	81.87	8.63	48	170	186
11	50.541	44	73.03	10.13	45	168	168
12	37.388	45	87.66	14.03	56	186	192
13	44.754	45	66.45	11.12	51	176	176
14	47.273	47	79.15	10.60	47	162	164
15	51.855	54	83.12	10.33	50	166	170
16	49.156	49	81.42	8.95	44	180	185
17	40.836	51	69.63	10.95	57	168	172
18	46.672	51	77.91	10.00	48	162	168
19	46.774	48	91.63	10.25	48	162	164
20	50.388	49	73.37	10.08	76	168	168
21	39.407	57	73.37	12.63	58	174	176
22	46.080	54	79.38	11.17	62	156	165
23	45.441	52	76.32	9.63	48	164	166
24	54.625	50	70.87	8.92	48	146	155
25	45.118	51	67.25	11.08	48	172	172
26	39.203	54	91.63	12.88	44	168	172
27	45.790	51	73.71	10.47	59	186	188
28	50.545	57	59.08	9.93	49	148	155
29	48.673	49	76.32	9.40	56	186	188
30	47.920	48	61.24	11.50	52	170	176
31	47.467	52	82.78	10.50	53	170	172

SAS Institute Inc., *SAS User's Guide: Statistics, 1982 Edition.* (Cary, North Carolina: SAS Institute Inc., 1982), p. 106.

TABLE 4.14 **Top eight subset models in terms of PRESS information, C_p, and s^2 for oxygen data**

Subset	PRESS	$\sum_{i=1}^{n} \lvert e_{i,-i} \rvert$	s^2	C_p	R^2
$(x_1, x_2, x_3, x_5, x_6)$	181.633	54.034	5.1763	5.106	0.848002
(x_1, x_3, x_5, x_6)	188.599	59.966	5.3435	4.880	0.836818
$(x_1, x_2, x_3, x_4, x_5, x_6)$	192.788	56.314	5.3682	7.000	0.848672
$(x_1, x_3, x_4, x_5, x_6)$	202.402	62.655	5.5499	6.846	0.837031
(x_1, x_3, x_5)	205.125	61.233	5.9567	6.960	0.811094
(x_1, x_2, x_3, x_5)	212.272	60.162	6.0090	8.104	0.816493
(x_3, x_5, x_6)	212.862	64.000	5.9916	7.135	0.809988
(x_2, x_3, x_5, x_6)	213.158	62.764	6.0301	8.206	0.815849

TABLE 4.15 **Prediction residuals and prediction performance criteria for the validation sample for the eight models of Table 4.14**

Subject	$(x_1, x_2, x_3, x_5, x_6)$	(x_1, x_3, x_5, x_6)	$(x_1, x_2, x_3, x_4, x_5, x_6)$	$(x_1, x_3, x_4, x_5, x_6)$
3	−2.8309	−3.2949	−2.8111	−3.3177
7	1.0714	1.5911	1.9597	2.4048
8	−0.3667	−0.3451	0.5895	0.4846
12	0.1801	0.4063	−0.5217	−0.1817
14	−0.2554	−0.4486	−0.5071	−0.6831
17	−5.8211	−5.3843	−5.4315	−5.0094
20	2.5691	2.5806	3.9505	3.7775
22	−1.6429	−1.7250	−0.5946	−0.8246
23	−4.0239	−4.2648	−3.9391	−4.2122
29	−0.1814	−0.1570	0.1358	0.1198
30	1.3265	2.4522	1.0409	2.3021
$\sum (y-\hat{y})^2$	70.5627	76.7253	74.6991	80.6239
$\sum \lvert y-\hat{y}\rvert$	20.2695	22.6498	21.4815	23.3173

Subject	(x_1, x_3, x_5)	(x_1, x_2, x_3, x_5)	(x_3, x_5, x_6)	(x_2, x_3, x_5, x_6)
3	−2.0577	−1.7833	−2.8383	−2.4712
7	2.2468	2.0451	2.8103	2.5689
8	0.6028	0.6421	0.6718	0.7564
12	2.9924	3.0236	0.9789	0.8778
14	−1.3629	−1.3226	0.1031	0.2913
17	−5.1459	−5.3317	−5.6171	−5.9433
20	1.2262	1.1509	2.5070	2.4918
22	−0.3594	−0.2516	−1.9289	−1.8919
23	−5.2498	−5.1915	−4.7547	−4.6354
29	−0.1936	−0.2066	−1.2496	−1.3737
30	3.5142	3.0586	2.6425	1.8792
$\sum (y-\hat{y})^2$	88.5185	84.8305	90.0832	86.1508
$\sum \lvert y-\hat{y}\rvert$	24.9518	24.0075	26.1021	25.1809

EXERCISES FOR CHAPTER 4

4.1 Consider the squid data of Example 3.2. There are 32 possible models involving all subsets.

(a) Select a model using stepwise regression with a significance level of 0.05 for entry and removal.

(b) Select a model using stepwise regression with a significance level of 0.25 for entry and .05 for removal.

(c) Select a model using backward elimination with a significance level of 0.10.

(d) Compute R^2, PRESS, $\sum_{i=1}^{n} |y_i - \hat{y}_{i,-i}|$, s^2, and C_p for all possible regressions. On the basis of this information, reduce your pool of models from 32 to a small subset for further consideration.

4.2 For the forestry data of Example 3.5, compute all possible regressions and compute PRESS, s^2, C_p, and $\sum_{i=1}^{n} |y_i - \hat{y}_{i,-i}|$. Comment on your choice of a small subset of models whose performance seems to be superior to the rest. Does one model stand out?

4.3 Refer to Exercise 3.11. Compute PRESS and C_p for all eight subset regressions.

4.4 In a chemical engineering experiment dealing with heat transfer in a shallow fluidized bed, data are collected on the following four regressor variables.

x_1: Fluidizing gas flow rate, lb/hr
x_2: Supernatant gas flow rate, lb/hr
x_3: Supernatant gas inlet nozzle opening, mm.
x_4: Supernatant gas inlet temperature, °F

The responses measured are

y_1: Heat transfer coefficient
y_2: Thermal efficiency

The data are as follows:

Observation	y_1	y_2	x_1	x_2	x_3	x_4
1	41.852	38.75	69.69	170.83	45	219.74
2	155.329	51.87	113.46	230.06	25	181.22
3	99.628	53.79	113.54	228.19	65	179.06
4	49.409	53.84	118.75	117.73	65	281.30
5	72.958	49.17	119.72	117.69	25	282.20
6	107.702	47.61	168.38	173.46	45	216.14
7	97.239	64.19	169.85	169.85	45	223.88
8	105.856	52.73	169.85	170.86	45	222.80
9	99.348	51.00	170.89	173.92	80	218.84
10	111.907	47.37	171.31	173.34	25	218.12
11	100.008	43.18	171.43	171.43	45	219.20
12	175.380	71.23	171.59	263.49	45	168.62
13	117.800	49.30	171.63	171.63	45	217.58
14	217.409	50.87	171.93	170.91	10	219.92
15	41.725	54.44	173.92	71.73	45	296.60
16	151.139	47.93	221.44	217.39	65	189.14
17	220.630	42.91	222.74	221.73	25	186.08
18	131.666	66.60	228.90	114.40	25	285.80
19	80.537	64.94	231.19	113.52	65	286.34
20	152.966	43.18	236.84	167.77	45	221.72

Data analyzed for the Department of Chemical Engineering by the Statistical Consulting Center, Virginia Polytechnic Institute and State University, Blacksburg, Virginia, October, 1983.

Consider the model for predicting the heat transfer coefficient response

$$y_{1i} = \beta_0 + \sum_{j=1}^{4} \beta_j x_{ji} + \sum_{j=1}^{4} \beta_{jj} x_{ji}^2 + \sum\sum_{j<l} \beta_{jl} x_{ji} x_{li} + \varepsilon_i \qquad (i = 1, 2, \ldots, 20)$$

(a) Compute PRESS and $\sum_{i=1}^{n} |y_i - \hat{y}_{i,-i}|$ for the least squares regression fit to the above model.

(b) Fit a second order model with x_4 completely eliminated, i.e., deleting all terms involving x_4. Compute the prediction criteria for the reduced model. Comment on the appropriateness of x_4 for prediction of the heat transfer coefficient.

(c) Repeat parts (a) and (b) for thermal efficiency.

4.5 Consider the punting data in Exercise 2.11. Consider the *multiple* regression model with both left-leg strength and right-leg strength as regressor variables. Compare, in terms of performance, the following competing models for predicting punting distance:

(a) Single regressor: right-leg strength

(b) Single regressor: left-leg strength

(c) Two regressor variables: right-leg strength and left-leg strength.

In your comparison, use the C_p statistic, PRESS, and the sum of the absolute residuals.

4.6 In the Department of Fisheries and Wildlife at Virginia Tech., an experiment was conducted to study the effect of stream characteristics on fish biomass. The regressor variables are as follows:

x_1: Average depth (of 50 cells)

x_2: Area of instream cover (i.e., undercut banks, logs, boulders, etc.)

x_3: Percent canopy cover (average of 12)

x_4: Area $\geqslant$ 25 cm in depth

The response is y, the fish biomass. The data are as follows:

Observation	y	x_1	x_2	x_3	x_4
1	100	14.3	15.0	12.2	48.0
2	388	19.1	29.4	26.0	152.2
3	755	54.6	58.0	24.2	469.7
4	1288	28.8	42.6	26.1	485.9
5	230	16.1	15.9	31.6	87.6
6	0	10.0	56.4	23.3	6.9
7	551	28.5	95.1	13.0	192.9
8	345	13.8	60.6	7.5	105.8
9	0	10.7	35.2	40.3	0.0
10	348	25.9	52.0	40.3	116.6

Data analyzed for the Department of Fisheries and Wildlife by the Statistical Consulting Center, Virginia Polytechnic Institute and State University, Blacksburg, Virginia, October, 1983.

(a) Compute s^2, C_p, PRESS, and the sum of the absolute PRESS residuals for the model involving all four variables with response y.

(b) Compute s^2, C_p, PRESS, and the sum of the absolute PRESS residuals for the model x_1, x_2, x_4 with response y.

(c) Compare the appropriateness of the models in parts (a) and (b) for predicting fish biomass.

4.7 Show that for $k = 1$, the HAT diagonal is given by

$$h_{ii} = \frac{1}{n} + \frac{(x_i - \bar{x})^2}{\sum_{i=1}^{n} (x_i - \bar{x})^2}$$

4.8 Consider the data of Exercise 3.11. Compute all eight regression models using ln transformations on each regressor.

(a) Compute PRESS, C_p, s^2, and the sum of the absolute PRESS residuals. Does any single model stand out as being superior to the rest? Are you able to eliminate any models from consideration?

(b) Fit all eight models using the first *twenty firms*. Use the resulting eight regressions to predict compensation for the last thirteen firms. Use the data splitting information to choose the best model.

4.9 Show that in a multiple linear regression a ranking of competing models using adjusted R^2 is equivalent to a ranking of competing models with the use of the residual mean square, s^2.

4.10 Consider the material on model underspecification in Section 4.3. Using the following data set, the experimenter fits a simple linear regression model

$$y_i = \beta_0^* + \beta_1(x_{1i} - \bar{x}_1) + \varepsilon_i \qquad \text{(Model A)}$$

y	x_1	x_2
—	10	1
—	10	2
—	20	1
—	20	2

Suppose, however, that the *true* model can be written

$$y_i = \beta_0^* + \beta_1(x_{1i} - \bar{x}_1) + \beta_2(x_{2i} - \bar{x}_2) + \varepsilon_i \qquad \text{(Model B)}$$

(a) Give $E(b_1)$ where b_1 is the least squares slope of Model A, the underfitted model. Comment. Can you give an intuitive explanation? Give $E(b_0^*)$, where b_0^* is the estimated intercept in Model A.

(b) Do your results in part (a) imply that $\hat{y}_0 = b_0^* + b_1(x_{1,0} - \bar{x}_1)$ (a predicted y at an arbitrary $x_{1,0}$) is unbiased? If so, explain why. If not, demonstrate why not

4.11 Consider, again, the data of Exercise 4.10. Suppose we scale the regressor values as well as center them. In other words, we define

$$z_1 = \frac{x_1 - \bar{x}_1}{5}$$

$$z_2 = \frac{x_2 - \bar{x}_2}{0.5}$$

As a result, the data become

z_1	z_2
−1	−1
−1	1
1	−1
1	1

(a) Now compute the HAT diagonals at the four data points using the z's.
(b) Compute the HAT diagonals at the four data points using centered x's.
(c) Compute the HAT diagonals at the four data points using the raw x's. You may wish to use a software package for this.

(*Comment*: The HAT diagonals are measures of *standardized* squared distance.)

(d) Discuss the apparent effect of centering and scaling on the HAT diagonals.

4.12 Show that $R^2_{\text{Pred}} < R^2$.

4.13 In Chapter 3 we learned that the HAT diagonals obey the inequality

$$\frac{1}{n} \leqslant h_{ii} \leqslant 1$$

for the multiple linear regression model with an intercept. Construct a data set and a corresponding model in which the HAT diagonals are all 1.0.

4.14 Revise the inequality on the h_{ii} for the case of the multiple linear regression model forced through the origin.

4.15 Consider the data set of Exercise 3.22. Compute C_p for the following subset models and comment on the results.

(x_1, x_2) $\quad$ $(x_1, x_2, x_1 \cdot x_2)$
(x_1, x_1^2, x_2) $\quad$ $(x_1, x_1^2, x_2, x_2^2, x_1 \cdot x_2)$
(x_1, x_1^2, x_2, x_2^2)

4.16 Consider the grain radius data of Exercise 3.19.

(a) Compute the HAT diagonals for the multiple regression model discussed in Exercise 3.19(a). Are you surprised?
(b) Consider the computation of C_p values for the following subset models.

(x_1) $\quad$ (x_1, x_2)
(x_1, x_2, x_3) $\quad$ $(x_1, x_2, x_3, x_1 \cdot x_2)$
$(x_1, x_2, x_3, x_1 \cdot x_2, x_1 \cdot x_3)$ $\quad$ $(x_1, x_2, x_3, x_1 \cdot x_2, x_1 \cdot x_3, x_2 \cdot x_3)$
$(x_1, x_2, x_3, x_1 \cdot x_2, x_1 \cdot x_3, x_2 \cdot x_3, x_1 \cdot x_2 \cdot x_3)$

First you should consider the seven-variable model as the "full model." Comment on the difficulty that you have in extracting C_p values for this problem. Do you have any ideas for the alleviation of this difficulty?

4.17 In a study involving the Departments of Statistics and Architecture at Virginia Polytechnic Institute and State University, an experiment was conducted to study the effects of atrium size and configuration on amount of light at points within the atrium. It was of interest to develop a model to predict illumination as a function of atrium floor area, distance from the top of the atrium, and reflectivity of the wall surface. In addition, it was of interest to include the height of the building in the study. The following data were collected:

Illumination (y)	Distance (Stories) (x_1)	Area of Atrium sq ft/100 (x_2)	Reflectivity (%) (x_3)	Number of Stories in Building (x_4)
150	1	1600	20	5
266	3	1600	20	5
773	5	1600	20	5
23	2	1600	20	10
77	5	1600	20	10
779	10	1600	20	10
14	3	1600	20	15
19	6	1600	20	15
38	9	1600	20	15
163	12	1600	20	15
728	15	1600	20	15
411	1	6400	20	5
544	3	6400	20	5
824	5	6400	20	5
192	2	6400	20	10
254	5	6400	20	10
999	10	6400	20	10
101	3	6400	20	15
137	6	6400	20	15
256	9	6400	20	15
495	12	6400	20	15
868	15	6400	20	15
516	1	14400	20	5
722	3	14400	20	5
992	5	14400	20	5
307	2	14400	20	10
489	5	14400	20	10
984	10	14400	20	10
168	3	14400	20	15
242	6	14400	20	15
377	9	14400	20	15
670	12	14400	20	15
1124	15	14400	20	15
629	1	25600	20	5
856	3	25600	20	5
1061	5	25600	20	5
478	2	25600	20	10
686	5	25600	20	10
1139	10	25600	20	10
329	3	25600	20	15
366	6	25600	20	15
601	9	25600	20	15

TABLE Continued

Illumination (y)	Distance (Stories) (x_1)	Area of Atrium sq ft/100 (x_2)	Reflectivity (%) (x_3)	Number of Stories in Building (x_4)
766	12	25600	20	15
1135	15	25600	20	15
155	1	1600	60	5
315	3	1600	60	5
684	5	1600	60	5
47	2	1600	60	10
97	5	1600	60	10
708	10	1600	60	10
16	3	1600	60	15
55	6	1600	60	15
76	9	1600	60	15
205	16	1600	60	15
678	15	1600	60	15
421	1	6400	60	5
590	3	6400	60	5
785	5	6400	60	5
211	2	6400	60	10
315	5	6400	60	10
769	10	6400	60	10
119	3	6400	60	15
151	6	6500	60	15
250	9	6400	60	15
438	12	6400	60	15
797	15	6400	60	15
490	1	14400	60	5
682	3	14400	60	5
884	5	14400	60	5
323	2	14400	60	10
471	5	14400	60	10
837	10	14400	60	10
200	3	14400	60	15
252	6	14400	60	15
355	9	14400	60	15
546	12	14400	60	15
841	15	14400	60	15
583	1	25600	60	5
743	3	25600	60	5
912	5	25600	60	5
410	2	25600	60	10
558	5	25600	60	10
870	10	25600	60	10
288	3	25600	60	15
343	6	25600	60	15
458	9	25600	60	15
627	12	25600	60	15
810	15	25600	60	15

There was a light measuring device on each story of the building. As one can see from the data, the illumination was read on various floors of the building.

(a) Discuss the possible multiple linear regression model formulations. How are you going to handle "height of building"?

(b) Choose, from all possible subsets, the model that best predicts illumination. Use C_p, PRESS, and residual mean square.

(c) Go through the process again, but replace x_2 by $x_2^{1/2}$ and x_1 by $\ln x_1$. Comment. This exercise will be the subject of further exercises in Chapters 5, 6, and 7.

REFERENCES FOR CHAPTER 4

1. Aitkin, M.A. 1974. Simultaneous inference and the choice of variable subsets in multiple linear regression. *Technometrics* 16: 221–227.
2. Allen, D.M. 1971. The prediction sum of squares as a criterion for selecting predictor variables. Technical Report No. 23, Department of Statistics, University of Kentucky, Lexington, Kentucky.
3. Allen, D.M. 1971. Mean square error of prediction as a criterion for selecting variables. *Technometrics* 13: 469–475.
4. Allen, D.M. 1974. The relationship between variable selection and data augmentation and a method for prediction. *Technometrics* 16: 125–127.
5. Anderson, R.L., D.M. Allen, and F.B. Cady. 1972. Selection of predictor variables in linear multiple regression. *Statistical Papers in Honor of George Snedecor*, ed. T.A. Bancroft. Ames, Iowa: Iowa State University Press.
6. Beale, E.M.L. 1970. Note on procedures for variable selection in multiple regression. *Technometrics*, 12: 909–914.
7. Beale, E.M.L., and P.C. Hutchinson. 1974. Note on constrained optimum regression. *Applied Statistics* 23: 208–210.
8. Beale, E.M.L., M.G. Kendall, and D.W. Mann. 1967. The discarding of variables in multivariate analysis. *Biometrika* 54: 357–366.
9. Bendel, R.B., and A.A. Afifi. 1977. Comparison of stopping rules in forward "stepwise" regression. *Journal of the American Statistical Association* 72: 46–53.
10. Berk, K.N. 1978. Comparing subset regression procedures. *Technometrics* 20: 1–6.
11. Cochran, W.G. 1938. The omission or addition of an independent variate in multiple linear regression. *Journal of the Royal Statistical Society, Series B* 5: 171–176.
12. Cox, D.R. 1960. Regression analysis when there is prior information about supplementary variables. *Journal of the Royal Statistical Society, Series B* 22: 172–176.
13. Forsythe, A.B., L. Engelman, R. Jennrich, and P.R.A. May. 1973. A stopping rule for variable selection in multiple regression. *Journal of the American Statistical Association* 68: 75–77.
14. Furnival, G.M. 1971. All possible regressions with less computation. *Technometrics* 13: 403–408.

15. Furnival, G.M., and W. Wilson. 1974. Regression by leaps and bounds. *Technometrics* 16: 499–511.

16. Geisser, S. 1975. The predictive sample reuse method with applications. *Journal of the American Statistical Association* 70: 320–328.

17. Goldberger, A.S. 1961. Stepwise least-squares: residual analysis and specification. *Journal of the American Statistical Association* 56: 998–1000.

18. Gorman, J.W., and R.J. Toman. 1966. Selection of variables for fitting equations to data. *Technometrics* 8: 27–51.

19. Gunst, R.F., and R.L. Mason. 1979. Some considerations in the evaluation of alternate prediction equations. *Technometrics* 21: 55–63.

20. Helms, R.W. 1974. The average estimated variance criterion for the selection-of-variables problem in general linear models. *Technometrics* 16: 261–273.

21. Henderson, H.V., and P.F. Velleman. 1981. Building multiple regression models interactively. *Biometrics* 37: 391–411.

22. Hocking, R.R. 1972. Criteria for selection of a subset regression: which one should be used? *Technometrics* 14: 967–970.

23. Hocking, R.R. 1976. The analysis and selection of variables in linear regression. *Biometrics* 32: 1–51.

24. Hocking, R.R. 1974. Misspecification in regression. *American Statistician* 28: 39–40.

25. Hocking, R.R., and R.N. Leslie. 1967. Selection of the best subset in regression analysis. *Technometrics* 9: 531–540.

26. Kennard, R.W. 1971. A note on the C_p statistic. *Technometrics* 13: 899–900, corrections 15: 657 (1973).

27. Kennedy, W.J., and T.A. Bancroft. 1971. Model building for prediction in regression based upon repeated significance tests. *Annals of Mathematical Statistics* 42: 1273–1284.

28. LaMotte, L.R. 1972. The SELECT routines: a program for identifying best subset regressions. *Applied Statistics* 21: 92–93.

29. Lindley, D.V. 1968. The choice of variables in multiple regression. *Journal of the Royal Statistical Society, Series B* 30: 31–53, discussion 54–66.

30. Mallows, C.L. 1964. Choosing variables in a linear regression: A graphical aid. Paper presented at the Central Regional Meeting of the Institute of Mathematical Statistics, Manhattan, Kansas.

31. Mallows, C.L. 1973. Some Comments on C_p. *Technometrics* 15: 661–675.

32. Mantel, N. 1970. Why stepdown procedures in variable selection. *Technometrics* 12: 621–625.

33. Montgomery, D.C., and E.A. Peck. 1982. *Introduction to Linear Regression Analysis*. New York: John Wiley & Sons.

34. Park C.N., and A.L. Dudycha. 1974. A cross-validation approach to sample size determination for regression models. *Journal of the American Statistical Association* 69: 214–218.

35. Pope, P.T., and J.T. Webster. 1972. The use of an *F*-statistic in stepwise regression procedures. *Technometrics* 14: 327–340.

36. Rao, P. 1971. Some notes on misspecification in multiple regression. *The American Statistician* 25: 37–39.

37. Rencher, A.C., and F.C. Pun. 1980. Inflation of R^2 in best subset regression. *Technometrics* 22: 49–53.

38. SAS Institute Inc. 1987. *SAS Views: SAS Principles of Regression Analysis, 1987 Edition.* Cary, North Carolina: SAS Institute Inc.

39. Schatzoff, M., R. Tsao, and S. Fienberg. 1968. The efficient calculation of all possible regressions. *Technometrics* 10: 769–779.

40. Snee, R.D. 1977. Validation of regression models: Methods and Examples. *Technometrics* 19: 415–428.

41. Stone, M. 1974. Cross-validatory choice and assessment of statistical predictions. *Journal of the Royal Statistical Society, Series B* 36: 111–147.

CHAPTER 5

ANALYSIS OF RESIDUALS

In several locations in the text, e.g., Chapters 2, 3, and 4, the notion of residuals was presented and incorporated into the development. Some effort was made to present residuals as major components in the computer output of a regression analysis. The residuals are often used to detect and assess the degree of discrepancy between the model assumed (includes assumptions made) and the data observed. Computations made on residuals have become standard in many commercial regression computer packages. Thus it is important that the user of regression analysis know the tools that are available for analysis of residuals and recognize what type of information can be recovered from them.

In Section 2.12 we emphasized the use of residuals in the case of simple linear regression. Emphasis was placed on detection of model misspecification and violation of the assumption of normal errors. The reader should review that material before proceeding to the details given in this chapter. Model misspecification will resurface in the context of multiple linear regression in this chapter. Diagnostic plots that highlight violation of the homogeneous variance assumption will also be discussed and demonstrated. In Chapter 2 a standardized form of a residual called a studentized residual was discussed. A more thorough discussion of the use of studentized residuals will be given here. Diagnostics that are used for detection of outliers make use of studentized residuals. These diagnostics are discussed and illustrated. Statisticians have been very attentive in recent years to the issue of transforming experimental data prior to computing the regression. Methods that diagnose the need for transformations can be very helpful in determining what the true "metric" should be for the analysis of the data. In this chapter we discuss some diagnostic type plots that are used for this purpose. These plots can also illustrate graphically the role of each variable in the regression. In addition, these special residual plots may highlight the existence of high influence observations. This chapter is tied very closely with Chapter 6, which is solely dedicated to the diagnosis of high influence observations.

5.1

INFORMATION RETRIEVED FROM RESIDUALS

The reader can better understand the value of a set of residuals after reviewing their properties *under ideal conditions.* Recall from Chapter 3 that, under the assumptions given in Section 3.1 on the ε_i, that the ith residual, e_i, has mean zero with variance given by

$$\boxed{\operatorname{Var}(e_i) = \sigma^2(1 - h_{ii})} \tag{5.1}$$

where h_{ii} is the ith HAT diagonal defined in Section 3.9. In addition, the ith and jth residuals ($i \neq j$) have covariance given by

$$\operatorname{Cov}(e_i, e_j) = -h_{ij} \cdot \sigma^2 \tag{5.2}$$

where h_{ij} is the (i, j) element of the HAT matrix $\mathbf{H} = \mathbf{X}(\mathbf{X}'\mathbf{X})^{-1}\mathbf{X}'$.

These important results are easily verified by considering results in Section 3.9. From Eq. (3.38), the vector of residuals $\mathbf{e}$ can be written

$$\mathbf{e} = \mathbf{y} - \mathbf{X}\mathbf{b} = [\mathbf{I} - \mathbf{H}]\mathbf{y}$$

Now, the variance-covariance matrix of the residuals can be determined by using the rule given in Exercise 3.1. Namely,

$$\operatorname{Var} \mathbf{e} = (\mathbf{I} - \mathbf{H})(\sigma^2 \mathbf{I}_n)(\mathbf{I} - \mathbf{H}) = \sigma^2(\mathbf{I} - \mathbf{H})^2$$

and since $(\mathbf{I} - \mathbf{H})$ is idempotent, $(\mathbf{I} - \mathbf{H})^2 = \mathbf{I} - \mathbf{H}$, and

$$\boxed{\operatorname{Var} \mathbf{e} = \sigma^2(\mathbf{I} - \mathbf{H})}$$

Thus the results in (5.1) and (5.2) follow. It is interesting to note that (5.1) is a generalization of the variance of a residual given for simple linear regression in

Eq. (2.32). For the *special case of one regressor variable*, the quantity

$$\frac{1}{n} + \frac{(x_i - \bar{x})^2}{S_{xx}}$$

is the ith HAT diagonal. The user is encouraged to verify this by using the centered regression model for the case of a single regressor and then working out the ith diagonal of $\mathbf{X}(\mathbf{X}'\mathbf{X})^{-1}\mathbf{X}'$. Equation (2.32) is then easily verified.

We study residuals for the purpose of detecting violations of assumptions. The model assumptions state our beliefs about the behavior of the ε_i, the conceptual errors. Of course, the validity of much regression technology depends on these assumptions. Residuals are available to us to mirror the behavior of the ε_i. Now, the task of the analyst would be greatly simplified if the residuals behaved like the ε_i, which are assumed uncorrelated with common variance, σ^2. Unfortunately, this ideal behavior is not the case in general, as can be seen from Eq. (5.1) and (5.2). Indeed, (5.1) reveals an important result which, by now, should be intuitively reasonable to the reader. *The residual is a more precise estimator of zero if it is associated with a data point that is remote from the data center* (i.e., h_{ii} near 1.0) *than if it is associated with a point that is near the center.* Please refer to the discussion of the HAT diagonals in Section 3.9. Accordingly, we can say that the least squares regression equation is expected to be a better fit to a remote point than to an interior point. See Figure 4.1 as an illustration.

Despite their relatively complicated variance-covariance structure, the residuals contain much information that can be beneficial. Of course, the analyst must account for the "deviation from the ideal" in their properties. In addition, it is important that many of the techniques in residual analysis be viewed as *diagnostic* in nature, and not as part of formal statistical inference. The reader will often be reminded of this as we progress. We discussed assumptions in Chapters 2 and 3. It should never be the intention of the analyst to ignore or deny violations of assumptions. But there are ways of dealing with them and one must begin with *detection*. The bases for detection are the residuals. Among the phenomena that can be detected are model underspecification, departure from the homogeneous variance assumption, existence of suspect data points, departure from normality in the model errors, and isolated *high influence* data points.

5.2

PLOTTING OF RESIDUALS

In the case of multiple regressor variables, a simple plotting of ordinary residuals against the fitted values $\hat{y}_i$ is often beneficial in highlighting either model underspecification or a deviation from the homogeneous variance assumption. The "classical appearance" of such a residual plot for an ideal situation is depicted in Figure 5.1. The picture indicates a random pattern around zero with no detectable trend. This is identical to the concept discussed in Chapter 2 in the case of simple linear regression.

FIGURE 5.1 **Ideal residual plot**

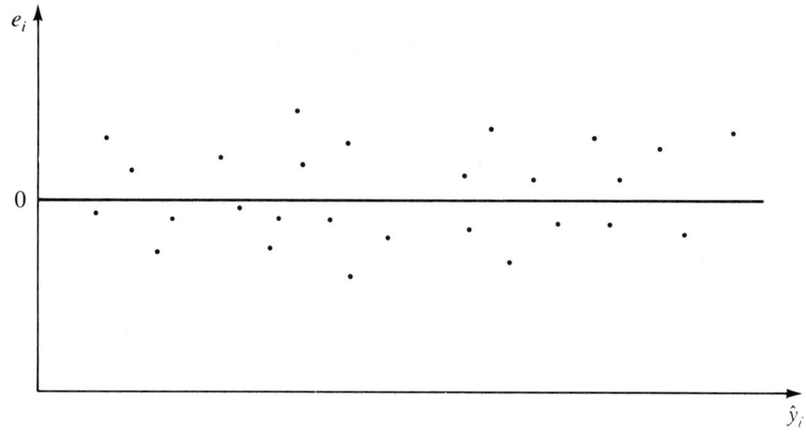

Figure 5.2 reveals residual patterns that give evidence of model under-specification (Figure 5.2(a)) and heterogeneous variance (Figure 5.2(b)). The situation in Figure 5.2(b) is reasonably clear. The *funnel* effect indicates that as the response variable gets larger, the deviations of the residuals from zero become greater. Hence one detects a condition whereby the error variance is not constant but increases as the measured response increases. In Figure 5.2(a), the systematic trend in the residuals usually indicates that a model term is missing, perhaps a quadratic term in one of the regressor variables.

FIGURE 5.2 **Residual plots indicating violation of assumptions:**

(a) Model should involve curvature ***(b)*** Heterogeneous variance

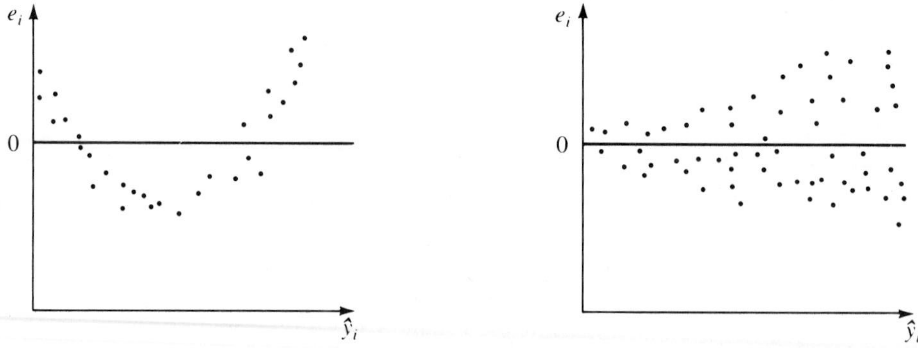

Example 5.1 ***Population Data***

The data in Table 5.1 give the population (in millions) of the U.S. from the year 1790 through the year 1970. Consider a simple linear regression fit to the 19 data points with a model of the type

$$y_i = \beta_0 + \beta_1 x_i + \varepsilon_i \qquad (i = 1, 2, \ldots, 19)$$

A least squares fit produces the regression

$$\hat{y} = -1958.366 + 1.078795x$$

with $R^2 = 0.9223$ and $s = 18.1274$. The listing below shows observed population y, fitted population $\hat{y}$ and residuals.

This is a class example in which a set of residuals depict a need for the regressor variable to enter in a curvilinear fashion. In fact, a mere inspection of the residuals indicates a rather curious set of "runs" of positive and negative values. Obviously the model underpredicts for the early and later years but overpredicts from the years 1830 through 1940.

Year	y (Population)	$\hat{y}$ (Fitted Population)	$e_i = y_i - \hat{y}_i$
1790	3.929	−27.324	31.253
1800	5.308	−16.536	21.844
1810	7.239	−5.748	12.987
1820	9.638	5.040	4.598
1830	12.866	15.828	−2.962
1840	17.069	26.616	−9.547
1850	23.191	37.404	−14.213
1860	31.443	48.192	−16.749
1870	39.818	58.980	−19.162
1880	50.155	69.767	−19.612
1890	62.947	80.555	−17.608
1900	75.994	91.343	−15.349
1910	91.972	102.131	−10.159
1920	105.710	112.919	−7.209
1930	122.775	123.707	−0.932
1940	131.669	134.495	−2.826
1950	151.325	145.283	6.042
1960	179.323	156.071	23.252
1970	203.211	166.859	36.352

Figure 5.3 shows a plot of the residuals against the $\hat{y}$ values. It is clear from the plot in Figure 5.3 that there are grave doubts about the model specification. The plot resembles in appearance the plot given in Figure 5.2(a). One would expect that the introduction of a quadratic term would produce a residual plot that would be

closer to the "ideal" appearance. A quadratic regression model fit to the data results in a fitted model given by

$$\hat{y} = 20{,}450.434 - 22.7806x + 0.0063456x^2$$

with $R^2 = 0.9983$ and $s = 2.781$. It would appear from the R^2 and the root residual mean square that the introduction of curvature via the quadratic term did provide an improved model. The residual plot given in Figure 5.4 for the revised model reveals a picture that is closer to the classic residual plot that appears in Figure 5.1.

However, there is still a need for further study. Two particular residuals appear considerably below the others. These two residuals represent years 1940 and 1950. We could conclude that the picture still depicts a possible model misspecification. Now, we do not, at this point, choose to conclude that these two residuals are "too far away from zero" in some statistical sense. We merely state that the pictorial display suggests that they are "suspect." If indeed they are suspect and if we are certain that the measured population is correct, then we interpret the problem as one of further model misspecification. There is an obvious explanation for the phenomenon seen here. The population during the period 1940 through 1950 was influenced by the war years, suggesting a need for a model shift, or a categorical variable. This was attempted by assigning a 1 to the categorical variable for the years 1940 and 1950 and a zero to the other years. (See Section 3.10.) The resulting regression equation is given by

$$\hat{y} = 20{,}982.754 - 23.36638x + 0.006507x^2 - 8.7415z$$

where z is 0 or 1 depending on the year. The R^2 and root residual mean square are given by $R^2 = 0.9998$ and $s = 0.9374$, an appreciable improvement over the previous model. The resulting residual plot is shown in Figure 5.5.

The plot is now much closer to the ideal picture in Figure 5.1. The point of the illustration here is the visual detection of model misspecifications (when they are obvious) from the residual plots. Of course, the analyst still needs to evaluate the models suggested by the residual plots. Clearly, an improvement was realized here.

TABLE 5.1 **Population data for the U.S. from 1790 through 1970**

Year (x)	Population (y)
1790	3.929
1800	5.308
1810	7.239
1820	9.638
1830	12.866
1840	17.069
1850	23.191
1860	31.443

TABLE 5.1 Continued

Population data for the U.S. from 1790 through 1970

Year (x)	Population (y)
1870	39.818
1880	50.155
1890	62.947
1900	75.994
1910	91.972
1920	105.710
1930	122.775
1940	131.669
1950	151.325
1960	179.323
1970	203.211

SAS Institute Inc., *SAS Views: Regression and ANOVA, 1981 Edition.* (Cary, North Carolina: SAS Institute Inc., 1981), pp. 1.2–1.3.

FIGURE 5.3 **Plot for the linear model fit to the population data of Table 5.1**

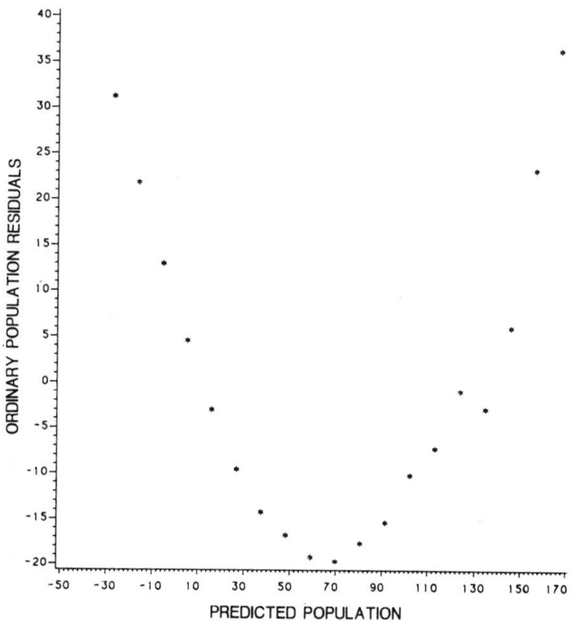

FIGURE 5.4 **Plot for the quadratic model fit to the population data of Table 5.1**

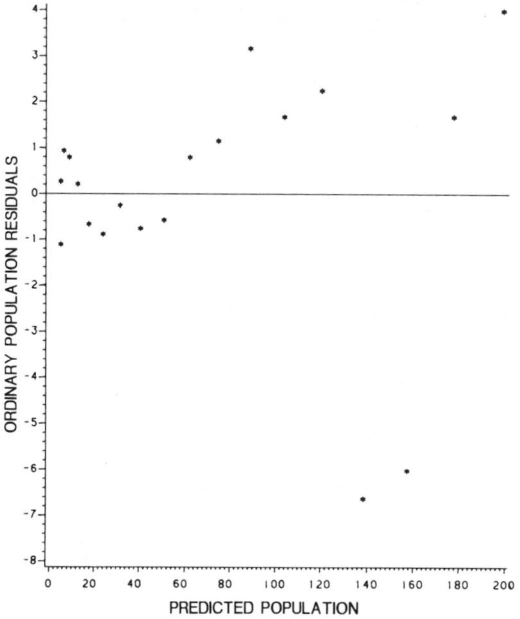

FIGURE 5.5 **Plot for the quadratic model with categorical variable fit to the population data of Table 5.1**

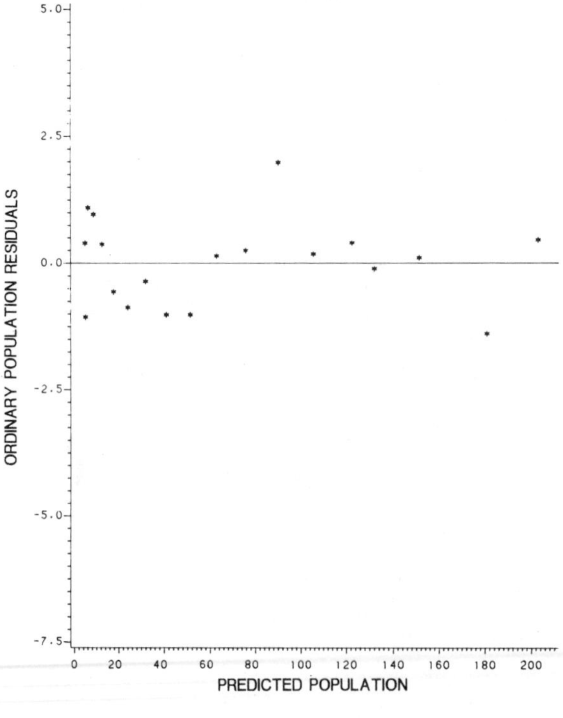

5.3

STUDENTIZED RESIDUALS

While plots like those described and illustrated in the foregoing can be very useful as diagnostic tools, random noise can often cloud the picture in many data sets. Surely the analyst cannot expect the classical pictures as shown in Figures 5.1 or 5.2 to appear with any degree of regularity. Experience is vital in becoming proficient at reading and extracting information from residual plots.

In many cases, plots of the *ordinary residuals* are not the optimum displays for revealing difficulties in the model and in the assumptions. Ideally, the quantities to be plotted should be uncorrelated with common variance to closely emulate the ε_i. We know that the residuals are not uncorrelated in general; from Eq. (5.1), it is clear that if there is a large variation in the HAT diagonals among the data points, there will be large differences in the variances of the residuals. We know that a HAT diagonal near unity defines a data point that is remote from the data center. The tendency for the regression fit to be good (i.e., a small residual) at such a point may completely mask any model inadequacy. As a result, deviations from assumptions are often best detected by working with residuals that have the same precision. Standardized residuals of the type $e_i/(\sigma\sqrt{1-h_{ii}})$ have zero mean and unit variance. Thus a *studentized residual* (s replaces σ)

$$r_i = \frac{e_i}{s\sqrt{1-h_{ii}}} \tag{5.3}$$

may be very useful to the analyst in residual diagnostics. The studentized residuals are scale-free, and the structure makes it t-like (it does not exactly follow a t-distribution). It is simpler to develop a crude yardstick to measure their size than it would be in the case of the ordinary residual. In summary, the advantage gained by using the studentized residual over the ordinary residual stems from the fact that the standardization eliminates the effect of the *location* of the data point in the regressor space (as measured by its h_{ii} value). Studentized residuals are rapidly becoming a standard part of the regression computer printout with most software packages.

Example 5.2 ***BOQ Data***

The U.S. Navy attempts to develop equations for estimation of manpower needs for manning installations such as Bachelor Officers Quarters. Regression equations are developed from data taken by measurement teams. The data in Table 5.2 were collected at 25 BOQ sites. The variable descriptions follow on page 218.

x_1: Average daily occupancy
x_2: Monthly average number of check-ins
x_3: Weekly hours of service desk operation
x_4: Square feet of common use area
x_5: Number of building wings
x_6: Operational berthing capacity
x_7: Number of rooms
y: Monthly man-hours

A multiple regression was fit to the data with the following results.

$$\hat{y} = 134.968 - 1.2838x_1 + 1.8035x_2 + 0.66915x_3 - 21.42263x_4 + 5.61923x_5 - 14.48025x_6 + 29.3248x_7$$

with $R^2 = 0.9613$ and $s = 455.167$.

TABLE 5.2 **Data on manpower and workload for U.S. Navy Bachelor Officers' Quarters**

Site	x_1	x_2	x_3	x_4	x_5	x_6	x_7	y
1	2.00	4.00	4	1.26	1	6	6	180.23
2	3.00	1.58	40	1.25	1	5	5	182.61
3	16.60	23.78	40	1.00	1	13	13	164.38
4	7.00	2.37	168	1.00	1	7	8	284.55
5	5.30	1.67	42.5	7.79	3	25	25	199.92
6	16.50	8.25	168	1.12	2	19	19	267.38
7	25.89	3.00	40	0	3	36	36	999.09
8	44.42	159.75	168	0.60	18	48	48	1103.24
9	39.63	50.86	40	27.37	10	77	77	944.21
10	31.92	40.08	168	5.52	6	47	47	931.84
11	97.33	255.08	168	19.00	6	165	130	2268.06
12	56.63	373.42	168	6.03	4	36	37	1489.50
13	96.67	206.67	168	17.86	14	120	120	1891.70
14	54.58	207.08	168	7.77	6	66	66	1387.82
15	113.88	981.00	168	24.48	6	166	179	3559.92
16	149.58	233.83	168	31.07	14	185	202	3115.29
17	134.32	145.82	168	25.99	12	192	192	2227.76
18	188.74	937.00	168	45.44	26	237	237	4804.24
19	110.24	410.00	168	20.05	12	115	115	2628.32
20	96.83	677.33	168	20.31	10	302	210	1880.84
21	102.33	288.83	168	21.01	14	131	131	3036.63
22	274.92	695.25	168	46.63	58	363	363	5539.98
23	811.08	714.33	168	22.76	17	242	242	3534.49
24	384.50	1473.66	168	7.36	24	540	453	8266.77
25	95.00	368.00	168	30.26	9	292	196	1845.89

Procedures and Analyses for Staffing Standards Development: Data/Regression Analysis Handbook (San Diego, California: Navy Manpower and Material Analysis Center, 1979).

Table 5.3 shows the residuals, HAT diagonal values, and the resulting studentized residuals. Figure 5.6 reveals a plot of the studentized residuals against $\hat{y}_i$. We detect a substantial increase in the studentized residuals for data points beyond a $\hat{y}$ of approximately 2,000 monthly man-hours. This information leads the analyst to conjecture that the homogeneous variance assumption may, in fact, be violated; or at least there is reason to believe that the model presumed becomes "strained" for the larger Naval installations, i.e., for those with monthly man-hours beyond the value of 2,000.

The results for installations 23 and 24 represent examples where the use of studentized residuals is clearly more informative than that for ordinary residuals. In the case of installation 23, the ordinary residual is -160.276, certainly not large in comparison to the other residuals. However, the HAT diagonal is 0.9885, characterizing the data point as being remote from the data center. The resulting studentized residual of -3.278 strongly suggests that under ideal conditions, the residual represents an unusual departure from zero, *given the location* of the point in the regressor space. The appearance in Figure 5.6 reveals a collection of large r_i values for the large installations, suggesting a possible model breakdown for the large installations. One may be tempted to interpret this as a failure of the homogeneous variance assumption.

TABLE 5.3 **Residuals and HAT diagonal values for data of Table 5.2**

Site	y_i	$\hat{y}_i$	e_i	r_i	h_{ii}
1	180.23	209.985	−29.755	−0.076	0.2573
2	182.61	213.796	−31.186	−0.075	0.1609
3	164.38	360.486	−196.106	−0.470	0.1614
4	284.55	360.106	−75.556	−0.181	0.1631
5	199.92	380.703	−180.783	−0.430	0.1475
6	267.38	510.373	−242.993	−0.582	0.1589
7	999.09	685.167	313.923	0.763	0.1829
8	1103.24	1279.299	−176.059	−0.483	0.3591
9	944.21	815.466	128.744	0.334	0.2808
10	931.84	891.846	39.994	0.094	0.1295
11	2268.06	1632.137	635.923	1.493	0.1241
12	1489.50	1305.177	184.323	0.453	0.2024
13	1891.70	1973.416	−81.716	−0.187	0.0802
14	1387.82	1397.786	−9.966	−0.023	0.0969
15	3559.92	4225.131	−665.211	−2.197	0.5576
16	3115.29	3134.895	−19.605	−0.056	0.4024
17	2227.76	2698.738	−470.978	−1.302	0.3682
18	4804.24	4385.778	418.462	1.236	0.4465
19	2628.32	2190.326	437.994	1.007	0.0868
20	1880.84	2750.910	−870.070	−2.401	0.3663
21	3036.63	2210.134	826.496	1.883	0.0704
22	5539.98	5863.847	−323.894	−1.536	0.7854
23	3534.49	3694.766	−160.276	−3.278	0.9885
24	8266.77	7853.505	413.265	2.580	0.8762
25	1845.89	1710.861	135.029	0.441	0.5467

FIGURE 5.6 Plot of studentized residuals against predicted man-hours data of Table 5.2

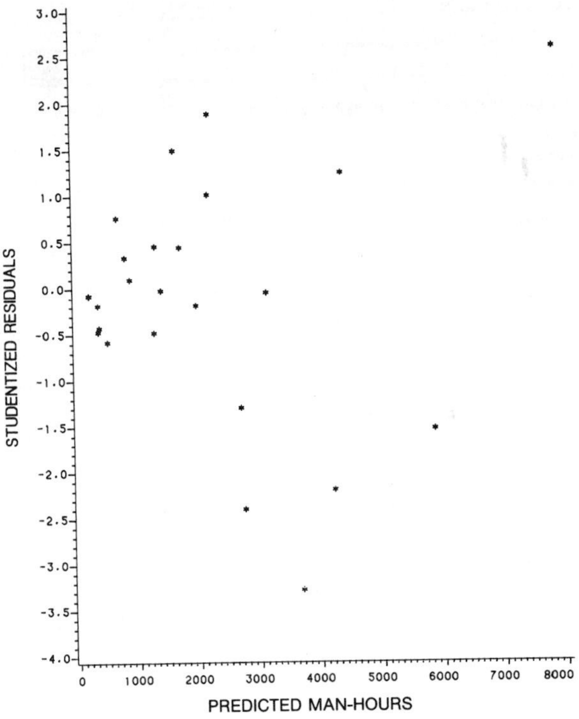

5.4

RELATION TO STANDARDIZED PRESS RESIDUALS

It is interesting to note the relationship between the standardized residual (or the studentized residual that one must use in practice) and the standardized PRESS residual. Up to this point in the text, we have used the PRESS residuals in the context of model selection as discussed in Chapter 4. However, their role in the diagnostic aspect of residual analysis is important from both a conceptual and analysis point of view. Recall from Eq. (4.6) that the ith PRESS residual is given by

$$e_{i,-i} = \frac{e_i}{1 - h_{ii}}$$

As a result, from Eq. (5.1) the variance of the ith PRESS residual can be written

$$\text{Var } e_{i,-i} = \frac{1}{(1 - h_{ii})^2}[\sigma^2(1 - h_{ii})] = \frac{\sigma^2}{1 - h_{ii}} \tag{5.4}$$

Thus we can *standardize* the PRESS residual and obtain

$$\frac{e_{i,-i}}{\sigma_{e_{i,-i}}} = \frac{e_i/(1-h_{ii})}{\sqrt{\sigma^2/(1-h_{ii})}}$$

$$= \frac{e_i}{\sigma\sqrt{1-h_{ii}}} \qquad \textbf{(5.5)}$$

which is identical to the *standardized ordinary residual.* This provides some interesting insight in dealing with tools for detection of suspect data points. This important topic is discussed in the following section.

5.5

DETECTION OF OUTLIERS

The regression analyst is unlikely to face a greater dilemma than that of how to cope with individual data points that do not fit the trend set by the balance of the data. These *suspect points* may arise for a number of reasons, most of which are classified into the *failure of assumption* category. As a result, it is natural and convenient to treat the problem with analysis of residuals. To completely understand the proper analytical device, the reader should note what model violations may produce a suspicious data point. The following are certainly possibilities:

1. There is a breakdown in the model at the ith point, producing a *location shift*, i.e., $E(\varepsilon_i) = \Delta_i \neq 0$. This is called the *mean shift* outlier model.
2. There is a breakdown in the model at the ith point and $\text{Var}(\varepsilon_i)$ exceeds the error variance at the other data locations.

There are often other explanations for observations that appear not to follow the proposed model fit to the balance of the data. It could be, of course, that there is no shift (bias) at the point in question or that the error variance is indeed stable, but that a large random disturbance is produced by chance. Either model breakdown, 1 or 2 above, will manifest itself in a residual that is expected to be large in magnitude *for a data point at that specific location in the regressor variables.* The size of the residual is an important portion of the diagnostic but, again, location of the point, as quantified by the HAT diagonal, must be taken into account. A statistic from which outlier detection evolves quite naturally is the PRESS residual,

$$e_{i,-i} = y_i - \hat{y}_{i,-i}.$$

If there is a model shift, Δ_i, associated with the ith point, then the mean of the ith PRESS residual is given by

$$E(y_i - \hat{y}_{i,-i}) = \Delta_i$$

Again, however, we need to standardize the PRESS residual to determine if the deviation of $e_{i,-i}$ from zero is more than mere chance. As a result, a diagnostic measure is given by (see Section 5.4)

$$e_{i,-i}/\sigma_{e_{i,-i}} = \frac{e_i}{\sigma\sqrt{1-h_{ii}}}$$

as given in Eq. (5.5). As a result, the residual and the PRESS residual, suitably standardized, give the same diagnostic for detection of outliers. Thus, for practical use, σ is replaced by s, and the studentized residual in Eq. (5.3) becomes a realistic diagnostic for outlier detection. Simply put, we use the studentized residual to determine if the residual (or the PRESS residual) is further from zero than we might expect under standard and ideal conditions. In some situations, there may be a slight preference for an alternative estimator of σ in Eq. (5.5). This is presented in the following subsection.

INTERNAL AND EXTERNAL STUDENTIZATION (THE R-STUDENT STATISTIC AS AN OUTLIER DIAGNOSTIC)

The studentized residual given by r_i in Eq. (5.3) is a diagnostic tool for outlier detection. However, a close inspection of this statistic reveals an alternative estimate for σ to use in the studentization. If the outlier results from a model breakdown as in the case of item 1, *the mean shift outlier*, then we know that s^2, the error mean square, described by Eq. (3.5) is biased upwards. This should be clear to the reader since the concept was discussed in Section 4.3 in regard to the bias of s^2 under model misspecification. The mean shift anomaly is merely a model misspecification. An alternate estimator is the root residual mean square calculated *without utilizing the* ith *observation.* One can make use of computational procedures similar to those used in Chapter 4 in the development of the PRESS statistic. (See Appendix B.5.) The desired estimate, denoted by s_{-i}, is given by

$$s_{-i} = \sqrt{\frac{(n-p)s^2 - e_i^2/(1-h_{ii})}{n-p-1}} \quad \textbf{(5.6)}$$

It is interesting that the residual sum of squares without the use of the ith observation differs from the residual sum of squares using all data by the quantity $e_i^2/(1-h_{ii})$.

The estimate in (5.6) is used in place of σ to produce an *externally studentized* residual, often called R-student, which is given by

$$t_i = \frac{y_i - \hat{y}_i}{s_{-i}\sqrt{1-h_{ii}}} \tag{5.7}$$

One should be aware of the difference between the two diagnostics, r_i and t_i. In many situations, r_i and t_i will give values that differ very little. The discrepancy between them will depend on the *influence* exerted on the results by the ith observation. Equation (5.6) reveals that a relatively large residual in combination with a HAT diagonal close to unity can produce a sizable discrepancy between s and s_{-i} and thus between r_i and t_i. (The notion of influence of individual data points is covered in depth in Chapter 6.) Often the R-student statistic will be more sensitive, i.e., become larger, in the presence of a discordant data point.

If we make standard assumptions on the ε_i, including normality, a single t_i value does indeed follow t_{n-p-1} degrees of freedom under the hypothesis that there is no mean shift. Thus the use of critical points from student's t-distribution is more reasonable for t_i than for r_i.

The R-student statistic allows a formal mechanism for detection of outliers through hypothesis testing. In the case of the location shift outlier, the hypothesis is given by

$$H_0: \Delta_i = 0$$
$$H_1: \Delta_i \neq 0$$

In the case where one attempts to detect an increase in error variance at the ith point, i.e., the model breakdown (2) described at the beginning of Section 5.5, the R-student is also appropriate. In this case, one supplements the model by postulating that at the ith data point, $\text{Var}(\varepsilon_i) = \sigma^2 + \sigma_i^2$, with σ_i^2 representing the increase in error variance due to an anomaly there. The hypothesis is given by

$$H_0: \sigma_i^2 = 0$$
$$H_1: \sigma_i^2 \neq 0$$

and the R-student statistic applies. In the case of either outlier model, a two-tailed t-test is appropriate.

Example 5.3 ***Coal-Cleansing Data***

An experiment[1] similar to that presented in Example 3.10 was conducted to gain some preliminary insight into the effect of three quantitative factors on the capability of a particular coal-cleansing operation. A polymer was used to clean the coal and the amount of suspended solids, y, (mg/1) in the "overflow" solution (following a cleansing operation) is a measure of the efficiency of the operation. The factors that influenced the suspended solids are

x_1: percent solids in the input solution

x_2: pH of the tank that holds the solution

x_3: flow rate of the cleansing polymer, ml/minute

All three factors were controlled in the experimental process. (The order of experimentation was random.) The data are as follows:

Experiment	x_1	x_2	x_3	y
1	1.5	6.0	1315	243
2	1.5	6.0	1315	261
3	1.5	9.0	1890	244
4	1.5	9.0	1890	285
5	2.0	7.5	1575	202
6	2.0	7.5	1575	180
7	2.0	7.5	1575	183
8	2.0	7.5	1575	207
9	2.5	9.0	1315	216
10	2.5	9.0	1315	160
11	2.5	6.0	1890	104
12	2.5	6.0	1890	110

A regression equation was constructed with the result given by

$$\hat{y} = 397.087 - 110.750x_1 + 15.5833x_2 - 0.058x_3$$

with $R^2 = 0.8993$ and $s^2 = 435.862$.

There was concern prior to the data analysis that the results from data point 9 were erroneous. Conditions were not held constant as was necessary. As a result, the engineers involved felt as if the project might benefit from deletion of the information. However, the analysis was initially conducted with the data point not removed in order to gain some insight into whether or not the analysis supports the

[1]Data were generated by the Mining Engineering Department and analyzed by the Statistical Consulting Center, Virginia Polytechnic Institute and State University, Blacksburg, Virginia, 1979.

conjecture regarding the data point. Table 5.4 shows the response, fitted response, residuals, HAT diagonal values, and R-student values associated with each of the 12 observations. Note that the residual associated with the 9th point is 32.2 mg/l. This is the largest residual in the data set. The R-student computation is designed to determine if the residual differs significantly from zero. The results give $t = 2.8695$ for data point 9, significant at below the 0.05 level. None of the other R-student values approach that of data point 9. The results would seem to confirm the suspicions regarding this observation. Thus deletion of observation 9 is a reasonable approach in the construction of the regression.

TABLE 5.4 **Results for the coal-cleansing example**

y_i	$\hat{y}_i$	e_i	h_{ii}	t_i
243	247.8	−4.8	0.4501	−0.2923
261	247.8	13.2	0.4501	0.8359
244	261.0	−17.0	0.4660	−1.1372
285	261.0	24.0	0.4660	1.7665
202	200.7	1.3	0.0838	0.0631
180	200.7	−20.7	0.0838	−1.0385
183	200.7	−17.7	0.0838	−0.8698
207	200.7	6.3	0.0838	0.2990
216	183.8	32.2	0.4501	2.8695
160	183.8	−23.8	0.4501	−1.7141
104	103.5	0.5	0.4660	0.0282
110	103.5	6.5	0.4660	0.4006

CRITICAL YARDSTICK FOR R-STUDENT

We have already indicated that the R-student given by Eq. (5.7) follows t_{n-p-1} under the appropriate outlier hypothesis. Thus, in the case of a single observation on which there is advance suspicion, the t-test with corresponding critical values is appropriate. However, if the analyst is simultaneously putting all observations under scrutiny, formal t-tests should not be used. Indeed, testing the observation that gives the largest t (without prior conviction) implies that n t-tests are essentially being made. Making these tests formally at the $\alpha = 0.05$ level is incorrect! For a sample size $n = 50$, the probability that the largest R-student exceeds the standard $t_{.05}$ critical value *is in the vicinity of* 0.9. Thus, multiple tests, treated formally as if they are separate tests, are entirely inappropriate. Useful, though conservative, critical values for the R-student statistic can be computed using the *Bonferroni inequality*. See Miller (1965) and Cook and Weisberg (1980). Suppose we wish an overall significance level of α for n tests. The use of the $(\alpha/n) \times 100\%$ point of the t_{n-p-1} distribution will produce a significance level that is *no larger than* α. The test certainly provides a guide in the situation where no suspect points are present prior to the analysis. Table C.4 in Appendix C provides critical values for making such outlier tests.

Example 5.4 ***BOQ Data***

Consider the data of Example 5.2. Suppose it is of interest to do an outlier analysis but, *a priori*, there are no data points under suspicion. Simply put, there is a need to determine if any data points are outliers using significance tests on all observations simultaneously. The following are the residuals, *R*-student values, and HAT diagonals for each of the 25 data points.

Site	e_i	t_i (*R*-student)	h_{ii}
1	−29.755	−0.0736	0.2573
2	−31.186	−0.0726	0.1609
3	−196.106	−0.4594	0.1614
4	−75.556	−0.1762	0.1631
5	−180.783	−0.4196	0.1475
6	−242.993	−0.5704	0.1589
7	313.923	0.7532	0.1829
8	−176.059	−0.4720	0.3591
9	128.744	0.3246	0.2808
10	39.994	0.0914	0.1295
11	635.923	1.5537	0.1241
12	184.323	0.4426	0.2024
13	−81.716	−0.1818	0.0802
14	−9.966	−0.0224	0.0969
15	−665.211	−2.5192	0.5576
16	−19.605	−0.0541	0.4024
17	−470.978	−1.3310	0.3682
18	418.462	1.2566	0.4465
19	437.994	1.0074	0.0868
20	−870.070	−2.8657	0.3663
21	826.496	2.0538	0.0704
22	−323.894	−1.6057	0.7854
23	−160.276	−5.2423*	0.9885
24	413.265	3.2093	0.8762
25	135.029	0.4299	0.5467

Given that there are 25 residuals being tested, the proper 0.05 level critical value for *R*-student (8 model parameters) is given by 3.69 (see Table C.4, Appendix C). As a result the only data point that can be classified as an outlier is that associated with installation 23 (note asterisk) in which the *R*-student value is −5.2423.

OUTLIER TESTS: DIAGNOSTIC OR FORMAL STATISTICAL INFERENCE?

Often the regression analyst will succumb to the temptation of eliminating data points in order to enhance the quality of fit. There is often no reason to believe that a suspect point should be deleted, even if the analysis confirms that it is an outlier. Surely the choice of *deletion or not* should not be made solely by the statistician, but rather in harmony with the subject matter scientist. As long as regression data are being collected for the purposes of imperfect empirical model building, there will be observations that do not satisfy the model entertained by the analyst.

An observation that is a *one in a thousand* occurrence, when compared to observations under equivalent conditions, has a different meaning to the scientist than an observation that merely does not *follow the trend* of the rest of the data and may be a result of a *flaw in modeling*. In the latter case, the detection of such a point supplies important information that would otherwise go unnoticed. Mistakes in modeling are much easier to understand in retrospect. The model may be necessarily altered after an analysis that involves outlier diagnosis. For example, the inclusion of a categorical variable (see Chapter 3) or a transformation (see Chapter 7) on a particular variable may be the necessary strategy resulting from such an exercise.

In a large number of instances, the diagnosis of outliers leads to further experimentation in the region where the outliers occur. The analyst must bear in mind that the detection of an outlier implies only that the observation does not agree with a specified model. Now, of course, if the curious observation is a result of a happening external to the system under study, e.g., a dirty test tube, a key punch error, or a computing error, then it should be deleted. Since it is impossible for the statistical calculations to determine if the anomaly is in the latter category, deletion should normally come from considerations that are external to the statistical analysis.

The modern trend in outlier regression analysis is to put less emphasis on the formal hypothesis testing framework described earlier in this section. Rather, one should consider the outlier test statistics as diagnostic in nature, suggesting data locations where further investigation is needed. The implication here is that tight critical values are not important. Only *crude cutoff* values are needed. For example, in the case of the BOQ data with R-student values given in Example 5.4, the only site that was formally detected as an outlier was site 23. The mechanism used was the conservative Bonferroni type test in which all sites are tested simultaneously. However, from the appearance of the other R-student values, one would suspect that the *deviation from ideal* is more complicated and certainly cannot be dismissed as merely a single outlier at site 23. One cannot ignore the fact that five sites produced an R-student exceeding 2.0 in magnitude, even though only one turned out to be significant at the 0.05 level. In addition, it seems noteworthy that all of the large R-student values occurred at relatively large sites. It is this type of *diagnostic view* of the R-student values (or studentized residuals as in the case of the presentation in Example 5.2) that can aid the analyst in searching for flaws in modeling. Also the reader must consider treatment of outliers to be very much related to the information presented in Chapter 6 on influence diagnostics. The total modern regression analysis involves an appropriate blending of these two topics.

USE OF INDICATOR VARIABLES IN OUTLIER DETECTION

Thus far in Section 5.5, we have focused on the detection of outliers through formal statistical inference or through the use of diagnostic information. The use of the *R*-student statistic can be used in the case of a single *prelabeled* observation, or for simultaneous testing of several or all observations through a conservative Bonferroni approach. For cases in which there are one or more prelabeled observations that are under test, there is an interesting application of indicator variables that can be implemented. Before we begin the discussion of outliers with indicator variables, we should caution the potential user of this methodology. Modeling outliers with indicator variables can be hazardous. It is important that an outlier or group of outliers be labeled *a priori*. Many students of regression view indicator variables as merely representing a model "shift in location" as discussed in Chapter 3. However, it should be understood that this is equivalent to our definition of the mean shift outlier model. That is, the presence of an outlier at the *i*th data point represents a shift in the intercept of the regression model at the *i*th point. As a result, the material in this subsection merely serves as a convenient extension of the formal *R*-student test statistic for cases in which there is suspicion associated with more than one data point. However, as in the case of many of the modern regression tools, there is considerable potential for abuse.

Consider a situation in which there is suspicion that a set of $r \geqslant 1$ observations $\mathbf{y}_1$ are outliers. As a result, if we consider the model in matrix notation we have, for the observation vector,

$$\mathbf{y} = \begin{bmatrix} \mathbf{y}_1 \\ --- \\ \mathbf{y}_2 \end{bmatrix} \begin{matrix} \}\, r \\ \\ \}\, n-r \end{matrix} \qquad \text{and} \qquad \begin{bmatrix} \mathbf{y}_1 \\ --- \\ \mathbf{y}_2 \end{bmatrix} = \mathbf{X}\boldsymbol{\beta} + \begin{bmatrix} \boldsymbol{\varepsilon}_1 \\ --- \\ \boldsymbol{\varepsilon}_2 \end{bmatrix}$$

and one is interested in testing

$$H_0\colon \boldsymbol{\Delta}_1 = \mathbf{0} \tag{5.8}$$

where $E(\boldsymbol{\varepsilon}_1) = \boldsymbol{\Delta}_1$. Now, the alternative hypothesis to Eq. (5.8) may take different forms depending on what the scientists' suspicions are about the outliers. It may be felt that (i) all r possible outliers are likely to have a *common* mean shift, (ii) all r outliers are likely to have different values for the mean shift, or (iii) some compromise between (i) and (ii). If it is felt that a collection of observations are questionable and the source of the difficulty is the same for each of them, then a common mean shift would be a reasonable alternative hypothesis. However, if a collection of possible outliers occur at apparently different magnitudes of error, then an alternative hypothesis similar to that described in (ii) should be used. In some applications the apparent outliers may even be divided into groups so that it can be assumed that all mean shifts within the group are similar.

COMMON MEAN SHIFT FOR A SET OF POSSIBLE OUTLIERS

Let us consider case (i) first. We have a labeled group of r observations $\mathbf{y}_1$, and we can write $E(\boldsymbol{\varepsilon}_1) = \boldsymbol{\Delta}_1 = \mathbf{j}\gamma$ where γ is a nonzero constant and $\mathbf{j}$ is a column of r ones. The hypothesis becomes

$$\begin{aligned} H_0&: \gamma = 0 \\ H_1&: \gamma \neq 0 \end{aligned} \tag{5.9}$$

The test of this hypothesis is conducted by fitting the regression model augmented with an indicator variable with two categories, one containing the r labeled observations and one containing the $n - r$ unaffected observations. The hypothesis of Eq. (5.9) is rejected or not according to whether or not the t-statistic for the indicator variable is significant.

It should be fairly simple for the reader to understand why the t-test for the indicator variable is a test of the outlier hypothesis of Eq. (5.9). The regression model for the situation in which there is a common mean shift for a set of possible outliers at $\mathbf{y}_1$ is given by

$$\begin{bmatrix} \mathbf{y}_1 \\ \text{---} \\ \mathbf{y}_2 \end{bmatrix} = \mathbf{X}\boldsymbol{\beta} + \begin{bmatrix} 1 \\ 1 \\ \vdots \\ 1 \\ \text{--} \\ 0 \\ 0 \\ \vdots \\ 0 \end{bmatrix} \gamma + \boldsymbol{\varepsilon} \tag{5.10}$$

where the vector of errors $\boldsymbol{\varepsilon}$ has the standard $N(\mathbf{0}, \sigma^2\mathbf{I})$ structure. It becomes clear from our development in Chapter 3 that this model is equivalent to what one achieves by adding an indicator variable with two categories, and the information regarding the significance of the mean shift lies in the t-test of the hypothesis of Eq. (5.9).

OUTLIER MODEL WITH NO COMMON SHIFT

The hypothetical outlier model in which the labeled outliers can be divided into groups is rare in practice. Often, even if the analyst does encounter observations that can be labeled as possible outliers, it is difficult to group them. If, for example, there are five observations that are *a priori* deemed to be suspicious, and if there is no cause for grouping them, then the significance of the outliers can be tested simultaneously through a model with an indicator variable with five categories. As an example,

$$\begin{matrix}\text{First outlier}\\ \text{Second outlier}\\ \text{Third outlier}\\ \text{Fourth outlier}\\ \text{Fifth outlier}\\ \\ \\ \\ \end{matrix}\begin{bmatrix}1 & 0 & 0 & 0 & 0\\ 0 & 1 & 0 & 0 & 0\\ 0 & 0 & 1 & 0 & 0\\ 0 & 0 & 0 & 1 & 0\\ 0 & 0 & 0 & 0 & 1\\ 0 & 0 & 0 & 0 & 0\\ \vdots & \vdots & \vdots & \vdots & \vdots\\ 0 & 0 & 0 & 0 & 0\end{bmatrix}$$

Example 5.5 Consider the BOQ data of Example 5.2. Suppose it was determined *a priori* that there is a threshold in size beyond which the governance and management of the larger BOQ render the installation quite different from the smaller ones. As a result, it is of interest to determine, by testing, if the large BOQ's, namely 15, 18, 22, 23, and 24 are, in fact, outliers in the context of the fitted model containing all of the BOQ's. This requires a model to be fit with an indicator variable containing five categories with a 1 in the position of the outlier and 0's elsewhere for each category.

Table 5.5 gives some pertinent results of a SAS PROC REG run in which the five large BOQ's are modeled as mean shift outliers. The model allows for separate shifts on each data point. Notice that variable D_4, signifying BOQ 23, is significant in this outlier model while BOQ 15, 18, 22, and 24 are not significant. This is not particularly surprising in light of the R-student computations in Example 5.4. The individual R-student values can be viewed as derivable from a model in which a single data point (BOQ in this case) is modeled *as a single outlier* as opposed to the case of modeling four outliers as in the present example. The reader should understand that this analysis of this example should not be viewed as any final analysis of these data. We use it here as only an illustration of the outlier model. The BOQ data become the subject of further analysis in Chapter 6, when we deal with influence diagnostics.

TABLE 5.5 **BOQ data, Example 5.5**

DEP VARIABLE: Y

ANALYSIS OF VARIANCE

SOURCE	DF	SUM OF SQUARES	MEAN SQUARE	F VALUE	PROB>F
MODEL	12	89725368.56	7477114.05	75.792	0.0001
ERROR	12	1183832.69	98652.72440		
C TOTAL	24	90909201.26			

ROOT MSE	314.0903	R-SQUARE	0.9870
DEP MEAN	2109.386	ADJ R-SQ	0.9740
C.V.	14.89013		

TABLE 5.5 Continued

BOQ data, Example 5.5

PARAMETER ESTIMATES

VARIABLE	DF	PARAMETER ESTIMATE	T FOR H0: PARAMETER=0	PROB > |T|
INTERCEP	1	142.15112		
X1	1	23.74365234		
X2	1	0.85304557		
X3	1	-0.20713669		
X4	1	9.56996481		
X5	1	12.76270486		
X6	1	-0.21055766		
X7	1	-6.07637585		
D1 (15)	1	723.57789	0.792	0.4440
D2 (18)	1	139.52092	0.228	0.8233
D3 (22)	1	-592.38447	-0.658	0.5229
D4 (23)	1	-15353.56021	-2.803	0.0160
D5 (24)	1	262.44392	0.189	0.8530

5.6

DIAGNOSTIC PLOTS

Plotting of residuals was discussed in Section 5.2. The analyst uses these plots to determine if model assumptions appear to be reasonable. In the case of a single regressor variable, a single plot of y versus x clearly reveals the picture of the relationship and can be an aid in determining if curvature is needed in the model or if any single data points are either suspicious or highly influential. Similar plots are not so informative if one is dealing in multiple regression. That is, for, say, k variables, plots of y versus x_1, y versus x_2, etc. do not truly highlight the role of the individual regressor variable since the interdependency among the x's, i.e., multicollinearity, has not been taken into account. A similar weakness exists if one attempts to gain information on plots of residuals against the individual regressor variables. Diagnostic plots can be used to display the impact of each individual regressor in the fitted model. In addition, plots are available that reveal which regressors should be transformed. Both of these types of plots involve the use of residuals. However, these types of residuals are not what many students of regression are accustomed to using.

Consider the linear regression model in matrix notation introduced in Chapter 3. That is,

$$\mathbf{y} = \mathbf{X}\boldsymbol{\beta} + \boldsymbol{\varepsilon}$$

Here $\mathbf{X}$ is of dimension $n \times p$.

Now, suppose we want to determine pictorially the role of the regressor x_j. We develop the machinery by partitioning $\mathbf{X}$ as

$$\mathbf{X} = [\mathbf{x}_j \vdots \mathbf{X}_{-j}] \quad (5.11)$$

Here $\mathbf{x}_j$ is the column of measurements on x_j, placed in the initial column of $\mathbf{X}$. The matrix $\mathbf{X}_{-j}$ contains measurements on all regressor variables *apart from* x_j. As a result, $\mathbf{X}_{-j}$ is of dimension $n \times (p-1)$. Suppose we were to regress $\mathbf{y}$ against the variables in $\mathbf{X}_{-j}$, obtaining the vector of residuals

$$\mathbf{y} - \mathbf{X}_{-j}(\mathbf{X}'_{-j}\mathbf{X}_{-j})^{-1}\mathbf{X}'_{-j}\mathbf{y} = \mathbf{e}_{y|\mathbf{x}_{-j}} \tag{5.12}$$

In addition, consider the regression of $\mathbf{x}_j$ against the variables in $\mathbf{X}_{-j}$, and the corresponding vector of residuals

$$\mathbf{x}_j - \mathbf{X}_{-j}(\mathbf{X}'_{-j}\mathbf{X}_{-j})^{-1}\mathbf{X}'_{-j}\mathbf{x}_j = \mathbf{e}_{x_j|\mathbf{x}_{-j}} \tag{5.13}$$

The types of residuals presented in Eq. (5.12) and (5.13), along with the standard set of residuals,

$$\mathbf{e} = \mathbf{y} - \mathbf{X}\mathbf{b} = \mathbf{e}_{y|\mathbf{X}}$$

will be used to generate the type of residual plots that can highlight the role of a prechosen regressor x_j in the regression involving all k regressors. It is interesting and instructive to note how these three sets of residuals are written in terms of HAT matrices; we already know that

$$\boxed{\mathbf{e} = \mathbf{e}_{y|\mathbf{X}} = (I - \mathbf{H})\mathbf{y}}$$

where, of course, $H = \mathbf{X}(\mathbf{X}'\mathbf{X})^{-1}\mathbf{X}'$. For the sets of residuals in Eq. (5.12) and (5.13),

$$\mathbf{e}_{y|\mathbf{X}_{-j}} = (I - \mathbf{H}_{-j})\mathbf{y} \qquad \text{and} \qquad \mathbf{e}_{x_j|\mathbf{X}_{-j}} = (\mathbf{I} - \mathbf{H}_{-j})\mathbf{x}_j$$

where $\mathbf{H}_{-j} = \mathbf{X}_{-j}(\mathbf{X}'_{-j}\mathbf{X}_{-j})^{-1}\mathbf{X}'_{-j}$.

There are four types of residual plots that can provide the type of diagnostic information necessary about the regressor variable x_j. These types of plots are as follows.

(i) Residual against predictor plots
(ii) Partial regression plots (added variable plots)
(iii) Component plus residual plots (partial residual plots)
(iv) Augmented partial residual plots

In what follows, we proceed to develop all four types of residual plots and discuss their strengths and weaknesses.

RESIDUAL AGAINST PREDICTOR PLOTS

The residual against predictor plots are quite straightforward and, in fact, special cases have been discussed in Chapter 2 and in Section 5.2. It is merely a scatter plot of

$$\boxed{\mathbf{e}_{y|\mathbf{X}} \quad \text{against} \quad x_j}$$

This is the type of plot around which diagnosis of model misspecification was illustrated with the population data in Example 5.1. As we indicated there and in Chapter 2, the plot should depict random scatter with a slope of zero. Model misspecification in the regressor x_j is indicated by a discernible trend. This type of plot *is not effective* for illustrating the impact of the model term $b_j x_j$ in the regression.

PARTIAL REGRESSION PLOTS (ADDED VARIABLE PLOTS)

The partial regression plots are available in many computer packages, including SAS. They provide a plot of

$$\boxed{\mathbf{e}_{y|\mathbf{X}_{-j}} \quad \text{against} \quad \mathbf{e}_{x_j|\mathbf{X}_{-j}}}$$

Here, of course, $\mathbf{e}_{y|\mathbf{X}_{-j}}$ is a set of residuals in which the linear dependency of y on all regressors *apart from* x_j *has been removed.* In a similar fashion $\mathbf{e}_{x_j|\mathbf{X}_{-j}}$ is a measure of x_j in which its *linear dependency with the other regressors has been removed.* Thus we have "y adjusted for the regressors in $\mathbf{X}_{-j}$" plotted against x_j, similarly adjusted.

The least squares slope of the partial regression plot is, indeed, b_j, the regression coefficient in the complete regression model. This plot nicely depicts the *role of* x_j in the regression. To provide further understanding, consider the complete model

$$\mathbf{y} = \mathbf{X}_{-j}\boldsymbol{\beta}_{-j} + \mathbf{x}_j\beta_j + \boldsymbol{\varepsilon} = \mathbf{X}\boldsymbol{\beta} + \boldsymbol{\varepsilon}$$

Now, premultiply by $\mathbf{I} - \mathbf{H}_{-j}$ and obtain

$$(\mathbf{I} - \mathbf{H}_{-j})\mathbf{y} = (\mathbf{I} - \mathbf{H}_{-j})\mathbf{X}_{-j}\boldsymbol{\beta}_{-j} + (\mathbf{I} - \mathbf{H}_{-j})\mathbf{x}_j\beta_j + (\mathbf{I} - \mathbf{H}_{-j})\boldsymbol{\varepsilon}$$

Noting that $(\mathbf{I} - \mathbf{H}_{-j})\mathbf{X}_{-j} = 0$, we have

$$\mathbf{e}_{y|\mathbf{X}_{-j}} = \beta_j \mathbf{e}_{x_j|\mathbf{X}_{-j}} + \boldsymbol{\varepsilon}^*$$

where $\boldsymbol{\varepsilon}^* = (\mathbf{I} - \mathbf{H}_{-j})\boldsymbol{\varepsilon}$. This development suggests that the partial regression plot has slope β_j. Thus, as we indicated, the least squares slope is b_j, an unbiased estimator of β_j.

The student of statistics should note from the above development that inference on individual coefficients in multiple regression *can actually be reduced to dealing with simple linear regression* (through the origin) in which the dependent and independent variables are special types of residuals, i.e., residuals in which both y and x_j have been adjusted for the balance of the regressors. Exercise 5.1 reveals more information along the same line. From what we learned in Chapter 2 (Eq. (2.16)), the coefficient b_j in this simple regression through the origin can be written

$$b_j = \frac{(\mathbf{e}_{y|\mathbf{X}_{-j}})'(\mathbf{e}_{x_j|\mathbf{X}_{-j}})}{(\mathbf{e}_{x_j|\mathbf{X}_{-j}})'(\mathbf{e}_{x_j|\mathbf{X}_{-j}})}$$

This happens to be, identically, the least squares estimator of β_j in the *multiple* linear regression involving all regressors. See Exercise 5.1.

Returning to the utility of the plot, if x_j enters the regression in a linear fashion, the partial regression plot should reflect a linear relationship through the origin. Variation around this plot reflects the strength of the model term $b_j x_j$. Also, the plots may uncover pictorially which (if any) single data points have a disproportionate amount of influence on the regression results. In particular, the plots will determine which regression coefficients are being most effected by these influential data points. The following example provides an illustration.

Example 5.6

Navy Manpower Data

In a study[2] to develop an equation for manning of a particular type of Naval functional activity, the following set of data was collected from ten sites.

Site	y (monthly man-hours)	x_1	x_2
1	1015.09	1160.33	9390.83
2	1105.18	1047.67	14942.33
3	1598.09	4435.67	14189.08
4	1204.65	5797.72	4998.58
5	2037.35	15409.60	12134.67
6	2060.42	7419.25	20267.75
7	2400.30	38561.67	11715.00
8	3183.13	36047.17	18358.29
9	3217.26	40000.00	20000.00
10	2776.20	35000.00	15000.00

SAS Institute Inc., *SAS User's Guide: Statistics, Version 5* (Cary, North Carolina: SAS Institute Inc., 1985).

The variables x_1 and x_2 represent workload factors that have a substantial influence on man-hours. A least squares fit to the data produced the multiple linear regression model

$$\hat{y} = 512.287 + 0.03875x_1 + 0.05894x_2$$

with $R^2 = 0.9468$ and $s = 216.672$. The set of residuals are as follows:

y_i	$\hat{y}_i$	$e_i = y_i - \hat{y}_i$	y_i	$\hat{y}_i$	$e_i = y_i - \hat{y}_i$
1015.09	1110.762	−95.672	2060.42	1994.405	66.015
1105.18	1433.611	−328.431	2400.30	2697.087	−296.787
1598.09	1520.501	77.589	3183.13	2991.215	191.915
1204.65	1031.578	173.072	3217.26	3241.156	−23.896
2037.35	1824.660	212.690	2776.20	2752.693	23.507

[2] *Procedures and Analyses for Staffing Standards Development: Data/Regression Analysis Handbook* (San Diego, California: Navy Manpower and Material Analysis Center, 1979).

The t-statistics for the coefficients of x_1 and x_2 are given by 8.278 and 3.594 respectively. This would seem to reflect the strength of each of the workload variables in explaining the man-hour variation. However, to gain some further pictorial insight into what role each variable plays in the regression, partial regression plots can be produced from the SAS PROC REG procedure. They appear in Figures 5.7 and 5.8.

The plot in Figure 5.7 involving x_1 appears to flatten at the top, suggesting that perhaps x_1 should enter the model in a curvilinear way. For the case of x_2, the plot reveals a linear trend, an appearance that illustrates the importance of x_2 as revealed in the t-statistic. The appearance of the plot on x_1 suggests the need for a transformation that produces a trend depicted by the figure. In this case a cube root transformation was attempted. The resulting regression was generated with the fitted model given by

$$\hat{y} = -387.072 + 9.9021x_1^{1/3} + 0.0596x_2$$

with

$$R^2 = 0.9776$$

and

$$s = 140.518$$

The residuals produced by this model are as follows:

y_i	$\hat{y}_i$	e_i
1015.09	896.565	118.525
1105.18	1203.146	−97.966
1598.09	1590.561	7.529
1204.65	1148.598	56.052
2037.35	2050.637	−13.287
2060.42	2164.544	−104.124
2400.30	2638.857	−238.557
3183.13	2982.994	200.136
3217.26	3161.152	56.108
2776.20	2760.616	15.584

The cube root transformation on x_1 resulted in an improved model with R^2 experiencing an increase and the value of s decreasing. In addition, the general complexion of the residuals improved. A reconstruction of the partial regression plots reveals a linear appearance on $x_1^{1/3}$ and x_2. These plots can be found in Figures 5.9 and 5.10.

FIGURE 5.7 **Partial regression plot of y against x_1; Naval manpower data**

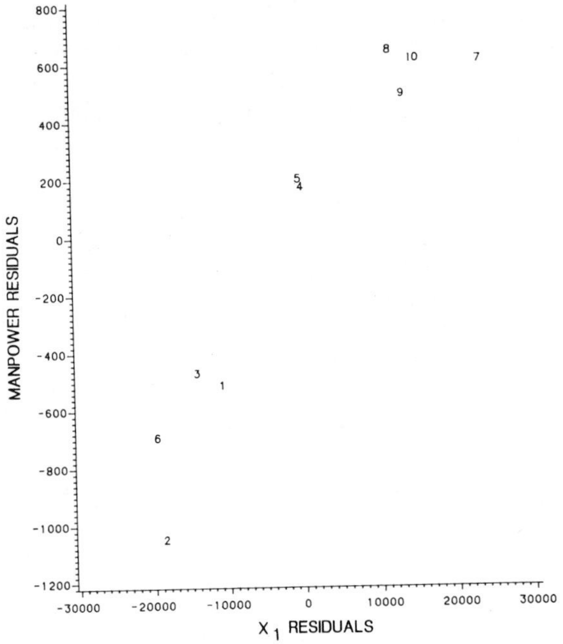

FIGURE 5.8 **Partial regression plot of y against x_2; Naval manpower data**

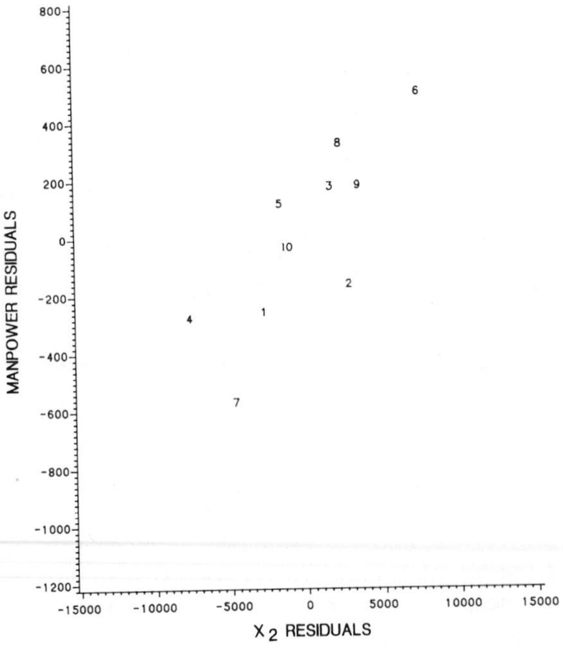

FIGURE 5.9 Partial regression plot of y against $x_1^{1/3}$ for the model involving x_2; Naval manpower data

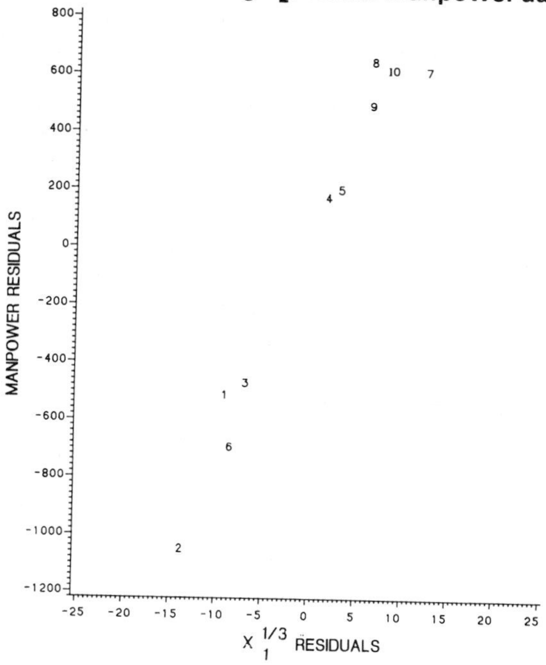

FIGURE 5.10 Partial regression plot of y against x_2 for the model involving $x_1^{1/3}$; Naval manpower data

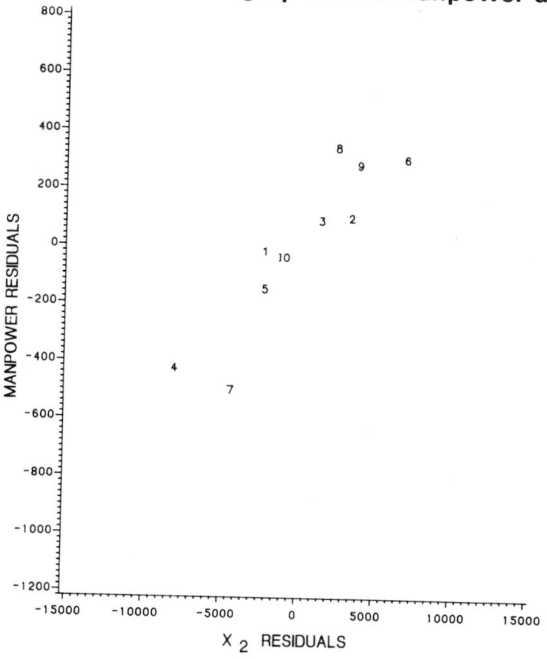

This example is not intended as a diagnostic to determine *exactly* what transformation is appropriate. However, it can aid the analyst in determining that a transformation will be helpful. In the example, there is no reason to believe that the *cube root* is the optimum choice. More on transformations, including hazards and advantages, appears in Chapter 7.

Some experienced regression analysts find that the most important utility in the partial regression plots is their value in displays. There are certainly nontechnical people who read reports describing the model-building procedure. However, they cannot understand or identify with regression coefficients, standard errors, *t*-statistics, etc., but need to ascertain the contribution from or role of each regressor. The partial plots provide this information in simple form. They represent the only way that a lay person can see the multiple regression mechanism simplified in two-dimensional pictures.

The partial regression plots were introduced by Mosteller and Tukey (1977). There are data sets in which partial regression plots may, in fact, not be effective in uncovering the need to transform one or more regressor variable. At times the component-plus-residual plot and augmented partial residual plots may be more effective.

COMPONENT-PLUS-RESIDUAL PLOTS (PARTIAL RESIDUAL PLOTS)

The component-plus-residual plot is generated by plotting the

$$\mathbf{e}_{y|\mathbf{X}} + x_j b_j \quad \text{against} \quad x_j$$

The title *component plus residual* describes what is plotted on the ordinate in this plot. Larsen and McCleary (1972) recommended this plot for usage. As in the case of the partial regression plot, the least squares slope of the scatter plot is b_j. One feature which distinguishes the two is that the abscissa is x_j, the metric on which the regressor under study is measured. In some cases this can result in a more effective detection of nonlinearity in the regressor x_j.

Example 5.7 **Navy Functional Activity Data**

Consider the data of Example 5.6. Figure 5.11 shows the component-plus-residual plot for x_1. Note, again, that the picture suggests that x_1 enters the multiple regression in a nonlinear fashion. Note the comparison to Figure 5.7, the partial regression plot. The plot for x_2, given in Figure 5.12, suggests that x_2 enters in a linear fashion.

FIGURE 5.11 Component-plus-residual plot for Naval manpower data

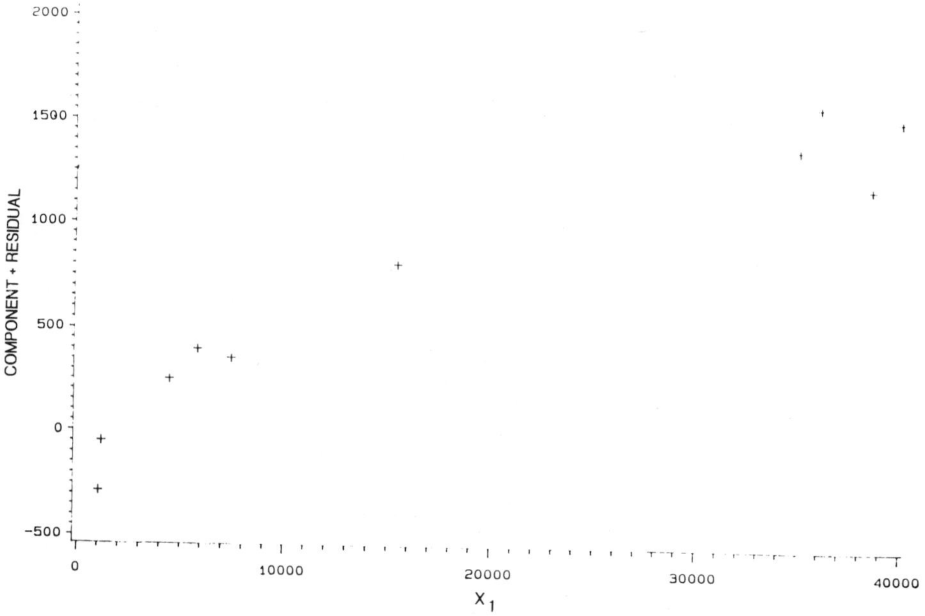

FIGURE 5.12 Component-plus-residual plot for Naval manpower data

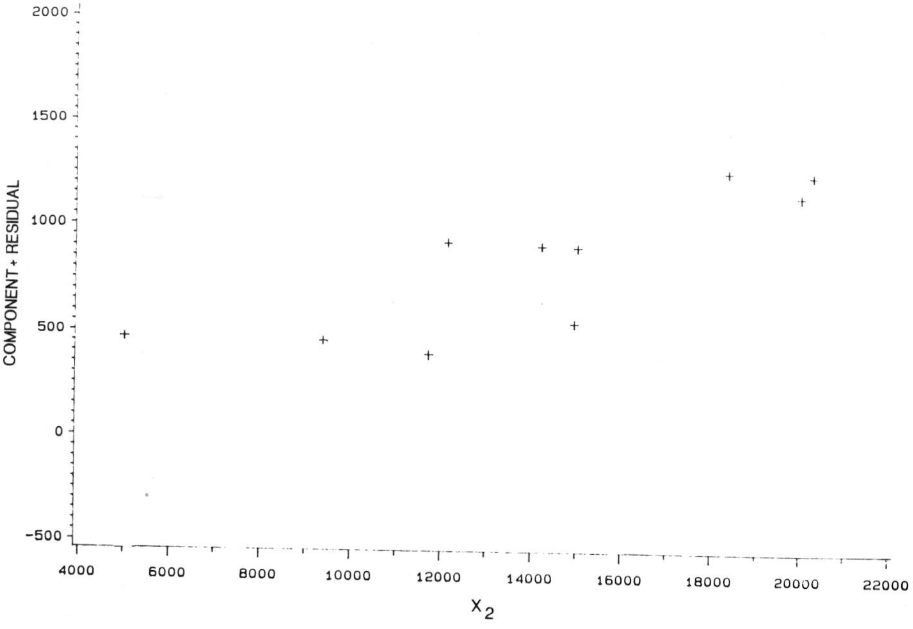

AUGMENTED PARTIAL RESIDUAL PLOTS

The diagnostic plots in (i), (ii), and (iii) on page 232 are designed to highlight nonlinearities in a particular regressor. Which is most effective depends on the particular application. Mallows (1986) introduced a new procedure that is a variation on the diagnostic in (iii), the partial residual plot. The plot appears to be more sensitive to the existence of nonlinear terms, i.e., the need to transform the regressor variables. As we indicated earlier, the partial regression plot suffers from the fact that the horizontal axis is that of a residual and thus is not the regressor x_j itself. In the case of the partial residual plot, Mallows conjectures that often the nonlinear contribution in the regressor, x_j, may be correlated with other regressors in the model. This collinearity with the other regressors does not allow nonlinearity to be exposed in the residuals for the complete model, i.e., the residuals in $\mathbf{e}_{y|\mathbf{X}}$. It is certainly true that, in methods (i), (ii), and (iii), the user is relying on the nonlinear contribution in x_j to be visible through the residuals, which are either all or a portion of what is plotted on the horizontal axis. However, it is certainly clear that the most effective way of exposing the need for a missing term or a transformed regressor *is to use it in the model.* However, rather than include the various possible transformations on the regressor x_j and thus induce undue collinearity, Mallows suggests the augmentation of the partial plot with a *quadratic term,* where this term is used both in the fitted model and in the plot. Thus the analyst fits a model that includes both a linear and quadratic term in the variable x_j. We then plot

$$\boxed{e_{y|\mathbf{X},x_j^2} + x_j b_j + x_j^2 b_{jj} \quad \text{against} \quad x_j}$$

where b_j is the coefficient of the linear term, b_{jj} is the coefficient of the quadratic term, and $e_{y|\mathbf{X},x_j^2}$ indicates the residual for the fitted model

$$\mathbf{y} = \mathbf{X}\boldsymbol{\beta} + \beta_j \mathbf{x}_j + \beta_{jj}\mathbf{x}_j^2 + \boldsymbol{\varepsilon}$$

One should not get the impression that the augmented partial residual plot is intended only to detect the need for a quadratic term. The introduction of the quadratic term is intended to provide a truncated approximation of the potential transform of x_j. If this quadratic term shows through the noise created by residuals, there is certainly reason to believe that the model will benefit from a transformation on x_j.

Example 5.8 ***Navy Functional Activity Data***

Consider again the data of Example 5.6. A model containing a linear and quadratic term in x_1 as well as a linear term in x_2 was fit for the purpose of plotting augmented partial residuals. The augmented partial residuals were formed from the fitted model and plotted against x_1. A similar plot was constructed for x_2. The plots are shown in Figures 5.13 and 5.14. As in the case of the partial residual plots, the augmented plots suggest that a transformation on x_1 may prove beneficial.

FIGURE 5.13 **Augmented partial residual plot for Naval manpower data**

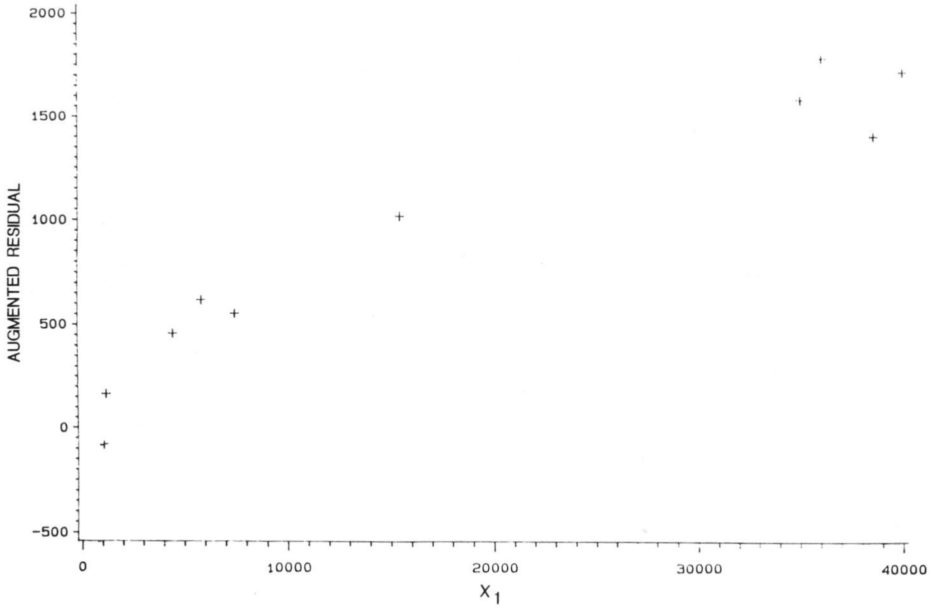

FIGURE 5.14 **Augmented partial residual plot for Naval manpower data**

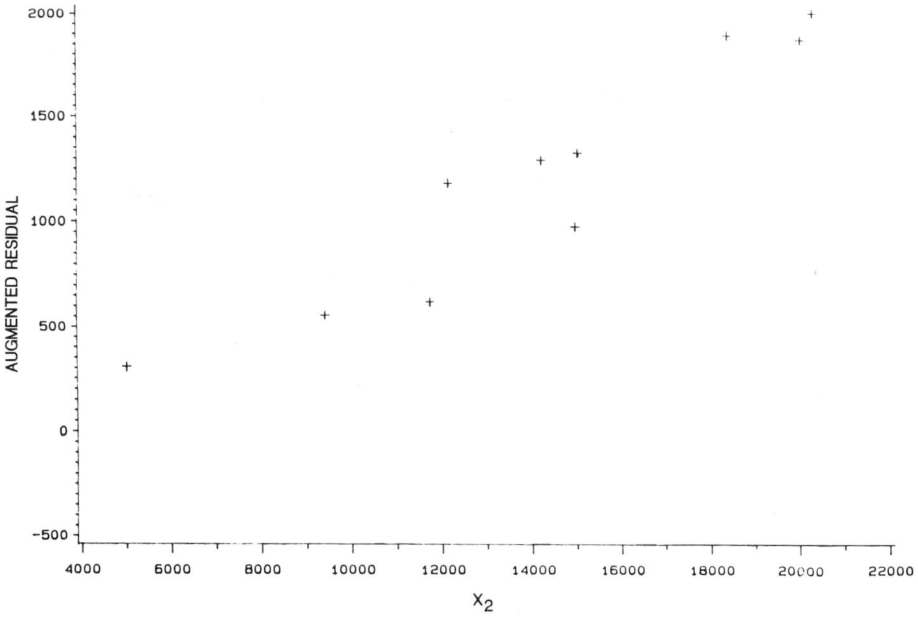

SUMMARY OF PERFORMANCE OF DIAGNOSTIC PLOTS

The diagnostic plots discussed here are designed to highlight graphically various aspects of the fitted regression. Among these are the need to transform, the relative importance of the individual regressors, and the obvious presence of *highly influential observations.* While detection of influential observations is discussed at length in Chapter 6, it becomes clear by our examples that the plots can indeed point to how some observations can be dominant in describing the role of each variable in the fitted model. For example, in the naval manpower data, observation 7 clearly is a large reason why x_1 should be transformed. See Figure 5.7. Even if the model is to remain untransformed, these plots can show which variables are important and which observations are influential.

DETECTION OF NONLINEARITIES

The augmented partial plots are perhaps the best at showing nonlinearities in the regressors. In many cases, the partial residual plots may be more effective than the partial regression plots. The residual against predictor plots may also highlight the need for transformation.

VISUAL DETECTION OF IMPORTANCE OF VARIABLES

The partial regression plots show the scatter of individual points around the line with the slope being the multiple regression coefficient. This scatter property depicts the sampling variation around b_j and shows a graphic picture of the standard error of b_j. As a result, it is a good summary of the relative strengths of individual regressors as they enter the multiple linear regression.

DETECTION OF HIGH INFLUENCE OBSERVATIONS

The partial regression plots are probably the best plots for highlighting influential observations. They are a nice graphic supplement to the influence diagnostics that are covered in the next chapter.

5.7 NORMAL RESIDUAL PLOTS

Residuals can produce information regarding the validity of the normality assumption on the model errors. Any reader who is prepared to study the details of the diagnostic procedure should first review why it is important to detect such departures from normality. Many acknowledge, either directly or obliquely, that there is a need to "check for normality." But why is it important? Recall that in

Chapters 2 and 3 an emphasis was placed on assumptions. Normality of the ε_i is required for the validity of hypothesis testing and confidence interval estimation. The prediction criteria (PRESS, C_p, etc.) used to compare models do not in any way require the assumption of normality. The usefulness of *model-fitting performance* criteria such as R^2, s^2, etc. is maintained whether or not the model errors are Gaussian. The normality assumption is not required to perform least squares estimation. However, the least squares estimators maintain better performance under normality than under conditions of nonnormality on the ε_i.

As a result, the analyst should understand that the least squares estimator being used in the nonnormal situation is not necessarily the optimal one, and perhaps can be improved upon. Indeed, where non-Gaussian errors are encountered, alternatives to least squares are available. These procedures, labeled *Robust Procedures* are discussed in Chapter 7.

In Chapter 2 we developed the methodology for plotting residuals for the purpose of detecting departures from normality of the model errors. The same technique is used in multiple linear regression, namely the ranked studentized residuals given by Eq. (5.3) are plotted against their expected values. An ideal plot is one that passes through the origin with a slope of 1.0. A substantial deviation from a straight line indicates a violation of the normality assumption on the ε_i. We do not choose to repeat the technical details here. However, the brief treatment that follows describes how fragile the residuals can be in transferring information about the model errors.

A deviation from the *normal* straight line can result from a model misspecification rather than from nonnormal errors. The validity of the normal error plots depends on *correct model specification.* In addition, if the sample size is small, it may be extremely difficult to detect any deviation from normality even though the deviation exists. Again, the structure of the residuals is such that a specific e_i does not yield ideal information required about its specific counterpart, ε_i. It is quite simple to illustrate that e_i, the ith residual, is a function of all of the ε's, not merely ε_i. Consider the vector of residuals:

$$\begin{aligned}\mathbf{y} - \mathbf{Xb} &= [\mathbf{I} - \mathbf{H}]\mathbf{y} \\ &= [\mathbf{I} - \mathbf{H}][\mathbf{X}\boldsymbol{\beta} + \boldsymbol{\varepsilon}] \\ &= \boldsymbol{\varepsilon} + \mathbf{X}\boldsymbol{\beta} - \mathbf{HX}\boldsymbol{\beta} - \mathbf{H}\boldsymbol{\varepsilon} \\ &= \boldsymbol{\varepsilon} - \mathbf{H}\boldsymbol{\varepsilon}\end{aligned}$$

Suppose we denote by h_{ij} the (i, j) element of the HAT matrix $\mathbf{H}$. Thus e_i can be written

$$e_i = \varepsilon_i - \sum_{j=1}^{n} h_{ij}\varepsilon_j \tag{5.14}$$

From Eq. (5.14), we see that a specific residual, e_j, is not merely a function of ε_j, but rather a linear combination of all errors. For small samples, the h_{ij} for $i \neq j$ will not be negligible. As n becomes larger, however, the e_i in (5.14) becomes dominated by the model error ε_i. (See Appendix B.6.) In the case of small and moderate size samples, there is a tendency for the right-hand side of (5.14) to be near normal even if the individual ε_i are not normal (see Gnadesikan (1977)). As a result, the user should not be optimistic about the success of detecting deviations from normality in the case of small samples.

5.8

FURTHER COMMENTS ON ANALYSIS OF RESIDUALS

The foregoing stressed the need for patience and caution on the user's part. It is unlikely that the reader will experience an immediate rash of uses for this warehouse of tools. In the case of the residual plots in Section 5.2, a user may investigate many plots for many data sets before encountering one that reveals a phenomenon that was not already obvious. Plots can be extremely informative. However, you cannot expect these plots to always produce dramatic (or even near dramatic) results. There are test cases (like the ones in this text) where positive results are shown. But we have deliberately used extreme examples so that the results appear clear and sharp. The user must be patient and continue to use residual analyses; there is no substitute for experience here. All of the analyses and plots discussed appear in commercial regression computer packages, and our presentation supplied some fundamental information regarding their use.

Much of the material in Chapters 6 and 7 relate to the presentation in this chapter. As indicated earlier, Chapter 6 deals with influence diagnostics, and Chapter 7 deals with nonideal conditions. Involved in the latter is the notion of relative importance of assumptions and transformations. The focus is on where the analyst should turn if, indeed, assumptions appear to be violated. Chapter 6 deals with tools that allow the analyst to determine the extent of the influence of outliers and high leverage data points. Again, residuals play an important role.

EXERCISES FOR CHAPTER 5

5.1 Consider the partial regression leverage plots in Section 5.6. Show that the least squares slope of the elements of $\mathbf{e}_{y|\mathbf{X}_{-j}}$ regressed against those of $\mathbf{e}_{x_j|\mathbf{X}_{-j}}$ is b_j, the slope of x_j in the multiple regression of y on $\mathbf{X}$.

5.2 Refer to the population data of Example 5.1. Form the residuals for the quadratic regression model as discussed in the text. Compute the R-student values for the years 1940 and 1950, and draw conclusions.

5.3 The following data represent the amount of money contributed by the alumni of a southern university for 14 graduating classes. The regressor variable is x, the number of years since graduation.

Observation	y	x
1	812.52	1
2	822.50	2
3	1212.00	3
4	1348.00	4
5	1301.00	8
6	2568.00	9
7	2527.00	10
8	2755.00	11
9	4391.00	12
10	5582.00	13
11	5548.00	14
12	6086.00	15
13	5764.00	16
14	8903.00	17

Data analyzed by the Statistical Consulting Center, Virginia Polytechnic Institute and State University, Blacksburg, Virginia, 1979.

(a) Fit a simple linear regression. Plot the residuals against x, the regressor variable.
(b) Compute the studentized residuals. Plot them against x.
(c) Do the plots in parts (a) and (b) suggest a need for any model alteration? Explain.

5.4 For the complete model of Exercise 4.6, generate the ordinary residuals and R-student values. Conduct formal outlier tests using Bonferroni critical values, and draw conclusions.

5.5 Consider the squid data of Example 3.2. For the "best" model selected in Exercise 4.1d, compute residuals, values of the HAT diagonals, and R-student values.

5.6 Often the appearance of the residuals is very much a function of the subset model that is being fit. Consider Example 5.2. These data were used as an illustration of a diagnosis of possible heterogeneous variance as well as outlier detection in Example 5.4. Fit the regression using only regressors x_2, x_4, and x_6. Do any dramatic changes occur in the complexion of the residuals? Would an outlier diagnosis produce the same or a different conclusion than that produced by the full model?

5.7 For the squid data of Example 3.2, compute R-student values and HAT diagonals using the full model. Does an outlier diagnosis produce the same or different conclusions than those produced by the model of Exercise 5.5?

5.8 Consider the sales data in Example 4.1. Generate the partial regression plots. From these plots, which one of the regressor variables appears to be the least important?

5.9 Consider the hospital data of Example 3.8. Using the full model, compute studentized residuals. With this residual analysis, comment on whether or not there appear to be "suspect" data points, i.e., hospitals for which data should be checked.

5.10 It is said that a subregion in the regressor/model space where prediction is expected to be poor is a subregion in which the fit of the model (from the relative size of the residuals) is expected to be relatively good. We alluded to this in Chapter 4. Using properties of residuals, explain what this means. Also make use of Var $\hat{y}_i$, Var$(y_i - \hat{y}_i)$, and properties of the HAT diagonals.

5.11 Consider the data of Exercise 4.17. Generate HAT diagonals and R-student values for the model with all terms linear in the regressors. Then generate the same diagnostics information with x_2 replaced by $x_2^{1/2}$ and x_1 replaced by ln x_1. Comment on the change in the complexion of these diagnostics.

5.12 Consider the data of Exercise 2.19. For a model that is linear in head diameter (simple linear regression model), construct an *augmented residual plot* against head diameter. Any comments? Does it suggest the need for a change in the model?

5.13 Consider the data of Exercise 3.19. Compute the standard error of the residuals at all eight data locations. Any comment? (*Comment*: These data are from a designed experiment, i.e., a 2^3 factorial experiment. One advantage is that all residuals have the same precision, i.e., the Var$(y_i - \hat{y}_i)$ are equal at all eight data points. This suggests that the quality, i.e., sensitivity of outlier detection is the same at all data points.)

5.14 Consider Exercise 5.13. The question deals in how the size of the HAT diagonal impacts the sensitivity of the test on an outlier. *For students who are relatively advanced in the field of linear models*, how does the size of h_{ii} impact the *power* of the test of

$$H_0: \Delta_i = 0$$
$$H_1: \Delta_i \neq 0$$

discussed in Section 5.5? (*Hint*: Consult Section 5.4 and begin with the PRESS residuals.)

5.15 Consider a multiple linear regression model

$$y_i = \mathbf{x}_i'\boldsymbol{\beta} + \varepsilon_i \qquad \text{where} \quad E(\varepsilon_i) = \Delta_i \neq 0$$

That is, there is a mean shift outlier at the ith data point. Find $E(e_i)$. Find $E(e_{i,-i})$.

5.16 The material in Chapter 5 dealing in outlier detection deals almost solely with the mean shift outlier model discussed in Section 5.5. The variance inflation outlier model can be written

$$y_i = \mathbf{x}_i'\boldsymbol{\beta} + \varepsilon_i \qquad (i = 1, 2, \ldots, n)$$

with the model errors being normal with

$$E\varepsilon_j^2 = \sigma^2 + \sigma_\Delta^2 \qquad \text{and} \qquad E\varepsilon_i^2 = \sigma^2 \text{ for all } i \neq j$$

and all model errors retain the property that $E(\varepsilon_i) = 0$. It turns out that the appropriate test statistic for testing $H_0: \sigma_\Delta^2 = 0$ is the R-student statistic. Thus the use of R-student is proper no matter which outlier model is prevalent.

(a) For students who are reasonably advanced in statistics, show that $(R\text{-student})_j$ has a t-distribution under H_0 above.

(b) Develop and display the distribution of $(R\text{-student})_j^2$ under $H_0: \sigma_\Delta^2 \neq 0$.

5.17 Consider a data set with the following two sets of diagnostics for two specific observations, i and j.

$$(R\text{-student})_i = 3.8 \qquad h_{ii} = 0.15$$
$$(R\text{-student})_j = 2.3 \qquad h_{jj} = 0.72$$

Will observation i or j have the largest amount of *influence* on the regression? Explain.

5.18 Consider the following data set:

Observation	x_1	x_2	x_3	y
1	12.980	0.317	9.998	57.702
2	14.295	2.028	6.776	59.296
3	15.531	5.305	2.947	56.166
4	15.133	4.738	4.201	55.767
5	15.342	7.038	2.053	51.722
6	17.149	5.982	−0.055	60.466
7	15.462	2.737	4.657	50.715
8	12.801	10.663	3.048	37.447
9	13.172	2.039	8.738	55.270
10	16.125	2.271	2.101	59.289
11	14.340	4.077	5.545	54.027
12	12.923	2.643	9.331	53.199
13	14.231	10.401	1.041	41.896
14	15.222	1.220	6.149	63.264
15	15.740	10.612	−1.691	45.798
16	14.958	4.815	4.111	58.699
17	14.125	3.153	8.453	50.086
18	16.391	9.698	−1.714	48.890
19	16.452	3.912	2.145	62.213
20	13.535	7.625	3.851	45.625
21	14.199	4.474	5.112	53.923
22	16.565	8.546	8.974	56.741
23	13.322	8.598	4.011	43.145
24	15.949	8.290	−0.248	50.706
25	14.123	0.578	−0.543	56.817

It is known *a priori* that in observations 10, 16, and 17 there were some difficulties in measuring the response y. Using the mean shift outlier model test simultaneously

$$H_0\colon \Delta_{10} = 0$$
$$\Delta_{16} = 0$$
$$\Delta_{17} = 0$$

and make conclusions. You may wish to use indicator variables.

REFERENCES FOR CHAPTER 5

1. Andrews, D.F. 1972. Plots of high-dimensional data. *Biometrics* 28: 125–136.
2. Andrews, D.F. 1971. Significance tests based on residuals. *Biometrika* 58, 139–148.
3. Andrews, D.F., and D. Pegibon. 1978. Finding the outliers that matter. *Journal of the Royal Statistical Society, Series B* 40: 85–93.
4. Anscombe, F.J. 1961. Examination of residuals. *Proceedings of the Fourth Berkeley Symposium on Mathematical Statistics and Probability* I: 1–36.
5. Anscombe, F.J., and J.W. Tukey. 1963. The examination and analysis of residuals. *Technometrics* 5: 141–160.
6. Beckman, R.J., and R.D. Cook. 1983. Outlier.....s. *Technometrics* 25: 119–145.
7. Beckman, R.J., and H.J. Trussell. 1974. The distribution of an arbitrary studentized residual and the effects of updating in multiple regression. *Journal of the American Statistical Association* 69: 199–201.
8. Behnken, D.W., and N.R. Draper. 1972. Residuals and their variance patterns. *Technometrics* 14: 101–111.
9. Belsley, D.A., E. Kuh, and R.E. Welsch. 1980. *Regression Diagnostics: Identifying Influential Data and Sources of Collinearity.* New York: John Wiley & Sons.
10. Cook, R.D., and S. Weisberg. 1982. *Residuals and Influence in Regression.* New York: Chapman and Hall.
11. Cox, D.R., and E.J. Snell. 1971. On test statistics calculated from residuals. *Biometrika* 58: 589–594.
12. Gentleman, J.F., and M.B. Wilk. 1975. Detecting outliers in a two-way table: I. Statistical behavior of residuals. *Technometrics* 17: 1–14.
13. Gentleman, J.F., and M.B. Wilk. 1975. Detecting outliers. II. Supplementing the direct analysis of residuals. *Biometrics* 31: 387–410.
14. Gnanadesikan, R. 1977. *Methods for Statistical Analysis of Multivariate Data.* New York: John Wiley & Sons.
15. Larsen, W.A., and S.J. McCleary. 1972. The use of partial residual plots in regression analysis. *Technometrics* 14: 781–790.
16. Mallows, C.L. 1896. Augmented partial residuals. *Technometrics* 28: 313–319.
17. Miller, R. 1965. *Simultaneous Inference.* New York: McGraw-Hill.
18. Mosteller, F., and J.W. Tukey. 1977. *Data Analysis and Regression.* Reading, Massachusetts: Addison-Wesley.
19. Prescott, P. 1975. An approximate test for outliers in linear regression. *Technometrics* 17: 129–132.
20. Prescott, P. 1979. Critical values for a sequential test for many outliers. *Applied Statistics* 28: 36–39.
21. Snee, R.D. 1971. A note on the use of residuals for examining the assumptions of covariance analysis. *Technometrics* 13: 430–437.
22. Tietjen, G.L., R.H. Moore, and R.J. Beckman. 1973. Testing for a single outlier in simple linear regression. *Technometrics* 15: 717–721.
23. Weisberg, S. 1980. *Applied Linear Regression.* New York: John Wiley & Sons.

CHAPTER 6

INFLUENCE DIAGNOSTICS

In Chapter 5, we focused considerable attention on studying residuals in order to shed light on possible violation of assumptions. Among the techniques presented was the so-called outlier analysis, which is designed to highlight suspect data points. Outliers are detected for further checking by the analyst, the lab technician, recorder, or whomever is involved in the data-taking process. A natural concern is that an erroneous observation will exert an undue amount of influence on the regression results—influence that is counterproductive. However, if the data point does reveal a serious model deficiency, then it is quite likely that much can be learned by it.

The conditions detected in the outlier diagnostics in Chapter 5 are errors in the *y-direction*, i.e., model shifts that produce anomalies in the measured response. The symptom, of course, is a residual, which is larger than would realistically be produced by chance. The influence on regression statistics from such an observation is the "fallout" that results from the regression being "pulled" toward the errant measured response. The analyst needs to be able to identify these observations and determine the extent to which predicted values, estimated regression coefficients, performance criteria, etc. are influenced by them. The methodology exists in standard regression computer packages for both identification and assessment of the extent of the influence. The diagnostic material presented in this chapter produces criteria that, in harmony with the least squares procedure, simultaneously build and criticize the model being considered.

6.1

SOURCES OF INFLUENCE

In certain types of data sets, it is quite common for a single observation or a small subset to determine the regression coefficients almost completely. In these cases, the majority of the data may have little impact. What causes a regression data point to be a high influence observation? First it may be an outlier as described previously. However, all high influence observations do not deserve indictment. Indeed, all high influence observations are not due to errors in the y-direction. An influential observation may be a legitimate and very important part of the data set. Influence can occur when a single observation is extreme in the x-direction; i.e., it is a disproportionate distance away from the data centroid in the x's, even though it is a proper observation and does not necessarily represent evidence of a model fallacy. For example, consider Figure 6.1(a) for the case of a single regressor. The single *high leverage* observation determines the slope of the regression almost entirely by itself. However, there is no evidence that this observation is an outlier, *in the sense of model fallacy*. On the other hand, consider Figure 6.1(b). The single observation that appears "off the trend" set by the rest of the data presumably was produced by a model shift or perhaps heterogeneous variance. Its impact on the estimated slope will be minimal compared to the isolated observation in Figure 6.1(a), though the intercept may be severely influenced by the observation.

FIGURE 6.1 ***(a)*** **Single influential observation remote from center** ***(b)*** **Single observation with error in *y*-direction**

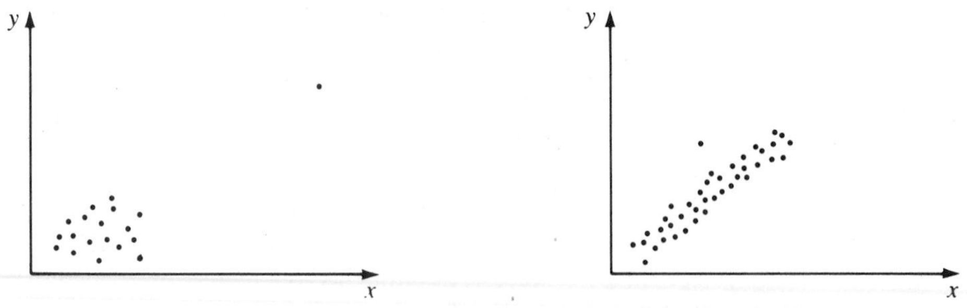

Figure 6.1 represents examples of what can occur, to different degrees, in practice. It is important that both be identified since both are instructive to the

analyst. In the case of Figure 6.1(a), additional data to "fill the gap" would be helpful. In the absence of supplemental information, the single high influence observation should be checked carefully using all possible resources. There is a clear danger involved with allowing one piece of information to totally dictate. For Figure 6.1(b), the comments made in Chapter 5 concerning the outlier analysis apply. Note that for the outlier in Figure 6.1(b), the symptom and thus the diagnostic information lies in the residual, vis-a-vis the *R*-student statistic. In Figure 6.1(a), one expects the residual to be relatively small. The diagnostic process in this case begins with the residual and the HAT diagonal value.

6.2

DIAGNOSTICS: RESIDUALS AND THE HAT MATRIX

Here we focus on diagnostic information that determines what individual data points exert disproportionate influence on the regression. We begin by recalling the **X** matrix

$$\mathbf{X} = \begin{bmatrix} \mathbf{x}_1' \\ \mathbf{x}_2' \\ \vdots \\ \mathbf{x}_n' \end{bmatrix}$$

and the vector of responses

$$\mathbf{y} = \begin{bmatrix} y_1 \\ y_2 \\ \vdots \\ y_n \end{bmatrix}$$

where the *i*th data point is given by $[\mathbf{x}_i' \vdots y_i]$. The diagnostics evolve, for the most part, from residuals $e_i = y_i - \hat{y}_i$ and elements of the HAT matrix $\mathbf{H} = \mathbf{X}(\mathbf{X}'\mathbf{X})^{-1}\mathbf{X}'$. The standardization of the residual via the *R*-student value (see Chapter 5), given by

$$t_i = \frac{e_i}{s_{-i}\sqrt{1 - h_{ii}}} \tag{6.1}$$

is a natural diagnostic used to detect data points that exert influence created by errors in the *y*-direction. In Chapter 5 some emphasis was placed on *R*-student as an outlier test statistic in a hypothesis-testing framework. Here the total focus is on its diagnostic value.

A large value of *R*-student (in magnitude) provides a signal that an unusually large error in fit has occurred at the *i*th data point. *R*-student does not assess the extent of that observation's influence nor does it determine what statistics are being influenced. A second type of signal, one that provides diagnostic information regarding what data points exert *high leverage*, is the HAT diagonal (see Section 3.9)

$$h_{ii} = \mathbf{x}_i'(\mathbf{X}'\mathbf{X})^{-1}\mathbf{x}_i \qquad \textbf{(6.2)}$$

Since the HAT diagonal provides a measure of standardized distance from the point $\mathbf{x}_i$ to $\bar{\mathbf{x}}$, which is the data center in the x's, it is a natural measure of leverage. It will highlight observations that are *extreme in the* x's. The reader should note that h_{ii} does not involve the y's. Thus it is intended as a "red flag" to expose observations that *potentially* exert undue influence on at least one regression coefficient as well as performance criteria.

A point that possesses a large HAT diagonal, but also *very closely* follows the trend in the model set by the other data points, will not exert undue influence *on the regression coefficients*. Contrast, for example, Figure 6.1(a) with Figure 6.2. In Figure 6.2, it is clear that the remote observation, unlike that in Figure 6.1(a), is not dominating the trend. This remote observation will have negligible impact on the slope and intercept. However, it will have a very positive effect on the standard error of both slope and intercept. Thus the data point in Figure 6.2 provides *reinforcement* of the regression already established and thus enhances the performance of the fitted model.

FIGURE 6.2 **Large HAT diagonal but not influential observation**

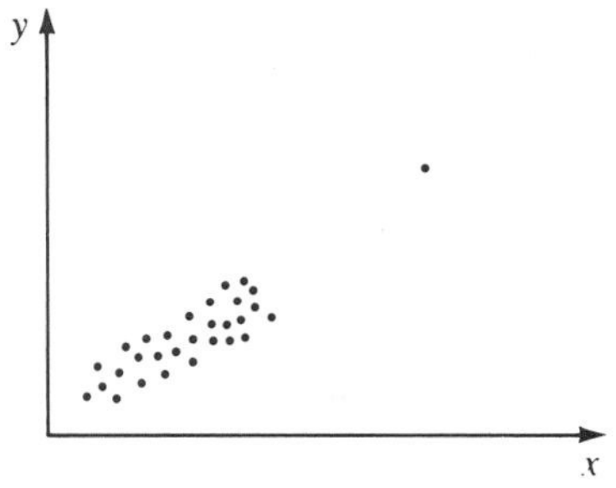

The figures shown here are certainly an oversimplification of what one faces in the case of more than one regressor. However, the R-student values and h_{ii} values remain the proper diagnostics to isolate data points that are potentially exerting disproportionate influence. The reader may view the influence as being the offspring of a collaboration between the HAT diagonal (leverage) and the nature of the fit of the model to the point in question (the residual). Obviously there will be a tendency for an observation with a large HAT diagonal to experience a small residual. The extent of the influence exerted by the high leverage point will be a function of how well the observation follows the model dictated by the other data points. Certainly, the combination that produces the greatest influence is a data point that contains a large HAT diagonal accompanied by a relatively large residual.

WHY IS IT IMPORTANT TO DETECT INFLUENTIAL OBSERVATIONS?

The reader should have noticed by now that the material in this chapter is very much related to that of Chapter 5 dealing in the detection of outliers. Both the concept of leverage and an outlying observation describe an unusual condition in an observation. A high leverage observation $\mathbf{x}_i$ (strictly involving measurements on the regressors and the model terms) is remote from the rest of the data. Outliers are observations that are not fit well by the model in question. *Which data points are high leverage points and which data points are outliers may change if the model formulation is changed.* Obviously, which observations are highly influential (if any) may change if the model formulation is changed. Now, are all high leverage observations influential and are all outliers influential? Clearly if a high leverage observation is one at which the model fits poorly, its presence in the data set does highly influence some aspect of the regression. However, consider Figure 6.3. It is obvious that point B is an influential observation. If it were removed, the slope of the regression line would change dramatically. On the other hand, the influence exerted by point A is not so great. Yet it certainly does appear as if A is an outlier. Point B and the remote point in Figure 6.2 are high leverage observations. But, in terms of effect on slope and intercept of the regression, point B is influential while the remote point in Figure 6.2 is not. Do we say that point B is an outlier? Note that a fitted line with B included in the data set will result in a fairly small residual at B.

FIGURE 6.3 **Point *B* is clearly influential**

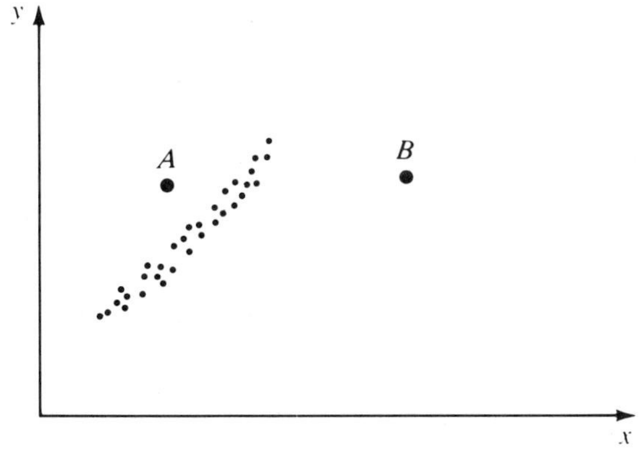

Some summary statements are as follows:

Outliers are not necessarily influential (depends on leverage)

High leverage observations are not necessarily influential (note Figure 6.2)

Influential observations are not necessarily outliers (note Figure 6.1(a))

The utility of influence diagnostics should be obvious to the reader. As we shall soon see by illustration, there are times when the analyst observes rather strange results, i.e., results that are difficult to interpret. There may be a negative coefficient that makes no scientific sense. An important regressor may be statistically insignificant, or a constructed model that makes good scientific sense may have poor performance properties. These unreasonable conditions of the regression may be due to the influence of a single observation or a few observations. The detection of "high impact" observations can often uncover the source of the problem.

HAT *DIAGONAL, R-STUDENT; HOW LARGE?*

To detect potentially high influence observations, we begin by noting the data points that have a large HAT diagonal, a large R-student value (in magnitude), or both. In Chapter 5, guidelines were given for what size R-student should be before it deserves special attention. However, much of that was clothed in the context of hypothesis testing. Here, we are isolating data points that *require further inspection*, or isolating areas in the regressor variables for further experimental exploration. As a result, the analyst should not be held to strict "yardstick values." Of course, one must be aware that the ± 2 guidelines implies that the *residual deviates from zero by two estimated standard errors*, and this certainly suggests that we should relegate the point to the category of further diagnostic analysis.

In the case of the HAT diagonal, we make use of the fact that

$$\sum_{i=1}^{n} h_{ii} = p \tag{6.3}$$

where p is the number of model parameters. (See Eq. (3.43).) As a result, the average h_{ii}, namely p/n, provides a norm. Surely *any* $h_{ii} > 2p/n$ *has potential for exerting strong influence on the results.* The relative magnitudes of p and n must be taken into account. Unfortunately there will be data sets in which $2p/n > 1$, and thus no HAT diagonal can exceed $2p/n$. Of course, in these cases, the guideline does not apply. In Example 6.1, we illustrate how important regression results and conclusions can be influenced dramatically by single observations. It should then become clear why it is important that these results be identified.

Example 6.1 ***BOQ Data***

Consider Example 5.2 with the BOQ workload data for the $n = 25$ observations appearing in Table 5.2. The purpose of that illustration was to show the utility of studentized residuals to highlight violation of specific assumptions. In particular, the homogeneous variance assumption was questioned based on the nature of the studentized residuals for the large installations. As the initial phase of an influence diagnostic study, we redisplay the residuals, HAT diagonal values, and the *R*-student values as follows:

Installation	e_i	h_{ii}	t_i (*R*-student)
1	−29.755	0.2573	−0.0736
2	−31.186	0.1609	−0.0726
3	−196.106	0.1614	−0.4594
4	−75.556	0.1631	−0.1762
5	−180.783	0.1475	−0.4196
6	−242.993	0.1589	−0.5704
7	313.923	0.1829	0.7532
8	−176.059	0.3591	−0.4720
9	128.744	0.2808	0.3246
10	39.994	0.1295	0.0914
11	635.923	0.1241	1.5537
12	184.323	0.2024	0.4426
13	−81.716	0.0802	−0.1818
14	−9.966	0.0969	−0.0224
15	−665.211	0.5576	−2.5192
16	−19.605	0.4024	−0.0541
17	−470.978	0.3682	−1.3310
18	418.462	0.4465	1.2566
19	437.994	0.0868	1.0074
20	−870.070	0.3663	−2.8657
21	826.496	0.0704	2.0538
22	−323.894	0.7854	−1.6057
23	−160.276	0.9885	−5.2423
24	413.265	0.8762	3.2093
25	135.029	0.5467	0.4299

We may find it informative to determine the influence exerted by one or more of the large installations where large *R*-student values are present. From a quick diagnosis, it appears that installations 23 and 24 are very influential. Both *R*-student values exceed 2.0 in magnitude, and the HAT diagonal values indicate strong leverage, especially in the case of installation 23 where the HAT diagonal is 0.9885, far exceeding the guideline of $2p/n = 16/25 \cong 2/3$. Based on the other *R*-student and h_{ii} values, other data points that *potentially* exert strong influence are installations 15, 20, 21, and 22. On the other hand, one might suspect that installation 21 is not as influential as some of the others since the HAT diagonal is very small. Installation 22 cannot be categorized as an outlier, but the HAT diagonal is 0.7854.

Some curious results occurred regarding the signs of the regression coefficients. In this type of application, positive signs on regression coefficients are expected. Yet, the coefficients of variables x_1, x_4, and x_6 are negative. (See regression equation in Example 5.2.) We will see subsequent evidence that the data of installation 23 has a profound influence on all three of these coefficients.

To illustrate the extent of the influence of data point 23 on the results, we present Table 6.1, which contains the set of coefficients, estimated standard errors, R^2, and s *with and without* data point 23.

The difference between these two sets of results is dramatic. The coefficients b_1 and b_4 become positive when data point 23 is eliminated. In addition, b_3 and b_5 change from positive to negative. The conclusions regarding the roles of the variables change considerably. Before setting aside the point, the coefficient b_1 is negative and experiences a t-value for H_0: $\beta_1 = 0$ of -1.595, a result that tempts the analyst to conclude that x_1 provides a marginal explanation of the response in the presence of the other variables. After removing data point 23, b_1 is positive and seemingly enjoys a strong role with a t-value of about 4.91. The coefficient b_4, which is *significant and negative* before the removal of data point 23, becomes positive and insignificant after removal. The coefficient b_6 remains negative but becomes statistically insignificant on the basis of the t-statistic. The change in coefficient b_7 also depicts a dramatic alteration in the conclusions drawn about it.

TABLE 6.1 **Results with and without installation 23**

With Installation 23		Without Installation 23	
$R^2 = 0.9613$	$s_{b_0} = 237.8140$	$R^2 = 0.9854$	$s_{b_0} = 148.8617$
$s = 455.1670$	$s_{b_1} = 0.8047$	$s = 284.6035$	$s_{b_1} = 4.2890$
$b_0 = 134.9680$	$s_{b_2} = 0.5162$	$b_0 = 171.4734$	$s_{b_2} = 0.3307$
$b_1 = -1.2838$	$s_{b_3} = 1.8464$	$b_1 = 21.0456$	$s_{b_3} = 1.1635$
$b_2 = 1.8035$	$s_{b_4} = 10.1716$	$b_2 = 1.4263$	$s_{b_4} = 8.4384$
$b_3 = 0.6692$	$s_{b_5} = 14.7461$	$b_3 = -0.0893$	$s_{b_5} = 9.4528$
$b_4 = -21.4226$	$s_{b_6} = 4.2202$	$b_4 = 7.6503$	$s_{b_6} = 3.3020$
$b_5 = 5.6192$	$s_{b_7} = 6.3659$	$b_5 = -5.3023$	$s_{b_7} = 6.8140$
$b_6 = -14.4803$		$b_6 = -4.0747$	
$b_7 = 29.3248$		$b_7 = 0.3319$	
$t_{b_0} = 0.5675$		$t_{b_0} = 1.1520$	
$t_{b_1} = -1.5954$		$t_{b_1} = 4.9069$	
$t_{b_2} = 3.4938$		$t_{b_2} = 4.3130$	
$t_{b_3} = 0.3624$		$t_{b_3} = -0.0768$	
$t_{b_4} = -2.1061$		$t_{b_4} = 0.9066$	
$t_{b_5} = 0.3811$		$t_{b_5} = -0.5609$	
$t_{b_6} = -3.4312$		$t_{b_6} = -1.2340$	
$t_{b_7} = 4.6065$		$t_{b_7} = 0.0487$	

Example 6.1 illustrates how a well-meaning observational study or experimental effort can be confusing or fraught with considerable uncertainty because of a single observation. We saw evidence in Chapter 5 that installation 23 qualified as an outlier. However, at that point, we had no measure of its impact on the results. Here we see that our conclusions regarding the roles of several regressors are strongly impacted by installation 23. In Chapter 5 the assumption of homogeneous variance was questioned on the basis of plots of studentized residuals. One may question which assumption is violated, but there can be no doubt that the residuals at the large activities are considerably larger than for the smaller ones. In addition, we believe there is a serious problem involving installation 23. The possible reasons for the gross inconsistency at this installation are numerous. At this point the analyst is certainly obliged to investigate rather than allow the conclusions to be false because of an erroneous observation or, perhaps, faulty modeling.

One should not feel committed to "delete and recompute" as we have done in Example 6.1. In what follows diagnostics are studied that allow the analyst to determine what is being influenced and to what extent by every observation, with the amount of work involved being merely the *computation of one regression.*

6.3

DIAGNOSTICS THAT DETERMINE EXTENT OF INFLUENCE

The R-student (or studentized residual) and HAT diagonal values reveal which individual observations have potential for exerting excessive influence. A series of conceptually meaningful statistics can be used to determine the extent of the influence. Standard regression computer software allows the user to observe the change that would be experienced in certain key statistics *if the* ith *data point were set aside from the rest of the data.* This then allows easy diagnosis of which statistics are influenced and *to what degree.* We saw in Example 6.1 that if we set aside a single high influence observation, the result can seriously alter the conclusions drawn by the analyst. In this section, readily accessible diagnostics are outlined that provide a *with* $\mathbf{x}_i$ contrasted against a *without* $\mathbf{x}_i$. We can compute the diagnostics without physically removing the ith point. (We removed observation 23 for illustrative purposes in Example 6.1.) In other words, information is available regarding the changes that occur if each observation were deleted from the data set.

The wealth of information that we can generate is beneficial when used wisely and with restraint by the analyst. These diagnostics are inexpensive and have rapidly become a major component of modern regression analysis. The computational ease is very much related to the simplicity with which the PRESS residuals (Chapter 4) were obtained. Recall that the expression for the ith PRESS residual, given by

$$e_{i,-i} = \frac{e_i}{1 - h_{ii}}$$

is developed in Appendix B.4. For the student interested in the development of the diagnostics discussed in this chapter, Appendix B.7 contains derivations of results.

INFLUENCE ON THE FITTED VALUE (DFFITS)

One may gain some insight into what influence observation i has on the predicted value or fitted value $\hat{y}_i$. A proper diagnostic is given by

$$\boxed{(\text{DFFITS})_i = \frac{\hat{y}_i - \hat{y}_{i,-i}}{s_{-i}\sqrt{h_{ii}}}} \tag{6.4}$$

The "DF" prefix means the *difference* between the result with $\mathbf{x}_i$ and without $\mathbf{x}_i$. In this case, it is the difference between the fitted value $\hat{y}_i$ and the predicted value $\hat{y}_{i,-i}$, i.e., the predicted response at $\mathbf{x}_i$ where the regression is used without the benefit of the ith point. The denominator in (6.4) merely provides a standardization since $\text{Var}\ \hat{y}_i = \sigma^2 h_{ii}$. The "$-i$" implies, as always, that the ith observation is not involved in the computation. Thus the value of $(\text{DFFITS})_i$ for the ith point represents the *number of estimated standard errors that the fitted value $\hat{y}_i$ changes* if the ith point is removed *from the data set*. The computational procedure for the value for $(\text{DFFITS})_i$ is interesting and instructive. It turns out that $(\text{DFFITS})_i$ is computed as follows:

$$\begin{aligned}(\text{DFFITS})_i &= \left[\frac{e_i}{s_{-i}\sqrt{1-h_{ii}}}\right]\left[\frac{h_{ii}}{1-h_{ii}}\right]^{1/2} \\ &= (R\text{-student})_i \left[\frac{h_{ii}}{1-h_{ii}}\right]^{1/2}\end{aligned} \tag{6.5}$$

(For details, see Appendix B.7.) Every term in (6.5) is computed from a single regression. (Recall that s_{-i} is computed by using Eq. (5.6).) Notice that DFFITS is essentially the R-student value, *magnified or shrunk* according to the leverage measure $[h_{ii}/(1-h_{ii})]^{1/2}$. Clearly if the data point is an outlier (larger R-student in magnitude) or is a high leverage point (h_{ii} close to 1.0), DFFITS will tend to be large. However, if $h_{ii} \cong 0$, the effect of R-student will certainly be moderated. On the other hand, a point with extremely high leverage will produce a small $(\text{DFFITS})_i$ if a near-zero residual leads to an unusually small $(R\text{-student})_i$. Thus, as expected, the diagnostic in Eq. (6.5) is produced by the impact of leverage and errors in the y-direction.

INFLUENCE ON THE REGRESSION COEFFICIENTS

For each regression coefficient, the influence diagnostics provide a statistic, which gives the number of *standard errors that the coefficient changes* if the ith observation were set aside. Specifically, we define

$$\boxed{(\text{DFBETAS})_{j,i} = \frac{b_j - b_{j,-i}}{s_{-i}\sqrt{c_{jj}}}} \tag{6.6}$$

where c_{jj} is the jth diagonal element of $(\mathbf{X}'\mathbf{X})^{-1}$. As a result, the denominator is an estimate of the standard error of b_j, the jth regression coefficient. The statistic $b_{j,-i}$ denotes the jth regression coefficient, computed without the use of the ith observation. A large value (in magnitude) of $(\text{DFBETAS})_{j,i}$ indicates that the ith observation has a sizable impact on the jth regression coefficient. The sign of $(\text{DFBETAS})_{j,i}$ may also be meaningful. Suppose, for example, that the jth regression coefficient b_j is negative in a setting in which a negative coefficient is meaningless and thus uninterpretable. From Eq. (6.6), we see that if $(\text{DFBETAS})_{j,i}$ is negative and relatively large in magnitude, it is likely that the negative coefficient *can be attributed to the* ith *observation.* This situation clearly suggests a need to do as much checking on the ith observation as possible. Quite possibly, a condition as fundamental as a wrong sign of a coefficient can be a result of one erroneous observation or perhaps a model fallacy in the region of the observation.

As in the case of $(\text{DFFITS})_i$, the computation of the $(\text{DFBETAS})_{j,i}$ is very interesting, though it is not as straightforward. Consider the $p \times n$ matrix

$$\mathbf{R} = (\mathbf{X}'\mathbf{X})^{-1}\mathbf{X}'$$

with the (q, s) element denoted by $r_{q,s}$. It turns out that the elements of the $\mathbf{R}$ matrix play an important role. In fact, one may view the n elements of the jth row of $\mathbf{R}$ as producing the leverage that the n observations exert on the coefficient b_j. Suppose we denote by $\mathbf{r}'_j$, the jth row of $\mathbf{R}$. Then

$$\begin{aligned}(\text{DFBETAS})_{j,i} &= \frac{r_{j,i}}{\sqrt{\mathbf{r}'_j\mathbf{r}_j}}\,\frac{e_i}{s_{-i}(1-h_{ii})} \\ &= \frac{r_{j,i}}{\sqrt{\mathbf{r}'_j\mathbf{r}_j}}\,\frac{1}{\sqrt{1-h_{ii}}}\,(R\text{-student})_i \end{aligned} \tag{6.7}$$

Again, the diagnostic represents the combination of leverage measures and the impact of errors in the y-direction. The value of $r_{j,i}/\sqrt{\mathbf{r}'_j\mathbf{r}_j}$ is a *normalized* measure of the impact of the ith observation on the jth coefficient. This value is swelled by the usual leverage measure, h_{ii}, of the ith observation. As expected, $(R\text{-student})_i$ also plays a role. Appendix B.7 provides the details of the derivation of the above result.

We use $(\text{DFBETAS})_{j,i}$ to ascertain which observations influence specific regression coefficients. As a result, the analyst must observe $n \times p$ statistics in assessing influence on the regression coefficients. In addition, associated with each data point is a single, essentially a composite, measure of the influence on the set of coefficients. The statistic is called *Cook's Distance*, or Cook's D, and is given by the scalar quantity

$$D_i = \frac{(\mathbf{b} - \mathbf{b}_{-i})'(\mathbf{X}'\mathbf{X})(\mathbf{b} - \mathbf{b}_{-i})}{ps^2} \tag{6.8}$$

Cook's Distance measure has become a fairly standard influence measure and has been incorporated in certain commercial computer packages. However, it is a bit difficult to understand for analysts who are not highly trained in statistics and matrix algebra. To understand the meaning of Cook's D, the user should first consider the vector

$$\mathbf{d}_i = \mathbf{b} - \mathbf{b}_{-i}$$

where $\mathbf{b}$ is the vector of coefficients and $\mathbf{b}_{-i}$ is the vector of coefficients with the ith observation set aside. Now, a composite measure of influence of the ith observation on the coefficients in $\mathbf{b}$ must be expressed as a scalar quantity that somehow standardizes $\mathbf{d}_i$. A proper standardization is to produce a quadratic form in which $\mathbf{d}_i$ is standardized by the inverse of the variance-covariance matrix. Since

$$\text{Var}(\mathbf{b}) = \sigma^2(\mathbf{X}'\mathbf{X})^{-1}$$

the quantity in Eq. (6.8) becomes a standardized version of $\mathbf{d}_i$. Actually, Cook's Distance is a distance measure, representing the standardized distance between the vector of least squares coefficients $\mathbf{b}$ and $\mathbf{b}_{-i}$. Cook's D is a positive quantity since $\mathbf{X}'\mathbf{X}$ is a positive definite matrix. A large value of D_i implies that the ith observation exerts undue influence on the *set* of coefficients. To determine which specific coefficients are affected, one must direct attention to the $(\text{DFBETAS})_{j,i}$.

Again, like the $(\text{DFFITS})_i$ and $(\text{DFBETAS})_{j,i}$, D_i is related to the residuals and data point leverage measures in a very interesting way. In fact, D_i is computed very simply from one regression (see Appendix B.7) as

$$D_i = \left(\frac{e_i^2}{(1-h_{ii})^2}\right)\left(\frac{h_{ii}}{s^2 p}\right) = \left(\frac{r_i^2}{p}\right)\left(\frac{h_{ii}}{1-h_{ii}}\right) \tag{6.9}$$

where r_i is the ith studentized residual. (Actually, if s_{-i} is used in place of s in the denominator of D_i, then r_i is replaced by t_i, the R-student value.) As before, D_i becomes large with either a poor fit (large r_i^2) at the ith point or high leverage (h_{ii} close to 1.0), or both. Also note the similarity between D_i and $(\text{DFFITS})_i$ in Eq. (6.5). From this, we see that an observation, which exerts strong influence on the fitted value, will, in turn, exert heavy influence on at least one regression coefficient.

HOW LARGE IS LARGE ON DFFITS, DFBETAS, AND COOK'S D?

The purpose of the collection of influence diagnostics is to aid the analyst in identifying which data points are most crucial. A practical question inevitably surfaces here regarding what magnitude in these measures should serve as a signal. For example, is there a level of DFFITS that defines "moderate influence" and another that defines "heavy influence," etc.?

In an excellent text on regression diagnostics, Belsley, Kuh and Welsch (1980) discuss "cutoffs," or levels that provide a means of categorizing data points as influential. These cutoffs are given later in this section, though the reader is referred to their text for a more detailed rationale on their use. From Eq. (6.6) and (6.4), we see that the diagnostics DFBETAS and DFFITS are formed by computing a difference between two statistics and providing an appropriate standardization. As we indicated earlier, these two measures are *t*-like. However, $\hat{y}_i$ and $\hat{y}_{i,-i}$ are dependent random variables, and the *t*-distribution does not provide an appropriate formal yardstick. Surely any analyst who is familiar with the concept of a standard error knows that if $(\text{DFBETAS})_{j,i}$ exceeds 2.0 in magnitude, the influence of the data point is unquestioned. But why should a value of 2.0 be viewed as a *minimal* requirement for classifying a point as influential? In the case of large data sets (say $n > 100$), DFFITS or DFBETAS values of 2.0 are extremely rare. However, even if n is large, it is not unusual to experience situations in which a single observation exerts sufficient influence to warrant some type of inspection.

There are various schools of thought regarding what levels of the diagnostics one should view as critical. The following are three *contrasting views*:

1. The analyst may hope that, for small and moderate sized samples, any observation can be identified which, if removed, would alter appreciably the results of the regression. In this case, a yardstick of approximately ± 2 on DFFITS and DFBETAS may be reasonable.
2. The analyst may hope that the yardstick used is one which would detect any existing influential observations, regardless of sample size. Mathematical yardsticks can give the analyst an impression of what data points represent substantial deviations from the *ideal*, the latter taken to be the situation in which there is uniform influence over all data points. The yardstick should be a function of sample size.
3. The analyst does not need a formula for a critical value or for a cutoff. Armed with awareness of what the criteria actually mean and with some knowledge of the system from which the data were produced, a user often can arrive at a yardstick value based on experience.

In the first case, a specific yardstick is used that is independent of sample size. The second case, suggests that formal cutoff values be put forth which take sample sizes into account. If one chooses to adopt view 1, a ± 2 may be a guideline; but the user who realizes the implication of ± 2 in the case of, say, DFBETAS, is tempted to adopt a *tougher* yardstick value of less than 2.0. There is no doubt that if yardsticks are necessary, they should depend on n. However, it is very difficult to produce critical values, based on n, that are appropriate for all cases. Belsley, Kuh, and Welsch (1980) suggest the cutoffs of $2/\sqrt{n}$ for the $(\text{DFBETAS})_{j,i}$ and $2\sqrt{p/n}$ for the $(\text{DFFITS})_i$. The reader is referred to their text for details and assumptions on which these levels are based. Surely there will be many instances in which these values will expose more data points than the analyst would care to check. However, these values tend to be reasonable for large samples.

The analyst should gain the experience necessary to *adopt the view given in* 3 *above*. Though attempting to generate cutoffs is admirable, one must constantly be aware that the purpose of these diagnostic statistics is not to do significance testing. Therefore, there are no natural *critical values* called for. In addition, one always has enough information available to determine the impact on the regression of an observation with, say, a specific $(\text{DFBETAS})_{j,i}$ or $(\text{DFFITS})_i$. For example, if $(\text{DFFITS})_i = 1.7$, this can always be translated into the actual units of the response (by making use of the standard error of prediction) to determine just how $\hat{y}_i$ is altered if the data point is removed. In addition, if a value of $(\text{DFBETAS})_{j,i}$ is, say, 1.4, an analyst can determine (by making use of the estimated standard error of the coefficient) to what degree the coefficient changed in the units of the problem. Thus an assessment can be made on whether a single observation is responsible for the significance or insignificance of a coefficient, or responsible for the coefficient attaining a sign which is uninterpretable. The analyst should not ignore the guidelines suggested by Belsley, Kuh, and Welsch; but some effort should be spent on gaining insight into what impact the observation has directly on a fitted value or a coefficient, in the context of the problem.

Cook's D, given by Eq. (6.8) is an F-like statistic with degrees of freedom p and $n-p$. However, a critical yardstick based on the F-distribution is no more appropriate than is the t-distribution for DFFITS and DFBETAS. In attempting to evaluate the overall influence on the coefficients, the user interprets a specific value of Cook's D as follows: Suppose the value of Cook's D is approximately equal to the 50% point of the $F_{p,n-p}$ distribution. One can say that deletion of the ith point moves the vector of coefficients from the center of the confidence region to the 50% confidence ellipsoid. However, many analysts will find discomfort with this interpretation and rely more heavily on the DFBETAS values.

Example 6.2 **BOQ *Data***

Consider again the Naval BOQ data of Table 5.2. In Example 6.1, we learned that data point 23 has a strong influence on at least three regression coefficients as well as the quality of fit. We actually eliminated data point 23 to provide the illustration. Though we concentrated on data point 23, there was also conjecture that data points 15, 20, 21, 22, and 24 may also be influential on the basis of R-student values and/or HAT diagonal values. Table 6.2 shows the DFFITS, Cook's D, and DFBETAS values, illustrating the influence of *all* 25 *installations without removing any of them.*

TABLE 6.2 **Cook's *D*, DFFITS, and DFBETAS values for the data of Table 5.2**

Installation	R-Student	h_{ii}	Standard Error of Prediction	DFFITS	Cook's D
1	−0.0736	0.2573	230.877	−0.0433	0.000
2	−0.0726	0.1609	182.567	−0.0318	0.000
3	−0.4594	0.1614	182.870	−0.2016	0.005

TABLE 6.2 Continued

Cook's *D*, DFFITS, and DFBETAS values for the data of Table 5.2

Installation	*R*-Student	h_{ii}	Standard Error of Prediction	DFFITS	Cook's *D*
4	−0.1762	0.1631	183.829	−0.0778	0.001
5	−0.4196	0.1475	174.801	−0.1745	0.004
6	−0.5704	0.1589	181.437	−0.2479	0.008
7	0.7532	0.1829	194.651	0.3563	0.016
8	−0.4720	0.3591	272.756	−0.3533	0.016
9	0.3246	0.2808	241.202	0.2029	0.005
10	0.0914	0.1295	163.824	0.0353	0.000
11	1.5537	0.1241	160.373	0.5849	0.039
12	0.4426	0.2024	204.778	0.2230	0.007
13	−0.1818	0.0802	128.903	−0.0537	0.000
14	−0.0224	0.0969	141.699	−0.0073	0.000
15	−2.5192	0.5576	339.885	−2.8282	0.761
16	−0.0541	0.4024	288.719	−0.0444	0.000
17	−1.3310	0.3682	276.208	−1.0162	0.123
18	1.2566	0.4465	304.143	1.1286	0.154
19	1.0074	0.0868	134.111	0.3106	0.012
20	−2.8657	0.3663	275.477	−2.1787	0.417
21	2.0538	0.0704	120.765	0.5652	0.034
22	−1.6057	0.7854	403.373	−3.0715	1.079
23	−5.2423	0.9885	452.533	−48.5179	115.041
24	3.2093	0.8762	426.057	8.5373	5.889
25	0.4299	0.5467	336.558	0.4722	0.029

Installation	DFBETAS Intercept	DFBETAS b_1	DFBETAS b_2	DFBETAS b_3
1	−0.0433	−0.0006	−0.0044	0.0345
2	−0.0310	−0.0006	−0.0025	0.0204
3	−0.1971	−0.0037	−0.0113	0.1318
4	0.0040	0.0026	0.0103	−0.0504
5	−0.1662	0.0041	0.0037	0.1231
6	0.0135	0.0115	0.0517	−0.1597
7	0.3234	−0.0274	−0.0700	−0.2195
8	−0.0055	−0.0160	−0.0577	−0.1285
9	0.1271	0.0070	−0.0108	−0.1376
10	−0.0029	−0.0031	−0.0110	0.0226
11	−0.0790	0.0289	−0.2033	0.2201
12	0.0069	0.0092	0.1441	0.0886
13	0.0085	0.0068	0.0209	−0.0277
14	0.0004	0.0010	−0.0002	−0.0045
15	−0.0598	0.9457	−1.7852	0.0088

TABLE 6.2 Continued

TABLE 6.2 Continued

Cook's *D*, DFFITS, and DFBETAS values for the data of Table 5.2

Installation	DFBETAS Intercept	DFBETAS b_1	DFBETAS b_2	DFBETAS b_3
16	0.0065	0.0124	0.0268	−0.0061
17	0.1531	0.2449	0.7725	−0.1946
18	−0.0431	−0.0822	0.7506	−0.1382
19	−0.0243	−0.0054	0.1403	0.1157
20	0.0223	−0.0035	−0.4404	−0.1233
21	−0.0929	−0.0984	−0.1058	0.2598
22	0.2010	0.2396	0.2936	0.3357
23	−0.2455	−44.3791	1.1685	0.6569
24	0.7391	−1.6514	0.5958	−1.2140
25	−0.0413	0.0586	−0.0851	0.0377

Installation	DFBETAS b_4	DFBETAS b_5	DFBETAS b_6	DFBETAS b_7
1	0.0036	−0.0020	−0.0007	0.0013
2	0.0046	−0.0020	−0.0008	0.0017
3	0.0374	−0.0025	0.0048	−0.0018
4	0.0261	−0.0006	0.0032	0.0003
5	−0.0128	0.0048	−0.0018	0.0020
6	0.0985	−0.0006	0.0165	−0.0119
7	−0.1375	−0.0160	−0.0576	0.0820
8	0.1666	−0.2715	−0.0783	0.1213
9	0.1242	−0.0116	0.0206	−0.0217
10	−0.0135	0.0020	−0.0033	0.0034
11	0.1826	−0.2019	0.2104	−0.1149
12	−0.0080	0.0454	0.0129	−0.0671
13	0.0051	−0.0037	0.0106	−0.0113
14	0.0022	0.0002	0.0013	−0.0007
15	−0.5855	1.2009	1.3983	−0.9411
16	−0.0034	0.0275	0.0285	−0.0343
17	0.0217	0.6030	0.4888	−0.6953
18	0.7150	0.1604	−0.0147	−0.2197
19	0.0965	0.0305	−0.0209	−0.0446
20	−0.5381	−0.1963	−1.7040	1.4204
21	0.0756	−0.0191	−0.1389	0.1178
22	0.1652	−2.0577	−0.0658	0.1008
23	−4.5712	1.1845	−3.9433	7.2839
24	−5.1382	−0.8793	−0.6539	2.1090
25	0.2267	−0.0205	0.3618	−0.2759

The influence of installation 23 is very evident. In fact, the $(\text{DFBETAS})_{1,23}$ value of −44.3791 is a graphic illustration. If data point 23 is removed, the regression coefficient b_1 increases by 44.3791 *estimated standard errors*. Its influence on coefficients b_4, b_6, and b_7 is also illustrated with DFBETAS values of −4.5712,

-3.9433, and 7.2839, respectively. Thus, the fact that the presence of data point 23 alters the signs of four coefficients as well as the conclusions regarding their role in the model is very evident here *without the need to actually compute a regression with data point* 23 *removed.* For further illustration of the influence of installation 23, the reader obviously notices the striking value of 115.041 for Cook's D and -48.5179 for DFFITS. The implication of the latter is that the inclusion of data point 23 forces the predicted response (monthly man-hours) to decrease by 48.5179 estimated standard errors. In case the analyst is interested in transforming this to the units of the response in the problem, the estimated standard error of prediction at data point 23 is required. This standard error usually appears in any regression package that displays the diagnostics. From Table 6.2, the appropriate value for data point 23 is 452.533 man-hours. (This, of course, is the estimate that makes use of s instead of s_{-i}.) The standard error of prediction computed with the use of s_{-i} is given by

$$452.533\left(\frac{s_{-i}}{s}\right) = (452.533)\left(\frac{284.6035}{455.1670}\right) = 282.9565$$

man-hours. The presence of data point 23 decreases the predicted response at that location by $(48.5179)(282.9565) = 13{,}728.4552$ man-hours. This dramatic change underscores the strong effect that installation 23 has on the results. Surely, all efforts should be made to verify the results there, or to determine if there is a nonstatistical reason why the model should change so dramatically in the presence of installation 23.

Of course, data point 23 is not the only installation whose data should be carefully checked. Data point 24 must be considered influential no matter what yardstick one uses for DFFITS, Cook's D, and the DFBETAS. Its impact on coefficients b_4 and b_7 cannot be ignored. Its DFFITS value of 8.5373 implies an extremely large increase in predicted man-hours at installation 24 due to its inclusion in the data set. Data points 15, 20, and 22 are also influential, though to a lesser extent than 23 and 24. Surely the influence that data point 22 exerts on b_5, and that data point 20 exerts on b_6, cannot be ignored. Data point 21 also provides an interesting illustration. The R-student value of 2.0538 signals an appreciably large residual induced by a relatively poor fit at the point. But is it really influential? A very small HAT diagonal ($h_{ii} = 0.0704$) leads one to expect that the influence might be mild, and, indeed, an inspection of the diagnostics reveals relatively light influence.

In the present example, the influence diagnostics highlighted several data points that need to be investigated. The diagnostics appear to reveal that whatever underlying model generates the data at the other sites, i.e., the noninfluential ones, does not prevail at sites 23, 24, 15, 20 and 21. This would certainly support the notion that a more complicated model may provide a better explanation. Additional data might be gathered at other large sites in order to study the possibility of a separate model for large installations.

Example 6.3 ***Coal-Cleansing Data***

Consider the coal-cleansing data of Example 5.3. There are five regressors, and the data were used to illustrate the outlier detection diagnostics. Recall that data point 9 was found, *a priori*, to be suspect, and the *R*-student statistic provided additional evidence that perhaps the observation should be removed from the data set. Before removing the outlying observation, it may be of interest to assess its influence on the result. The essential diagnostics for determining the influence of all 12 observations follow:

Experiment	e_i	*R*-Student	h_{ii}	DFFITS	Cook's *D*
1	−4.8	−0.2923	0.4501	−0.2644	0.020
2	13.2	0.8359	0.4501	0.7563	0.149
3	−17.0	−1.1372	0.4660	−1.0624	0.272
4	24.0	1.7665	0.4660	1.6503	0.538
5	1.3	0.0631	0.0838	0.0191	0.000
6	−20.7	−1.0385	0.0838	−0.3142	0.024
7	−17.7	−0.8698	0.0838	−0.2631	0.018
8	6.3	0.2990	0.0838	0.0905	0.002
9	32.2	2.8695	0.4501	2.5963	0.885
10	−23.8	−1.7141	0.4501	−1.5508	0.484
11	0.5	0.0282	0.4660	0.0264	0.000
12	6.5	0.4006	0.4660	0.3743	0.039

Experiment	DFBETAS Intercept	DFBETAS b_1	DFBETAS b_2	DFBETAS b_3
1	−0.2461	0.1394	0.1394	0.1347
2	0.7039	−0.3986	−0.3986	−0.3853
3	0.3905	0.5502	−0.5502	−0.5669
4	−0.6066	−0.8547	0.8547	0.8806
5	0.0028	0.0000	0.0000	−0.0015
6	−0.0460	−0.0000	−0.0000	0.0244
7	−0.0385	−0.0000	−0.0000	0.0205
8	0.0132	0.0000	0.0000	−0.0070
9	−0.4802	1.3682	1.3682	−1.3225
10	0.2869	−0.8172	−0.8172	0.7900
11	−0.0065	0.0136	−0.0136	0.0141
12	−0.0920	0.1938	−0.1938	0.1997

The *R*-student value of 2.8695 at data point 9 is accompanied by a HAT diagonal value of 0.4501. The latter is above the average of $p/n = 1/3$ but is not greater than the yardstick of $2p/n = 0.66$. However DFFITS and the DFBETAS, associated with coefficients b_1, b_2, and b_3, are larger than the corresponding diagnostics for any other observation. Surely a comparison with the yardstick values of $2/\sqrt{n}$ for DFBETAS and $2\sqrt{p/n}$ for DFFITS would suggest that the magnitude of the influence is appreciable. The influence of data point 9 on b_3, the coefficient of x_3, flow rate, is particularly crucial. The coefficient takes on a value of -0.058 with a standard error of 0.0256. (See Example 5.3.) The DFBETAS value of -1.3225 suggests that without the 9th observation the regression coefficient increases by 1.3225 standard errors, rendering it statistically insignificant on the basis of a t-test. Thus any conclusion about the impact of flow rate in this coal-cleansing operation is very much affected by the result of the 9th run. The regression diagnostics allowed us to evaluate the impact of the outlier without actually rerunning the regression.

6.4

INFLUENCE ON PERFORMANCE

The diagnostics $(\text{DFFITS})_i$ and $(\text{DFBETAS})_{j,i}$ highlight those observations whose presence in the data set strongly impacts the results. The values of these diagnostics reflect influence but do not give an indication of whether the influence is directed at achieving better performance on the part of the regression equation. An unusually large value of $(\text{DFBETAS})_{j,i}$ merely signifies that the ith observation is prominent in describing b_j, the jth regression coefficient. The value provides no focus on whether or not the presence of the ith observation appreciably sharpened the estimation of the coefficient.

A single statistic that provides a convenient scalar measure of the variance-covariance properties of the coefficients is the *generalized variance* (GV) of the regression coefficients, given by

$$\text{GV} = |\text{Var}\,\mathbf{b}| = |(\mathbf{X}'\mathbf{X})^{-1}\sigma^2| \tag{6.10}$$

The expression in (6.10) has some simple and interesting interpretations. See Graybill (1976). The analyst strives for small values of the determinant to obtain a quality estimation of the set of regression coefficients. The generalized variance, apart from σ^2, is a function of the conditioning of the $\mathbf{X}$ matrix. To capture the role of the ith observation when determining the generalized variance, we define the ratio (called COVRATIO) of the property *without i* to the property *with i*. The estimators s_{-i}^2 and s^2 are used in place of σ^2 in the numerator and denominator respectively. The result is given by

$$(\text{COVRATIO})_i = \frac{|(\mathbf{X}'_{-i}\mathbf{X}_{-i})^{-1}s_{-i}^2|}{|(\mathbf{X}'\mathbf{X})^{-1}s^2|} \tag{6.11}$$

where $\mathbf{X}_{-i}$ denotes the $(n-1) \times p$ data matrix with the ith observation eliminated. The COVRATIO in (6.11) has no standard error type scaling. However, it is clear that a value exceeding 1.0 implies that the ith point provides an improvement—a reduction in the *estimated generalized variance* of the coefficient over what would be produced without the data point. A value less than 1.0 reveals that the inclusion of the ith point results in an increase in the estimated generalized variance. From (6.12), one can readily see what prescription produces a $(\text{COVRATIO})_i$ that exceeds 1.0.

$$(\text{COVRATIO})_i = \frac{(s_{-i})^{2p}}{s^{2p}}\left(\frac{1}{1-h_{ii}}\right) \tag{6.12}$$

The reader who is interested in the development of (6.12) should refer to Belsley, Kuh, and Welsch (1980). The rather simple term $(1-h_{ii})^{-1}$ in (6.12) is the ratio of $|(\mathbf{X}'_{-i}\mathbf{X}_{-i})^{-1}|$ to $|(\mathbf{X}'\mathbf{X})^{-1}|$. Strong leverage ($h_{ii} \cong 1.0$) induces COVRATIO to be large. This is an anticipated result since we expect an extreme point to substantially sharpen the trend set by the other data, assuming that the point is not an outlying observation in the y-direction. Now, if the ith observation is, indeed, an outlier, s_{-i}^{2p}/s^{2p} will be considerably less than 1.0. (See Eq. (5.6).) Thus, as in the case of all the diagnostics we have discussed, the combination of leverage (distance $\mathbf{x}_i$ is away from the data center) and the error in fit at the point $\mathbf{x}_i$ work together to produce the diagnostic result. In the case of $(\text{COVRATIO})_i$, the combination of high leverage and a small residual at $\mathbf{x}_i$ result in a point that enhances the dispersion properties of the regression coefficients.

As in the cases of the previous diagnostics, yardstick values for COVRATIO are very difficult to determine. The problem is further complicated by the fact that, to most users, the notion of influence on a generalized variance is more abstract than the case of a fitted response or an estimated regression coefficient. Belsley, Kuh, and Welsch (1980) suggest a rough yardstick that is reasonable to use for large samples. In essence, the yardstick suggests that we consider that the ith data point exerts an unusual amount of influence on the generalized variance if $(\text{COVRATIO})_i > 1 + 3p/n$ or $(\text{COVRATIO})_i < 1 - 3p/n$. Clearly, the lower bound of this guideline applies only when $n > 3p$.

Example 6.4 ***Squid Data***

Consider Example 3.2, in which data were taken on 22 squid and a model was sought that related the weight of the squid to various beak dimensions. The data appear in Table 3.2. Table 6.3 gives the important diagnostic information. We include the COVRATIO for the first time. This example is very instructive. We offer some summary comments as follows.

The greatest leverage is associated with observation 19 ($h_{ii} = 0.6111$). The HAT diagonal value exceeds our crude guideline of $2p/n = 12/22$. However, the R-student value is extremely low ($t_{19} = -0.0200$), which indicates a very good fit at that point. Cook's D ($\cong 0$) and DFFITS (-0.0251) reflect that essentially no influence is exerted by data point 19. However, the analyst should not consider the point *inert* since it does give reinforcement of the regression provided in its absence. This conclusion is reflected in $(\text{COVRATIO})_{19}$, which takes on the value 3.7868. Thus observation 19 is extremely important in that it enhances the properties of the regression. A similar conclusion could be drawn about data point 3, though it is not as dramatic. Data points 2, 13, and 18 are influential observations. Point 13 is not an outlier, but there is sufficient leverage ($h_{ii} = 0.5578$) to provide influence on at least two of the regression coefficients. Its leverage also provides a moderate enhancement of the properties of the regression coefficients (COVRATIO = 1.4599). Data points 2 and 18 are somewhat similar. The R-student value for point 2 is large in magnitude, suggesting a sizable error in fit. This combines with high leverage ($h_{ii} = 0.5647$), to cause the point to exert strong influence on three coefficients. Data point 18 carries less leverage ($h_{ii} = 0.3068$) and hence has somewhat less influence. Note also that both data points 2 and 18 cause the estimated generalized variance to increase (COVRATIO < 1), though the increase is more pronounced in data point 18. This result occurs because data point 2 has more leverage.

TABLE 6.3 **Influence diagnostics for squid data of Table 3.2**

Observation	e_i	t_i (R-Student)	h_{ii}	COVRATIO	Cook's D	DFFITS
1	−0.24	−0.3627	0.1320	1.6103	0.004	−0.1414
2	−0.96	−2.3390	0.5647	0.5237	0.924	−2.6639
3	−0.07	−0.1109	0.3098	2.1236	0.001	−0.0743
4	0.86	1.3576	0.1441	0.8586	0.049	0.5571
5	0.28	0.4298	0.1855	1.6803	0.007	0.2051
6	0.30	0.4629	0.1949	1.6802	0.009	0.2278
7	0.58	1.0501	0.3888	1.5745	0.116	0.8374
8	−0.11	−0.1586	0.1387	1.6930	0.001	−0.0636
9	0.32	0.4829	0.1395	1.5603	0.007	0.1944
10	−0.37	−0.5420	0.0984	1.4542	0.006	−0.1791
11	−0.59	−0.8655	0.0661	1.1774	0.009	−0.2303
12	−0.86	−1.3612	0.1497	0.8612	0.052	−0.5711
13	0.67	1.4868	0.5578	1.4599	0.432	1.6698
14	−0.65	−1.0114	0.1751	1.2019	0.036	−0.4659
15	0.53	1.0246	0.4596	1.8165	0.148	0.9449
16	−0.61	−0.9517	0.1809	1.2650	0.034	−0.4472
17	0.65	1.0135	0.1662	1.1872	0.034	0.4525
18	−1.26	−2.4738	0.3068	0.2728	0.342	−1.6459
19	−0.01	−0.0200	0.6111	3.7868	0.000	−0.0251
20	0.42	0.6600	0.2084	1.5670	0.020	0.3386
21	0.46	0.8202	0.3748	1.8107	0.069	0.6351
22	0.65	1.2747	0.4471	1.4372	0.211	1.1462

TABLE 6.3 Continued

TABLE 6.3 Continued

Influence diagnostics for squid data of Table 3.2

Observation	DFBETAS Intercept	DFBETAS b_1	DFBETAS b_2	DFBETAS b_3	DFBETAS b_4	DFBETAS b_5
1	0.0182	−0.0711	0.0402	−0.0329	0.0071	0.0820
2	0.1941	−0.0395	−2.4969	1.2841	1.2751	0.1606
3	−0.0706	0.0303	−0.0019	0.0418	−0.0194	−0.0551
4	0.3482	0.0340	0.0559	−0.1082	−0.1086	0.0431
5	0.1699	−0.0951	−0.0186	−0.0698	0.1097	0.0752
6	0.0170	0.0101	−0.0610	0.1568	−0.0302	−0.1106
7	−0.0581	0.4199	0.1273	0.2421	−0.6715	−0.2819
8	−0.0033	0.0007	0.0157	−0.0176	−0.0241	0.0327
9	0.1498	−0.0540	−0.0057	−0.0096	0.0128	0.0412
10	0.0096	−0.0784	−0.0173	0.0105	0.0342	0.0833
11	−0.0833	−0.0102	0.0354	0.0126	−0.0442	0.0267
12	0.1001	−0.1497	0.3032	−0.3856	−0.0471	0.3502
13	0.2473	0.6737	−0.1559	−0.9827	−0.3556	0.9868
14	−0.1814	0.0469	−0.0251	0.3435	−0.1762	−0.2416
15	−0.4737	0.3209	−0.5877	0.3466	0.3613	−0.4199
16	0.2966	−0.2082	−0.0342	−0.2017	0.2516	0.2121
17	0.0651	−0.2586	0.1048	−0.1401	0.2249	0.1527
18	−0.4478	0.4260	1.0236	−0.1744	−0.7310	−0.8926
19	−0.0012	0.0095	−0.0017	−0.0097	0.0051	−0.0061
20	−0.0112	−0.2043	−0.0329	0.1299	0.1319	0.0339
21	−0.4252	0.4083	0.1203	−0.0605	−0.3289	−0.1347
22	−0.0817	−0.6815	0.4965	−0.0496	0.4741	−0.1599

6.5

WHAT DO WE DO WITH HIGH INFLUENCE POINTS?

The analyst should always be aware that the diagnostics available *do not represent a set of independent influence measures.* For example, one can rest assured that if Cook's D_i produces an unusually high result, at least one of the $(\text{DFBETAS})_{j,i}$ will suggest strong influence on a specific regression coefficient. This same condition should reflect an impact on $(\text{DFFITS})_i$ and $(\text{COVRATIO})_i$. Thus there is much overlap in the information presented by the diagnostics. Indeed, an experienced analyst will quickly assess, with a mere inspection of the *R*-student and HAT-diagonal values, those data points that have influence. However, the analyst should look at all available diagnostics. For example, if the conclusions drawn rely heavily on interpretation of coefficients, the DFBETAS can provide extra important information.

Diagnostics are designed to offer the analyst *signals*, i.e., signs, that if resources exist for reinvestigating some data points for accuracy, the influential ones should be given a thorough investigation. This is particularly important if an unexpected result occurs due to a single observation. Is there ever a need to eliminate a high influence observation? The analyst's attitude toward a high influence data point can be no more harsh than that taken toward an outlier (which may itself exert substantial influence). If a serious problem is found through a reevaluation of a high influence point, then the point's presence should surely be questioned. However, if reevaluation verifies that an influential point is a valid observation, there can be no reasonable justification for its removal. In some cases a high influence observation may represent the most important information gathered. It may provide the primary support for a model that is postulated. On the other hand, it may be evidence that is counter to a postulated model, but this too can be crucial to the total analysis. Now, the most desirable leverage condition in a data set is the achievement of a uniform distribution of leverage. This occurs when all HAT diagonals take on the value of p/n, and potential influence derived from leverage is evenly divided among the data points. However, an uneven allotment of leverage (and hence influence) is not an uncommon occurrence in many types of data sets. The presence of this condition does not automatically imply that a regression cannot be salvaged. Influential points should not be reason for permanently deflecting the analyst from the model-building task. In summary, the diagnostics produce important results with little computational effort. However, the information from the diagnostics often forces the analyst to probe more. Hence, the path to the eventual goal of building an effective model may be altered somewhat.

The user of regression influence diagnostics must put the information in the context of ordinary regression analysis. The diagnostic procedures are a welcome addition to the user's arsenal, *simply because traditional hypothesis testing and model fitting so often fail at building usable models.* When a model has been fit and rejected in some traditional sense, the standard procedures (variable screening, hypothesis testing, etc.) do not provide suitable illumination as to why the model failed. Nor do they always hint at what a suitable model is. By the same token, if a model is not discarded by traditional procedures, there is no assurance that the model has been given a complete examination. The diagnostics are designed to fill these voids. Traditional analyses do not look at a model performance through a microscope as the diagnostics do.

The relationship to the material in Chapter 5 is obvious, and we have discussed it. There is also a very clear relationship to the material in Chapter 7, particularly the material that deals with transformations. The presence of an outlier or set of high influence data points may be a signal for a transformation. It would be an oversight if we did not relate the diagnostic information to the progression of steps we suggested in Chapter 4 for building a model. We described the ideal exercise of using all possible regressions to reduce the number of candidate models to a manageable number. We can then inspect these candidates more closely with emphasis placed on residuals, PRESS residuals, standard errors of prediction, etc. Concerning the ordering of operations, there is a sound argument for observing influence diagnostics on the entire data set (all of the regressors) early in the procedure. Decisions influenced by this inspection may alter appreciably the model-building process. We have already illustrated, by example, how the roles of the variables are driven by single observations.

There is a point that suggests caution. Many software packages provide the single observation deletion (set aside) results we have described. The mathematics is readily available for easy access to multiple observation diagnostics. The material in *this* chapter is difficult enough for the experienced user. One must make a slow approach to the explosion brought about by these obvious extensions, lest the resulting volumes of computer printout bring discouragement and terminal exhaustion.

EXERCISES FOR CHAPTER 6

6.1 Consider the hospital data introduced initially in Example 3.8. Errors in measurement can be abundant in data of these types. As a result, a *post analysis* remeasurement is often necessary.

(a) For the model containing all 5 regressor variables, compute the following influence diagnostics:

(i) DFBETAS (ii) Cook's D

(iii) DFFITS (iv) HAT diagonals

for all 17 hospitals.

(b) Discuss your interpretation of the results in part (a), and state which hospitals should be reexamined or remeasured.

6.2 Consider the forestry data of Example 3.5. Compute HAT diagonals, Cook's D values, and DFBETAS values for each of the 20 observations. From these results, determine whether or not any of the 20 observations exert a disproportionate influence on the results.

6.3 Consider the data of Exercise 3.11. Use the influence diagnostics in Chapter 6 to determine if any of the 33 firms studied should be reexamined or measured once again. That is, are any individual firms exerting undue influence on the regression results?

6.4 Cook's D given in Eq. (6.8) represents a composite measure of influence on the set of regression coefficients. It is rather difficult to produce a yardstick on this measure. One interpretation of Cook's D is related to the joint confidence region on the coefficients on $\boldsymbol{\beta}$. Consult Chapter 3, and discuss the relationship between Cook's D and the joint confidence region. As a result, discuss an interpretation of Cook's D that is based on this confidence region.

6.5 Consider the experiment depicted in Exercise 4.10. Suppose that the y observations are taken and that the residuals are all extremely small. Thus the fit of the model to the data is quite good. What do you expect the nature of the influence diagnostics to be? Will any observation have a disproportionate influence on the results? Explain.

6.6 Consider the data with Exercise 5.18. In particular, consider observations 22 and 25.

These two data points look reasonably innocent even though their leverage is somewhat higher than the norm. However, at times, influence can come from observations in sets and not necessarily only individual observations. With regard to this data set, answer the following questions.

(a) Fit the multiple linear regression using the entire data set, and note regression coefficients and standard errors. Also note variance inflation factors on coefficients.

(b) Remove 22 and 25, and look at the same information. Comment.

6.7 Consider the data of Exercise 3.19. Compute all influence diagnostics. Do any points exert a disproportionate amount of influence on the regression results?

6.8 Influential observations and multicollinearity in a data set often go hand in hand. Consider, for example, the data discussed in Exercises 6.5 and 6.8. All HAT diagonals are equal and thus the "potential" influence is quite proportionate among observations. Also note that the variance inflation factors are all 1.0 in these two factorial experiments. On the other hand, consider the following data set, which appears in Section 3.8.

y	x_1	x_2
—	10.0	10.0
—	11.0	11.4
—	11.9	12.2
—	12.7	12.5
—	13.3	13.2
—	14.2	13.9
—	14.7	14.4
—	15.0	15.0

The variance inflation factors are both 63.94. Compute the HAT diagonals, and discuss the dispersion in HAT diagonals as opposed to what you notice in Exercises 6.5 and 6.8.

REFERENCES FOR CHAPTER 6

1. Atkinson, A.C. 1983. Diagnostic regression for shifted power transformations. *Technometrics* 25: 23–33.
2. Atkinson, A.C. 1986. Masking unmasked. *Biometrika* 73: 533–541.
3. Atkinson, A.C. 1982. Regression diagnostics, transformations and constructed variables. *Journal of the Royal Statistical Society, Series B* 44: 1–36.
4. Belsley, D.A., E. Kuh, and R.E. Welsch. 1980. *Regression Diagnostics: Identifying Influential Data and Sources of Collinearity*. New York: John Wiley & Sons.
5. Chatterjee, S., and A.S. Hadi. 1986. Influential observations, high leverage points, and outliers in linear regression (with discussion). *Statistical Science* 1: 379–416.
6. Chatterjee, S., and A.S. Hadi. 1988. *Sensitivity Analysis in Linear Regression*. New York: John Wiley & Sons.

7. Cook, R.D. 1986. Assessment of local influence (with discussion). *Journal of the Royal Statistical Society, Series B* 48: 133–169.

8. Cook, R.D. 1977. Detection of influential observations in linear regression. *Technometrics* 19: 15–18.

9. Cook, R.D. 1979. Influential observations in linear regression. *Journal of the American Statistical Association* 74: 169–174.

10. Cook, R.D., and S. Weisberg. 1982. *Residuals and Influence in Regression*. New York and London: Chapman and Hall.

11. Draper, N.R., and J.A. John. 1981. Influential observations and outliers in regression. *Technometrics* 23: 21–26.

12. Gray, J.B. 1985. Graphics for regression diagnostics. *American Statistical Association Proceedings of the Statistical Computing Section*, pp. 102–107. American Statistical Association, Washington, D.C.

13. Gray, J.B., and R.F. Ling. 1984. *K*-clustering as a detection tool for influential subsets in regression. *Technometrics* 26: 305–330.

14. Graybill, F.A. 1976. *Theory and Application of the Linear Model*. Boston, Massachusetts: Duxbury Press.

15. Hocking, R.R., and O.J. Pendleton. 1983. The regression dilemma. *Communications in Statistics (Theory and Methods)* 12: 497–527.

16. Velleman, P.F., and R.E. Welsch. 1981. Efficient computing of regression diagnostics. *American Statistician* 35: 234–242.

17. Weisberg, S. 1981. A statistic for allocating C_p to individual cases. *Technometrics* 23: 27–31.

C H A P T E R 7

NONSTANDARD CONDITIONS, VIOLATIONS OF ASSUMPTIONS, AND TRANSFORMATIONS

Chapters 3 and 4 dealt with standard statistical inference associated with the multiple linear regression model. Also presented were criteria and algorithms for finding the best subset model from a group of candidate models. Underlying all of this material was the *method of least squares* for estimation of model coefficients. Implicit in the least squares procedure are the assumptions:

1. $E(\varepsilon_i) = 0$.
2. The ε_i are uncorrelated with homogeneous variance σ^2.

In addition, *normality* on the ε_i is required in order for the estimators to attain the property of minimum variance of the class of unbiased estimators. Chapter 5 focused at length on procedures for detection of violations of one or more of these assumptions. Techniques that exploit residuals provide graphical and analytical methods for these diagnoses.

Chapter 6 can be viewed as an extension of Chapter 5. The nature of the residuals for a data set are combined with leverage measures to create influence diagnostics. In a sense, Chapters 5 and 6 provide machinery for *model/data criticism*, an activity that should be carried out for any and every model that deserves strong consideration. Part of the model criticism aspect of the analysis (particularly that derived from Chapter 5) is intended to uncover violation of assumptions, particularly model misspecification, heterogeneous variance, nonnormal errors, and the presence of outliers.

In this chapter we turn our attention toward procedures for negotiating model building under some of the same nonstandard conditions detected through the diagnostics developed in Chapter 5. The nonstandard conditions covered in this chapter include

1. Heterogeneous variance
2. Model misspecification that necessitates data transformation
3. Correlated errors
4. Outliers in the data or nonnormal errors
5. Regression data in which model errors are obviously nonnormal
6. Measurement errors in the regressor variables

Heavy emphasis is placed on data transformation. It is important for the analyst to know which *metric* is most appropriate for the response or the regressor variables. At times the decision can and should be made by the subject matter scientist. But there are circumstances where the analyst requires diagnostic information that is designed to point out transformations that are desirable. The reader should note the connection to material in Chapter 5 in which various types of plots highlighted the need for data transformation, but they provided no indication of which transformation to use.

Violation of the normality assumption is discussed extensively in this chapter. Estimation methods that are resistant or *robust* to lack of normality are discussed and illustrated. These robust estimation procedures are insensitive to both deviations from normality and outliers in the data. In addition, deviations from normality that involve special classes of response variables will be covered. In many applications in the biological, health, and physical sciences, the responses are counts. They may be binary numbers (0 or 1) or counts of the form 1, 2, 3, ... In other applications the model errors are continuous, but it may be known that distributional assumptions other than normal are more appropriate. In this chapter we discuss *generalized linear models*, a unified approach that accommodates a large class of nonnormal errors. Examples of applications include binomial regression and Poisson regression models.

The reader who is interested in applications to his or her own field should naturally link this material to that in Chapter 5. The methodology presented here should be considered as alternatives to fundamental least squares if procedures in Chapter 5 suggest that a particular assumption has been violated.

In real applications, violation of assumptions is often the rule rather than the exception. It is likely that the experienced reader is already privy to this fact. The analyst must be able to know when an assumption violation is too subtle or unimportant to be the subject of a major alteration in standard procedures. Data analysts often overreact to a violation. One can overdo manipulations such as data transformation, weighting of data, and deletion of apparent outlying observations. In the proper place, the available strategies to accommodate or to counter violations are an integral part of the analyst's machinery. But it should not be a subconscious accommodation, not one without thought. It takes patience and some deliberation to make these strategies accomplish their purpose.

7.1

HETEROGENEOUS VARIANCE: WEIGHTED LEAST SQUARES

The assumption of homogeneous error variance is, unfortunately, often violated in practical situations. It is almost endemic that in scientific measurements, as numbers become larger, either in the regressor variables or in the response, variation around the trend or fitted model becomes larger. We saw evidence of this in residual diagnostics in Chapter 5. From a *theoretical* point of view, it is quite simple to alter the least squares procedure to accommodate the heterogeneous variance difficulty. Practically speaking, however, it is not always clear what should be done. There is a sense of importance associated with the problem because the least squares procedure is, indeed, relatively sensitive to violations of the homogeneous variance assumption (Seber (1977)). When a strategy is being mapped by the analyst for combating heterogeneous variance, initial consideration should be given to *weighted least squares.*

Let us recall the general linear model

$$\mathbf{y} = \mathbf{X}\boldsymbol{\beta} + \boldsymbol{\varepsilon}$$

with the ordinary least squares estimator given by

$$\mathbf{b} = (\mathbf{X}'\mathbf{X})^{-1}\mathbf{X}'\mathbf{y} \tag{7.1}$$

The estimators in (7.1) are quite appropriate under ideal conditions. Suppose, however, we relax the assumption that $\text{Var}(\boldsymbol{\varepsilon}) = \sigma^2 \mathbf{I}_n$ and assume instead that there is a positive definite matrix $\mathbf{V}$ (see Appendix A.2) for which

$$\text{Var}(\boldsymbol{\varepsilon}) = \mathbf{V} \tag{7.2}$$

In stating Eq. (7.2), we invoke a more generalized approach to least squares estimation. The matrix $\mathbf{V}$ is a variance-covariance matrix, and thus, by introducing it, we account for a possible deviation from common error variance as well as nonzero covariances among the errors. We may wish to consider

$$\mathbf{V} = \text{diag}[\sigma_1^2, \sigma_2^2, \ldots, \sigma_n^2] \tag{7.3}$$

in which case we are *assuming uncorrelated errors* with *error variances that vary from observation to observation.* The reader can gain some understanding of the impact of heterogeneous variance by considering that the individual observations, the y_i, have variances that vary from observation to observation. As a result, it would seem reasonable that the proper estimator of $\boldsymbol{\beta}$ should take this into account by weighting the observations in some way that allows for the differences in the precision of the results. The condition we describe here produces information concerning the model from all observations, but quite simply, the pieces of information do not have equal precision.

The appropriate estimator of $\boldsymbol{\beta}$ in the general linear model when the ε_i have variance-covariance structure given by Eq. (7.2) is *not* the ordinary least squares estimator given by Eq. (7.1). Rather, the appropriate estimator is the *generalized least squares estimator* given by

$$\boxed{\boldsymbol{\beta}^* = (\mathbf{X}'\mathbf{V}^{-1}\mathbf{X})^{-1}\mathbf{X}'\mathbf{V}^{-1}\mathbf{y}} \tag{7.4}$$

The following are important characteristics of this generalized least squares estimator.

1. The estimator $\boldsymbol{\beta}^*$ is unbiased, i.e., $E(\boldsymbol{\beta}^*) = \boldsymbol{\beta}$. See Exercise 7.1.
2. $\boldsymbol{\beta}^*$ is a maximum likelihood estimator under normality conditions on $\boldsymbol{\varepsilon}$, i.e., if $\boldsymbol{\varepsilon} \sim N(\mathbf{0}, \mathbf{V})$. See Appendix B.3.
3. The estimators in $\boldsymbol{\beta}^*$ achieve minimum variance of all unbiased estimators under the condition $\boldsymbol{\varepsilon} \sim N(\mathbf{0}, \mathbf{V})$.
4. If the assumption of normality is relaxed, a more generalized Gauss-Markoff Theorem applies (see Section 3.3); namely, the estimators in $\boldsymbol{\beta}^*$ achieve the minimum variance of all linear unbiased estimators.

From 1–4, it is clear that the estimator in (7.4) attains the same properties as the ordinary least squares estimator when conditions are ideal, i.e., when $\mathbf{V} = \sigma^2\mathbf{I}$. In the latter situation $\boldsymbol{\beta}^*$ reduces to $\mathbf{b}$ in Eq. (7.1).

WHAT IS BEING MINIMIZED BY β^*

The notion of generalized least squares implies that $\boldsymbol{\beta}^*$ is minimizing some type of sum of squares. The function being minimized depends on $\mathbf{V}$ and is given by (see Exercise 7.2)

$$SS_{\mathrm{Res},\mathbf{V}} = (\mathbf{y} - \mathbf{X}\boldsymbol{\beta}^*)'\mathbf{V}^{-1}(\mathbf{y} - \mathbf{X}\boldsymbol{\beta}^*) \tag{7.5}$$

Those readers interested in the maximum likelihood approach can verify that the minimization of the result in (7.5) does indeed maximize the likelihood under the condition that $\boldsymbol{\varepsilon} \sim N(\mathbf{0}, \mathbf{V})$. The resulting estimator given by (7.4) appeals nicely to intuition when the estimator is viewed as a weighted least squares estimator.

WEIGHTED LEAST SQUARES

Suppose we assume, as mentioned earlier, that the model errors are uncorrelated but the homogeneous variance assumption does not hold. In other words, the error variances at the n data points are $\sigma_1^2, \sigma_2^2, \ldots, \sigma_n^2$. The $\mathbf{V}$ matrix is given by

$$\mathbf{V} = \text{diag}[\sigma_1^2, \sigma_2^2, \ldots, \sigma_n^2] \tag{7.6}$$

A graphical illustration of the heterogeneous variance situation is given in Figure 7.1. Suppose, for example, that a regression data set is governed by the model

$$y_{ij} = \beta_0 + \beta_1 x_i + \varepsilon_{ij} \qquad (i = 1, 2, \ldots, m;\ j = 1, 2, \ldots, n)$$

for which the ε_{ij} are uncorrelated with mean zero but unequal variances. That is, $\text{Var}(\varepsilon_{ij}) = \sigma_i^2$. In Figure 7.1, there is an increase in variability as x increases. Obviously as x increases, the information that is available for estimating slope and intercept becomes lessened due to the larger variances. For small values of x, the small error variances, i.e., high reproducibility of the experiment, produce a greater amount of information. It is easy to verify that for this very important special case the generalized least squares estimator of $\boldsymbol{\beta}$ minimizes (from Eq. (7.5))

$$SS_{\text{Res(weighted)}} = \sum_{i=1}^{n} w_i(y_i - \hat{y}_i)^2 \tag{7.7}$$

where $w_i = 1/\sigma_i^2$. That is, it is a least squares estimator, but each residual is first weighted by the reciprocal of the error standard deviation. The appropriate weighted least squares estimator is then given by Eq. (7.4) with $\mathbf{V}$ given by Eq. (7.6).

The criterion of minimization of (7.7) is a very logical extension of standard least squares. The implication is that residuals at data locations with large error variances *should not count as much.* The contribution to the residual sum of squares should be greatest at data locations where the model is *sharper* or *more precise.*

FIGURE 7.1 **Simple linear regression with heterogeneous variance**

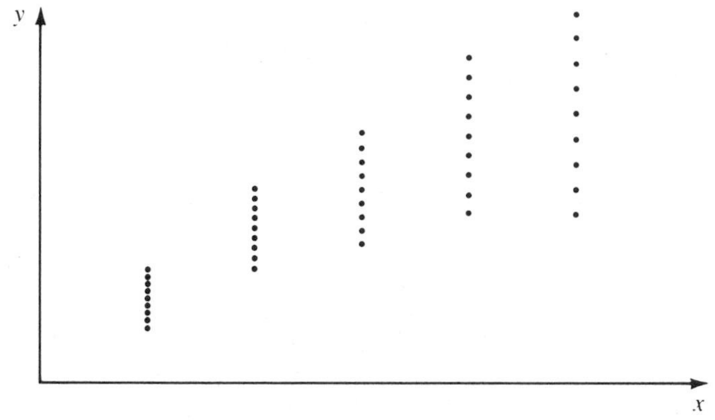

Most commercial regression computer packages contain a weighted regression routine. The user is often free to insert the weights. However, the estimator is not truly the generalized least squares estimator unless the weights depend on the error variances as indicated here. In fact, *robust regression* methods presented later in this chapter exploit weighted regression as an analytical tool. However, in the presence of normality, independence, and heterogeneous variance, weighted regression with $w_i = 1/\sigma_i^2$ is the optimal procedure.

Practical Difficulties

If the analyst has evidence that heterogeneous variance is a problem, then clearly weighted regression is an option that must be considered. However, in many cases, a practical difficulty surfaces. *Where does one obtain* σ_i?

In Figure 7.1 sample variances can be used for estimating the individual error variances. Thus the weights can be estimated. However, in most applications where the regressor variables do not arise from a designed experiment, only one y observation is recorded at each regressor combination; thus estimates of the σ_i are unlikely to be available. As a result, weighted least squares for producing the optimal estimator is often not practicable. Estimation of the weights is essentially confined to experimental situations in which repeated runs on the response can be made at each regressor combination. From these repeated runs, a simple estimate of error variance can be computed. The analyst should be cautious however; the use of estimates in place of the σ_i may not be appropriate if they are based on limited information. Poorly estimated weights implemented in weighted regression may indeed produce results that are more unsatisfactory than merely not weighting at all. A crude guideline on the use of so-called *estimated* generalized least squares (estimating weights) is that estimated weights should not be used unless they are each based on a sample of size approximately nine. (See Deaton *et al.* (1983).) This is not intended as a firm rule of thumb—only a guideline. Since many experimental plans do not involve nine or more replicated runs at each data point, ignoring the weights may often be the most effective course of action. For more information on the use of estimated weights, the reader is referred to Williams (1967) and Deaton, *et al.* (1983).

VARIANCE-COVARIANCE PROPERTIES OF THE GENERALIZED LEAST SQUARES ESTIMATOR

The weighted least squares estimator, of course, is a special case of the generalized least squares estimator. The former is designed to produce an optimal unbiased estimator of $\boldsymbol{\beta}$ for the heterogeneous variance situation. However, as we indicated in the previous section, the optimality only holds if the true σ_i and thus the proper weights $w_i = 1/\sigma_i^2$ are known. If estimates are substituted for the w_i, the weights become random variables, and the properties of the estimator of $\boldsymbol{\beta}$ become very complicated. It is important, however, for the student of linear regression to know the properties of the generalized least squares estimator (or the weighted regression special case) for known **V**.

As we indicated earlier, the estimator $\boldsymbol{\beta}^*$ in (7.4) is unbiased. The variance-covariance matrix is easily developed. Consider

$$\begin{aligned}\text{Var}(\boldsymbol{\beta}^*) &= \text{Var}(\mathbf{X}'\mathbf{V}^{-1}\mathbf{X})^{-1}\mathbf{X}'\mathbf{V}^{-1}\mathbf{y}\\ &= (\mathbf{X}'\mathbf{V}^{-1}\mathbf{X})^{-1}\mathbf{X}'\mathbf{V}^{-1}[\text{Var }\mathbf{y}]\mathbf{V}^{-1}\mathbf{X}(\mathbf{X}'\mathbf{V}^{-1}\mathbf{X})^{-1}\end{aligned}$$

Since Var $\mathbf{y} = \mathbf{V}$, we have

$$\boxed{\text{Var }\boldsymbol{\beta}^* = (\mathbf{X}'\mathbf{V}^{-1}\mathbf{X})^{-1}} \tag{7.8}$$

Note that if $\mathbf{V} = \sigma^2\mathbf{I}$, the expression in (7.8) reduces to $(\mathbf{X}'\mathbf{X})^{-1}\sigma^2$, the variance-covariance matrix of the ordinary least squares estimator under standard conditions.

Example 7.1, which follows, presents an analysis of a data set in which there is a violation of the homogeneous variance assumption. Weighted regression is used with weights that are estimated from the data.

Example 7.1 ***Transfer Efficiency Data***

Consider the following set of transfer efficiency data in which two factors, air velocity, x_1, and voltage, x_2, influence the efficiency of a particular electrostatic type of spray paint equipment. Four combinations of the factors were chosen in a designed experiment and ten experimental runs were made at each of the combinations. Air velocity is in feet per minute and voltage in kilovolts.

		Voltage (x_2) 50		Voltage (x_2) 70	
Air Velocity (x_1)	60	87.5	88.2	77.4	68.1
		88.1	87.3	70.7	65.3
		89.5	89.2	67.0	61.0
		86.2	85.9	71.7	81.7
		90.0	87.0	79.2	60.3
	120	82.5	81.3	61.2	50.7
		81.6	80.7	67.2	52.3
		77.4	79.3	55.9	68.6
		81.5	82.0	52.0	69.5
		79.7	79.2	63.5	70.1

With the multiple runs available at each factor combination, one can estimate the error variance at each of the four factor combinations. These estimates are given by $\frac{1}{9}\sum(y-\bar{y})^2$ at each location. The results are given by

Voltage	Air Velocity	Estimate of Variance	Weights
50	60	1.8899	0.5291
50	120	2.5018	0.3997
70	60	54.3204	0.0184
70	120	60.6933	0.0165

From the variance estimates given, it would seem that the homogeneous variance assumption is suspect. As a result, we use weighted regression with the weights computed from the data as the reciprocals of the variance estimates. The estimator is given by Eq. (7.4) with **V** given by Eq. (7.6) with the estimates substituted for the population variances. The larger variances at the high voltage suggest that the model precision reduces considerably at the high voltage. The following represents the computed weighted regression:

$$\hat{y} = 141.5612 - 0.1239x_1 - 0.9242x_2$$

with the observed efficiency, fitted values, and residuals given in Table 7.1. The residuals clearly reflect the heterogeneous variance problem. There are more large residuals occurring at the high voltage than those observed at the low voltage.

TABLE 7.1 **Observed value, predicted value, and residuals for transfer efficiency data**

Velocity (x_1)	Voltage (x_2)	Efficiency (y)	Fitted Value ($\hat{y}$)	Residual ($y-\hat{y}$)
60	50	87.50	87.918	−0.418
60	50	88.10	87.918	0.182
60	50	89.50	87.918	1.582
60	50	86.20	87.918	−1.718
60	50	90.00	87.918	2.082
60	50	88.20	87.918	0.282
60	50	87.30	87.918	−0.618
60	50	89.20	87.918	1.282
60	50	85.90	87.918	−2.018
60	50	87.00	87.918	−0.918
120	50	82.50	80.483	2.017
120	50	81.60	80.483	1.117
120	50	77.40	80.483	−3.083
120	50	81.50	80.483	1.017

TABLE 7.1 Continued

Observed value, predicted value, and residuals for transfer efficiency data

Velocity (x_1)	Voltage (x_2)	Efficiency (y)	Fitted Value ($\hat{y}$)	Residual ($y-\hat{y}$)
120	50	79.70	80.483	−0.783
120	50	81.30	80.483	0.817
120	50	80.70	80.483	0.217
120	50	79.30	80.483	−1.183
120	50	82.00	80.483	1.517
120	50	79.20	80.483	−1.283
60	70	77.40	69.435	7.965
60	70	70.70	69.435	1.265
60	70	67.00	69.435	−2.435
60	70	71.70	69.435	2.265
60	70	79.20	69.435	9.765
60	70	68.10	69.435	−1.335
60	70	65.30	69.435	−4.135
60	70	61.00	69.435	−8.435
60	70	81.70	69.435	12.265
60	70	60.30	69.435	−9.135
120	70	61.20	62.000	−0.800
120	70	67.20	62.000	5.200
120	70	55.90	62.000	−6.100
120	70	52.00	62.000	−10.000
120	70	63.50	62.000	1.500
120	70	50.70	62.000	−11.300
120	70	52.30	62.000	−9.700
120	70	68.60	62.000	6.600
120	70	69.50	62.000	6.500
120	70	70.10	62.000	8.100

FURTHER TECHNIQUES IN WEIGHTED REGRESSION

To this point in Section 7.1, we have paid considerable attention to justifying the use of the weighted regression estimator given by Eq. (7.4) in cases where it is known that the homogeneous variance assumption does not hold. Example 7.1 illustrates the computed weighted regression, but what about other traditional tests and computational procedures that we exploit in ordinary weighted regression? Do they apply and what adjustments must be made? The major difficulties that surface revolve around the assumed variance-covariance structure associated with the model errors. Standard tests, computation of C_p, PRESS, etc., described in Chapters 3 and 4 depend on the form of the estimator being as shown in Eq. (7.1) or on the variance-covariance matrix having the form $(\mathbf{X}'\mathbf{X})^{-1}\sigma^2$. As a result, some extensions to the weighted regression case quite naturally involve the **V** matrix, and in fact, their success depends on **V** and thus quality estimation of the weights. We consider here further procedures in weighted regression. However, in much of the foregoing, we assume that the weights given by $w_i = 1/\sigma_i^2$ are known.

Standard Errors and Tests

If we assume that $\mathbf{V}$ is known, i.e., not a random variable, then the standard errors of the weighted regression coefficients are quite easily computed. One merely computes variances of the coefficients as

$$c_{jj} = j\text{th diagonal element of } (\mathbf{X}'\mathbf{V}^{-1}\mathbf{X})^{-1}$$

and thus the standard error is computed as

$$\sigma_{\beta_j^*} = \sqrt{c_{jj}}$$

Tests of significance on model coefficients can be made very simply by using the statistic

$$t = \frac{\beta_j^*}{\sqrt{c_{jj}}}$$

where β_j^* is the coefficient of x_j from weighted regression. The statistic is designed to test

$$H_0: \beta_j = 0$$
$$H_1: \beta_j \neq 0$$

We use the "t" notation here but, of course, if we assume $\mathbf{V}$ is known, then the standard error is not an estimate. As a result, the appropriate test is a two-tailed test with the critical region being that of a standard normal distribution.

As in the case of standard unweighted least squares, a sum of squares identity becomes very useful in explaining the partition of variation. It also affords a *variation explained* or regression sum of squares as in the standard unweighted case. In matrix notation, this sum of squares identity is given by

$$\boxed{\mathbf{y}'\mathbf{V}^{-1}\mathbf{y} = \boldsymbol{\beta}^{*\prime}\mathbf{X}'\mathbf{V}^{-1}\mathbf{y} + (\mathbf{y} - \mathbf{X}\boldsymbol{\beta}^*)'\mathbf{V}^{-1}(\mathbf{y} - \mathbf{X}\boldsymbol{\beta}^*)}$$

or

$$\text{Total weighted } SS = \text{Weighted regression } SS + \text{Weighted residual } SS$$

One easily recognizes the weighted residual sum of squares

$$(\mathbf{y} - \mathbf{X}\boldsymbol{\beta}^*)'\mathbf{V}^{-1}(\mathbf{y} - \mathbf{X}\boldsymbol{\beta}^*)$$

as the now familiar weighted sum of squares in Eq. (7.7).

The weighted residual sum of squares maintains a role in variance estimation if the model assumption is altered slightly. Now, suppose we are not willing to assume knowledge of the error variances, but rather we assume we know a diagonal matrix $\mathbf{V}$, where

$$\text{Var}(\boldsymbol{\varepsilon}) = \mathbf{V}\sigma^2$$

In other words, we assume the variances are known apart from the constant σ^2. Now, the procedure for estimation of the regression coefficients is identical to the weighted least squares described earlier in this section. The fact that σ^2 is unknown does not alter the weights; the weights remain the diagonal elements of $\mathbf{V}^{-1}$. In this case, the weighted residual sum of squares is used in estimation of σ^2. In fact, the estimator, s^2, is given by

$$s^2 = \sum_{i=1}^{n} \frac{w_i(y_i - \hat{y}_i)^2}{(n-p)}$$

where p remains the number of model parameters.

Example 7.2 ***Transfer Efficiency Data***

Consider the transfer efficiency data of Example 6.1. Using the weights in that example, the variance-covariance matrix of coefficients is given by

$$(\mathbf{X}'\mathbf{V}^{-1}\mathbf{X})^{-1} = \begin{bmatrix} \sum_{i=1}^{n} w_i & \sum_{i=1}^{n} w_i x_{1i} & \sum_{i=1}^{n} w_i x_{2i} \\ & \sum_{i=1}^{n} w_i x_{1i}^2 & \sum_{i=1}^{n} w_i x_{1i} x_{2i} \\ \text{symmetric} & & \sum_{i=1}^{n} w_i x_{2i}^2 \end{bmatrix}^{-1}$$

$$= \begin{bmatrix} 9.637329 & 827.9550 & 488.8426 \\ 827.9550 & 79643.13 & 42014.09 \\ 488.8436 & 42014.09 & 24930.58 \end{bmatrix}^{-1}$$

$$= \begin{bmatrix} 19.9777 & -0.0093446 & -0.37597 \\ -0.0093446 & 0.00011750 & -0.000014791 \\ -0.37597 & -0.000014791 & 0.00743730 \end{bmatrix}$$

The elements of $\mathbf{X}\mathbf{V}^{-1}\mathbf{y}$ are given by

$$\mathbf{X}'\mathbf{V}^{-1}\mathbf{y} = \begin{bmatrix} 488.8436 \\ 42014.09 \\ 24930.58 \end{bmatrix}$$

As a result, the estimates (from Eq. (7.4)) and standard errors (from Eq. (7.8)) are given by

$$\beta_0^* = 141.5612 \qquad s_{\beta_0^*} = 4.4696$$
$$\beta_1^* = -0.1239 \qquad s_{\beta_1^*} = 0.0108$$
$$\beta_2^* = -0.9242 \qquad s_{\beta_2^*} = 0.0862$$

Then, if we consider test statistics to determine if the coefficients differ significantly from zero, we have for β_0^*,

$$t = \frac{141.5612}{4.4696} = 31.67$$

and for β_1^*,

$$t = \frac{-0.1239}{0.0108} = -11.47$$

and for β_2^*,

$$t = \frac{-0.9242}{0.0862} = -10.72$$

From the above ratios of coefficients to their standard errors, it would appear that conclusions can be drawn that all regression coefficients (including the constant term) differ significantly from zero. Both voltage and velocity seem to have an important effect on transfer efficiency.

TRANSFORMATIONS TO STABILIZE VARIANCE

In many fields of application of regression analysis, failure of the homogeneous variance assumption is very natural and expected. The source of the problem is that the error variance is often *not independent* of the mean $E(y)$. As we indicated earlier, as regressor values and responses in the data become larger, variance around the regression tends to grow.

It is unlikely that the analyst would know exactly the functional form relating error variance to $E(y)$. However, there can be cases where evidence in the residuals or evidence from replication variance would give some insight regarding the structure. Knowledge of how error variance fluctuates with the mean response can often suggest a transformation on the response that is designed to *stabilize the error variance.* The analyst would then work with the transformed y in building the regression.

A variance-stabilizing transformation, which is used quite often, is the *natural log* transformation, i.e., $\ln y$. This is an appropriate transformation when the *error standard deviation* is proportional to $E(y)$. Another is the inverse transformation, i.e., $1/y$. This transformation may be appropriate in the rather extreme situation in which the error standard deviation appears to be a quadratic function of the mean response. That is, σ is proportional to $[E(y)]^2$.

The purpose of the transformations discussed here is to make the very important homogeneous variance assumption more palatable. A transformation may be very effective. However, the analyst must develop an awareness that correcting one violation (by, say, the transforming of data) may cause another violation to surface. In the present context, a transformation to stabilize error variance clearly changes the assumption regarding the functional form of the model. One must avoid achieving stable error variance with the "fallout" being a serious deterioration of quality of fit and prediction. If a transformation is made, say the natural log transformation, a prudent analyst will look very carefully into what was the effect of the transformation on the fit and the properties of the prediction of y, not ln y. Further information regarding the impact of transformations will be given and illustrated in Section 7.3.

7.2

PROBLEM WITH CORRELATED ERRORS (AUTOCORRELATION)

In Section 7.1 we introduced the general methodology that allows for estimation of regression coefficients when the variance-covariance conditions on the model errors are not ideal. Emphasis was placed on the use of weighted least squares for the situation in which the analyst is faced with the heterogeneous variance problem. On the other hand, the method of generalized least squares with the estimators of regression coefficients given by Eq. (7.4) was introduced as a general methodology if one knows the variance-covariance structure and has an extremely good estimate of the matrix **V**. However, more often than not the elements of **V** are not known.

There are specific types of applications of regression analysis in which there is a natural effect on results *over time*. The data involved are often called *time series data*. When regression data are time dependent, often it is not advisable to assume that the errors are uncorrelated. Rather, one can often assume that the errors have so-called serial correlation. In other words, $E(\varepsilon_i \cdot \varepsilon_{i+j}) \neq 0$. We say that the errors are *autocorrelated*.

The concept of autocorrelation due to the effect of time is quite easy to understand. Suppose a scientist is interested in building a model that relates the demand of a particular product to the price of the product. However, suppose that the data are taken over time, e.g., demand and price are average or monthly readings. Other variables that are influenced by time and likely influence demand for the product are not included in the model. Variables such as seasonal influences and consumer population size are essentially synonymous with the variable *time*. As a result, autocorrelation can often be viewed as a problem of *model misspecification*. Suppose we postulate a simple linear regression model

$$y_i = \beta_0 + \beta_1 x_i + z_i \qquad (i = 1, 2, \ldots, n)$$

when in fact the error in the model can be written

$$z_i = \varepsilon_i + \sum_j \beta_j w_{ji}$$

where ε_i is a more conventional model error and $\sum_j \beta_j w_{ji}$ represents the accumulation of time dependent regressors that were not included in the model. Obviously, the z_i, dominated by the *hidden variables*, will be correlated from one time period to another.

IMPACT AND DETECTION OF AUTOCORRELATION

The existence of autocorrelated errors implies, of course, that the variance-covariance matrix of errors is not $\sigma^2 I$ and thus conditions are not ideal. The presence of the autocorrelation causes difficulty in the estimation of error variance and, as a result, in tests of hypotheses and confidence interval estimation. The existence of positively correlated errors can result in an estimate of σ^2 that is a substantial underestimate. This, of course, tends to inflate t-statistics on coefficients and deflate the width of confidence intervals on coefficients.

Detection of autocorrelation can be accomplished through diagnostic plots or through a formal test. Simple plots of residuals can be helpful. If the residuals from a standard least squares analysis are plotted against time, and unusually large numbers of residuals with the same sign tend to cluster together, then there is an indication that the errors are governed by a positive autocorrelation. On the other hand, rapid changes in sign signify a negative correlation. Figure 7.2 provides an illustration.

FIGURE 7.2 ***(a)*** **Residual plot for which errors are positively autocorrelated** ***(b)*** **Residual plot for which errors are negatively autocorrelated**

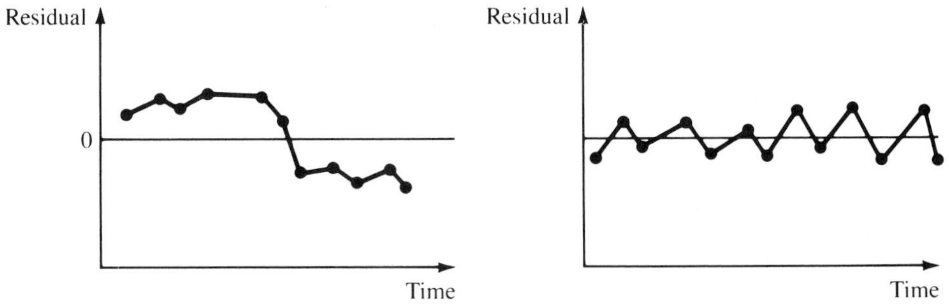

There are formal tests for detection of autocorrelation. Perhaps the best known test is the Durbin–Watson test (D-W). It appears in many commercial regression packages. The D-W test is designed to detect positive autocorrelation. Despite its wide usage, it is often criticized for being restrictive in the type of correlation structure that is assumed. It is assumed that for equally spaced time periods we have

$$\varepsilon_t = \rho\varepsilon_{t-1} + u_t \quad \textbf{(7.9)}$$

where ε_t is the error at time period t for the regression model, ρ is the autocorrelation, and u_t is a random disturbance, assumed to be $N(0, \sigma_u)$. Of course, the parameter ρ cannot exceed 1.0 in magnitude. The error structure in (7.9) describes a *first-order autoregressive* error structure. Now, since the model errors are related to each other through (7.9), it is instructive to focus a bit closer on the properties of a given error ε_t. If we successively use (7.9), we have

$$\varepsilon_t = \sum_{j=0}^{\infty} \rho^j u_{t-j} \tag{7.10}$$

Thus the model error is a linear combination of independent normally distributed errors in the expression of (7.9). The u values that have the greatest impact on ε_t are those disturbances that are the recent ones in time. We can write

$$E(\varepsilon_t) = 0$$

$$\mathrm{Var}(\varepsilon_t) = \sigma_u^2 \sum_{j=0}^{\infty} \rho^{2j}$$

The expression $\sum_{j=0}^{\infty} \rho^{2j}$ can actually be written as $1/(1 - \rho^2)$. Thus

$$\mathrm{Var}\ \varepsilon_t = \sigma_u^2 \left(\frac{1}{1 - \rho^2} \right) \tag{7.11}$$

In a similar fashion it is not difficult to show that

$$\mathrm{Cov}(\varepsilon_t, \varepsilon_{t+j}) = \rho^{|j|} \sigma_u^2 \left(\frac{1}{1 - \rho^2} \right) \tag{7.12}$$

As a result, the autocorrelation structure given by Eq. (7.9) results in autocorrelation between model errors unless $\rho = 0$. In addition, the autocorrelation, as expected, is more severe for errors that are close together in time.

The D-W test is designed to test

$$H_0: \rho = 0$$
$$H_1: \rho > 0$$

The reader should suspect that the test statistic should somehow be based on the magnitude of the differences *between adjacent residuals.* If the errors exhibit positive autocorrelation, one would expect adjacent residuals to be numerically similar. The D-W test statistic is given by

$$d = \frac{\sum_{t=2}^{n} (e_t - e_{t-1})^2}{\sum_{t=1}^{n} e_i^2} \tag{7.13}$$

Here, of course, the e_i are the ordinary residuals from the least squares analysis of the data. Obviously if $\rho > 0$, one would expect the numerator of d to be fairly small. Hence, H_0 is rejected on small values of d. Bounds given by Durbin and Watson (1951) appear in Table C.6 for various values of n and α. These bounds are denoted by d_L and d_U. The D-W test is to

Reject H_0: $\rho = 0$ if $d < d_L$.
Do not reject H_0: $\rho = 0$ if $d > d_U$.
Declare test inconclusive if $d_L < d < d_U$.

Example 7.3 In Chapter 5 we illustrated the use of residual plots with the data of Example 5.1 in which the population of the U.S. was regressed against the year. The data appear in Table 5.1. Residual plots for various model specifications appear in Figures 5.3, 5.4, and 5.5. The example was shown at that point in the text in order to illustrate how model misspecification can be detected through residual plots. However, since the population data are obviously time dependent, the true model misspecification may be hidden model misspecification that often accompanies autocorrelation. Certainly it is of interest to test for positive autocorrelation for at least one model that was fit to the data. The model containing a linear term and a quadratic term in the regressor variable, year, as well as an indicator variable isolating the 1940 and 1950 years is given by

$$\hat{y} = 20{,}982.784 - 23.36638x + 0.006507x^2 - 8.7415z$$

where x is the year and z is the indicator variable described in Section 5.1. Table 7.2 redisplays the ordinary residuals. The computed D-W value is given by

$$d = \frac{\sum_{i=2}^{17} (e_i - e_{i-1})^2}{\sum_{i=1}^{17} e_i^2} = 1.3825$$

We see from Table C.7 that the critical values for sample size 19 do not appear. However, if one interpolates between values $n = 15$ and $n = 20$, it appears as if the test is inconclusive.

Detection of autocorrelated errors in time-dependent data can often be made by merely observing residual plots. For example, the residual plots for Figures 5.3, 5.4, and 5.5 would certainly bring about suspicion, though when the quadratic term is added the residuals do take on a more orthodox appearance. In the population example, time, namely *year*, was the primary regressor variable. Thus it is natural to suspect some form of autocorrelation. In general, when data are time dependent, one should investigate residuals and look for a hidden regressor that may help explain the variability in response and thus remove the autocorrelation. It must be

TABLE 7.2 Table of residuals for population data of Example 5.1. (Model includes intercept, linear term, quadratic term, and indicator variable.)

Year	Residual $e_i = y_i - \hat{y}_i$
1790	31.253
1800	21.844
1810	12.987
1820	4.598
1830	−2.962
1840	−9.547
1850	−14.213
1860	−16.749
1870	−19.162
1880	−19.612
1890	−17.608
1900	−15.349
1910	−10.159
1920	−7.209
1930	−0.932
1940	−2.826
1950	6.042
1960	23.252
1970	36.352

emphasized, however, that autocorrelation may appear even if time is not the regressor variable. Any regression data set that is time dependent may exhibit autocorrelated errors. The use of the Durbin–Watson statistic is helpful in detection.

If one does detect autocorrelated errors, there are several methods that can be used to remove the autocorrelation. Initially, of course, the subject matter scientist should search for an additional regressor variable or variables. In this text we do not come close to exhausting the methods for handling autocorrelation. The reader is referred to an excellent text by Box and Jenkins (1976) for a thorough discussion and development.

METHODS FOR HANDLING AUTOCORRELATION

One procedure for removing autocorrelation is to reformulate the model by using *lagged values* of the response variable. When autocorrelation is present in time-dependent data, it is reasonable to assume that the value of the response y_t should be empirically dependent on y_{t-1}, the value at the adjacent time period. Thus, a fitted model of the type

$$y_t = \beta_0 + \beta_1 x_t + \beta_2 y_{t-1} + \varepsilon_t$$

should be considered. A reinvestigation of residuals as well as performance criteria discussed in Chapter 4 should be used to assess the value of such a model. The Durbin–Watson statistic should not be used for a model containing lagged variables.

A fairly simple method for estimating parameters in a model containing autocorrelation was developed by Cochrane and Orcutt (1949). This method is based on transforming both response and regressor variables so that standard least squares regression assumptions hold. Consider the simple linear regression model

$$y_t = \beta_0 + \beta_1 x_t + \varepsilon_t \qquad (t = 1, 2, \ldots, n)$$

with the errors satisfying the autoregressive structure of (7.9). Consider the transformed response

$$y_t^* = y_t - \rho y_{t-1} \tag{7.14}$$

Making use of Eq. (7.9) and (7.14), we have

$$\begin{aligned} y_t^* &= \beta_0 + \beta_1 x_t + \varepsilon_t - \rho(\beta_0 + \beta_1 x_{t-1} + \varepsilon_{t-1}) \\ &= \beta_0(1-\rho) + \beta_1(x_t - \rho x_{t-1}) + (\varepsilon_t - \rho\varepsilon_{t-1}) \quad (t = 1, 2, \ldots, n) \\ &= \beta_0^* + \beta_1^* x_t^* + u_t \end{aligned} \tag{7.15}$$

Notice that the errors in model (7.15) are the errors of the autoregressive structure in Eq. (7.9) and are normal and independent with mean zero and variance σ_u^2. As a result, the standard least squares procedure can be used *if* the parameter ρ is known. Of course, it is unlikely in practice that ρ would be known. A point estimate of ρ can be used in producing an estimate of β_0 and β_1. From the first-order autoregressive structure of Eq. (7.9), a simple estimate of ρ is easily obtained. Viewing (7.9) as a simple linear regression through the origin, we have the following estimator

$$\boxed{\hat{\rho} = \frac{\sum_{t=2}^{n} e_t \cdot e_{t-1}}{\sum_{t=1}^{n} e_t^2}} \tag{7.16}$$

where the e_t are the residuals from an ordinary least squares regression of y_t on x_t.

Thus a single iteration estimation procedure for β_0 and β_1 is to:

(i) Obtain $\hat{\rho}$ after an ordinary least squares fit.

(ii) Form the y_t^* and x_t^* and do an ordinary least squares regression of y_t^* on x_t^*. Find β_0^* and β_1^*.

(iii) The estimators of the intercept and slope are $\beta_0^*/(1-\rho)$ and β_1^*.

Additional iterations of the procedure are often required. An application of the Durbin–Watson statistic to the residuals of the model of Eq. (7.15) may be used as a guide. A second iteration may be fruitful if the D-W statistic indicates that positive autocorrelation is still apparent. A second iteration involves computing $\hat{\rho}$ from Eq. (7.16) using the residuals of the regression of model (7.15). This produces a new y_t^* and x_t^* and hence a new β_0^* and β_1^*.

7.3 TRANSFORMATIONS TO IMPROVE FIT AND PREDICTION

Transformation of data is sometimes an effective alternative that produces a better fitting, or perhaps better predicting model. Transformation or *reexpression* of data is used in order to make an assumption more reasonable. Thus far, in this chapter, we have made use of transformations to stabilize variance and to eliminate autocorrelation in regression data. Here the motivation revolves around another assumption, that of *model form*.

In the following subsections, procedures are discussed for determining *alternative model forms*, or *transformations*, and situations under which the use of these transformations might prove successful. In some cases the need for some type of transformation is obvious. Plots of data in the single regressor case may show a definite curvilinear appearance.

TRANSFORMATION IN THE CASE OF A SINGLE REGRESSOR

Consider a regression situation with a single regressor variable. A transformation to change the model structure is normally required when the data reflect curvature. A deviation from the ordinary straight line regression model

$$y = \beta_0 + \beta_1 x + \varepsilon$$

may, of course, be detected from residual plots (see Section 5.2) or from a simple plotting of the data. If a plot reflects a trend that has a specific curvilinear appearance, one may be able to alter the model to accommodate the trend.

Parabola

The first model alteration we consider is more than a mere transformation but rather *the addition of a quadratic model term.* The model is given by

$$y = \beta_0 + \beta_1 x + \beta_2 x^2 + \varepsilon$$

The nature of the plot produced by data generated by the parabolic model depends on the signs and magnitudes of the coefficients β_0, β_1, and β_2. Two typical examples are given in Figure 7.3.

FIGURE 7.3

(a) **Parabola for β_0, β_1, and $\beta_2 > 0$**

(b) **Parabola for $\beta_0 > 0$, $\beta_1 > 0$, $\beta_2 < 0$**

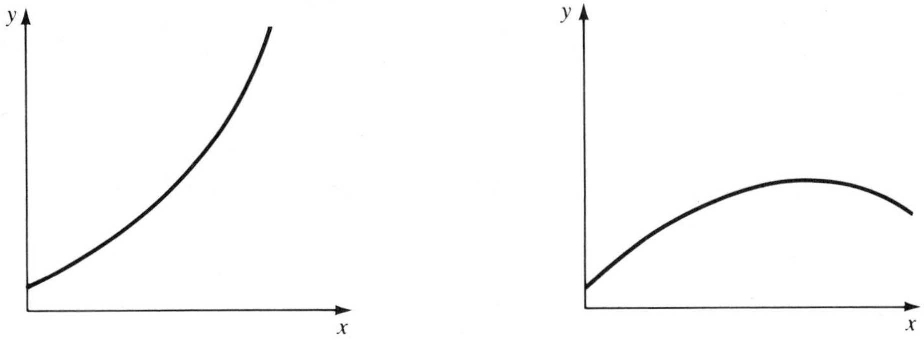

Hyperbola (Inverse Transformation on y and x)

Applications in areas of biology, economics, and certain other fields lead to the use of a hyperbolic function, which can be produced by transformations on both the response variable y and the regressor variable x. Typical plots that suggest the use of the hyperbola are given in Figure 7.4. The true functional form of the hyperbola is *nonlinear* in the model coefficients. The equation is given by $y = x/(\alpha + \beta x)$. The linearized form involves the inverse transformation on both variables; i.e., one regresses $1/y$ against $1/x$, and thus adopts the model structure (assuming observations x_i, y_i, $i = 1, 2, \ldots, n$)

$$\frac{1}{y_i} = \beta_0 + \beta_1\left(\frac{1}{x_i}\right) + \varepsilon_i$$

FIGURE 7.4

(a) **Hyperbola with negative curvature**

(b) **Hyperbola with positive curvature**

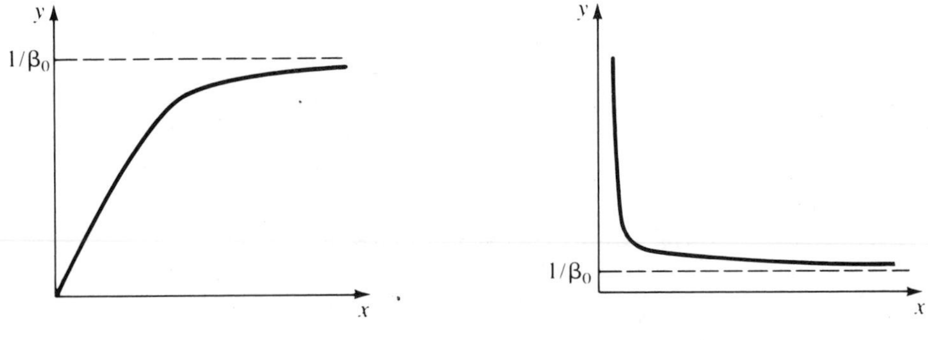

It is easily verified that $\beta_0 = \beta$ and $\beta_1 = \alpha$. The asymptote that may be of interest to the analyst is indicated by the dotted lines in Figure 7.4. Negative and positive curvature are, respectively, produced when $\beta_1 > 0$ and $\beta_1 < 0$.

Exponential Function; Natural Log Transformation on y

In Section 7.1 we discussed the log transformation on the response as a mechanism that may be useful for countering heterogeneous variance in certain situations. The transformation may also be useful to produce a reasonable model assumption when the data's appearance suggests curvature of a certain type. If the picture is not a straight line but, rather, looks like that depicted in either part of Figure 7.5, the true structure may be of the form $y = \alpha e^{\beta x}$. Thus a fitted model of the type

$$\ln(y_i) = \beta_0 + \beta_1 x_i + \varepsilon_i$$

may be appropriate. Here, of course, $\beta_0 = \ln \alpha$ and $\beta_1 = \beta$. Figure 7.5(a) illustrates the situation in which $\beta > 0$. Figure 7.5(b) reveals a case in which $\beta < 0$.

FIGURE 7.5 (a) Exponential function: use ln *y* transform; (b) Exponential function: use ln *y* transform

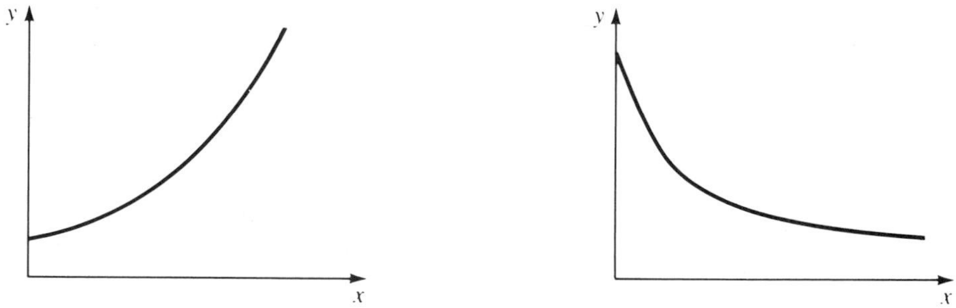

Power Functions (Natural Log Transformations on y and x)

At times, the curvature that the analyst sees in the plot of y against x suggests a mechanism of the type $y = \alpha x^{\beta}$, a power function that is clearly linearized by using the log transformation on both variables. Thus the fitted model is given by

$$\ln(y_i) = \beta_0 + \beta_1(\ln x_i) + \varepsilon_i$$

with the coefficients β_0 and β_1 estimated by standard least squares procedures. The actual appearance of the power function plot depends on the sign and magnitude of the constant β_1. One set of pictures appears in Figure 7.6(a). Another set appears in Figure 7.6(b).

FIGURE 7.6

(a) **Power function: use natural log transformation on *y* and *x***

(b) **Power function: β_1 negative; use natural log transformation on *y* and *x***

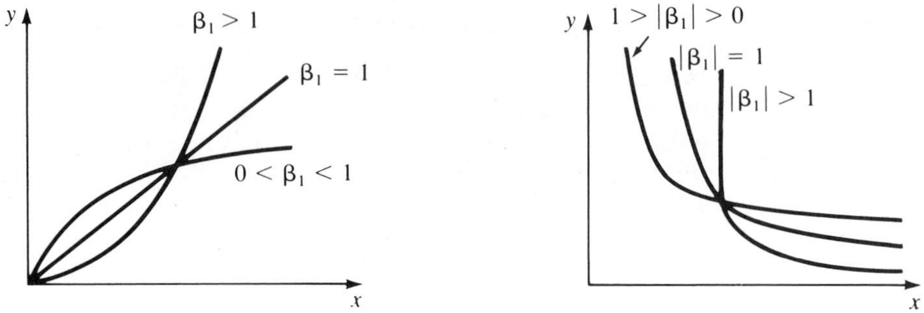

Inverse Exponential (Natural Log Transformation on y; Inverse Transformation on x)

There are many scientific phenomena that are exponential in nature but do not fall into a family presented earlier. Often the exponential portion of the mechanism is proportional to the *inverse* of x instead of x; this results in pictures that do not resemble those presented in Figure 7.5. The mechanism is $y = \alpha e^{\beta/x}$. The fitted model, reflecting the transformation is given by

$$\ln(y_i) = \beta_0 + \beta_1\left(\frac{1}{x_i}\right) + \varepsilon_i$$

Figure 7.7 reveals the appearance of the y against x plot.

The purpose of this section is to give the user alternative models that may provide more success in fitting and predicting than that of the straight line simple

FIGURE 7.7

(a) **Inverse exponential: use natural log transformation on *y* and inverse transformation on *x***

(b) **Inverse exponential: use natural log transformation on *y* and inverse transformation on *x***

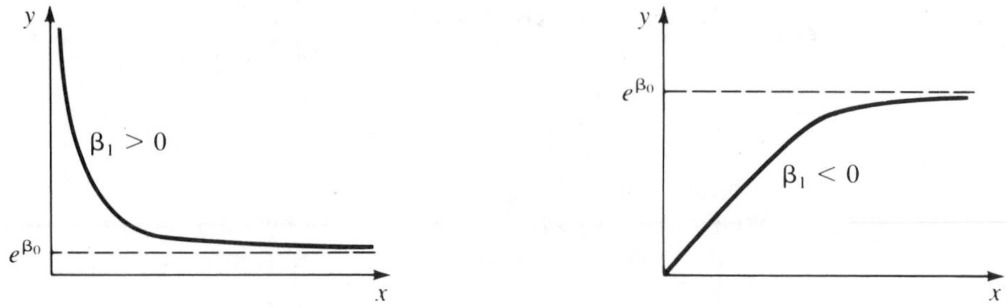

linear regression. In the case of a single regressor variable, if curvature in a plot reveals an appearance similar to the one depicted here, the user should employ the indicated transformation and determine if an improvement results. The improvement may take the form of (i) more attractive pattern of residuals, or (ii) better fit and prediction statistics in the original, natural variables. In the case of (ii), the analyst should form residuals in the original y-units and thus produce an error mean square (residual sum of squares divided by residual degrees of freedom). In addition, PRESS residuals can be found for the y variable in its original untransformed units, and thus a PRESS statistic can be calculated. The user has no dearth of criteria with which to determine if the transformed model is a more reasonable one. Example 7.4 will illustrate this.

What Happens to Model Structure Under Transformation

Transformations are an integral part of statistical data analysis. We attempt to formalize them here so the user can understand their purpose. However, the data analyst must know what the total structure of the model is, once the decision is made to transform the data. A transformation would seem to be a mere stroke of a pen or a change in a computer package model statement, with the motivation being that the data appear to follow the new model more closely. However, there is more involved, and the analyst should be aware. We can best illustrate with an example transformation. Suppose a plot of y against x reveals curvature that resembles Figure 7.6(a). One may assume that a power function of the type $y = \alpha x^{\beta}$ generated the data (apart from random error) and that a model of the type

$$\ln y_i = \beta_0 + \beta_1 \ln x_i + \varepsilon_i \tag{7.17}$$

should be fit. Indeed, all indications from the resulting fit may point to evidence of an improvement. But, where did the transformation truly come from, and what has the analyst done? The transformation works because the true mechanism is the power function $y = \alpha x^{\beta}$. If we ignore errors in the model (which truly we cannot), then, by taking the natural log of the power function, we are led to $\ln y = \beta_0 + \beta_1 \ln x$. However, if we allocate the usual additive error terms to the two relationships, a log transformation of the power function model does not produce Eq. (7.17). One should ask, then, the obvious question: What is the assumption that must be made in order that the model of Eq. (7.17) (with usual homogeneous variance assumption) be valid?

We can obtain Eq. (7.17), the model of the transformed data, by beginning with the model

$$y = \alpha x^{\beta}(1 + \varepsilon^*) = \alpha x^{\beta} + \varepsilon^{**} \tag{7.18}$$

Here, we might assume that ε^* has the usual properties, i.e., $E(\varepsilon^*) = 0$ and homogeneous variance. If these assumptions hold, then $\text{Var}\,\varepsilon^{**} = \sigma^2[E(y)]^2$, which varies with $E(y)$. We say, in this case, that there is a *multiplicative error structure*. Now, under these conditions, a natural log transformation to both sides of (7.18) results in

$$\ln y = \ln \alpha + \beta \ln x + \ln(1 + \varepsilon^*) \tag{7.19}$$

Now, let us closely inspect the model of Eq. (7.19). It is of the linear regression form. But, what about the resulting error term? Suppose $E[\ln(1 + \varepsilon^*)] = \alpha^*$; then we can write

$$\ln(y) = (\ln \alpha + \alpha^*) + \beta \ln x + \varepsilon \tag{7.20}$$

where $E(\varepsilon) = 0$ and $\text{Var } \varepsilon = \sigma^2$, say, is homogeneous, not varying with $E(y)$. Now, the model of Eq. (7.20) is of the form of that in (7.17). As a result, a transformation that produces (7.17) truly began with the multiplicative error assumption of the model in (7.18).

Suppose the model of (7.18) is unrealistic, and instead, the error structure is *additive*. That is,

$$y = \alpha x^\beta + \varepsilon^* \tag{7.21}$$

where ε^*, as before, has the standard assumptions. Taking natural logs here *does not* produce Eq. (7.17) with standard error assumptions. Suppose we write (7.21) as

$$y = \alpha x^\beta \left\{1 + \frac{\varepsilon^*}{E(y)}\right\}$$

where, of course, $E(y) = \alpha x^\beta$. We can now write the model in order to accommodate the log transformation, namely

$$y = \alpha x^\beta (1 + \theta)$$

Now, $\theta = \varepsilon^*/E(y)$ and the log transformation results in

$$\ln y = \ln \alpha + \beta \ln x + \ln(1 + \theta) \tag{7.22}$$

In this case the term $\ln(1 + \theta)$ plays the role of the model error of Eq. (7.17). However, the variance of $1 + \theta$ varies with $E(y)$, and thus the standard least squares assumptions *do not hold.* The upshot of this development is that a transformation on the data implicitly *alters the error structure.* While the transformation may be structurally advantageous, it is quite possible that it may produce a violation of either the homogeneous variance assumption or the normality assumption. Thus we may be faced with the tradeoff of eliminating an assumption violation but producing a second violation. In the case of normality, for example, if ε^* in Eq. (7.18) is normally distributed, then ε of Eq. (7.20) cannot be normal. In addition, if ε^* of Eq. (7.21) is normal, then $\ln(1 + \theta)$ of Eq. (7.22) cannot be normal.

The disadvantages of transformations outlined here should in no way discourage their use. However, it is sound precautionary advice to make routine the careful inspection of studentized residuals when a transformation has been used. The reader should also be aware that model structures such as that given in (7.21), i.e., models nonlinear in the coefficients, where model errors are additive, fall into the category of *nonlinear regression* models. The estimation of the parameters in nonlinear models by least squares is more involved and is covered in Chapter 9. One who deals regularly in data containing curvature should learn about nonlinear regression. In certain instances, which occur on a regular basis, it may be preferable not to transform the data but rather to assume an additive model and do nonlinear regression.

Example 7.4 **Surgical Services Data**

In an effort to develop a preliminary manpower equation for estimation of man-hours per month expended in surgical services at Naval hospitals, the U.S. Navy collected data on y (man-hours per month) and x (surgical cases) from 15 hospitals. The data are as follows:

y (Man-hours per Month)	x (Surgical Cases)
1,275	230
1,350	235
1,650	250
2,000	277
3,750	522
4,222	545
5,018	625
6,125	713
6,200	735
8,150	820
9,975	992
12,200	1,322
12,750	1,900
13,014	2,022
13,275	2,155

Procedures and Analyses for Staffing Standards Development: Data/Regression Analysis Handbook (San Diego, California: Navy Manpower and Material Analysis Center, 1979).

A plot of the data is shown in Figure 7.8. From the appearance of the plot, it would seem that a transformation of the data might be in order. Several models were attempted. The results are discussed on page 300.

FIGURE 7.8 Plot of surgical case data

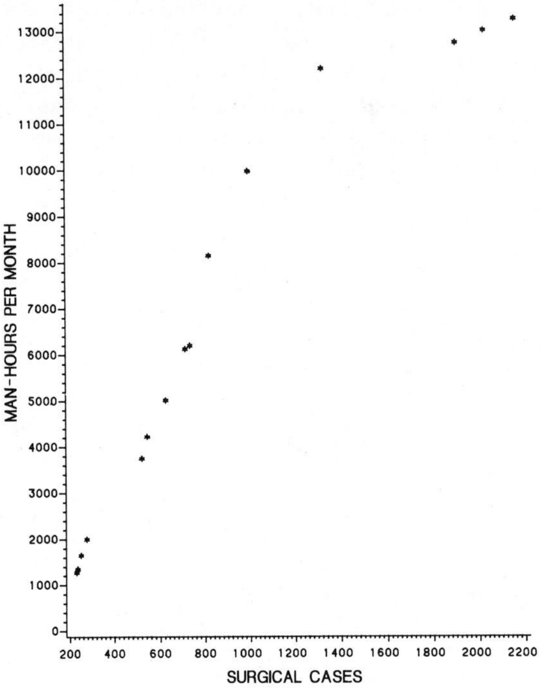

Simple Linear Model

As one might expect the simple linear regression model produces results that cannot be considered satisfactory. However, it serves as a basis or starting point for comparing with transformed models. The least squares simple regression model is given by

$$\hat{y} = 936.926 + 6.51279x$$

with the following performance criteria:

$$R^2 = 0.9076$$
$$s = 1427.565$$
$$\text{PRESS} = 35{,}927{,}142.72$$
$$\sum_{i=1}^{n} |y_i - \hat{y}_{i,-i}| = 19{,}064.7$$

The residuals and PRESS residuals are shown below:

y (Man-hours/Mo)	$\hat{y}$ (Man-hours/Mo)	$e_i = y_i - \hat{y}_i$	$y_i - \hat{y}_{i,-i} = e_{i,-i}$
1275	2434.9	−1159.9	−1344.9
1350	2467.4	−1117.4	−1294.1
1650	2565.1	−915.1	−1055.9
2000	2741.0	−741.0	−849.6
3750	4336.6	−586.6	−643.7
4222	4486.4	−264.4	−289.3
5018	5007.4	10.6	11.5
6125	5580.5	544.5	586.5
6200	5723.8	476.2	512.3
8150	6277.4	1872.6	2008.0
9975	7397.6	2577.4	2766.6
12200	9546.8	2653.2	2938.7
12750	13311.2	−561.2	−731.8
13014	14105.8	−1091.8	−1507.5
13275	14972.0	−1697.0	−2524.3

The inappropriateness of the simple linear model in the presence of the curvature is reflected in the pattern of the residuals. The regression overestimates the response for low values of x; it underestimates in the middle portion of the range and overestimates at the upper portion. In the next section, transformations to account for curvature in the data are illustrated and compared.

Inverse Exponential Function

The inverse exponential function was used to fit the surgical case data with the resulting fitted function given by

$$\widehat{\ln(y)} = 9.6652 - 590.912\left(\frac{1}{x}\right)$$

The value of the coefficient of determination, i.e., the proportion of variation in the $\ln y_i$ explained by the above model is 0.9675. However, any comparison in performance with the linear regression model should also be made with criteria that involve *original* units, i.e., man-hours, not log man-hours. In fact, the PRESS statistic and an analog to s^2 can be computed for comparative purposes from the residuals and PRESS residuals transformed back to natural y units. Thus fitted values are obtained from taking the antilog of the $\widehat{\ln y}$ values from the fitted transformed model. Results are given in the table shown at the top of page 302.

y (Observed Man-hours)	$\widehat{\ln y}$	$\hat{y} = \text{antilog} \widehat{\ln y}$	Residual $= y_i - \hat{y}_i$ (Natural Units)
1275	7.096	1207.2	67.8
1350	7.151	1275.0	75.0
1650	7.302	1482.6	167.4
2000	7.532	1866.7	133.3
3750	8.533	5080.6	−1330.6
4222	8.581	5329.2	−1107.2
5018	8.720	6122.6	−1104.6
6125	8.836	6880.4	−755.4
6200	8.861	7053.3	−853.3
8150	8.945	7666.2	483.8
9975	9.070	8686.5	1288.5
12200	9.218	10079.1	2120.9
12750	9.354	11547.2	1202.8
13014	9.373	11766.0	1248.0
13275	9.391	11980.1	1294.9

From the residuals,

$$s(\text{natural}) = \sqrt{\sum_{i=1}^{n} \frac{(y_i - \hat{y}_i)^2}{n-2}} = 1{,}131.53$$

where the residuals are in natural units. The appropriate PRESS residuals can be obtained from the PRESS residuals of the transformed model. The latter, of course, are obtained as usual from a standard regression output. One can compute the antilog of $\widehat{\ln y_{i,-i}}$, and thus the ith PRESS residual, as $y_i - \text{antilog}(\widehat{\ln y_{i,-i}})$. The sum of the squares of these PRESS residuals is then a PRESS statistic in natural units. The results are as follows:

y_i	$\widehat{\ln y_{i,-i}}$	$\text{antilog}(\widehat{\ln y_{i,-i}})$	$y_i - \text{antilog}(\widehat{\ln y_{i,-i}})$
1275	7.0750	1182.1	92.9
1350	7.1304	1249.4	100.6
1650	7.2712	1438.2	211.8
2000	7.5179	1840.8	159.2
3750	8.5549	5192.0	−1442.0
4222	8.5977	5419.0	−1197.0
5018	8.7348	6215.5	−1197.5
6125	8.8460	6946.8	−821.8
6200	8.8721	7130.4	−930.4
8150	8.9390	7623.2	526.8
9975	9.0548	8559.4	1415.6
12200	9.1934	9832.1	2367.9
12750	9.3386	11368.5	1381.5
13014	9.3567	11575.6	1438.4
13275	9.3740	11777.7	1497.3

The result is a PRESS of 20,684,140 and the sum of absolute PRESS residuals,

$$\sum_{i=1}^{n} |y_i - \text{antilog}(\widehat{\ln y_{i,-i}})| = 14{,}780.7$$

On the basis of s and the PRESS information, it should be apparent that the inverse exponential model is preferable to the simple linear model. The values for the performance criteria in the natural units will serve as bases for comparison with other transformed models.

Hyperbola

The hyperbolic model was fit to the surgical case data with the resulting least squares equation given by

$$(\widehat{1/y_i}) = -0.0000550675 + 0.174331(1/x_i)$$

The coefficient of determination (variation explained in $1/y$) is given by 0.9693. Again, an analog to s^2 can be computed from residuals in the natural response variable. The results are as follows:

y_i	$(\widehat{1/y_i})^{-1}$	$y_i - (\widehat{1/y_i})^{-1}$
1275	1422.7	−147.7
1350	1456.1	−106.1
1650	1557.0	93.0
2000	1741.3	258.7
3750	3585.5	164.5
4222	3776.3	445.7
5018	4467.0	551.0
6125	5278.8	846.2
6200	5491.0	709.0
8150	6347.9	1802.1
9975	8287.1	1687.9
12200	13020.5	−820.5
12750	27258.6	−14508.6
13014	32103.1	−19089.1
13275	38716.8	−25441.8

The resulting root residual mean square in natural units is

$$s(\text{natural}) = \sqrt{\sum_{i=1}^{n} (y_i - \hat{y}_i)^2/(n-2)} = 9{,}730.21$$

The PRESS residuals in the natural units are computed for the hyperbola in a manner similar to the approach used for the inverse exponential model. The PRESS residuals are in the final column listed as follows:

y_i	$\widehat{(1/y_i)}_{-i}$	$[\widehat{(1/y_i)}_{-i}]^{-1}$	$y_i - [\widehat{(1/y_i)}_{-i}]^{-1} = e_{i,-i}$
1275	.0006716	1488.9	−213.9
1350	.0006676	1497.8	−147.8
1650	.0006525	1532.5	117.5
2000	.0005894	1696.7	303.3
3750	.0002798	3574.3	175.7
4222	.0002668	3748.0	474.0
5018	.0002257	4430.2	587.8
6125	.0001916	5219.3	905.7
6200	.0001839	5438.5	761.5
8150	.0001607	6221.5	1928.5
9975	.0001228	8140.2	1834.8
12200	.0000761	13135.4	−935.4
12750	.0000301	33208.3	−20458.3
13014	.0000238	42088.7	−29074.7
13275	.0000176	56781.5	−43506.5

with PRESS = 3,166,829,641 and $\sum_{i=1}^{n} |y_i - [\widehat{(1/y_i)}_{-i}]^{-1}| = 101{,}425$.

Parabola

A parabola was fit to the surgical case data with the least squares regression fitted model given by

$$\hat{y} = -2{,}292.774 + 15.1086x - 0.003680x^2$$

with $R^2 = 0.9830$ and $s = 637.172$. The residuals and PRESS residuals are as follows:

y_i	$\hat{y}_i$	$y_i - \hat{y}_i$	$y_i - \hat{y}_{i,-i}$
1275	987.5	287.5	375.5
1350	1054.5	295.5	383.2
1650	1254.4	395.6	502.6
2000	1610.0	390.0	479.8
3750	4591.2	−841.2	−926.1
4222	4848.4	−626.4	−690.0
5018	5712.6	−694.6	−772.7
6125	6608.9	−483.9	−550.3
6200	6824.1	−624.1	−714.5
8150	7621.9	528.1	622.1
9975	9073.7	901.3	1124.3
12200	11249.5	950.5	1236.4
12750	13129.1	−379.1	−495.3
13014	13211.6	−197.6	−287.5
13275	13176.7	98.3	187.9

with PRESS = 7,057,691.54 and $\sum_{i=1}^{n} |y_i - \hat{y}_{i,-i}| = 9{,}348.06$.

Based on the values of PRESS, sum of the absolute PRESS residuals, and root residual mean square, the transformed models, apart from the hyperbola, are superior to the simple linear regression. The prediction criteria were based on the capability of the transformed model to predict the response in the original untransformed units. The parabolic model appears preferable to all alternative models. Several results are noteworthy. If one were to confine the computations to regression results in the transformed units, the hyperbola appears promising with a coefficient of determination of 0.9693, seemingly a substantial increase over that of the simple linear regression. However, when one considers the performance of the model back in the untransformed man-hours per month, the hyperbola is quite disappointing. A serious problem occurs in the residuals for the last three observations, i.e., the three largest hospitals. When fitting errors are transformed back into deviations in the natural y units, unusually large errors occur. These residuals appear in the column $y_i - (\widehat{1/y_i})^{-1}$. This result stems from the fact that, for large y, a modest error in $1/y$ can lead to huge errors in y. A similar problem occurs in the PRESS residuals and essentially renders the hyperbola inferior, even in comparison to the simple linear regression. This illustrates the hidden difficulties with transformations if one does not investigate model performance in the untransformed units. If prediction is important, usually in application the analyst is obligated to use the model in the untransformed units.

In this example, better results might be expected in the case of the hyperbola if a nonlinear model is fit of the type

$$y = \frac{x}{\beta_1 + \beta_0 x} + \varepsilon$$

One may bypass the linearization that led to the hyperbola and fit the above model using nonlinear procedures discussed in Chapter 9. Here we minimize the residual sum of squares on y as opposed to $1/y$, and hence the huge residuals when y is large may be avoided. (See Exercise 9.4.)

The parabola appeared to produce the most attractive model from among the candidates. Troublesome patterns in the residuals, however, remain for all models. Even in the case of the parabola, underestimation is followed by consistent overestimation, and then by underestimation once again. This data set will be the subject of further analysis in this section and in Chapter 9.

TRANSFORMATIONS ON MULTIPLE REGRESSORS

The need to transform the data is often quite apparent in the single regressor case since two-dimensional plots often depict the curvature. In the case of multiple regressors, curvature may be more difficult to detect since collinearity often renders two-dimensional plots misleading. Of course, we learned in Chapter 5 that partial leverage plots can often offer evidence that transformations on the regressor variables are needed. In what follows, suggestions are given regarding model alterations in the multiple regressor case. These alterations may help produce an improved fit or prediction.

Interaction in the Model

A very simple model alteration to account for interaction among the regressor variables can often result in an improved model. The reader should recall the implication of interaction in the discussion of indicator variables in Chapter 3. The reader should be cautioned against confusing interaction among variables with *multicollinearity*, which is discussed in Chapters 3 and 8. Interaction between x_1 and x_2 represents a situation in which the *rate of change of response with respect to a variable, say x_1, depends on the level of x_2*. In other words, it is a deviation from the additivity reflected in the usual multiple linear regression model. For example, in the case of three regressor variables, when all three interact, the model form may be taken to be as follows:

$$y = \beta_0 + \beta_1 x_1 + \beta_2 x_2 + \beta_3 x_3 + \beta_{12} x_1 x_2 + \beta_{13} x_1 x_3 + \beta_{23} x_2 x_3 + \varepsilon \quad \textbf{(7.23)}$$

Equation (7.23) contains first-order regression terms and three *linear by linear interaction terms*. To verify that the model does accommodate what we have described as interaction, we should consider the rate of change

$$\frac{\partial y}{\partial x_1} = \beta_1 + \beta_{12} x_2 + \beta_{13} x_3$$

Note from the above that the slope of the regression in the x_1 direction depends on x_2 and x_3. Figure 7.9 represents interaction pictorially for the case of two regressors.

FIGURE 7.9

(a) **No interaction x_1 and x_2** ***(b)*** **Interaction between x_1 and x_2**

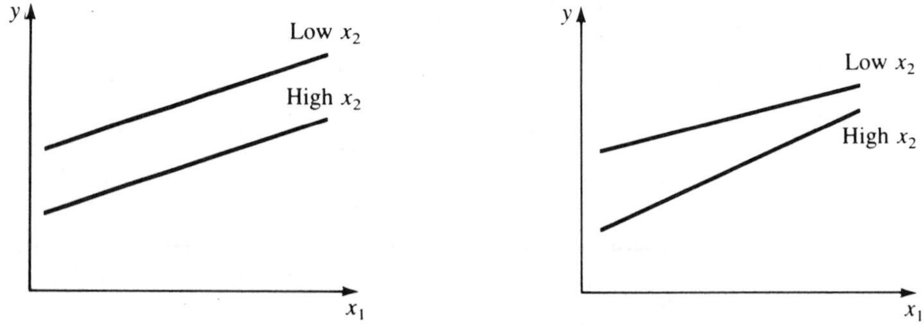

Power Transformations on the Regressor Variables (Box–Tidwell Procedure)

Earlier in this section, we discussed transformations in the case of a single regressor. The general appearance of a plot of the data suggests the type of transformation. Unfortunately this does not always provide a simple answer. The "textbook" appearances of the plots often are clouded by noise in the data, and of course, several of the various nonlinear plots have a similar appearance. In the case of multiple regressors, multicollinearity prohibits these two-dimensional plots (y versus x_1, y versus x_2, etc.) from revealing the true role of the individual regressors. While partial plots might uncover the need for a transformation, they will not always indicate what the transformation (or transformations) should be. As a result, a data analyst should have additional techniques available.

Box and Tidwell (1962) proposed a procedure for estimating exponents $\alpha_1, \alpha_2, \alpha_3, \ldots, \alpha_k$ in a model of the type

$$y = \beta_0 + \beta_1 w_1 + \cdots + \beta_k w_k + \varepsilon \tag{7.24}$$

where

$$w_j = \begin{cases} x_j^{\alpha_j} & \text{if } \alpha_j \neq 0 \\ \ln(x_j) & \text{if } \alpha_j = 0 \end{cases}$$

The method accommodates exponents on one or more of the regressor variables. The methodology of Box and Tidwell is quite easy to use, given that the analyst has access to a multiple linear regression computer software package.

The technique is based on a Taylor Series expansion (see Appendix B.9) of the model function in Eq. (7.24). Suppose one begins with starting values, i.e., an initial guess of $\boldsymbol{\alpha} = (\alpha_1, \alpha_2, \ldots, \alpha_k)$. These beginning values are usually taken to be 1.0. The Taylor Series expansion around $\boldsymbol{\alpha}_0 = (\alpha_{1,0}, \alpha_{2,0}, \ldots, \alpha_{k,0})$ is given by

$$\begin{aligned} E(y) \cong [f(\alpha_1, \alpha_2, \ldots, \alpha_k)]_{\boldsymbol{\alpha}=\boldsymbol{\alpha}_0} &+ (\alpha_1 - \alpha_{1,0})\left[\frac{\partial f}{\partial \alpha_1}\right]_{\boldsymbol{\alpha}=\boldsymbol{\alpha}_0} \\ &+ (\alpha_2 - \alpha_{2,0})\left[\frac{\partial f}{\partial \alpha_2}\right]_{\boldsymbol{\alpha}=\boldsymbol{\alpha}_0} + \cdots + (\alpha_k - \alpha_{k,0})\left[\frac{\partial f}{\partial \alpha_k}\right]_{\boldsymbol{\alpha}=\boldsymbol{\alpha}_0} \end{aligned}$$

where $[f(\alpha_1, \alpha_2, \ldots, \alpha_k)]_{\boldsymbol{\alpha}=\boldsymbol{\alpha}_0}$ is merely

$$\beta_0 + \beta_1 x_1^{\alpha_{1,0}} + \beta_2 x_2^{\alpha_{2,0}} + \cdots + \beta_k x_k^{\alpha_{k,0}}$$

Using the value 1.0 for all starting values, we have

$$\begin{aligned} E(y) \cong \beta_0 + \beta_1 x_1 + \cdots + \beta_k x_k &+ (\alpha_1 - 1)\beta_1 x_1 \ln x_1 \\ &+ (\alpha_2 - 1)\beta_2 x_2 \ln x_2 + \cdots + (\alpha_k - 1)\beta_k x_k \ln x_k \end{aligned} \tag{7.25}$$

Thus the model of (7.25) reduces to

$$E(y) \cong \beta_0 + \beta_1 x_1 + \cdots + \beta_k x_k + \gamma_1 z_1 + \gamma_2 z_2 + \cdots + \gamma_k z_k \tag{7.26}$$

where

$$\left.\begin{aligned} \gamma_j &= (\alpha_j - 1)\beta_j \\ z_j &= x_j \ln x_j \end{aligned}\right\} \qquad j = 1, 2, \ldots, k$$

A step-by-step procedure for estimation of the α_j is as follows:

1. Do a multiple linear regression with the model

$$E(y_i) = \beta_0 + \beta_1 x_{1i} + \cdots + \beta_k x_{ki}$$

   Denote parameter estimates by $b_0, b_1, \ldots, b_k$.
2. Do a regression of y on $x_1, x_2, \ldots, x_k, z_1, z_2, \ldots, z_k$. Thus estimate $\gamma_1, \gamma_2, \ldots, \gamma_k$, the coefficients of the z's. Denote the estimates by $\hat{\gamma}_1, \hat{\gamma}_2, \ldots, \hat{\gamma}_k$.
3. Estimate $\alpha_1, \alpha_2, \ldots, \alpha_k$ by

$$\hat{\alpha}_j = \frac{\hat{\gamma}_j}{b_j} + 1 \qquad (j = 1, 2, \ldots, k) \tag{7.27}$$

The results given by Eq. (7.27) may be viewed as an updated estimate of α_j. Often a one-step computation is sufficient, though in practice the analyst should monitor the residual sum of squares. As a second phase, the following steps may be considered:

1. Use $w_1^* = x_1^{\hat{\alpha}_1}$, $w_2^* = x_2^{\hat{\alpha}_2}$, $\ldots$, $w_k^* = x_k^{\hat{\alpha}_k}$, and fit the model

$$E(y) = \beta_0 + \beta_1 w_1^* + \cdots + \beta_k w_k^*$$

   with estimates $\hat{\beta}_0, \hat{\beta}_1, \ldots, \hat{\beta}_k$.
2. Define $z_1^* = w_1^* \ln w_1^*$, $z_2^* = w_2^* \ln w_2^*$, $\ldots$, $z_k^* = w_k^* \ln w_k^*$.
3. Fit a regression of y on $w_1^*, w_2^*, \ldots, w_k^*, z_1^*, z_2^*, \ldots, z_k^*$ with new coefficients of $z_1^*, z_2^*, \ldots, z_k^*$ denoted by $\hat{\gamma}_1, \hat{\gamma}_2, \ldots, \hat{\gamma}_k$.
4. Compute the updated $\hat{\alpha}_j$ by

$$\text{New } \hat{\alpha}_j = \left(\frac{\hat{\gamma}_j}{\hat{\beta}_j} + 1\right)(\text{Current value of } \hat{\alpha}_j) \tag{7.28}$$

Essentially, the second iteration described above proceeds like the first, but the x_j is replaced by $x_j^{\hat{\alpha}_j}$, where $\hat{\alpha}_j$ is from the previous iteration and the expression $((\hat{\gamma}_j/\hat{\beta}_j) + 1)$ is an estimate of the exponent of $(x_j^{\hat{\alpha}_j})$. Thus the updated term in x_j is $(x_j^{\hat{\alpha}_j})^{((\hat{\gamma}_j/\hat{\beta}_j)+1)}$. This brings about the result of Eq. (7.28).

The user should gain some understanding of the role of the Taylor Series expansion here. The form of Eq. (7.26) involves the original model (linear in the x's) augmented with a portion in the z's. The augmented portion accounts for deviation from the original postulation of linearity in the x's. As a result, large γ estimates are evidence of the need of a transformation. Note that a γ_j value near zero implies that $(\alpha_j - 1) \cong 0$, and hence a transformation is not indicated. The procedure may detect the need for a quadratic term ($\alpha_j \cong 2$), a natural log transformation ($\alpha_j \cong 0$), a reciprocal transformation ($\alpha_j \cong -1$), etc.

Example 7.5 ***Surgical Services Data***

Consider the surgical case data of Example 7.4. The Box–Tidwell procedure was attempted on that data set and is demonstrated here. Recall that y is man-hours per month expended and the regressor variable is the number of surgical cases. In order to provide an appropriate transformation on x, we consider the model

$$y = \beta_0 + \beta_1 w + \varepsilon$$

where $w = x^\alpha$ and α is to be estimated. Following the procedure outlined earlier in this section, we first fit a regression

$$\hat{y} = b_0 + b_1 x$$

with $b_0 = 936.926$, $b_1 = 6.51278$, $R^2 = 0.9076$ and $s = 1427.565$. Next, a regression of y on x and $z = x \ln x$ was fit, with the result

$$\hat{y} = -4{,}247.759 + 57.1398x - 6.3689z$$

Then by (7.27), the first estimate of α is

$$\hat{\alpha} = \frac{-6.3689}{6.51278} + 1.0 = 0.022099$$

The second stage is accomplished by regressing y against $w^* = x^{0.022099}$. The result is given by

$$\hat{y} = -251{,}309 + 223{,}370w^*$$

A regression is then fit of y on w^* and $z^* = w^* \ln w^*$ with the result

$$\hat{y} = 3{,}001{,}426 - 2{,}998{,}989w^* + 2{,}815{,}748z^*$$

From the results, the next stage computation of $\hat{\alpha}$ is given by Eq. (7.28) and is

$$\hat{\alpha} = \left[\frac{2{,}815{,}748}{223{,}370} + 1\right][0.022099] = 0.300677$$

Third and fourth iterations were negotiated with the resulting estimates of α being 0.26465 and 0.26949 respectively. The latter value was adopted, and thus the fitted model is given by

$$y_i = \beta_0 + \beta_1 x^{0.26949} + \varepsilon_i$$

The estimated regression is given by

$$\hat{y} = -14{,}581.49 + 3{,}600.649x^{0.26949}$$

with $R^2 = 0.9654$ and $s = 873.967$. The PRESS statistic is given by 12,394,896, and the sum of the absolute PRESS residuals is 11,348.4. The model is clearly superior to the simple linear regression model but does appear to be inferior, in terms of performance criteria, to the parabola used earlier in this section.

BOX–COX TRANSFORMATION ON THE RESPONSE

In many cases much can be accomplished by performing a power transformation on the response rather than estimating multiple exponents as in the case of the Box–Tidwell procedures. In this way only one parameter needs to be estimated and, in essence, the analyst has access to an important family of transformations. The Box–Cox (1964) transformation makes use of the response y^λ, where λ is estimated from the data. Formally, we define the response to be w, where w is given by

$$w = \begin{cases} \dfrac{y^\lambda - 1}{\lambda} & \text{if } \lambda \neq 0 \\ \ln y & \text{if } \lambda = 0 \end{cases}$$

The purpose of the Box–Cox procedure is to simultaneously estimate λ and parameters $\beta_0, \beta_1, \ldots, \beta_k$ in the model

$$w_i = \beta_0 + \beta_1 x_{1i} + \beta_2 x_{2i} + \cdots + \beta_k x_{ki} + \varepsilon_i \qquad (i = 1, 2, \ldots, n) \qquad \textbf{(7.29)}$$

This power transformation on the response has many nice features. It covers a wide variety of relationships that can exist between the response and the regressor variables. It does not rule out the adoption of transformations on the raw regressors. For example, squares and cross products in the regressor variables could be included in the model of Eq. (7.29). Notice that if we simply consider the range $[-2, 2]$ in λ, the Box–Cox procedure allows transformations on the response that accommodate negative or positive square root ($\lambda = 1/2$, $\lambda = -1/2$), square ($\lambda = 2$), log($\lambda = 0$), inverse square ($\lambda = -2$), and, of course, other fractional powers between -2 and 2.

The fitting of the model of Eq. (7.29) involves, of course, simultaneous estimation of the β's and the important parameter λ. The use of $\ln y$ as the proper transformation for $\lambda = 0$ is the result of the fact that the limit of $(y^\lambda - 1)/\lambda$ approaches $\ln y$ as λ approaches zero. This assignment of $\ln y$ for $\lambda = 0$ allows for continuity in λ. Many of the procedures for estimation center around plots that are designed to provide this joint estimation of λ and the regression coefficients. In addition, these plots can provide useful diagnostic information for the analyst as he or she attempts to determine the *proper metric* for the response. An important method to consider for the estimation process is the method of maximum likelihood.

Maximum Likelihood Method for Estimating β's and λ

The reader should recall that the maximum likelihood estimation procedure was introduced in Chapter 2. We first begin with the likelihood function for the model of Eq. (7.29) where the errors are normal, independent with common variance, σ^2.

The appropriate likelihood function is given by

$$L(\boldsymbol{\beta}, \lambda, \sigma^2) = \frac{1}{(2\pi)^{n/2}(\sigma^2)^{n/2}} e^{(-1/2\sigma^2)(\mathbf{w}-\mathbf{X}\boldsymbol{\beta})'(\mathbf{w}-\mathbf{X}\boldsymbol{\beta})} \cdot J(\lambda, \mathbf{y}) \tag{7.30}$$

This is very much like the likelihood function described in Chapter 2 and in Appendix B. However, the likelihood involves an extra parameter, λ, and the **Jacobian**

$$J(\lambda, \mathbf{y}) = \prod_{i=1}^{n} \frac{\partial w_i}{\partial y_i}$$

The likelihood is based on the linear model in the form

$$\mathbf{w} = \mathbf{X}\boldsymbol{\beta} + \boldsymbol{\varepsilon}$$

where $\mathbf{w}$ is the $n \times 1$ vector of responses in Eq. (7.29).

The analyst should view this power transformation in the context of that described in Eq. (7.29), with the value of λ being dictated by the data. We need to account in the likelihood for the change in scale created by changes in λ. For example when $\lambda \cong 0$ (response is $\ln y$), the scale on the transformed response is quite different than, say, when $\lambda = 2$ (response is y^2). The Jacobian $J(\lambda, \mathbf{y})$ accounts for the change in scale. This adjustment is necessary since the criterion for choice of λ depends on the likelihood. The Jacobian of the transformation is given by

$$J(\lambda, \mathbf{y}) = \prod_{i=1}^{n} \frac{\partial w_i}{\partial y_i} = \prod_{i=1}^{n} y_i^{\lambda - 1}$$

The use of the scaled likelihood in evaluating various alternatives in the family of transformations is easily accomplished by considering a slightly different form of the model. Now suppose that we rewrite the transformation in standardized form at the outset. Let us redefine the transformation as

$$\boxed{\begin{aligned} z_i &= \frac{w_i}{\{J(\lambda, \mathbf{y})\}^{1/n}} = \frac{w_i}{\left\{\left(\prod_{i=1}^{n} y_i\right)^{1/n}\right\}^{\lambda-1}} \\ &= \frac{y_i^{\lambda} - 1}{\lambda(\tilde{y})^{\lambda-1}} \qquad (\lambda \neq 0) \\ &= \frac{\ln y_i}{(\tilde{y})^{\lambda-1}} \qquad (\lambda = 0) \end{aligned}} \tag{7.31}$$

The quantity $\tilde{y} = \left(\prod_{i=1}^{n} y_i\right)^{1/n}$, the geometric mean of the y's, provides a standardization of the transformed response, i.e., the w's. Working with the z's as responses, the maximum of the log likelihood becomes

$$\max[\ln L(\boldsymbol{\beta}, \lambda)] = -\frac{n}{2}\ln(2\pi) - \frac{n}{2}\ln(\hat{\sigma}^2) - \frac{1}{2\hat{\sigma}^2}(\mathbf{z} - \mathbf{X}\hat{\boldsymbol{\beta}})'(\mathbf{z} - \mathbf{X}\hat{\boldsymbol{\beta}})$$

where $\hat{\sigma}^2$ is the maximum likelihood estimator of σ^2. Now, from Chapter 3 we know that for any fixed λ, $\hat{\sigma}^2$ is given by $\hat{\sigma}^2 = SS_{\text{Res}}/n$. Thus, the maximum of the log likelihood reduces to (apart from constants)

$$\boxed{\max[\ln L(\beta, \lambda)] = -\frac{n}{2}\ln \hat{\sigma}^2} \tag{7.32}$$

Maximization with respect to λ is tantamount to minimization of $\hat{\sigma}^2$ and thus *minimization of the residual sum of squares.* This leads to the following step-by-step procedure for determining the λ value for the transformation.

(i) Select a grid of λ values between -2 and 2.

(ii) For each λ, fit the model

$$\mathbf{z} = \mathbf{X}\boldsymbol{\beta} + \boldsymbol{\varepsilon} \tag{7.33}$$

where $\mathbf{z}$ is defined in (7.31).

(iii) Plot $\max[\ln L(\boldsymbol{\beta}, \lambda)]$, in (7.32), against λ.

(iv) Choose as the estimator for $\hat{\lambda}$ that which produces the largest value of $\max[\ln L(\boldsymbol{\beta}, \lambda)]$.

The $\lambda = \hat{\lambda}$ chosen through this procedure describes the transformation that one should make on the response y. If the optimum λ is significantly different from 1.0, the implication is that a transformation of the response will result in a better fit. In the interval $[-2.0, 2.0]$, usually 10–20 values are sufficient for plotting purposes. Obviously there are key values in this interval which result in models that often describe scientific phenomena. If λ is close to 2, squaring the response would seem to be a reasonable approach; if $\lambda \cong -1$, a reciprocal transformation is indicated. If one experiences an optimum λ of, say, 0.15, a log transformation may be used rather than using $y^{0.15}$ as the response. One should keep in mind that we can actually plot the residual sum of squares rather than the expression in Eq. (7.32). The choice of λ would be that which minimizes the residual sum of squares. In order to determine if the estimated λ (keep in mind that this is a random variable) differs significantly from some particular value, we need a mechanism for hypothesis testing or confidence interval estimation.

Confidence Interval on λ

Important evidence of the need for a transformation and the appropriate λ is derived from a confidence interval on λ. An approximate $100(1-\alpha)\%$ confidence interval (see Atkinson (1982), Draper and Smith (1981)) is based on the theory dealing with the *likelihood ratio.* Consider an arbitrary value $\lambda = \lambda_0$. Whether or not the value λ_0 falls inside a $100(1-\alpha)\%$ confidence interval depends on the difference

$$\max[\ln L(\boldsymbol{\beta}, \lambda)] - \max[\ln L(\boldsymbol{\beta}, \lambda_0)] = -\frac{n}{2}\ln \hat{\sigma}^2(\hat{\lambda}) - \left[-\frac{n}{2}\ln \hat{\sigma}^2(\lambda_0)\right] \tag{7.34}$$

where $\hat{\sigma}^2(\hat{\lambda})$, as before, is the maximum likelihood estimate of error variance for the model of Eq. (7.33) in which λ has been set to $\hat{\lambda}$, the value which minimizes SS_{Res} in the Box-Cox procedure. As we indicated earlier, this provides the maximum likelihood estimator of λ and thus results in the largest value for the likelihood over all values of λ. Now, $\hat{\sigma}^2(\lambda_0)$ is computed as

$$\hat{\sigma}^2(\lambda_0) = \frac{SS_{\text{Res}}(\lambda_0)}{n}$$

from the same model, i.e., Eq. (7.33), but the transformation to z for the model is constrained by $\lambda = \lambda_0$. The quantity on the left-hand side of Eq. (7.34) is positive, and thus

$$-\frac{n}{2}\ln \hat{\sigma}^2(\hat{\lambda}) - \left[-\frac{n}{2}\ln \hat{\sigma}^2(\lambda_0)\right] \geqslant 0$$

In the constrained case, one fewer parameter is being estimated. Evidence indicating the reasonableness of λ_0 as a parameter of the transformation lies in this difference. This approach is called the *likelihood ratio approach* and the asymptotic distribution involved is the chi-squared distribution. Values of λ_0 that fall inside a $100(1-\alpha)\%$ confidence interval are given by those satisfying

$$-\frac{n}{2}\ln \hat{\sigma}^2(\hat{\lambda}) - \left[-\frac{n}{2}\ln \hat{\sigma}^2(\lambda_0)\right] \leqslant \frac{1}{2}\chi^2_{\alpha,1} \qquad \textbf{(7.35)}$$

where $\chi^2_{\alpha,1}$ is the upper αth percent point of the χ^2 distribution with 1 degree of freedom. Values of the percent points of the χ^2 distribution are given in Table C.6.

A helpful procedure for showing graphically the reasonable values for λ can be derived from the plot of $(-n/2)\ln \sigma^2(\lambda)$ against λ. One can then depict approximate confidence bounds according to the inequality

$$\boxed{-\frac{n}{2}\ln \hat{\sigma}^2(\lambda) \geqslant -\frac{n}{2}\ln \hat{\sigma}^2(\hat{\lambda}) - \frac{1}{2}\chi^2_{1,\alpha}}$$

An illustration is given by Figure 7.10. The vertical lines shown depict the approximate $100(1-\alpha)\%$ confidence interval and hence display values of the transformation parameter that are not unreasonable.

Example 7.6 ***Box–Cox Transformation for* BOQ *Data***

Consider the BOQ data of Example 5.2. The reader should recall that the data were used to illustrate the detection of outliers and high influence observations in Chapter 5. The Box–Cox procedure is used to determine if a transformation is needed. The model in Eq. (7.33) was used with the response **z** defined in Eq. (7.31). Values of λ from -2 to 2 were used, and the plot of $(-n/2)\ln \hat{\sigma}^2(\lambda)$ against λ was generated. The plot is shown in Figure 7.11. The reader should bear in mind that

FIGURE 7.10 **Typical plot of $-n/2 \ln \sigma^2(\lambda)$ against λ**

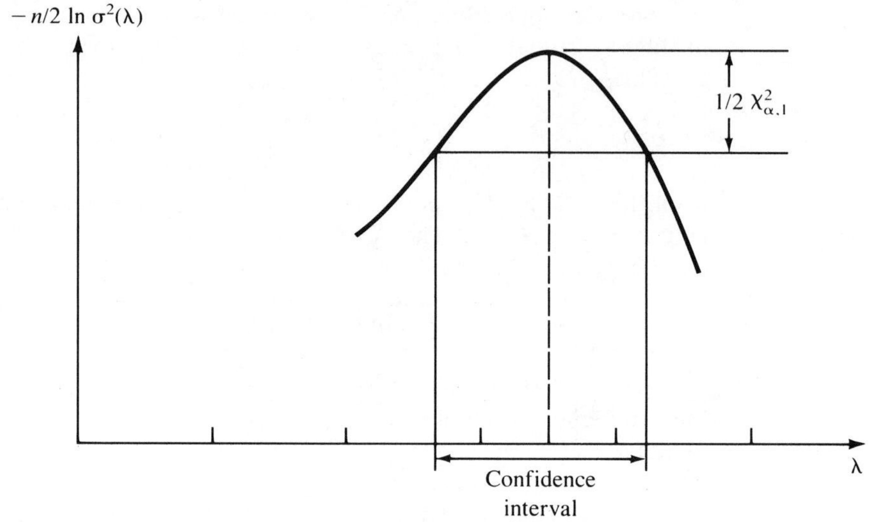

FIGURE 7.11 **Box–Cox plot for the BOQ data**

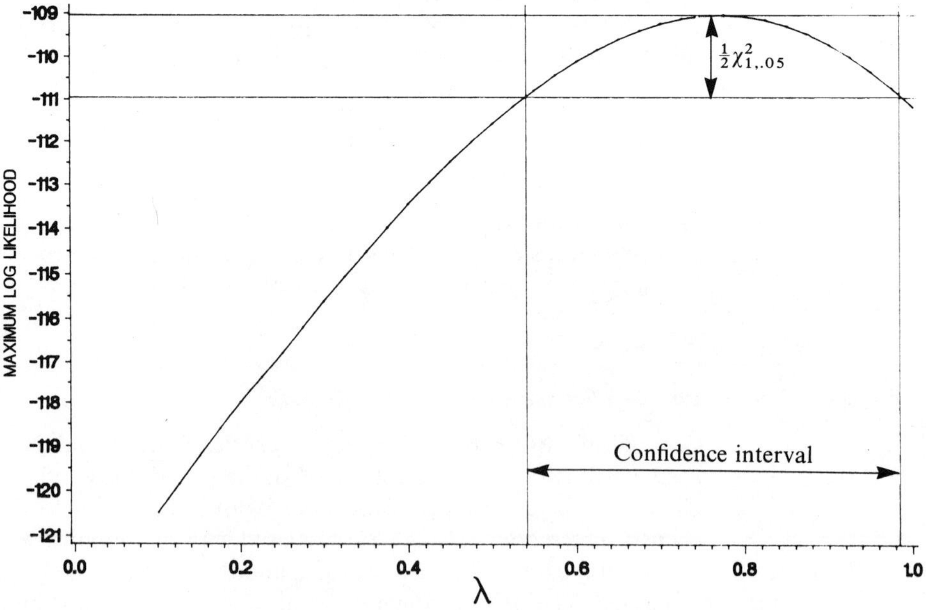

the $\hat{\lambda}$ value, which maximizes $(-n/2)\ \ln\ \sigma^2(\lambda)$, also minimizes SS_{Res}. A horizontal line that depicts a 95% confidence on λ is shown. The value $\hat{\lambda} = 0.775$ is that which maximizes the likelihood. Notice that the value $\lambda = 1.0$ is not inside the confidence interval though it is very close to the border. The fitted transformed model is given by

$$y^{.775} = 49.9517 - 0.1312x_1 + .2247x_2 + .3687x_3 - 1.0035x_4 + .0458x_5 - 1.8534x_6 + 3.6630x_7$$

7.4

REGRESSION WITH A BINARY RESPONSE

In many practical situations, the response is basically binary; i.e., in the outcome there are two possibilities, and these may be assigned the values 0 or 1. The motivation behind the model-building exercise is very much like that of the case of a continuous response. The analyst may need to determine the role of a set of regressor variables $x_1, x_2, \ldots, x_k$ on the binary response. In addition, there may be a need to predict or, rather, *estimate the probability* of one of the two specific response outcomes at a certain combination $x_1, x_2, \ldots, x_k$. Applications occur in many fields, including the engineering, biological, and health sciences. In recent years, weapons research has become an area where this type of modeling is used extensively. The material presented here may be applicable in any situation in which the response on the *experimental unit* is in either one of two categories and the category depends on the levels of the regressor variables. A numerical example will be presented subsequently.

It should come as no surprise that where the response variable is not continuous, special difficulties naturally occur, which should invite attention from the analyst. Before we cite these problems one by one, we will attempt to explain what this *discrete response* model really means. For n regression data points, consider the model

$$y_i = \beta_0 + \beta_1 x_{1i} + \beta_2 x_{2i} + \cdots + \beta_k x_{ki} + \varepsilon_i \quad \begin{cases} (i = 1, 2, \ldots, n) \\ y_i = \{0, 1\} \end{cases} \tag{7.36}$$

If we assume the usual $E(\varepsilon_i) = 0$, we have

$$E(y_i) = \beta_0 + \beta_1 x_{1i} + \beta_2 x_{2i} + \cdots + \beta_k x_{ki} \tag{7.37}$$

Now, the reader may view $E(y_i) = P_i$ as the *population proportion* of observations at $x_{1i}, x_{2i}, \ldots, x_{ki}$ for which $y = 1$. In other words,

$$\left.\begin{aligned} P_i &= \text{Prob}(y_i = 1) \\ Q_i = 1 - P_i &= \text{Prob}(y_i = 0) \end{aligned}\right\} (i = 1, 2, \ldots, n) \tag{7.38}$$

Thus for n distinct data points there are n probabilities $P_1, P_2, \ldots, P_n$, each of which is a parameter of a *Bernoulli distribution.*

If we consider the error term of the model in (7.36), it becomes clear that ε_i cannot be continuous since only two values are possible; namely

$$\varepsilon_i = y_i - [\beta_0 + \beta_1 x_{1i} + \beta_2 x_{2i} + \cdots + \beta_k x_{ki}]$$

which can either be $1 - P_i = Q_i$ or $0 - P_i = -P_i$. As a result, there can be no assumption of normality on the model errors. A second assumption, which is clearly violated, is the homogeneous variance assumption. Since $E(\varepsilon_i) = 0$,

$$\begin{aligned} \text{Var}(\varepsilon_i) &= E(\varepsilon_i)^2 \\ &= (Q_i)^2 \text{ Prob}[y_i = 1] + (-P_i)^2 \text{ Prob}[y_i = 0] \\ &= Q_i^2 P_i + P_i^2 Q_i \\ &= P_i Q_i (Q_i + P_i) \\ &= P_i Q_i \end{aligned} \tag{7.39}$$

Since P_i varies with the levels of regressor variables, the error variance is not homogeneous. In fact, from (7.39)

$$\begin{aligned} \text{Var}(\varepsilon_i) = &(\beta_0 + \beta_1 x_{1i} + \beta_2 x_{2i} + \cdots + \beta_k x_{ki}) \\ &\times (1 - \beta_0 - \beta_1 x_{1i} - \beta_2 x_{2i} - \cdots - \beta_k x_{ki}) \end{aligned}$$

As a result, the following are two difficulties with the use of ordinary least squares in the model of (7.36):

1. Distribution of the ε_i is discrete and thus not normal.
2. Error variance is not homogeneous.

In the case of both assumptions, a violation is a result of the nonideal condition naturally brought about by the binary response situation. One solution is to use weighted least squares, taking advantage of Eq. (7.39).

In Section 7.1 we discussed weighted least squares as a procedure for estimating parameters in the case of nonhomogeneous variance. Equation (7.39) gives the error variance at the ith data location. However, the P_i and hence $\text{Var}(\varepsilon_i)$ $(i = 1, 2, \ldots, n)$ are not known in practice and, at best, can be estimated from the data. One procedure is the following:

1. Estimate the regression function with ordinary least squares.
2. Use the fitted values $\hat{y}_1, \hat{y}_2, \ldots, \hat{y}_n$ to estimate weights. That is,

$$\hat{w}_i = \frac{1}{\hat{\sigma}_{\varepsilon_i}^2} = \frac{1}{\hat{y}_i(1 - \hat{y}_i)} \qquad (i = 1, 2, \ldots, n)$$

Here, $\hat{y}_i(1 - \hat{y}_i)$ represents an estimate of $P_i Q_i$, the error variance at the ith location.

3. Reestimate the parameters using Eq. (7.4); i.e.,

$$\boldsymbol{\beta}^* = (\mathbf{X}'\mathbf{V}^{-1}\mathbf{X})^{-1}\mathbf{X}'\mathbf{V}^{-1}\mathbf{y}$$

where

$$\mathbf{V} = \text{diag}[\hat{y}_1(1-\hat{y}_1), \hat{y}_2(1-\hat{y}_2), \ldots, \hat{y}_n(1-\hat{y}_n)]$$

In some cases, the user may wish to produce a second iteration, i.e., use the fitted values from the estimation in 3 to form new weights in 2.

While weighted least squares is not an unreasonable approach to the regression situation with a binary response, a more popular approach is to model the mean response against the regressors through a logistic function. This important concept is discussed in the following subsection.

LOGISTIC REGRESSION

The probability of a success is postulated to be a function of a set of regressor variables $x_1, x_2, \ldots, x_k$. Suppose there exists a set of data that takes the form

$$\left.\begin{matrix} n_1 & r_1 & x_{11} & x_{21} & \cdots & x_{k1} \\ n_2 & r_2 & x_{12} & x_{22} & \cdots & x_{k2} \\ \vdots & \vdots & \vdots & \vdots & & \vdots \\ n_s & r_s & x_{1,s} & x_{2,s} & \cdots & x_{k,s} \end{matrix}\right\} (s > k)$$

Here, $r_1, r_2, \ldots, r_s$ represents the number of successes in $n_1, n_2, \ldots, n_s$ trials respectively. For each of the s set of trials, there are changes in the regressor variables. The regressors may be dosages of a combination of drugs and a success represents a cure of a particular disease. In an engineering type application, a regressor may be a load type variable that is subject to failure. For the ith load n_i specimen are tested, and r_i represents the number that fail (or do not fail). Applications occur frequently in the social sciences where people or actions of people often fall into *one of two categories.* For example, we might model a marketing data set in which the response is whether or not an advertising coupon was redeemed by an individual and the regressor variable is the price discount. A sample of n_i ($i = 1, 2, \ldots, s$) was taken, and r_i is the number redeemed. The purpose is to determine the relationship between redemption proportion and discount.

A very popular and useful model to accommodate this binary response situation is the *logistic regression model* given by

$$\boxed{P(\mathbf{x}_i) = \frac{1}{1 + e^{-\mathbf{x}_i'\boldsymbol{\beta}}} \qquad (i = 1, 2, \ldots, s)} \tag{7.40}$$

This model relates the *probability of occurrence* $P(\mathbf{x}_i)$ against the regressor variables. Here, the quantity

$$\mathbf{x}_i'\boldsymbol{\beta} = \beta_0 + \beta_1 x_{1i} + \cdots + \beta_{ki} x_{ki}$$

which is the *multiple linear regression contribution.* At a given value of i, $n_i \geqslant 1$ observations occur, giving rise to r_i successes. The logistic function lies between zero and one and, of course, takes on an interesting S-shape as depicted in Figure 7.12.

FIGURE 7.12 **A plot depicting the logistic function**

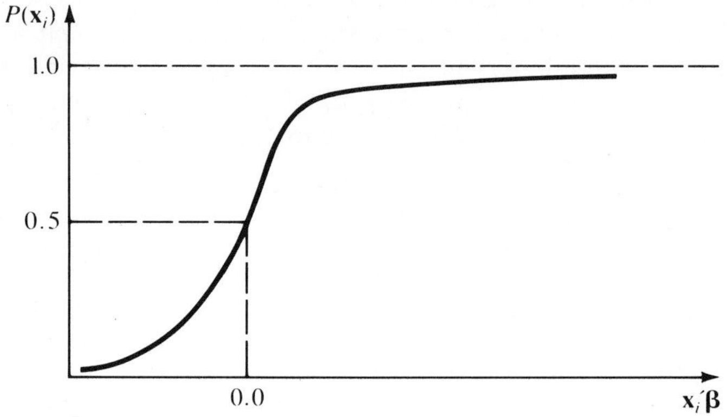

We will eventually consider maximum likelihood estimation of the parameters in the model of Eq. (7.40). However, it is instructive to consider the *logit transformation*, a linearization of the logistic model. Namely, from (7.40) we can write

$$\ln\left[\frac{P(\mathbf{x}_i)}{(1 - P(\mathbf{x}_i))}\right] = \beta_0 + \beta_1 x_{1i} + \cdots + \beta_k x_{ki} \tag{7.41}$$

Now, the information at each combination allows for estimation of the $P(\mathbf{x}_i)$ probability values. As one would expect, the estimates are given by the sample proportion of successes, namely

$$\hat{P}(\mathbf{x}_i) = \frac{r_i}{n_i} \qquad (i = 1, 2, \ldots, s)$$

So, we impose these estimates on the model of (7.41) and *regress*

$$\ln\left(\frac{\hat{P}(\mathbf{x}_i)}{1 - \hat{P}(\mathbf{x}_i)}\right) \quad \text{against} \quad x_1, x_2, \ldots, x_k$$

Thus the user may formally view the postulated model as follows:

$$\boxed{\ln\left(\frac{\hat{P}(\mathbf{x}_i)}{1 - \hat{P}(\mathbf{x}_i)}\right) = \beta_0 + \beta_1 x_{1i} + \beta_2 x_{2i} + \cdots + \beta_k x_{ki} + \varepsilon_i \quad (i = 1, 2, \ldots, s)} \tag{7.42}$$

The *transformed logistic regression model* of (7.42) can now be considered a least squares candidate model. However, as in the previous section, we must be attentive to the error variance. At a fixed combination $\mathbf{x}_i$,

$$\text{Var} \ln\left(\frac{\hat{P}(\mathbf{x}_i)}{1-\hat{P}(\mathbf{x}_i)}\right) \cong \frac{1}{n_i P(\mathbf{x}_i)(1-P(\mathbf{x}_i))}$$

As a result, weighted regression is an approach that one must consider. Since the weights are the reciprocals of variances, then an estimate of the weight applied at the ith data point is given by

$$w_i = n_i[(\hat{P}(\mathbf{x}_i))(1-\hat{P}(\mathbf{x}_i)] \qquad (i = 1, 2, \ldots, s) \tag{7.43}$$

Thus the estimator of the vector of regression coefficients is given by Eq. (7.4) with $\mathbf{V}$ being a diagonal matrix with ith diagonal element given by

$$\frac{1}{[n_i \hat{P}(\mathbf{x}_i)(1-\hat{P}(\mathbf{x}_i))]}$$

The preceding description of the use of weighted least squares to fit a logistic regression model is reasonable when the number of observations at the individual $\mathbf{x}_i$ is not small. There are two reasons for this restriction. First, as we indicated earlier in this chapter, weighted regression should be avoided when weights are estimated with a relatively small amount of information. Secondly, the variance of

$$\ln\left(\frac{\hat{P}(\mathbf{x}_i)}{1-\hat{P}(\mathbf{x}_i)}\right)$$

is only an approximation, and the result is most accurate when the n_i are large.

Once the estimation procedure has been accomplished by least squares, the use of the logistic regression for prediction or, rather, estimation of the probability of one of two outcomes, is very simple. Specifically, if one requires estimation of P_0 (probability of a success) at $x = x_0$, computation is given by

$$\hat{P}(\mathbf{x}_0) = \frac{1}{1+e^{-\mathbf{x}_0'\boldsymbol{\beta}^*}}$$

where $\boldsymbol{\beta}^*$ is the vector of weighted least squares estimates from the model of Eq. (7.42).

Example 7.7 ***Dose Response Data***

From a set of streptonigrin dose-response data, an experimenter desires to develop a relationship between the proportion of lymphoblasts sampled that contain aberrations and the dosage of streptonigrin. Five dosage levels were applied to the rabbits used for the experiment. The data are shown in Table 7.3.

A logistic regression model was postulated. With relatively large values in the n_i, the model form of Eq. (7.42) was considered. We thus have

$$\ln\left(\frac{P(x_i)}{1-P(x_i)}\right) = \beta_0 + \beta_1 x_i + \varepsilon_i \qquad (i = 1, 2, 3, 4, 5)$$

Weighted regression is appropriate to use here with weights given by Eq. (7.43). The transformed response, dosage values, and computed weights are as follows:

$\ln(\hat{P}(x_i)/(1-\hat{P}(x_i))$	Dosage (x_i)	Weight (w_i)
−3.6636	0	14.625
−1.4371	30	77.568
−0.7908	60	128.794
−0.6946	75	66.633
−0.0680	90	74.913

The weighted least squares logistic regression equation is given by

$$\ln\left(\frac{\hat{P}(x)}{1-\hat{P}(x)}\right) = -2.56488 + 0.02806x$$

As a result, the logistic regression function is given by

$$\hat{P}(x) = \frac{1}{1 + e^{-[-2.56488 + 0.025806x]}}$$

where $\hat{P}$ is interpreted as the probability of an aberration at a given dosage level, x.

TABLE 7.3 **Streptonigrin dose response data**

Dose (mg/kg)	Number of Lymphoblasts (n_i)	Number with Aberrations (r_i)	Proportion with Aberrations ($\bar{P}_i$)
0	600	15	.025
30	500	96	.192
60	600	187	.312
75	300	100	.333
90	300	145	.483

Data taken from E. L. Frome, "Regression Methods for Binomial and Poisson Distributed Data." Paper presented at the American Association of Physicists in Medicine First Midyear Topical Symposium, March, 1984, Mobile, Alabama.

Maximum Likelihood Estimation for Grouped Data

As we indicated earlier, logistic regression is used a great deal in many scientific areas. Part of this interest stems from the fact that many software packages contain options for doing logistic regression. For example, SAS (1983) offers PROC

LOGIST to which we will refer later in this section. A more general package, GLIM (see Baker and Nelder (1978)), can also be used for doing logistic regression. While the maximum likelihood estimation of coefficients in the model of (7.40) does not allow a "closed form" solution, it is relatively easy to use iterative procedures that produce the coefficients.

Consider the model of Eq. (7.40) with, once again, n_i subjects or observations associated with the ith group. The response is truly discrete (success or failure), and hence, *at the* ith *group*, the likelihood is the product of the probabilities, or the joint probability of occurrence of r_i successes in a sample of n_i. As a result, the likelihood at the ith group is given by

$$[P(\mathbf{x}_i)]^{r_i}[1 - P(\mathbf{x}_i)]^{n_i - r_i} = \left[\frac{1}{1 + e^{-\mathbf{x}_i'\boldsymbol{\beta}}}\right]^{r_i}\left[\frac{e^{-\mathbf{x}_i'\boldsymbol{\beta}}}{1 + e^{-\mathbf{x}_i'\boldsymbol{\beta}}}\right]^{n_i - r_i}$$

Denoting $e^{-\mathbf{x}_i'\boldsymbol{\beta}}$ as η_i, we have as the likelihood of the entire sample

$$L(\boldsymbol{\beta}, \mathbf{x}_i) = \prod_{i=1}^{s}\left[\frac{1}{1 + \eta_i}\right]^{r_i}\left[\frac{\eta_i}{1 + \eta_i}\right]^{n_i - r_i} \tag{7.44}$$

As usual, we work with the log of the likelihood in our attempt to find maximum likelihood estimates of the parameters in $\boldsymbol{\beta}$. We have

$$\ln L(\boldsymbol{\beta}, \mathbf{x}_i) = \sum_{i=1}^{s} \{r_i[-\ln(1 + \eta_i)] + [n_i - r_i][\ln \eta_i - \ln(1 + \eta_i)]\}$$

Simplifying, we have

$$\begin{aligned} \ln L(\boldsymbol{\beta}, \mathbf{x}_i) &= \sum_{i=1}^{s} \{n_i[\ln \eta_i - \ln(1 + \eta_i)] - r_i[\ln \eta_i]\} \\ &= \sum_{i=1}^{s} \{n_i[-\mathbf{x}_i'\boldsymbol{\beta} - \ln(1 + e^{-\mathbf{x}_i'\boldsymbol{\beta}})] - r_i[-\mathbf{x}_i'\boldsymbol{\beta}]\} \end{aligned}$$

For estimation of the parameters in $\boldsymbol{\beta}$, we require setting derivatives with respect to $\boldsymbol{\beta}$ equal to zero.

$$\frac{\partial \ln L(\boldsymbol{\beta}, \mathbf{x}_i)}{\partial \boldsymbol{\beta}} = -\sum_{i=1}^{s} n_i\left[1 - \frac{e^{-\mathbf{x}_i'\boldsymbol{\beta}}}{(1 + e^{-\mathbf{x}_i'\boldsymbol{\beta}})}\right]\mathbf{x}_i + \sum_{i=1}^{s} r_i\mathbf{x}_i = \mathbf{0}$$

As a result, the maximum likelihood estimates are given as solutions $\hat{\boldsymbol{\beta}}$ to the following set of equations

$$\boxed{\sum_{i=1}^{s} n_i\left[1 - \frac{e^{-\mathbf{x}_i'\hat{\boldsymbol{\beta}}}}{(1 + e^{-\mathbf{x}_i'\hat{\boldsymbol{\beta}}})}\right]\mathbf{x}_i = \sum_{i=1}^{s} r_i\mathbf{x}_i} \tag{7.45}$$

It is clear that the equations (p equations if $\mathbf{x}_i'\boldsymbol{\beta}$ contains p parameters) in (7.45) are not linear in $\boldsymbol{\beta}$ and cannot be solved directly. However, there are many iterative methods for obtaining the estimates and thus estimating probabilities as a function of the regressor variables. Equation (7.45) represents the equation that must be solved when the data are grouped; i.e., for each combination of the regressor variables, several subjects or experimental units are observed as in the case of the dose response data of Example (7.7). However, in observational studies in which experiments cannot be planned, the data are not grouped. While we will eventually illustrate with examples of both types of situations, it is important to display the likelihood structure for ungrouped data.

THE LIKELIHOOD FUNCTION FOR THE UNGROUPED CASE

For the case of ungrouped data, we assume that there are n observations, and n_1 of them are characterized as a success while $n - n_1$ are characterized as failures. However, there are not necessarily any groups of experimental units with equivalent values of the regressor variables. Denoting the first n_1 observations as those with "success," we have for the likelihood function,

$$\begin{aligned} L(\boldsymbol{\beta}, \mathbf{x}_i) &= \prod_{i=1}^{n_1} \left(\frac{1}{1+\eta_i}\right) \prod_{i=n_1+1}^{n} \left[1 - \left(\frac{1}{1+\eta_i}\right)\right] \\ &= \prod_{i=1}^{n_1} \left(\frac{1}{1+\eta_i}\right) \prod_{i=n_1+1}^{n} \left(\frac{\eta_i}{1+\eta_i}\right) \\ &= \prod_{i=1}^{n_1} \left(\frac{1}{1+e^{-\mathbf{x}_i'\beta}}\right) \prod_{i=n_1+1}^{n} \left(\frac{e^{-\mathbf{x}_i'\beta}}{1+e^{-\mathbf{x}_i'\beta}}\right) \end{aligned}$$

It can easily be verified through simple algebra that the likelihood can be written in the fairly simple form

$$L(\boldsymbol{\beta}, \mathbf{x}_i) = \frac{\prod_{i=1}^{n_1} (e^{\mathbf{x}_i'\boldsymbol{\beta}})}{\prod_{i=1}^{n} (1 + e^{\mathbf{x}_i'\boldsymbol{\beta}})} \tag{7.46}$$

Again, maximization of Eq. (7.46) with respect to $\boldsymbol{\beta}$ involves the maximization of a function nonlinear in $\boldsymbol{\beta}$ as in the grouped data case.

Likelihood Ratio as a Basis for Inference on Coefficients

Several commercial packages produce estimates of the β's for the logistic regression model as well as information that leads to the statistical inference regarding the roles of the variables in the logistic regression. In addition, estimates of the asymptotic variance-covariance matrix of the coefficients are often displayed. This allows us to make use of standard errors of coefficients. In this section we highlight such methods and follow with two examples, one containing grouped data and one containing ungrouped data.

Measures of lack of fit of the logistic model can be determined by making use of known features of maximum likelihood estimation. The reader should recall that in the previous section on the Box–Cox transformation procedure we made use of our knowledge of the maximum likelihood estimation of the transformation parameter λ to develop a confidence region on λ. This confidence interval was based on the *likelihood ratio criterion.* Similar concepts allow for measures of lack of fit and tests of hypotheses in the case of logistic regression when maximum likelihood is used. For purposes of this development, *we shall consider the ungrouped data case.* One measure of the quality of the logistic regression fit is to compare the *maximum of the likelihood* in Eq. (7.46) with the maximum that would be experienced if one were to produce a *perfect fit* of the success-failure data in the ungrouped case. As one would expect, the more complicated the model being fit, the larger will be the likelihood. This is analogous to the minimum SS_{Res} in the standard regression context discussed in Chapters 2, 3, and 4. The most complicated model, i.e., that containing the greatest number of parameters, and hence the largest likelihood is simply

$$y_i = P_i + \varepsilon_i \qquad (i = 1, 2, \ldots, n)$$

where y_i is the basic 0 or 1 response observed (failure or success) and P_i is the probability of a success. As a result, the regressor variables are being ignored and, as one would expect, we must estimate n parameters, $P_1, P_2, \ldots P_n$, and the likelihood is given by

$$L(\mathbf{P}) = \prod_{i=1}^{n} [P_i]^{y_i}[1 - P_i]^{1-y_i}$$

Here, by $\mathbf{P}$, we denote the vector of probabilities of success. This model essentially states that there is nothing that allows the regressors to link to estimation of the probabilities. Taking the log of the likelihood function, we have

$$\ln L(\mathbf{P}) = \sum_{i=1}^{n} [y_i \ln P_i + (1 - y_i) \ln(1 - P_i)] \tag{7.47}$$

Though Eq. (7.47) is a bit tricky, one can see by inspection that if we consider a specific probability parameter, say P_i, its contribution to the likelihood is given by

$$y_i \ln P_i + (1 - y_i) \ln(1 - P_i)$$

and if $y_i = 0$, the log likelihood is largest when $P_i = 0$; if $y_i = 1$ (a success), the log likelihood is maximized when $P_i = 1.0$. As a result, the maximum likelihood estimates are as expected, namely,

$$\begin{aligned} \hat{P}_i &= 1.0 \qquad \text{if } y_i = 1.0 \\ &= 0 \qquad\;\;\, \text{if } y_i = 0 \end{aligned}$$

One can view this model as being an overfitted model in which the residuals, $y_i - \hat{y}_i$, *are zero.* While this model is senseless if the regression variables are having any impact, it provides a "perfect fit" basis for determining if the regressors as a whole are effective. Thus the lack of fit test will provide a mechanism that determines if the logistic regression model is *significantly different from the perfect fit model.*

We now have two likelihoods, the largest that the likelihood can be, given by

$$L(\hat{\mathbf{P}}) = \prod_{i=1}^{n} (y_i)^{y_i}(1 - y_i)^{(1-y_i)} \tag{7.48}$$

and

$$L(\hat{\boldsymbol{\beta}}) = \frac{\prod_{i=1}^{n_1} (e^{\mathbf{x}_i'\hat{\boldsymbol{\beta}}})}{\prod_{i=1}^{n} (1 + e^{\mathbf{x}_i'\hat{\boldsymbol{\beta}}})} \tag{7.49}$$

where $\hat{\boldsymbol{\beta}}$ is the maximum likelihood estimator maximizing (7.49) through a computerized iterative routine. Now, the log of the ratio of these likelihoods is an important measure of the goodness of fit of the logistic model. The statistic

$$\boxed{\lambda(\boldsymbol{\beta}) = -2 \ln\left[\frac{L(\hat{\boldsymbol{\beta}})}{L(\hat{\mathbf{P}})}\right]} \tag{7.50}$$

is called a *likelihood ratio statistic* and is used to determine if the logistic model is a worthwhile description of the data as compared to the fit described by the more complete model $y_i = P(\mathbf{x}_i) + \varepsilon_i$. If the ratio $L(\hat{\boldsymbol{\beta}})/L(\hat{\mathbf{P}})$ is sufficiently close to 1.0, one can reason that since $L(\hat{\boldsymbol{\beta}})$ is not significantly less than $L(\hat{\mathbf{P}})$, there is *no significant lack of fit* with the logistic regression model. The test statistic $\lambda(\boldsymbol{\beta})$ in Eq. (7.50) is called the *deviance* associated with the fitted logistic regression. If we wish to formulate the hypothesis

$$H_0\colon P(x_i) = \frac{1}{1 + e^{-\mathbf{x}_i'\boldsymbol{\beta}}}$$

the appropriate test information lies in $\lambda(\boldsymbol{\beta})$. In fact, under H_0, $\lambda(\boldsymbol{\beta})$ is approximately distributed as a χ^2 random variable with $n - k - 1$ degrees of freedom; thus H_0 is not rejected; i.e., the lack of fit is not significant if $\lambda(\boldsymbol{\beta})$ is sufficiently close to zero. As a result we reject H_0 on values in the *upper tail* of the χ^2 distribution. That is,

$$\boxed{\text{Reject } H_0 \text{ at the level of significant } \alpha \text{ if } \lambda(\boldsymbol{\beta}) > \chi^2_{\alpha,n-k-1}}$$

The reader should study Eq. (7.50) and draw the connection between it and the material that produced the confidence interval on λ in the Box–Cox procedure. They both are of the general form

$$\ln L(\text{unrestricted}) - \ln L(\text{restricted})$$

where *unrestricted* means that we are substituting maximum likelihood estimators for all parameters (or for more parameters than in the restricted case). Notice that in the Box–Cox situation the factor of "2" used in Eq. (7.50) was incorporated into the χ^2 percentage point in Eq. (7.35) in the Box–Cox procedure.

In the case where the data are grouped, the concept is the same. However, the maximum likelihood estimate under the most complete model is $\hat{P}_i = r_i/n_i$ and hence (for s groups)

$$L(\hat{P}) = \prod_{i=1}^{s} \left(\frac{r_i}{n_i}\right)^{r_i} \left(\frac{n_i - r_i}{n_i}\right)^{(n_i - r_i)}$$

and

$$L(\hat{\boldsymbol{\beta}}) = \prod_{i=1}^{s} \left[\frac{1}{1 + e^{-\mathbf{x}_i'\hat{\boldsymbol{\beta}}}}\right]^{r_i} \left[\frac{e^{-\mathbf{x}_i'\hat{\boldsymbol{\beta}}}}{1 + e^{-\mathbf{x}_i'\hat{\boldsymbol{\beta}}}}\right]^{n_i - r_i}$$

where $\hat{\boldsymbol{\beta}}$ contains the maximum likelihood estimates for the grouped case. The statistic $\lambda(\boldsymbol{\beta})$ has an approximate χ^2 distribution *with $s - k - 1$ degrees of freedom.*

As we have seen, the logistic model is found to be inappropriate if the deviance is sufficiently large. In this sense, the deviance plays a role not unlike that of the residual sum of squares in the ordinary regression case. (See Exercise 7.9.) We can extend the use of deviance to allow tests of other hypotheses. In fact, contributions to the deviance can be produced that are analogous to contributions to the residual sum of squares (or regression sum of squares) in the ordinary regression case. Inference concerning any regressor or subset can be computed by determining how much the presence of each regressor contributes to the reduction in deviance. Consider, for example, the contribution of the jth regressor x_j in a setting where there are k regressors. Let us define

$$\boxed{\lambda(\beta_j \mid \beta_1, \beta_2, \ldots, \beta_{j-1}, \beta_{j+1}, \ldots, \beta_k) = \lambda(\beta_1, \beta_2, \ldots, \beta_{j-1}, \beta_{j+1}, \ldots, \beta_k) - \lambda(\boldsymbol{\beta})}$$

The notation in this equation is quite consistent with the $R(\cdot)$ notation in Chapter 3. The term $\lambda(\boldsymbol{\beta})$ is the deviance calculated with all parameters in the model while $\lambda(\beta_1, \beta_2, \ldots, \beta_{j-1}, \beta_{j+1}, \ldots, \beta_k)$ is the deviance computed with $\beta_j x_j$ not included in the model. The difference, the amount of reduction in deviance attributed to $\beta_j x_j$, *adjusted for the other regressors*, is computed as

$$\lambda(\beta_j \mid \beta_1, \beta_2, \ldots, \beta_{j-1}, \beta_{j+1}, \ldots, \beta_k) = -2 \ln\left[\frac{L(\hat{\beta}_1, \hat{\beta}_2, \ldots, \hat{\beta}_{j-1}, \hat{\beta}_{j+1}, \ldots, \hat{\beta}_k)}{L(\hat{\boldsymbol{\beta}})}\right] \tag{7.51}$$

The numerator likelihood in Eq. (7.51) is the maximum likelihood with $\beta_j x_j$ removed. To verify this important result we need only consider

$$\begin{aligned}
\lambda(\beta_j \mid \beta_1, \beta_2, \ldots, \beta_{j-1}, \beta_{j+1}, \ldots, \beta_k) &= -2 \ln\left[\frac{L(\hat{\beta}_1, \hat{\beta}_2, \ldots, \hat{\beta}_{j-1}, \hat{\beta}_{j+1}, \ldots, \hat{\beta}_k)}{L(\hat{\mathbf{P}})}\right] \\
&\quad + 2 \ln\left[\frac{L(\hat{\boldsymbol{\beta}})}{L(\hat{\mathbf{P}})}\right] \\
&= -2 \ln\left[\frac{L(\hat{\beta}_1, \hat{\beta}_2, \ldots, \hat{\beta}_{j-1}, \hat{\beta}_{j+1}, \ldots, \hat{\beta}_k)}{L(\hat{\boldsymbol{\beta}})}\right]
\end{aligned}$$

We can view the left-hand side of Eq. (7.51) as expressing the difference in deviance, which is an analog of the *difference in variation explained*. Variation explained is expressed as the difference in 2 ln L *between the full and reduced model*. Now, the log likelihood ratio statistic

$$\lambda(\beta_j \mid \beta_1, \beta_2, \ldots, \beta_{j-1}, \beta_{j+1}, \ldots, \beta_k)$$

has an approximate χ^2-distribution with 1 degree of freedom under the appropriate hypothesis on β_j. This is consistent in concept with $\lambda(\boldsymbol{\beta})$. Namely, the degrees of freedom is the difference in the number of parameters being estimated in the denominator and numerator of the log likelihood ratio.

The above formulation allows for *partial χ^2 tests* on individual coefficients. That is, to test

$$H_0: \beta_j = 0$$
$$H_1: \beta_j \neq 0$$

we reject H_0 if $\lambda(\beta_j \mid \beta_1, \beta_2, \ldots, \beta_{j-1}, \beta_{j+1}, \ldots, \beta_k)$ exceeds the appropriate upper tail χ_1^2 percent point

That is, reject H_0 if $\lambda(\beta_j \mid \beta_1, \beta_2, \ldots, \beta_{j-1}, \beta_{j+1}, \ldots, \beta_k) > \chi^2_{\alpha,1}$.

The same χ^2 procedure can be used for testing subsets of coefficients. Again, the importance of a subset $\boldsymbol{\beta}_1$, where $\boldsymbol{\beta}' = [\boldsymbol{\beta}_1', \boldsymbol{\beta}_2']$, depends on how much the deviance is reduced when $\boldsymbol{\beta}_1$ is introduced into the model in the presence of $\boldsymbol{\beta}_2$. Assume that $\boldsymbol{\beta}_1$ contains $r < k + 1$ parameters. We reject H_0 in the hypothesis

$$H_0: \boldsymbol{\beta}_1 = \mathbf{0}$$
$$H_1: \boldsymbol{\beta}_1 \neq \mathbf{0}$$

if $\lambda(\boldsymbol{\beta}_1 \mid \boldsymbol{\beta}_2) > \chi^2_{r,\alpha}$.

Here, of course,

$$\begin{aligned}\lambda(\boldsymbol{\beta}_1 \mid \boldsymbol{\beta}_2) &= \lambda(\boldsymbol{\beta}_2) - \lambda(\boldsymbol{\beta}) \\ &= -2 \ln\left[\frac{L(\hat{\boldsymbol{\beta}}_2)}{L(\hat{\boldsymbol{\beta}})}\right] \\ &= 2 \ln L(\hat{\boldsymbol{\beta}}) - 2 \ln L(\hat{\boldsymbol{\beta}}_2)\end{aligned}$$

The notion of the likelihood ratio procedure can be very helpful in determining the importance of model terms in logistic regression. In fact, stepwise type procedures similar to those discussed in Chapter 4 can be implemented for situations containing large numbers of regressors. (See SAS (1983).)

Standard Errors of Coefficients

The use of asymptotic theory allows us to determine approximate variances and covariances of regression coefficients. Estimates of variances and covariances can be obtained from the matrix of second partials of the log likelihood. This allows for the use of standard errors of coefficients for testing individual parameters in the $\boldsymbol{\beta}$

vector. Thus we can display an analog to the t-tests that are used in ordinary regression. Since we are not using standard linear regression, the $\mathbf{X}'\mathbf{X}$ matrix does not play the major role. Instead, we form the $\mathbf{C}$ matrix from the derivatives

$$c_{ii} = \frac{-\partial^2 \ln L(\hat{\boldsymbol{\beta}})}{\partial \hat{\beta}_i^2} \qquad (i = 1, 2, \ldots, k+1)$$

and

$$c_{ij} = \frac{-\partial^2 \ln L(\hat{\boldsymbol{\beta}})}{\partial \hat{\beta}_i (\partial \hat{\beta}_j)} \qquad (i \neq j)$$

The estimate of the variance-covariance matrix of the coefficients is the matrix $\mathbf{C}^{-1}$. As a result, an approximate χ_1^2 statistic for testing

$$H_0\colon \beta_j = 0$$
$$H_1\colon \beta_j \neq 0$$

can be used by computing

$$\chi^2 = \frac{b_j^2}{c^{jj}} \qquad (j = 0, 1, \ldots, k)$$

where the c^{jj} are the diagonal elements of $\mathbf{C}^{-1}$. The standard errors of the coefficients are square roots of the c^{jj}. Thus we have two methods for testing individual coefficients, one method that makes use of the deviances and the second method that makes use of the approximate variance-covariance matrix. In both cases, the distribution involved is a χ_1^2. However, the numerical results will not be equivalent. Both are asymptotically correct. For more information regarding the asymptotic theory that leads to the estimated variance-covariance matrix, the reader is referred to McCullagh and Nelder (1983).

Measures of Performance of the Logistic Model

In addition to the hypothesis testing material developed in the previous subsection, the analyst has access to a modest number of performance criteria for building the logistic regression model when the method of maximum likelihood is used to estimate the parameters. A convenient analog to R^2 is certainly available through consideration of the log likelihoods. We have demonstrated that if we have a logistic regression model with parameters $\beta_0, \beta_1, \beta_2, \ldots, \beta_k$, the analog to the regression sum of squares explained by $x_1, x_2, \ldots, x_k$, in the presence of the constant β_0, is given by the following k degree of freedom deviance

$$\begin{aligned}\lambda(\beta_1, \beta_2, \ldots, \beta_k \mid \beta_0) &= \lambda(\beta_0) - \lambda(\beta_0, \beta_1, \beta_2, \ldots, \beta_k) \\ &= 2 \ln L(\hat{\beta}_0, \hat{\beta}_1, \ldots, \hat{\beta}_k) - 2 \ln L(\hat{\beta}_0)\end{aligned}$$

The analog to the *total sum of squares* is the deviance experienced if one were to fit a logistic model only containing a constant term β_0. As a result, a reasonable *proportion χ^2 explained* is given by

$$R^2 = \frac{\lambda(\beta_1, \beta_2, \ldots, \beta_k \mid \beta_0)}{\lambda(\beta_0)} \tag{7.52}$$

An adjusted type of R^2, which has its roots in an analog to the C_p statistic, can also be formed. It provides an adjustment that results in a reduction if regressors are added to the model that do not result in sufficient reduction in deviance, or, correspondingly, increase in explained χ^2. This statistic is given by

$$\bar{R}^2 = \frac{\lambda(\beta_1, \beta_2, \ldots, \beta_k \mid \beta_0) - 2p}{\lambda(\beta_0)} \tag{7.53}$$

where $p = k + 1$ is the number of model parameters. Here, if $\lambda(\beta_1, \beta_2, \ldots, \beta_k \mid \beta_0)$ is less than $2p$, $\bar{R}^2$ is set to zero. If variables are added to the model that increase $\lambda(\beta_1, \beta_2, \ldots, \beta_k \mid \beta_0)$ but reduce $\bar{R}^2$, an analog to C_p is increased. (See Chapter 4.)

There are other rather creative performance measures and even outlier and influence diagnostics. However, the reader is referred to Pregibon (1981), Atkinson (1982), Cox (1970), and Walker and Duncan (1967) for further details in dealing in inference in logistic regression.

Two examples follow that illustrate many of the methods discussed here. The first example involves grouped data and the second involves observational data. Here we use maximum likelihood and illustrate the lack of fit test for the logistic model.

Example 7.8 ***Failure Data—Maximum Likelihood Estimation***

Consider the data of Table 7.4. Varying numbers of fabric type specimens were exposed to load (x) in lb/in^2. Also listed is the number of specimens that failed. A logistic model is chosen to fit to the data. The data are grouped. Here we choose to use maximum likelihood estimation with groups and group sample sizes shown in Table 7.4. Table 7.5 represents the results of the analysis. There is strong evidence that β_0 and β_1 both differ significantly from zero. Notice that the χ^2-values for testing β_1 do not give identical results. The procedures are not equivalent. The R^2-like statistic can be calculated as

$$\begin{aligned} R^2 &= \frac{\lambda(\beta_1 \mid \beta_0)}{\lambda(\beta_0)} = 1 - \frac{\lambda(\beta_0, \beta_1)}{\lambda(\beta_0)} \\ &= 1 - \frac{36.22}{319.27} \\ &= 0.886 \end{aligned}$$

Now, while the R^2-like statistic seems to imply that the model is reasonable, the lack of fit test with the statistic

$$\lambda(\beta_0, \beta_1) = -2 \ln\left[\frac{L(\beta_0, \beta_1)}{L(\hat{p})}\right] \quad (5 - 2 = 3 \text{ df})$$
$$= 36.22$$

is quite significant ($P < 0.001$). Thus we might benefit from a change in the linear portion of the model in the logistic function. Consider, as an alternative,

$$P(\mathbf{x}_i) = \frac{1}{1 + e^{-(\beta_0 + \beta_1 \ln x_i)}}$$

Table 7.6 shows the analysis.

TABLE 7.4 **Fabric failure data**

Load	Number of Specimens	Number of Failures
5	600	13
35	500	95
70	600	189
80	300	95
90	300	130

TABLE 7.5 **Analysis of failure data using maximum likelihood estimation**

Estimate	Standard Error	χ^2	P
$\hat{\beta}_0 = -2.995$	0.1459	421.10	0.0001
$\hat{\beta}_1 = 0.0307$	0.0021	215.84	0.0001
Model $\chi^2 = \lambda(\beta_0) - \lambda(\beta_0, \beta_1)$		$\lambda(\beta_0, \beta_1) = 36.22$	(3 df)
$= 283.06$	(1 df)	$\lambda(\beta_0) = 319.27$	(4 df)

TABLE 7.6 **Analysis of failure data using log load**

Estimate	Standard Error	χ^2	P
$\hat{\beta}_0 = -5.578$	0.3682	229.49	0.0001
$\hat{\beta}_1 = 1.1400$	0.0892	163.09	0.0001
Model $\chi^2 = \lambda(\beta_0) - \lambda(\beta_0, \beta_1)$		$\lambda(\beta_0, \beta_1) = 5.388$	(3 df)
$= 313.89$	(1 df)	$\lambda(\beta_0) = 319.27$	(4 df)

The coefficient $\hat{\beta}_1$ in Table 7.6 is the coefficient of $\ln x$. In this case, the proportion of explained variation

$$R^2 = 1 - \frac{5.388}{319.27} = 0.983$$

is considerably higher than what we experienced with only a linear term in load; in addition the deviance (3 df) of 5.388 is now insignificant, suggesting that this logistic model does not differ significantly from the model in which five parameters were estimated ignoring load. As a result, the user would certainly favor the model involving the log of load in the logistic structure. It is interesting that the coefficient χ^2-values are larger with the model containing the linear load term. However, the considerably smaller deviance favors the model containing log load. Further evidence can be observed from a table of *fitted values* using the two models.

Linear Load		Log Load	
Observed	Fitted	Observed	Fitted
13	33.09	13	13.87
95	64.04	95	89.33
189	180.79	189	194.43
95	110.92	95	107.47
130	133.15	130	116.90

It is clear that the log model is considerably better at the lower loads at estimating proportions, or number of failures of the material. For comparing competing models that appear in the exponential portion of the logistic model, the computed deviance is very beneficial.

In the following example, we turn to an illustration in which the data are not grouped. In fact the user has no capability for controlling the regressor variables. It is an observational study in which subjects were exposed to varying levels of rate and volume of air inspired on a transient vaso restriction on the skin of the digits. Only the occurrence or nonoccurrence of the restriction was observed. For more details, see Finney (1947).

Example 7.9 The data described above appear in Table 7.7. A logistic regression was fit to the data with the method of maximum likelihood used to estimate the parameters. The model is given by

$$P(\mathbf{x}_i) = \frac{1}{1 + e^{-(\beta_0 + \beta_1 x_{1i} + \beta_2 x_{2i})}}$$

where x_{1i} is the volume of air inspired and x_{2i} is the rate. Table 7.8 gives the analysis, showing the lack of fit test, the χ^2 statistic on the model, and χ^2 statistics on the individual parameters. The goal is to determine the appropriateness of the logistic model and to study the impact of the two regressor variables, volume and rate. As in the previous example, a computer software package was used to produce the maximum likelihood estimates.

The model deviance, i.e., that statistic which produces the lack of fit test, gives a χ^2-value of 29.772 with 36 degrees of freedom. The P-value is approximately 0.8. Thus the logistic model appears to be quite adequate. From the χ^2-values in Table 7.8, both volume and rate appear to contribute significantly in the logistic model. In this illustration we will not discuss any transformations on volume and rate.

TABLE 7.7 **Volume and rate data**

Volume	Rate	Response (occurrence signified by 1)
3.70	0.825	1
3.50	1.090	1
1.25	2.500	1
0.75	1.500	1
0.80	3.200	1
0.70	3.500	1
0.60	0.750	0
1.10	1.700	0
0.90	0.750	0
0.90	0.450	0
0.80	0.570	0
0.55	2.750	0
0.60	3.000	0
1.40	2.330	1
0.75	3.750	1
2.30	1.640	1
3.20	1.600	1
0.85	1.415	1
1.70	1.060	0
1.80	1.800	1
0.40	2.000	0
0.95	1.360	0
1.35	1.350	0
1.50	1.360	0
1.60	1.780	1
0.60	1.500	0
1.80	1.500	1
0.95	1.900	0

TABLE 7.7 Continued

TABLE 7.7 Continued

Volume and rate data

Volume	Rate	Response (occurrence signified by 1)
1.90	0.950	1
1.60	0.400	0
2.70	0.750	1
2.35	0.030	0
1.10	1.830	0
1.10	2.200	1
1.20	2.000	1
0.80	3.330	1
0.95	1.900	0
0.75	1.9000	0
1.30	1.625	1

TABLE 7.8

Results of the analysis of the volume and rate data

Coefficient	Standard Error	χ^2	P
$\hat{\beta}_0 = -9.530$	3.233	8.69	0.003
$\hat{\beta}_1 = 3.882$	1.429	7.38	0.007
$\hat{\beta}_2 = 2.649$	0.914	8.40	0.0038

$$\text{Model } \chi^2 = \lambda(\beta_0) - \lambda(\beta_0, \beta_1, \beta_2)$$
$$= 24.27 \quad (2 \text{ df})$$
$$P < 0.0001$$
$$\lambda(\beta_0, \beta_1, \beta_2) = 29.772 \quad (36 \text{ df})$$

In the section that follows we consider further the notion of regression with a discrete response. We move to the development of a regression model in which the response is Poisson in nature.

7.5

FURTHER DEVELOPMENT IN MODELS WITH A DISCRETE RESPONSE (POISSON REGRESSION)

Developments in the previous section revolved around situations in which the response is binary. Applications are abundant in medicine, economics, engineering, and biology. However, in many applications, the response is discrete but not necessarily binary. The number of counts are assumed to be a function of one or more regressor variables. For example, in a biological experiment, the number of insects on a specific type of plant may be a function of certain biological characteristics of the plant. In a metallurgy application, it may be important to model the number of deformities (bubbles, etc.) per square foot of material as a function of certain process variables. Many applications arise in the medical and industrial processing fields. In this section we will assume that the mean number of "incidents" is the parameter of a Poisson distribution. This Poisson mean, then, is a function of the regressor variables.

THE POISSON DISTRIBUTION

Poisson regression is, of course, based on the use of the Poisson distribution. The Poisson distribution models the probability of y "events" or incidents, according to the Poisson process with the probability given by

$$p(y;\mu) = \frac{e^{-\mu}\mu^y}{y!} \qquad (y = 0, 1, 2, \ldots)$$

The mean of the Poisson distribution is μ. It is important to note that the parameter μ is very much dependent on some specified unit or period of time, distance, area, volume, etc. The distribution is then used to model relatively rare events that occur during the chosen "basic period." For example, if μ is the mean incidents per unit time and t is the time period of interest, etc., in the application, then the mean y is μt. Thus, the probability of y events in time period t is given by

$$P(y;\mu) = \frac{e^{-\mu t}(\mu t)^y}{y!}$$

This expression for the probability of y incidents, and the mean number of incidents, μt, is based on the assumption that the mean number of incidents *per unit period is constant.* But often the mean number of incidents depends on levels of regressor variables that are changing in the data-taking process. Thus we use the terminology *Poisson regression.*

MODEL FOR POISSON REGRESSION

As in the case of logistic regression in Section 7.4, the model for Poisson regression essentially expresses the mean of the discrete distribution (in the case of the logistic, it was the binomial distribution) as a function of regressor variables. Suppose the data take the form

$$\begin{array}{ccccc} y_1 & x_{11} & x_{21} & \cdots & x_{k1} \\ \vdots & \vdots & \vdots & & \vdots \\ y_n & x_{1n} & x_{2n} & \cdots & x_{kn} \end{array}$$

The model is written

$$y_i = \mu_i + \varepsilon_i \qquad (i = 1, 2, \ldots, n)$$

Now, μ_i is the mean number of incidents in period t_i. We make use of the Poisson distribution and assume that the μ_i are not changing independently from data point to data point. Rather we model the μ_i as a function of the set of k regressor variables. Thus we have

$$p(y_i; \boldsymbol{\beta}) = \frac{e^{-t_i[\mu(\mathbf{x}_i, \boldsymbol{\beta})]}[t_i \mu(\mathbf{x}_i, \boldsymbol{\beta})]^{y_i}}{y_i!} \qquad (i = 1, 2, \ldots, n) \tag{7.54}$$

where $\mu(\mathbf{x}_i; \boldsymbol{\beta})$ is the Poisson mean and thus replaces μ in the standard Poisson case. The $\boldsymbol{\beta}$ vector represents a set of parameters to be estimated. The function $\mu(\mathbf{x}_i, \boldsymbol{\beta})$ can be chosen by the user. The function must always be nonnegative. One candidate, of course, is $e^{\mathbf{x}_i'\boldsymbol{\beta}}$ where $\mathbf{x}_i'\boldsymbol{\beta}$ is a linear function. Another is $\ln(\mathbf{x}_i'\boldsymbol{\beta})$, where $\mathbf{x}_i'\boldsymbol{\beta} > 1$. Another, of course, is the linear function itself, i.e., $\mathbf{x}_i'\boldsymbol{\beta}$ where $\mathbf{x}_i'\boldsymbol{\beta} > 0$. This function that relates the regressor variables to the distribution mean is called the *link function*. For example, in the logistic case, the link function results in the logistic function displayed in Eq. (7.40). Thus we can model the Poisson mean as

$$\mu_i = t_i \mu(\mathbf{x}_i, \boldsymbol{\beta}) \qquad (i = 1, 2, \ldots, n) \tag{7.55}$$

The random variable y_i has mean μ_i, and since the Poisson distribution applies, the variance of y_i is also μ_i which, of course, varies from data point to data point. More attention will be paid to the role of the link function in subsequent sections.

MAXIMUM LIKELIHOOD ESTIMATION IN POISSON REGRESSION

We now have Poisson regression in a structure similar to that of logistic regression. The mean is modeled as a function of a set of regressor variables. We need to set up the likelihood function and the likelihood equations that allow estimation of the elements of $\boldsymbol{\beta}$. For the Poisson distribution as described in Eq. (7.54), the likelihood is given by

$$\begin{aligned} L(\mathbf{y}, \boldsymbol{\beta}) &= \prod_{i=1}^{n} p(y_i, \boldsymbol{\beta}) \\ &= \prod_{i=1}^{n} \left\{ \frac{[t_i \mu(\mathbf{x}_i, \boldsymbol{\beta})]^{y_i} e^{-[t_i \mu(\mathbf{x}_i, \boldsymbol{\beta})]}}{y_i!} \right\} \\ &= \frac{\left\{ \prod_{i=1}^{n} [t_i \mu(\mathbf{x}_i, \boldsymbol{\beta})]^{y_i} \right\} e^{-\sum_{i=1}^{n} t_i \mu(\mathbf{x}_i, \boldsymbol{\beta})}}{\prod_{i=1}^{n} y_i!} \end{aligned} \tag{7.56}$$

Once the functional form $\mu(\mathbf{x}_i, \boldsymbol{\beta})$ is chosen, the maximization of (7.56) with the use of iterative techniques will produce the maximum likelihood estimators for the regression coefficients in $\boldsymbol{\beta}$. The maximization can be done through packages such as GLIM and CATMAX. (See Stokes and Koch (1983).) The procedure often recommended for finding maximum likelihood estimators for both logistic regression and Poisson regression is a method called iteratively reweighted least squares (IRWLS). IRWLS will be discussed in conjunction with robust regression later in this chapter. Actually, a SAS procedure called NLIN, which is illustrated in Chapter 9 when we deal with *nonlinear regression*, allows the use of IRWLS and thus can be used to do either logistic or Poisson regression.

Formally, of course, the method used to find maximum likelihood estimators is called upon to solve

$$\frac{\partial \ln L(\mathbf{y}, \boldsymbol{\beta})}{\partial \boldsymbol{\beta}} = \mathbf{0}$$

We have

$$\ln L(\mathbf{y}, \boldsymbol{\beta}) = \sum_{i=1}^{n} y_i \ln t_i \mu(\mathbf{x}_i, \boldsymbol{\beta}) - \sum_{i=1}^{n} t_i \mu(\mathbf{x}_i, \boldsymbol{\beta}) - \sum_{i=1}^{n} \ln(y_i)!$$

and thus the likelihood equations are

$$\boxed{\sum_{i=1}^{n} \left[\frac{y_i}{\mu(\mathbf{x}_i, \hat{\boldsymbol{\beta}})} - t_i\right]\left[\frac{\partial \mu(\mathbf{x}_i, \hat{\boldsymbol{\beta}})}{\partial \hat{\boldsymbol{\beta}}}\right] = \mathbf{0}} \tag{7.57}$$

Equation (7.57) is to be solved for $\hat{\boldsymbol{\beta}}$. Note that, unlike standard regression, even if $\mu(\mathbf{x}_i, \boldsymbol{\beta}) = \mathbf{x}_i'\boldsymbol{\beta}$, a linear function, the likelihood equations are nonlinear. The reader should note that the expression $\partial[\mu(\mathbf{x}_i, \boldsymbol{\beta})]/\partial\boldsymbol{\beta}$ is a column vector in the formulation in Eq. (7.57).

WHAT ARE THE IMPORTANT RESULTS IN POISSON REGRESSION?

In Chapters 2 and 3, we learned that for ordinary linear regression there is an implicit assumption of normality on the model errors. Then maximum likelihood estimation leads to least squares estimation. After estimation of coefficients, the analyst can use the fitted model

$$(\hat{E}y) = \hat{\beta}_0 + \hat{\beta}_1 x_1 + \hat{\beta}_2 x_2 + \cdots + \hat{\beta}_k x_k$$

to estimate mean response. Now, as in the case of logistic regression, Poisson regression is used when the mean response is a mean of a *particular discrete distribution*, whereas in the standard regression model above, the mean is the mean of a normal distribution. But in all three cases, we are modeling the mean y against regressor variables. For logistic regression that model is given by Eq. (7.40), while for the Poisson case, it is given in Eq. (7.55). Thus for the Poisson case, we have, after estimation,

$$\hat{\mu}_i = t_i \mu(\mathbf{x}_i, \hat{\boldsymbol{\beta}})$$

For example, if $\mu(\mathbf{x}_i, \boldsymbol{\beta})$ is chosen to be $e^{\mathbf{x}_i'\boldsymbol{\beta}}$, then for period t and regressors $\mathbf{x}$,

$$\hat{\mu} = te^{\mathbf{x}'\hat{\boldsymbol{\beta}}}$$

This then gives an estimate of the mean number of incidents for period t and for the set of regressor variables given by $\mathbf{x}$. Additionally though, one may wish to estimate the probability of obtaining, say, y incidents. This, of course, is found through the use of the Poisson distribution and the computation resides with Eq. (7.54).

For model building, inferences on coefficients, etc., we can still use likelihood ratio and the concept of deviance, etc., as we introduced for the case of the logistic model in Section 7.4. Equation (7.56) gives the user some insight into $L(\mathbf{y}, \boldsymbol{\beta})$ and $\ln L(\mathbf{y}, \boldsymbol{\beta})$ for the Poisson case. We choose to illustrate the use of inference procedures for Poisson regression with an example.

Example 7.10 Consider a data set in which counts were observed on the number of injuries or fractures that occur in the upper seam of mines in the coal fields of the Appalachian region in western Virginia. A total of 44 observations were collected on mines in this area. Four variables were collected, all of which are functions of the material in the land and the mining area. The data are shown in Table 7.9. The variables include

TABLE 7.9 **Mining Fracture Data**

Subject	y (number of fractures per subregion)	x_1	x_2	x_3	x_4
1	2	50	70	52	1.0
2	1	230	65	42	6.0
3	0	125	70	45	1.0
4	4	75	65	68	0.5
5	1	70	65	53	0.5
6	2	65	70	46	3.0
7	0	65	60	62	1.0
8	0	350	60	54	0.5
9	4	350	90	54	0.5
10	4	160	80	38	0.0
11	1	145	65	38	10.0
12	4	145	85	38	0.0
13	1	180	70	42	2.0
14	5	43	80	40	0.0
15	2	42	85	51	12.0
16	5	42	85	51	0.0
17	5	45	85	42	0.0
18	5	83	85	48	10.0
19	0	300	65	68	10.0
20	5	190	90	84	6.0
21	1	145	90	54	12.0

TABLE 7.9 Continued

Mining Fracture Data

Subject	y (number of fractures per subregion	x_1	x_2	x_3	x_4
22	1	510	80	57	10.0
23	3	65	75	68	5.0
24	3	470	90	90	9.0
25	2	300	80	165	9.0
26	2	275	90	40	4.0
27	0	420	50	44	17.0
28	1	65	80	48	15.0
29	5	40	75	51	15.0
30	2	900	90	48	35.0
31	3	95	88	36	20.0
32	3	40	85	57	10.0
33	3	140	90	38	7.0
34	0	150	50	44	5.0
35	0	80	60	96	5.0
36	2	80	85	96	5.0
37	0	145	65	72	9.0
38	0	100	65	72	9.0
39	3	150	80	48	3.0
40	2	150	80	48	0.0
41	3	210	75	42	2.0
42	5	11	75	42	0.0
43	0	100	65	60	25.0
44	3	50	88	60	20.0

Data analyzed by the Statistics Consulting Center, Virginia Polytechnic Institute and State University, 1988.

x_1(INB): Inner burden thickness (feet), which is the shortest distance between seam floor and lower seam

x_2(EXTRP): Percent of extraction of the lower previously mined seam

x_3(SEAMH): Lower seam height (in feet)

x_4(TIME): Time (years) that the mine has been opened

A Poisson regression model was fit to the data. The link function $\mathbf{x}_i'\boldsymbol{\beta} = \ln \mu_i$ was used and maximum likelihood estimators were found using Eq. (7.57). We set $t = 1$ since the "basis" is taken as a single mine. The GLIM software was used and one of the purposes of the Poisson regression analysis was to determine which variables actually have impact on the number of fractures. The "full" model can be written

$$\hat{\mu} = e^{(\beta_0 + \beta_1 x_1 + \beta_2 x_2 + \beta_3 x_3 + \beta_4 x_4)}$$

In order to gain some insight into which subset of model variables one should retain, we offer the table on page 338 of model deviance for several subset models.

Model	Deviance	Model	Deviance
x_1	71.84	x_2x_3	47.59
x_2^*	48.62	$x_2x_4^*$	41.63
x_3	73.84	x_3x_4	71.28
x_4	72.12	$x_1x_2x_3$	41.75
x_1x_2	42.09	$x_1x_2x_4^*$	38.03
x_1x_3	71.07	$x_1x_3x_4$	69.81
x_1x_4	70.43	$x_2x_3x_4$	41.02
		$x_1x_2x_3x_4^*$	37.86

The best single-variable model, of course, involves EXTRP; the best two-variable model involves EXTRP and time; the best three-variable model involves INB, EXTRP, and time. Clearly, x_3 (SEAMH) is not significant in the full model since

$$\begin{aligned}\lambda(\beta_3 \mid \beta_0, \beta_1, \beta_2, \beta_4) &= \lambda(\beta_0, \beta_1, \beta_2, \beta_4) - \lambda(\beta_0, \beta_1, \beta_2, \beta_3, \beta_4)\\ &= 38.03 - 37.86\\ &= 0.17\end{aligned}$$

which is, of course, not significant as a χ_1^2 random variable. Does the Poisson model need variable x_1 in the presence of x_2 and x_4? We compute

$$\begin{aligned}\lambda(\beta_1 \mid \beta_2, \beta_4) &= 41.63 - 38.03\\ &= 3.60 \quad (1 \text{ df})\end{aligned}$$

which is significant at the 0.057 level. *As a result, a reasonable model includes* x_1, x_2, *and* x_4. The following gives estimates, standard errors, and χ^2-values for each parameter estimate in this three-parameter model. The standard errors are asymptotic standard errors, and the methodology is identical to that discussed in the case of the logistic regression case in Section 7.4.

	Estimate	Standard Error	χ^2	P
$\hat{\beta}_0$	−3.721	0.9786	14.46	0.0014
$\hat{\beta}_1$	−0.001479	0.0008242	3.22	0.0727
$\hat{\beta}_2$	0.06270	0.01227	26.11	0.0001
$\hat{\beta}_4$	−0.03165	0.01631	3.76	0.0523

The χ^2-values for each model term support the use of this three-variable model. As a result, the final model is given by

$$\hat{\mu} = e^{[-3.721 - 0.001479x_1 + 0.06270x_2 - 0.03165x_4]}$$

where, of course, $\hat{\mu}$ is the estimate of the mean number of fractures. It is interesting that the mean number of fractures is increased with an increase in EXTRP, whereas an increase in INB or time decreases the mean number of fractures.

7.6

GENERALIZED LINEAR MODELS

In the two previous sections, we developed the procedures and illustrated the use of regression analysis when the response is discrete. It is noteworthy that this represented a serious departure in concept from all of the material that preceded this chapter. A large portion of what came before Chapter 7 made use of the presumption of normal errors on the linear model, either by assumption or certainly by implication. Please recall that material in both Chapters 2 and 3 emphasized that least squares estimation with the linear regression model is also a maximum likelihood estimator when the errors are normal. It should also be recalled from Chapter 3 that the properties of the least squares estimators are stronger (uniformly minimum variance unbiased) when the errors are normal than when they are not normal.

OTHER OBVIOUS NONNORMAL ERROR SITUATIONS

Certainly the binomial and Poisson regression situations developed in the previous two sections are obvious candidates where ordinary least squares does not apply. However, there certainly are others. In biological applications where the phenomena that produce counts are contagious in nature, and, in fact, the variance in the response is obviously greater than the mean, the *negative binomial distribution* may apply. There are situations where the response is continuous, and yet the Gaussian error assumption simply makes no sense. In reliability theory there are important applications of regression in reliability growth models where mean reliability of a system or the breaking strength of an alloy is a function of processing regressor variables. In this case, the errors are likely to be distributed as an exponential, gamma, or even a Weibull random variable. Statisticians have long been able to classify scientific processes into categories that determine the likely distribution of the data generated. But only recently has there been developed some interesting connective tissue between this knowledge and the notion of model fitting. This is important because in many applications (for example, reliability growth) the probabilities or other parameters that need to be computed are functions of regressor variables. Poisson and logistic regression are then only a special case of a more general family of linear models.

FORM OF THE GENERALIZED LINEAR MODEL (GLM)

An important unified approach to linear models, and hence regression, was introduced by Nelder and Wedderburn (1972). The emphasis was placed on a general family of error distributions for fitting linear regression models. The normal distribution of errors becomes a special case. The Poisson and binomial fall into this general class. Let us consider, in general, an experiment which is modeled by

(i) A set of observations y_i ($i = 1, 2, \ldots, n$) are independently distributed with mean $E(y_i) = \mu_i$. These data take on a *regression structure*, i.e., a set of regressors $x_{1i}, x_{2i}, \ldots, x_{ki}$ are also observed.

(ii) The density function (or probability function in the discrete case) is in the *exponential family* with the general form

$$\boxed{f(y_i, \phi, \theta_i) = \exp\{r(\phi)[y_i\theta_i - g(\theta_i)] + h(\phi, y_i)\}} \quad \textbf{(7.58)}$$

where ϕ is merely a nuisance parameter.

(iii) The parameters θ_i may vary from data point to data point. This is the parameter that *links* the density function to the regressor variables through the linear function $\mathbf{x}_i'\boldsymbol{\beta}$.

(iv) The quantity $\mathbf{x}_i'\boldsymbol{\beta}$ is a linear predictor of $\mu_i = E(y_i)$.

(v) The *link function* between μ_i and $\mathbf{x}_i'\boldsymbol{\beta}$ is accomplished through the parameter θ_i and is given by

$$\boxed{\mathbf{x}_i'\boldsymbol{\beta} = s(\mu_i)} \quad \textbf{(7.59)}$$

Now, θ_i is not necessarily the mean μ_i. However, in order to bring the regressors and the accompanying *linear model* into the formulation, we must relate θ_i to μ_i.

(vi) The mean μ_i is not necessarily linked to the regressors in a linear fashion. More often than not the link is nonlinear. However, it is assumed that $E(y_i)$ is a monotonic, differentiable function of $\mathbf{x}_i'\boldsymbol{\beta}$.

(vii) There is *no assumption* that the variance $\text{Var}(y_i)$ is homogeneous. However, it is assumed that the variance of y_i *varies with the x's through* μ_i *alone.*

The above seven items outline the fundamental principles of generalized linear models. It is important now to tie the important parameter θ_i to the distribution means and variance, i.e., $E(y_i)$ and the variance of y_i. Please keep in mind that *both mean and variance of* y_i *may vary with the regressor variables* from data point to data point.

THE MEAN AND VARIANCE OF y_i

Consider the exponential family of Eq. (7.58). From McCullagh and Nelder (1983), we know that

$$\boxed{\begin{aligned} E(y_i) = \mu_i &= \frac{\partial g(\theta_i)}{\partial \theta_i} \\ \text{Var}(y_i) = \sigma_i^2 &= \frac{\partial^2 g(\theta_i)/\partial \theta_i^2}{r(\phi)} \end{aligned}}$$

As a result, through the portion $g(\theta_i)$ of the density function and the link function, one is able to model the mean μ_i as a function of the linear expression $\mathbf{x}_i'\boldsymbol{\beta}$. The following example may be helpful. Suppose the distribution of y_i is *normal.* In the exponential family of Eq. (7.58), we can write the density function as

$$f(y_i, \mu_i, \sigma) = \exp\left\{\sigma^{-2}\left[\mu_i y_i - \frac{1}{2}\mu_i^2 - \frac{1}{2}y_i^2\right] + \ln\frac{1}{\sqrt{2\pi\sigma}}\right\}$$

Now, for this case

$$\theta_i = \mu_i$$
$$g(\theta_i) = \tfrac{1}{2}\mu_i^2$$
$$r(\phi) = \sigma^{-2}$$
$$h(y_i, \phi) = -\frac{1}{2}y_i^2 + \ln\frac{1}{\sqrt{2\pi\sigma}}$$

Thus, as we already know, the mean and variance are given by

$$\frac{\partial g(\theta_i)}{\partial\theta_i} = \mu_i$$
$$\frac{\partial^2 g(\theta_i)/\partial\theta_i^2}{r(\phi)} = \sigma^2 \qquad \text{(constant for all } i\text{)}$$

Thus as one might expect, *the homogeneous variance assumption is very much consistent with the assumption of normal errors, within the framework of generalized linear models.* The link function in Eq. (7.59) for the normal case is the *unity link*, i.e.,

$$\boxed{\mathbf{x}_i'\boldsymbol{\beta} = \mu_i}$$

and thus we have the *standard multiple linear regression model*

$$y_i = \mathbf{x}_i'\boldsymbol{\beta} + \varepsilon_i \qquad (i = 1, 2, \ldots, n) \tag{7.60}$$

with $\text{Var}(y_i) = \sigma^2$. Also we know from Chapters 2 and 3 that the maximum likelihood estimator of $\boldsymbol{\beta}$ is the least squares estimator via the above model.

POISSON AND LOGISTIC REGRESSION IN THE FRAMEWORK OF GLM

Consider the Poisson regression situation discussed in Section 7.5. The probability function for y_i, the number of "incidents" in period t_i is given in Eq. (7.54), namely

$$p(y_i, \boldsymbol{\beta}) = \frac{e^{-t_i[\mu(\mathbf{x}_i, \boldsymbol{\beta})]}[t_i\mu(\mathbf{x}_i', \boldsymbol{\beta})]^{y_i}}{y_i!}$$

It is easy to see that this probability function is a member of the exponential family since

$$p(y_i, \boldsymbol{\beta}) = \exp\{-t_i[\mu(\mathbf{x}_i'\boldsymbol{\beta})] + y_i \ln t_i\mu(\mathbf{x}_i, \boldsymbol{\beta}) - \ln(y_i)!\}$$

Thus, we have

$$r(\phi) = 1.0$$
$$\theta_i = \ln t_i[\mu(\mathbf{x}_i, \boldsymbol{\beta})]$$
$$g(\theta_i) = t_i\mu(\mathbf{x}_i, \boldsymbol{\beta}) = e^{\theta_i}$$
$$h(y_i, \phi) = -\ln(y_i)!$$

As a result,

$$\mu_i = \frac{\partial g(\theta_i)}{\partial \theta_i} = e^{\theta_i} = t_i\mu(\mathbf{x}_i, \boldsymbol{\beta})$$
$$\text{Var}(y_i) = \frac{\partial^2 g(\theta_i)}{\partial \theta_i^2} = e^{\theta_i} = t_i\mu(\mathbf{x}_i, \boldsymbol{\beta})$$

Recall that in Section 7.5, we linked the Poisson mean, $t_i\mu_i(\mathbf{x}_i, \boldsymbol{\beta})$, to the regressors through the *link function* $\mathbf{x}_i'\boldsymbol{\beta} = \ln(\mu_i/t_i)$, which leads to the equation

$$\boxed{\mu_i = t_i e^{\mathbf{x}_i'\boldsymbol{\beta}}}$$

As a result, we can once again write that the model to be fit for Poisson regression is given by

$$y_i = t_i e^{\mathbf{x}_i'\boldsymbol{\beta}} + \varepsilon_i \tag{7.61}$$

However, since $\text{Var}(y_i) = t_i e^{\mathbf{x}_i'\boldsymbol{\beta}}$ is not homogeneous from observation to observation, *standard least squares does not apply.* From our development and illustrations in Section 7.5, we know that maximum likelihood estimation methods are recommended for estimation of the β's in the Poisson regression model.

The GLM framework has essentially linked the *linear function*, $\mathbf{x}_i'\boldsymbol{\beta}$, of the regressors to the probability structure of the Poisson distribution. The contrast between the normal and Poisson is interesting. In the normal error case, we are led to the linear model of Eq. (7.60). In the Poisson case, we are led to the *nonlinear model* of Eq. (7.61) through the very reasonable link function $\mathbf{x}_i'\boldsymbol{\beta} = \ln(\mu_i/t_i)$. In the normal case, the unity link function $\mathbf{x}_i'\boldsymbol{\beta} = \mu_i$ leads to a linear model. In addition, normality leads to a condition in which $\text{Var}(y_i)$ is homogeneous and standard least squares produces maximum likelihood estimation of coefficients. Clearly *the first five chapters of the text make use of this type of model.*

The logistic regression as discussed in Section 7.4 falls into the GLM framework. The probability function is one that is in the exponential family. Suppose we consider the grouped data case, i.e., where a set of n_i observations are taken under the set of regressors $\mathbf{x}_i$, and r_i successes are measured. Here the probability of obtaining r_i successes is given by the binomial distribution. Here $r_i = y_i$, the basic response random variable. The probability of r_i successes is given by (see Walpole and Myers (1989))

$$p(r_i, \boldsymbol{\beta}, \mathbf{x}_i) = \binom{n_i}{r_i}[P(\mathbf{x}_i)]^{r_i}[1 - P(\mathbf{x}_i)]^{n_i - r_i} \qquad (r_i = 0, 1, 2, \ldots, n_i)$$

$$= \exp\left[\ln\binom{n_i}{r_i} + r_i \ln P(\mathbf{x}_i) + (n_i - r_i)\ln[1 - P(\mathbf{x}_i)]\right]$$

As before, $P(\mathbf{x}_i)$ is the probability of a success at the ith measurement on the regressors. In this case, from Eq. (7.58), we have

$$\theta_i = \ln P(\mathbf{x}_i) - \ln[1 - P(\mathbf{x}_i)]$$

$$g(\theta_i) = -n_i \ln[1 - P(\mathbf{x}_i)] = -n_i \ln\left[\frac{e^{-\theta_i}}{1 + e^{-\theta_i}}\right]$$

$$r(\phi) = 1$$

$$h(y_i, \phi) = \ln\binom{n_i}{r_i}$$

From this information we obtain, as expected,

$$\mu_i = \frac{\partial g(\theta_i)}{\partial \theta_i} = n_i P(\mathbf{x}_i)$$

and

$$\text{Var}(y_i) = \frac{\partial^2 g(\theta_i)}{\partial \theta_i^2} = n_i P(\mathbf{x}_i)[1 - P(\mathbf{x}_i)]$$

In Section 7.4 we introduced the logistic model, which relates the mean of y_i, i.e., $n_i P(\mathbf{x}_i)$ to the regressor variables. This essentially defines the link function, namely the *logistic link*. Again, for the grouped data case in which the basic observation is r_i, the number of successes, the logistic relationship is

$$n_i P(\mathbf{x}_i) = \frac{n_i}{1 + e^{-\mathbf{x}_i'\boldsymbol{\beta}}}$$

or, with $P(\mathbf{x}_i)$ being the mean of a single observation, the logistic model is written

$$\boxed{p(\mathbf{x}_i) = \frac{1}{1 + e^{-\mathbf{x}_i'\boldsymbol{\beta}}}} \tag{7.62}$$

Equation (7.62) represents the logistic regression for the case in which the measured response is binary. The formal model is, then

$$y_i = \frac{1}{1 + e^{-\mathbf{x}_i'\boldsymbol{\beta}}} + \varepsilon_i \tag{7.63}$$

Again, as in the case of Poisson regression, the model of Eq. (7.63) is nonlinear in the coefficients. Also, since

$$\mathrm{Var}(y_i) = n_i p(\mathbf{x}_i)[1 - p(\mathbf{x}_i)]$$

which varies with the regressor variables, least squares procedures are not appropriate. Maximum likelihood methods were used in Section 7.4 in conjunction with logistic regression.

WHERE DOES THE LINK FUNCTION COME FROM?

As we indicated earlier, the link function determines the model that relates $\mu_i = E(y_i)$ to the linear function $\mathbf{x}_i'\boldsymbol{\beta}$. Obviously the regression results depend on the choice of the link function. Those suggested here for the normal, binomial, and Poisson situations are certainly reasonable for each situation. In fact, any monotonic, differentiable function can be used. However, there are certain functions that are preferred. In the binomial case, obviously the logistic function has great appeal. In the Poisson case, a function that gives only positive values of μ_i is necessary. In each of the three cases discussed, *the link function is derived from*

$$\boxed{\mathbf{x}_i'\boldsymbol{\beta} = \theta_i} \tag{7.64}$$

where θ_i is described in Eq. (7.58). See Exercise 7.10. This type of link function is called a *canonical link function.* It is used as a mechanism for producing models in which the linear function of the regressors, $\mathbf{x}_i'\boldsymbol{\beta}$, is a monotonic function of a clearly important location parameter. As in these cases, the models are often simple and interpretable.

We have often reminded the reader that, apart from the case of the Gaussian distribution, standard least squares as we know it to this point cannot be used for analysis of the GLM. In the following section, we discuss the analysis of the generalized linear model and develop it in the context of the material in Sections 7.4 and 7.5.

ANALYSIS OF THE GLM

Apart from the case of the normal distribution, the model resulting from the link function is nonlinear, and the error variance is not homogeneous. Sections 7.4 and 7.5 emphasized the use of maximum likelihood for estimation of $\boldsymbol{\beta}$ for the binomial and Poisson case. The maximum likelihood procedure can be generalized to cover all of the distributions that fall into the exponential family of distributions. Before we discuss the general notion of maximum likelihood estimation for GLM type models, let us consider Table 7.10, which provides the probability structure, the appropriate canonical link function relating the mean to the regressors, and the variance for several distributions that are in the exponential family.

TABLE 7.10

Distributions, variances, and link functions for some distributions in the exponential family

Distribution	$r(\phi)$	Canonical Link Function	Mean	Variance	Model (Canonical Link)	θ_i	$g(\theta_i)$
Normal:							
$f(y_i, \mu_i, \sigma) = \frac{1}{\sqrt{2\pi}\sigma} e^{-1/2((y_i - \mu_i)/\sigma)^2}$ $\sigma > 0$	σ^{-2}	$\mathbf{x}_i'\boldsymbol{\beta} = \mu_i$	μ_i	σ^2 (constant)	$E(y_i) = \mathbf{x}_i'\boldsymbol{\beta}$	μ_i	$\frac{\theta_i^2}{2}$
Poisson:							
$f(y_i, \mu_i) = \frac{e^{-\mu_i}\mu_i^{y_i}}{y_i!}$ $y = 0, 1, 2, \ldots; \mu_i > 0$	1	$\mathbf{x}_i'\boldsymbol{\beta} = \ln \mu_i$	μ_i	μ_i	$E(y_i) = t_i e^{\mathbf{x}_i'\boldsymbol{\beta}}$	$\ln(\mu_i)$	e^{θ_i}
Binomial (Logistic): Grouped Data							
$f(y_i, n_i, P(\mathbf{x}_i)) = \binom{n_i}{y_i} P(\mathbf{x}_i)^{y_i}[1 - P(\mathbf{x}_i)]^{n_i - y_i}$ $y_i = 0, 1, 2, \ldots; 0 \leqslant P(\mathbf{x}_i) \leqslant 1$	1	$\mathbf{x}_i'\boldsymbol{\beta} = \ln\left[\frac{\mu_i}{n - \mu_i}\right]$	$n_i P(\mathbf{x}_i)$	$\mu_i(1 - P(\mathbf{x}_i))$	$E(y_i) = \frac{n_i}{1 + e^{-\mathbf{x}_i'\boldsymbol{\beta}}}$	$\ln \frac{P(\mathbf{x}_i)}{1 - P(\mathbf{x}_i)}$	$-n_i \ln\left[\frac{e^{-\theta_i}}{1 + e^{-\theta_i}}\right]$
Gamma:							
$f(y_i, r, \lambda_i) = \frac{\lambda_i^r}{\Gamma(r)} e^{-y_i\lambda_i} y_i^{r-1}$ $y_i, \lambda_i, r > 0$	$-r$	$\mathbf{x}_i'\boldsymbol{\beta} = \frac{1}{\mu_i}$	$\frac{r}{\lambda_i}$	$\frac{\mu_i^2}{r}$	$E(y_i) = \frac{1}{\mathbf{x}_i'\boldsymbol{\beta}}$	$\frac{\lambda_i}{r}$	$-\ln\left(\frac{1}{\theta_i}\right)$
Geometric:							
$f(y_i, P(\mathbf{x}_i)) = [P(\mathbf{x}_i)][1 - P(\mathbf{x}_i)]^{y_i}$ $y_i = 0, 1, 2, \ldots; 0 \leqslant P(\mathbf{x}_i) \leqslant 1$	1	$\mathbf{x}_i'\boldsymbol{\beta} = \ln\left[\frac{\mu_i}{1 + \mu_i}\right]$	$(1 - P(\mathbf{x}_i))/P(\mathbf{x}_i)$	$\mu_i(1 + \mu_i)$	$E(y_i) = \frac{1}{e^{-\mathbf{x}_i'\boldsymbol{\beta}} - 1}$	$\ln(1 - P(\mathbf{x}_i))$	$-\ln(1 - e^{\theta_i})$
Negative Binomial:							
$f(y_i, P(\mathbf{x}_i), \alpha) = \binom{y_i + \alpha - 1}{\alpha - 1}[P(\mathbf{x}_i)]^{\alpha}[1 - P(\mathbf{x}_i)]^{y_i}$ $y_i = 0, 1, 2, \ldots; \alpha > 0; 0 \leqslant P(\mathbf{x}_i) \leqslant 1$	1	$\mathbf{x}_i'\boldsymbol{\beta} = -\ln\left[\frac{\alpha + \mu_i}{\mu_i}\right]$	$\alpha(1 - P(\mathbf{x}_i))/P(\mathbf{x}_i)$	$\mu_i\left(1 + \frac{\mu_i}{\alpha}\right)$	$E(y_i) = \frac{\alpha}{e^{-\mathbf{x}_i'\boldsymbol{\beta}} - 1}$	$\ln(1 - P(\mathbf{x}_i))$	$-\alpha \ln(1 - e^{\theta_i})$

The reader should keep in mind that the canonical link function is not the only one that may be considered in practice. Also, the variance listed in Table 7.10 may be extremely helpful in the maximum likelihood analysis.

In Sections 7.4 and 7.5, we went through considerable detail in outlining the likelihood function; we then indicated that the solution may be obtained through a computer software package. The procedure is iterative. In this section we hope to shed a bit more light on the general nature of maximum likelihood for the GLM family of distributions. It turns out that if we proceed to use the density function in Eq. (7.58), and use the link function to incorporate the linear function $\mathbf{x}_i'\boldsymbol{\beta}$, we have the following set of equations that require solution for estimation of $\boldsymbol{\beta}$: (see Nelder and Wedderburn (1972))

$$\boxed{\mathbf{X}'\boldsymbol{\Delta}\mathbf{e} = \mathbf{0}} \tag{7.65}$$

where
$$\mathbf{e} = \begin{bmatrix} y_1 - \hat{\mu}_1 \\ y_2 - \hat{\mu}_2 \\ \vdots \\ y_n - \hat{\mu}_n \end{bmatrix}$$

a vector of *residuals* that contains $\hat{\boldsymbol{\beta}}$ in the $\hat{\mu}_i$. The matrix $\boldsymbol{\Delta}$ is a diagonal matrix that reflects the link function. We know that the link function connects $\mathbf{x}_i'\boldsymbol{\beta}$ to μ_i and also gives the connection between θ_i and $\mathbf{x}_i'\boldsymbol{\beta}$. The diagonal elements $\boldsymbol{\delta}_i (i = 1, 2, \ldots n)$ of $\boldsymbol{\Delta}$ are given by

$$\boldsymbol{\delta}_i = \frac{\partial \theta_i}{\partial \eta_i}$$

where $\eta_i = \mathbf{x}_i'\boldsymbol{\beta}$. Since the often used canonical link has $\theta_i = \eta_i$, $\boldsymbol{\Delta}$ *will be the identity matrix for this case.* The development of Eq. (7.65) is rather involved and will not be given here. Note that in the case of the canonical link, the equation

$$\mathbf{X}'\mathbf{e} = 0 \tag{7.66}$$

must be solved. (See Exercises 7.11 and 7.12.) The reader may gain some insight into the important Eq. (7.65) by considering the normal error case. In the normal error case with the canonical link, we have $\mathbf{e} = \mathbf{y} - \mathbf{X}\boldsymbol{\beta}$ (model is linear), and hence

$$\mathbf{X}'(\mathbf{y} - \mathbf{X}\boldsymbol{\beta}) = \mathbf{0} \qquad \text{and thus} \qquad (\mathbf{X}'\mathbf{X})\hat{\boldsymbol{\beta}} = \mathbf{X}'\mathbf{y}$$

which is our set of normal equations for estimation $\boldsymbol{\beta}$.

The solution to Eq. (7.65) or (7.66) does not appear, on the surface, to be formidable. However, keep in mind that the residuals themselves (apart from the normal, canonical link case) reflect the nonlinearity of the model. The $\hat{\boldsymbol{\beta}}$ vector surfaces in the residuals. A Newton–Raphson method can be used. (See McCullagh and Nelder (1983).) Considering the link function, there may be situations in which something other than the canonical link function is used. To illustrate, consider the normal error case in which we wish to assume that $\mu_i = \ln(\mathbf{x}_i'\boldsymbol{\beta})$. In the normal case

$$\theta_i = \mu_i$$

and thus

$$\frac{\partial \theta_i}{\partial(\mathbf{x}_i'\boldsymbol{\beta})} = \frac{1}{\mathbf{x}_i'\boldsymbol{\beta}}$$

so, the matrix $\boldsymbol{\Delta}$ in Eq. (7.65) would contain $1/(\mathbf{x}_i'\hat{\boldsymbol{\beta}})$ as diagonal elements in the solution for the maximum likelihood estimation of the β_1. In addition, of course, the model would be given by

$$y_i = \ln(\mathbf{x}_i'\boldsymbol{\beta}) + \varepsilon_i$$

As a result, the solution will be given by Eq. (7.65) with

$$e_i = y_i - \ln(\mathbf{x}_i'\hat{\boldsymbol{\beta}})$$

Thus, for any of the distributions discussed here, the canonical link function is not necessary. Based on the information in Table 7.10, one can construct the $\boldsymbol{\Delta}$ matrix and use Eq. (7.65) for solution of the likelihood equations.

FURTHER ANALYSIS OF GLM *(USE OF DEVIANCE)*

Recall that in Sections 7.4 and 7.5 we introduced the notion of *deviance* to illustrate the methodology of doing significance tests on coefficients and testing for lack of fit of the fitted model for the case of the Poisson and logistic model. The notion of deviance applies quite generally for any distribution that falls into the exponential family of Eq. (7.58). In fact, given the general formulation of Eq. (7.58), we denote the likelihood of the exponential family as $L(\boldsymbol{\beta})$ and, through the link function the parameter, $\boldsymbol{\beta}$ is replaced by its maximum likelihood estimator $\hat{\beta}$. As before, we have two likelihoods, the maximum likelihood for the model being fit, $L(\hat{\boldsymbol{\beta}})$, and the likelihood computed when the fit is perfect; i.e., $L(\hat{\boldsymbol{\mu}})$ is maximized by the estimator $\hat{\mu}_i = y_i$. The lack of fit measure is the deviance, defined as

$$\lambda(\boldsymbol{\beta}) = -2\ln[L(\hat{\boldsymbol{\beta}}) - L(\hat{\boldsymbol{\mu}})] \qquad \textbf{(7.67)}$$

which is asymptotically a χ^2 distribution with $n-k-1$ degrees of freedom under the condition that the model being fit is correct. Again, $k+1$ is the number of parameters in $\boldsymbol{\beta}$. Evidence of lack of fit of the model appears if $\lambda(\beta) > \chi^2_{\alpha,n-k-1}$. The deviance decreases with an increase in the number of parameters in $\boldsymbol{\beta}$. In addition, the hierarchical structure can be used for testing subsets or individual parameters. Again, this is much like that presented for the Poisson and logistic regression models in Sections (7.4) and (7.5). For further experience in dealing with the GLM, the reader should consult Exercises 7.11, 7.12, 7.13, and 7.14. A numerical exercise appears as Exercise 7.15.

7.7

FAILURE OF NORMALITY ASSUMPTION; PRESENCE OF OUTLIERS

In this section we continue to investigate procedures that are used when conditions are not ideal. Earlier we dealt with weighted least squares as an alternative when homogeneous variance fails to hold. We then presented information that dealt with transformations to accommodate the nonideal conditions: heterogeneous error variance and curvature in the model. We also introduced methodology to handle discrete responses and the more generalized linear model. In this section we reinvestigate normality as an assumption and consider alternative methods for estimating regression relationships under one of the following conditions:

1. Nonnormal errors
2. Variant data or outliers in the data set

In Chapter 5 we discussed computational techniques on residuals that allow for detection of severe deviations from normality in the errors. The reader should recall that the properties of the least squares estimators are more impressive (minimum variance of all unbiased estimators) under normality than when normality is sacrificed (minimum variance of all *linear* unbiased estimators). This implies that when the errors are clearly *not normal*, an alternative to the least squares procedure may be beneficial, an alternative that is *resistant to departure from normality*.

In many applications of regression analysis, one needs access to procedures that are resistant to the effect of *outliers* in the data set. In Chapter 5 we discussed the *detection* of outliers. In Chapter 6, techniques were supplied which diagnose the influence of outliers. Of course the signal that an outlier is present evolves from the value of its residual. By its very nature, least squares allows outliers to exert a disproportionate influence on regression results. This should come as no surprise to the user since the least squares criterion is the minimization of the sum of *squares* of residuals. So, while least squares estimation represents a noble and intuitively reasonable estimation methodology, it suffers in performance under certain nonideal situations, the presence of outliers and certain nonnormal error distributions.

Specific nonnormal error distributions that reduce the effectiveness of least squares are those with *heavy tails*. In practice, of course, one may not distinguish between the presence of model shifts (outliers) and a heavy-tailed error distribution. However, the end result of the two is the same when least squares is used; namely, there will be strong influence on the results created by the "pull" of the regression toward the deviant data points. A robust procedure is one that is resistant or insensitive to the nonideal condition. Here, in the regression situation, the focus is on the residuals. The useful robust procedures essentially "weight down" the influence of data points that produce residuals that are large in magnitude.

THE INFLUENCE FUNCTION

An interesting measure of how individual data points affect regression results is given by the influence function. This is particularly revealing in the case of least squares. An illustration of the influence function begins with the fitted function

$$y_i = \mathbf{x}_i'\mathbf{b} \qquad (i = 1, 2, \ldots, n) \tag{7.68}$$

where $\mathbf{b}$ is the vector of least squares estimators. The least squares methodology is to determine $\mathbf{b}$ for which $\sum_{i=1}^{n} (y_i - \hat{y}_i)^2$ is minimized. This produces $\mathbf{b}$ as the solution to

$$\sum_{i=1}^{n} e_i \mathbf{x}_i = \mathbf{0} \qquad \text{where} \qquad e_i = y_i - \hat{y}_i \tag{7.69}$$

Equation (7.69) provides a good illustration of the influence exerted by data points with large (in magnitude) residuals. Suppose we view a more general form of estimation as the solution to

$$\sum_{i=1}^{n} \psi\left[\frac{e_i}{\sigma}\right]\mathbf{x}_i = \mathbf{0} \tag{7.70}$$

The least squares procedure (Eq. (7.69)) is, then, a special case of the formulation in Eq. (7.70). The function $\psi(\cdot)$ is called an *influence function.* From Eq. (7.69), it is clear that in the least squares case, the influence exerted by the ith data point is *proportional to the residual* e_i. In other words, the influence function is *linear in* e_i, a result of minimization of residual sum of *squares.* A procedure resistant to outliers can be formulated from Eq. (7.70) by choosing the influence function $\psi(\cdot)$ so as to not allow data points with large residuals to exert undue influence.

THE M-ESTIMATOR

The estimates of the coefficients of the linear regression model appear, of course, in the $y_i - \hat{y}_i$ in Eq. (7.69) for the case of least squares estimation and in the more general case of Eq. (7.70). The solution for the regression coefficients in (7.70) is called an *M-estimator.* The influence function $\psi(\cdot)$ is generally chosen in a manner that down weights the influence of data points containing large errors in fit. It should be clear to the reader, at this point, that the least squares estimator with $\psi(e_i) = e_i$ is not robust to outliers. For the purpose of contrasting it with others, we provide a plot of the influence function for least squares in Figure 7.13.

One can readily see that a more reasonable approach might be to use an influence function that is bounded. A popular example is offered in Huber's (1973) influence function, which is intuitively appealing. Huber's function is given by

$$\begin{aligned} \psi(e_i^*) &= e_i^* & |e_i^*| &\leq r \\ &= r & e_i^* &> r \\ &= -r & e_i^* &< -r \end{aligned}$$

FIGURE 7.13 **The influence function for least squares estimation**

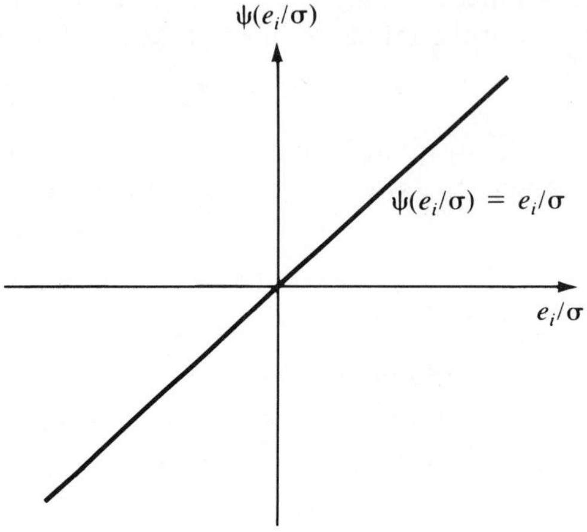

where $e_i^* = e_i/\sigma$. A pictorial display is given in Figure 7.14.

The Huber function does not allow influence produced by any residual to be any greater than that exerted by one at level, say, r. Reasonable values of r might be 1, 1.5, or perhaps 2.0. The implication is that, in the case of, say, $r = 1$, a residual that exceeds σ will exert no more influence than a residual with a value of σ.

The notion of an influence function allows the regression user to develop an estimator with a view toward formally reducing the influence of data points that result in large residuals. In principle, Eq. (7.70), with $\psi(\cdot)$ as described in Figure 7.14, is to be solved for the regression coefficients. The reader must keep in mind that the coefficients are involved in the e_i in $\psi(e_i/\sigma)$. In the following subsection, we discuss the computation of the regression via the solution of Eq. (7.70) for the coefficients.

FIGURE 7.14 **Plot of Huber's influence function**

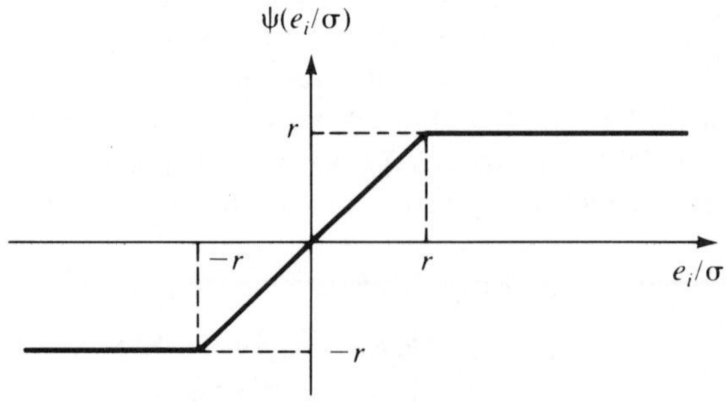

COMPUTATION OF THE M-ESTIMATOR (ITERATIVELY REWEIGHTED LEAST SQUARES)

The solution to the regression coefficients in Eq. (7.70) is the M-estimator, with the influence function $\psi(\cdot)$ chosen in order to down weight data points that appear to be outliers. Huber's influence function represents a reasonable selection of ψ, though there are other choices. The equations that must be solved in (7.70) are nonlinear and thus must be solved iteratively. Also, in practice, σ must be replaced by an estimate, say $\hat{\sigma}$. There are many choices for this estimate. However, one should choose to remain consistent with the spirit of robustness and use a *robust estimate of scale*; a reasonable choice is given by

$$\hat{\sigma} = 1.5 \operatorname{med}|e_i| \qquad (i = 1, 2, \ldots, n) \tag{7.71}$$

where $\operatorname{med}|e_i|$ is the median of the absolute residuals. For details on the estimate of σ, the reader is referred to Welsch (1975).

There are computer programs available for doing M-estimation. It is interesting to note how one can approach the solution iteratively by doing repeated *weighted least squares*, allowing the data to dictate the weights. Equation (7.70) can be written ($\hat{\sigma}$ replacing σ)

$$\sum_{i=1}^{n} \frac{\psi(e_i^*)}{(e_i^*)}(e_i^*)\mathbf{x}_i = \mathbf{0} \tag{7.72}$$

where e_i^* is the ith scaled residual; i.e., $e_i^* = e_i/\hat{\sigma}$. Equation (7.72) is of the form

$$\boxed{\sum_{i=1}^{n} w_i e_i^* \mathbf{x}_i = \mathbf{0}} \tag{7.73}$$

where $w_i = \psi(e_i^*)/(e_i^*)$. Now, Eq. (7.73) is also the solution to the minimization of $\sum_{i=1}^{n} w_i(y_i - \hat{y}_i)^2$, i.e., weighted least squares. Thus weighted regression can be used as a computational device for computing the M-estimator. The weights are dependent on the residuals which, in turn, depend on the coefficients. As a result, an iterative procedure is necessary. The following is a description of the methodology:

1. Obtain a vector of initial estimates $\mathbf{b}_0$, and from them obtain residuals, $e_{i,0}$.
2. From the initial residuals, compute $\hat{\sigma}_0$ and initial weights, $w_{i,0} = \psi(e_{i,0}^*)/(e_{i,0}^*)$.
3. Use weighted least squares to obtain new robust parameter estimates

   $$\mathbf{b}_{R_0} = (\mathbf{X}'\mathbf{W}_0\mathbf{X})^{-1}\mathbf{X}'\mathbf{W}_0\mathbf{y}$$

   where $\mathbf{W}_0$ is a diagonal matrix of weights with ith diagonal element $w_{i,0}$.
4. Let the parameter estimates from step 3 take the role of $\mathbf{b}_0$ in step 1 and obtain new residuals, a new value of $\hat{\sigma}$, and hence new weights.
5. Go back to step 3.

The methodology is to be continued until convergence is reached. The procedure is called *iteratively reweighted least squares* (IRWLS). It depends on a choice of an influence function $\psi(\cdot)$. In the case of Huber's function, one must choose r, the *tuning constant*. Clearly if r is large, say $r = 3$, the robust estimator may behave very much (perhaps too much) like least squares, depending on the distribution of the residuals. A reasonable compromise value for the tuning constant is $r = 1.0$ or 1.5. The method also requires starting values, $\mathbf{b}_0$, of the coefficients and $\hat{\sigma}_0$, which must be computed from starting values. Obviously least squares is the simplest mechanism for producing starting values. However, alternative selection for starting values is to use the coefficients that result from the minimization of the sum of the absolute residuals, i.e., minimization of $\sum_{i=1}^{n} |y_i - \hat{y}_i|$. Computer algorithms are common for this estimation problem. For example, the SAS system contains such a procedure.

The approach to M-estimation has been through emphasis on the influence function, particularly Huber's $\psi(\cdot)$ function. However, it is instructive to note the criterion used; i.e., what is being minimized with, say, Huber's function. The reader should realize by now that the influence function is a result of differentiating the criterion function. In the case of Huber's M-estimator, we choose $\mathbf{b}$ for which

$$\sum_{i=1}^{n} \rho(e_i^*) = \sum_{i=1}^{n} \rho\left(\frac{y_i - \hat{y}_i}{\hat{\sigma}}\right)$$

is minimized, where

$$\rho(e_i^*) = \frac{e_i^{*2}}{2} \qquad |e_i^*| \leq r$$

$$= r|e_i^*| - \frac{1}{2}(r^2) \qquad |e_i^*| > r$$

Thus, it is clear that M-estimation, via Huber's $\psi(\cdot)$ function, involves squaring "small" residuals as in ordinary least squares, but treating large residuals in such a way as to reduce their influence.

Example 7.11 ***BOQ Data***

Consider the data of Table 5.2 in which a regression was constructed, which relates manpower requirements in Bachelor Officers' Quarters with seven workload variables. In Chapter 5, the data were used to illustrate outlier detection with studentized residuals and R-student values. The fit of the regression appeared to be quite poor at several sites. While the coefficient of determination $R^2 = 0.9613$, it is quite likely that an improved regression might be experienced by allowing for a robust estimation procedure, thereby not allowing the suspect sites to exert undue influence on the results. Huber's influence function and the procedure of iteratively reweighted least squares was employed to generate a set of coefficients. A value of

$r = 1.0$ was chosen in computing the M-estimator. We used the value of $\hat{\sigma}$ from Eq. (7.71) and employed the ordinary least squares procedure to generate initial estimates and thus initial residuals. Table 7.11 gives the initial residuals and the weights applied in the initial IRWLS exercise.

The sum of the absolute residuals for the least squares estimation procedure is 7068.3 man-hours per month. The initial robust estimate of the scale parameter is $\hat{\sigma}_0 = 276.484$. Weighted least squares was then employed to generate new coefficients and thus new residuals. The new estimates and new residuals are shown in Table 7.12. The notation b_{j,R_0} denotes *robust estimate.*

The sum of the absolute residuals after the first IRWLS step is 6,531.39. The procedure was continued with new estimates of σ being computed at each step. The results for the ninth and final iteration are given in Table 7.13. Included are the weights that produced the final robust regression coefficients and the final set of residuals.

The sum of the absolute residuals at the final iteration is 6050.12, indicating a better overall quality of fit than that provided by ordinary least squares. Table 7.14 shows the values of the regression coefficients and the sum of the absolute residuals for OLS and at each of the nine iterations. Note the sites that produced *weights* substantially less than 1.0. The influence of sites 11, 15, 20, and 21 was reduced by the robust estimation procedure. By comparing these results with the outlier diagnostic results that appear in Example 5.4 with the same data, you see that these down-weighted sites are, indeed, those that have the largest least squares residuals. Also, in the case of sites 15, 20, and 21, sizable R-student values were experienced in the residual diagnostic analysis in Chapter 5. In the case of site 11, the residual was relatively large. However, site 11 has a HAT diagonal value of only 0.1241, and thus the R-student value is a relatively modest 1.5537.

Contradictions seem to occur for sites 23 and 24. For site 23, for example, the R-student value provided in Chapter 5 is -5.2423. In fact it is the site that is found to be an outlier when the formal outlier test is used with the Bonferroni critical value. Yet, in the M-estimation, this site carries a weight of 1.0. A closer look reveals that the residual is -160.276, which is not at all large in this data set, and the HAT diagonal is 0.9885! The *leverage* associated with this site produced the large R-student value. However, the M-estimation, as described here, weights according to the size of the residuals *without regard to the leverage* associated with the observations. We learned in Chapter 5 that when h_{ii} is close to unity, one expects the residual to be close to zero, even if the data point is an outlying observation. As a result, there is always a danger that a high leverage observation might undeservedly receive full weight because the residual is kept low due to the high leverage. This difficulty with standard M-estimation has generated interest in modifications, which fall into the general class entitled *bounded influence regression.*

The notion behind bounded influence, in the context of robust regression, allows for an adjustment in the residual. The adjustment accounts for the leverage induced

TABLE 7.11 **Initial residuals and weights in robust regression for manpower data of Table 5.2**

Site	$e_i = y_i - \hat{y}_i$	$w_{i,0}$	Site	$e_i = y_i - \hat{y}_i$	$w_{i,0}$
1	−29.755	1.0	14	−9.966	1.0
2	−31.186	1.0	15	−665.211	0.416
3	−196.106	1.0	16	−19.605	1.0
4	−75.556	1.0	17	−470.978	0.587
5	−180.783	1.0	18	418.462	0.661
6	−242.993	1.0	19	437.994	0.631
7	313.923	0.881	20	−870.070	0.318
8	−176.059	1.0	21	826.496	0.334
9	128.744	1.0	22	−323.894	0.854
10	39.994	1.0	23	−160.276	1.0
11	635.923	0.435	24	413.265	0.669
12	184.323	1.0	25	135.029	1.0
13	−81.716	1.0			

TABLE 7.12 **Regression coefficients and new residuals after the first IRWLS step**

Site	$e_i = y_i - \hat{y}_i$
1	−32.154
2	−24.277
3	−189.034
4	−33.933
5	−175.233
6	−194.624
7	347.311
8	−90.447
9	120.428
10	96.554
11	604.966
12	143.234
13	−33.553
14	6.567
15	−836.491
16	44.108
17	−401.617
18	301.215
19	418.094
20	−1033.070
21	849.106
22	−193.537
23	−74.238
24	271.858
25	15.748

Robust Regression Coefficients

$b_{0,\mathrm{Ro}} = 140.322$

$b_{1,\mathrm{Ro}} = -1.46765$

$b_{2,\mathrm{Ro}} = 2.07494$

$b_{3,\mathrm{Ro}} = 0.42421$

$b_{4,\mathrm{Ro}} = -18.7427$

$b_{5,\mathrm{Ro}} = 1.81296$

$b_{6,\mathrm{Ro}} = -13.4412$

$b_{7,\mathrm{Ro}} = 27.9085$

TABLE 7.13 **Results for the final IRWLS step**

Site	$e_i = y_i - \hat{y}_i$	w_i
1	2.354	1.0
2	7.350	1.0
3	−159.623	0.968
4	−14.355	1.0
5	−146.540	1.0
6	−165.955	0.934
7	407.889	0.380
8	−15.205	1.0
9	132.849	1.0
10	133.380	1.0
11	579.303	0.267
12	39.006	1.0
13	−18.165	1.0
14	−18.115	1.0
15	−1179.200	0.132
16	33.709	1.0
17	−368.485	0.421
18	68.143	1.0
19	337.678	0.456
20	−1147.110	0.135
21	827.024	0.187
22	−98.820	1.0
23	−35.440	1.0
24	100.589	1.0
25	−13.844	1.0

Robust Regression Coefficients

$b_{0,\text{Ro}} = 107.735$
$b_{1,\text{Ro}} = -1.73117$
$b_{2,\text{Ro}} = 2.46923$
$b_{3,\text{Ro}} = 0.531307$
$b_{4,\text{Ro}} = -15.7004$
$b_{5,\text{Ro}} = -3.77555$
$b_{6,\text{Ro}} = -14.1068$
$b_{7,\text{Ro}} = 28.2999$

TABLE 7.14
Regression coefficients and sum of absolute residuals for the nine IRWLS steps

Step	$b_{0,\text{Ro}}$	$b_{1,\text{Ro}}$	$b_{2,\text{Ro}}$	$b_{3,\text{Ro}}$	$b_{4,\text{Ro}}$	$b_{5,\text{Ro}}$	$b_{6,\text{Ro}}$	$b_{7,\text{Ro}}$	$\sum_{i=1}^{n} \lvert y_i - \hat{y}_i \rvert$
OLS	134.968	−1.28377	1.80351	0.66915	−21.4226	5.61923	−14.4803	29.3248	7068.30
1	140.322	−1.46765	2.07494	0.42421	−18.7427	1.81296	−13.4412	27.9085	6531.39
2	137.058	−1.56468	2.26161	0.41268	−17.8232	0.00805	−13.4863	27.7528	6335.89
3	128.904	−1.62762	2.36330	0.44254	−16.8489	−1.20189	−13.6902	27.8340	6224.06
4	120.791	−1.66808	2.41187	0.47519	−16.3509	−2.16349	−13.8502	27.9861	6146.24
5	115.921	−1.69041	2.43526	0.49435	−16.0887	−2.72010	−13.9426	28.0878	6103.46
6	112.429	−1.70580	2.44941	0.50636	−15.9189	−3.11633	−14.0043	28.1644	6079.56
7	109.869	−1.71689	2.45904	0.51484	−15.7998	−3.40541	−14.0482	28.2212	6062.74
8	108.213	−1.72516	2.46515	0.52583	−15.7360	−3.61901	−14.0815	28.2655	6055.78
9	107.735	−1.73117	2.46923	0.53131	−15.7004	−3.77555	−14.1068	28.2999	6050.12

by the data point. An example of an effective bounded influence procedure would be to do robust regression as described previously with the residuals replaced by studentized residuals. Another possibility is the use of PRESS residuals or even the DFFITS influence diagnostics discussed in Chapter 6. Clearly, if h_{ii} is near 1.0, the studentized residual

$$r_i = \frac{e_i}{\hat{\sigma}\sqrt{1 - h_{ii}}}$$

will produce a larger value than $e_i^* = e_i/\hat{\sigma}$, and hence it will be assigned a smaller weight than that assigned under standard robust regression described here. As a result, bounded influence regression involves a replacement of $e_i/\hat{\sigma}$ by something that better *uncovers* outlying data. Further discussion of bounded influence regression is found in Krasker and Welsch (1979). In addition, the reader should consider Exercise 7.7 at the end of this chapter.

FURTHER COMMENTS CONCERNING ROBUST REGRESSION

Convergence properties of the M-estimator depend on starting values. In addition, convergence is more precarious in the case of some influence functions than with others. Some influence functions redescend. See Hampel (1974), Andrews (1974), Beaton and Tukey (1974), and Birch (1980).

In many cases, the analyst may very well be satisfied with using the mean absolute residuals as a convergence criterion. Once the robust procedure begins, the mean absolute residuals will decrease (assuming some observations are down weighted). As in Example 6.6, one may monitor $\sum_{i=1}^{n} |e_i|$ as the iterative procedure progresses.

As we indicated earlier, the purpose of robust regression is to provide an alternative to least squares when outliers are present. One may also view robust regression as a method for outlier diagnosis. Clearly any point that receives a weight of less than unity must be considered under suspicion. But the primary function of the methodology is estimation, not diagnostic. The regression coefficients are superior to those of least squares in the case of heavy-tailed error distributions. In Chapter 5 we discussed the diagnosis of outliers at length. However, the diagnosis leaves no formula for what to do with these alleged maverick data. There are certainly many situations in which the detection of outliers turns out to be an instructive and useful exercise. However, the analyst may not be prepared to totally eliminate the outlier (or outliers), even though there is a danger that their influence on the results may be counterproductive. Robust regression has a diagnostic aspect and yet offers coefficient estimates superior in properties to that of least squares in nonideal situations. The reader should not consider this an exhaustive account of robust regression. A serious user should certainly investigate other influence functions as well as further details regarding bounded influence regression. For example, see Krasker and Welsch (1982).

7.8

MEASUREMENT ERRORS IN THE REGRESSOR VARIABLES

In this chapter we have dealt with model violations that encompass the areas of normality, homogeneous variance, and model specification. In this section we discuss with considerably less detail the effect of one additional model violation.

Early in Chapter 2, we emphasized that the stated properties of least squares estimators depended on the assumption that the regressor variables are not random. Later in Chapter 2, we treated the case where y and x are both random with the linear model being accompanied by the assumption of a bivariate normal distribution. However, we stated little or nothing in the fixed x case concerning the impact of measurement error in the regressor variables. We have no intention in this chapter of thoroughly treating the subject. Our purpose is to make the reader aware of why the problem generates difficulties in the least squares procedure. For important treatments the reader is referred to Seber (1977), Davies and Hutton (1975), Berkson (1950), and Mandansky (1959).

Consider the standard multiple regression model

$$y_i = \beta_0 + \beta_1 x_{1i} + \cdots + \beta_k x_{ki} + \varepsilon_i \qquad (i = 1, 2, \ldots, n)$$

In the traditional *fixed* x model, the x_{ji} are not random variables. However, suppose each regressor is observed with error, and as a result, the analyst observes

$$u_{ji} = x_{ji} + \delta_{ji} \tag{7.74}$$

The δ_{ji} are random measurement errors. It would be reasonable to assume that

$$E(\delta_{ji}) = 0$$
$$\text{Var}(\delta_{ji}) = \sigma_j^2$$

and that the δ_{ji} are uncorrelated with each other and with the ε_i. As a result, the regression model is written

$$\begin{aligned} y_i &= \beta_0 + \sum_{j=1}^{k} \beta_j(u_{ji} - \delta_{ji}) + \varepsilon_i \\ &= \beta_0 + \sum_{j=1}^{k} \beta_j u_{ji} + \left(\varepsilon_i - \sum_{j=1}^{k} \beta_j \delta_{ji}\right) \\ &= \beta_0 + \sum_{j=1}^{k} \beta_j u_{ji} + \xi_i \end{aligned} \tag{7.75}$$

Now, Eq. (7.75) represents the model in the observable regressors, namely the u_{ji}. The expected value of the new model error

$$\xi_i = \varepsilon_i - \sum_{j=1}^{k} \beta_j \delta_{ji}$$

is zero. The variance of ξ_i is easily seen to be

$$\text{Var}(\xi_i) = \sigma^2 + \sum_{j=1}^{k} \beta_j^2 \sigma_j^2 \quad \textbf{(7.76)}$$

From (7.76) it becomes clear that the variances of the measurement errors in the regressors are transmitted to the model error, thereby inflating the model error variance. The error variance, namely $\text{Var}(\xi_i)$ is thus made larger than what would be experienced if there were no measurement errors. Obviously the magnitude of the inflation depends in large part on the values of the σ_j^2, individual error variances.

An inflation of model error variance is by no means the only difficulty one must face in the presence of measurement error. The observed u_{ji}, the values that the analyst uses in the regression, are random variables according to Eq. (7.74). An inspection of the ξ_i and a specific u_{ji} reveals that

$$\begin{aligned} \text{Cov}(\xi_i, u_{ji}) &= E(\xi_i)(u_{ji} - x_{ji}) \\ &= E\left(\varepsilon_i - \sum_{r=1}^{k} \beta_r \delta_{ri}\right)(\delta_{ji}) \\ &= -\beta_j \sigma_j^2 \end{aligned} \quad \textbf{(7.77)}$$

So the model error ξ_i is correlated with each regressor variable. Thus, due to the random nature of the regressors and the nonzero covariance between the model error and regressor variables, the Gauss–Markoff Theorem quoted in Chapter 3 no longer applies. In fact, the reader should recognize that the methods applied in Chapter 3 for showing that the least squares estimators are unbiased cannot be used. Indeed, the least squares parameter estimators are biased. As might be expected, the bias incurred in the regression coefficients will depend on the measurement error variances. One may expect that the effect of measurement error on the regressors can be ignored with only minor consequence if the jth measurement error variance, the σ_j^2, is small in comparison to the variation $\sum_{i=1}^{n} (u_{ji} - \bar{u}_j)^2$ encountered in the jth regressor. For details on the bias in the coefficients and rules of thumb regarding the seriousness of the problem, see Davies and Hutton (1975) and Seber (1977).

EXERCISES FOR CHAPTER 7

7.1 Show that $\boldsymbol{\beta}^*$, the generalized least squares estimator given by Eq. (7.4), is unbiased for $\boldsymbol{\beta}$.

7.2 Show that the vector $\boldsymbol{\beta}^*$, which minimizes

$$(\mathbf{y} - \mathbf{X}\boldsymbol{\beta}^*)'\mathbf{V}^{-1}(\mathbf{y} - \mathbf{X}\boldsymbol{\beta}^*)$$

is given by the generalized least squares estimator in Eq. (7.4).

7.3 In order to estimate the angler harvest in terms of the total weight of channel catfish, it becomes necessary to study the length-weight relationship. Length (in millimeters) is easy to measure but weight (in grams) of live fish is difficult to determine. As a result, a length-weight relationship may be used to estimate the weight from the length. Twenty-three fish were captured in the Kanawha River in Charleston, West Virginia. The weights and lengths were measured. The data are given below in log (base 10) units:

Observation	Log of Weight	Log of Length	Observation	Log of Weight	Log of Length
1	1.973	2.338	13	2.405	2.496
2	1.973	2.367	14	2.626	2.558
3	2.064	2.398	15	2.713	2.559
4	2.152	2.401	16	2.665	2.582
5	2.158	2.412	17	2.737	2.585
6	2.326	2.459	18	2.746	2.600
7	2.262	2.465	19	2.825	2.615
8	2.299	2.473	20	2.892	2.631
9	2.362	2.474	21	2.975	2.639
10	2.338	2.484	22	2.940	2.646
11	2.449	2.484	23	3.025	2.670
12	2.384	2.493			

W.E. Ricker, "Compütation and Interpretation of Biological Statistics of Fish Populations," *Bull. Fish. Res. Board Can.* 191 (1975): 382.

We conjecture that the weight of a fish varies with length by the following approximate relationship

$$y \cong ax^b$$

where y is weight and x is length.

(a) Fit a simple linear regression

$$y_i = \beta_0 + \beta_1 x_i + \varepsilon_i$$

(b) Fit a model

$$\log y_i = \gamma_0 + \gamma_1(\log x_i) + \varepsilon_i$$

(c) Compare the two models. Assume interest is in prediction of y in natural units (not log units).

7.4 In order to study the growth characteristics of corn roots in the presence of a particular herbicide, an experiment was conducted in which the herbicide was applied to a particular type of soil (Acredale). Data were collected and percent of control, meaning the percent of the growth observed *without the herbicide*, was used as the response.

Concentration of Herbicide (x) (parts per billion)	Percent of Control (y)
0.5	95.8467
1.0	91.6561
2.0	81.5142
8.0	75.7477
32.0	68.7061
128.0	35.9895

Data analyzed for the Department of Plant Pathology and Physiology by the Statistical Consulting Center, Virginia Polytechnic Institute and State University, Blacksburg, Virginia, 1984.

(a) Fit a model of the type

$$y_i = \beta_0 + \beta_1 x_i + \beta_{11} x_i^2 + \varepsilon_i$$

Compute s^2, the PRESS residuals, the ordinary residuals, PRESS, and the sum of the absolute PRESS residuals.

(b) Fit the model

$$\ln(y_i) = \beta_0 + \beta_1 x_i + \varepsilon_i$$

Compute the same information as in part (a) for the natural response units.

(c) Make a comparison between the models in parts (a) and (b). Which is preferred?

7.5 Consider the data of Exercise 7.3. Fit a model of the type

$$y_i = \beta_0 + \beta_1 x_i^{\alpha} + \varepsilon_i$$

but first use the Box–Tidwell procedure to estimate α. Comment concerning the comparison of this model with the model in Exercise 7.3b. Make whatever computations are necessary for the Box–Tidwell model.

7.6 There are various methods for capturing insects. Some are very elaborate and time consuming while others are quick and require very little work. The following data were used to develop a relationship between a *drop net catch* procedure for catching grasshoppers and a *sweep net* procedure. The latter is quick but not as effective. Other factors influencing the drop net catch are as follows:

Average stage of development of insects, a 1–8 score

Mean plant height

Mean ground cover: proportion of the ground covered by vegetation

Four quadrants were used. The average values per quadrant are as follows:

y (Drop Net Catch)	x_1 (Sweep Net Catch)	x_2 (Stage of Development)	x_3 (Average Plant Height, Inches)	x_4 (Mean Ground Cover, %)
9.0000	49.8571	2.80285	52.705	0.850694
4.4375	24.2857	3.69100	42.069	0.768229
1.0000	1.9091	7.00000	34.766	0.420139
10.0000	27.9375	3.30478	27.622	0.937500
1.1875	3.0625	6.10294	45.879	0.984375
1.3750	6.8750	2.80303	97.472	0.975694
1.6667	8.4167	1.51240	102.062	0.962963
0.5000	1.6250	3.35294	97.790	0.954861
0.6667	1.4545	6.83333	88.265	0.973380
0.8750	1.3125	7.00000	58.737	0.935764
2.0625	6.7500	4.16312	42.386	0.826389
6.4375	29.4375	2.80139	31.274	0.888889
2.6875	5.4375	5.93846	31.750	0.677951
14.0000	80.2500	2.69363	35.401	0.820313
2.3750	10.4375	6.44878	64.516	0.894965
0.8750	1.7500	7.00000	25.241	0.593750
0.0667	0.1875	5.50000	36.354	0.567708

Data analyzed for the Department of Entomology by the Statistical Consulting Center, Virginia Polytechnic Institute and State University, Blacksburg, Virginia, 1982.

(a) Fit a multiple linear regression using the model

$$y_i = \beta_0 + \beta_1 x_{1i} + \beta_2 x_{2i} + \beta_3 x_{3i} + \varepsilon_i$$

Compute ordinary residuals and the sum of the absolute residuals.

(b) Use IRWLS to compute M-estimators for the model in part (a). Use a tuning constant $r = 1.0$. Compute the sum of the absolute residuals.

7.7 Consider the Navy BOQ data of Table 5.2. A robust regression was accomplished in Example 7.11. Do a *bounded influence regression* with the same data. Use studentized residuals rather than ordinary residuals in the IRWLS procedure. Use $r = 1.0$ and $\hat{\sigma}$ as given in Eq. (7.71).

7.8 For the data of Example 7.1, do an *ordinary least squares* fit to the model

$$y_i = \beta_0 + \beta_1 x_{1i} + \beta_2 x_{2i} + \varepsilon_i$$

Plot studentized residuals in this fit against $\hat{y}$. Is the heterogeneous variance problem evident?

7.9 Suppose we are dealing with a regression model of the type

$$\mathbf{y} = \mathbf{X}\boldsymbol{\beta} + \boldsymbol{\varepsilon}$$

where $\mathbf{X} = [\mathbf{X}_1 \,\vdots\, \mathbf{X}_2]$ and $\mathbf{X}_1$ is $n \times r$ and $\mathbf{X}_2$ is $n \times (p - r)$. We wish to test

$$H_0\colon \boldsymbol{\beta}_1 = \mathbf{0}$$

$$H_1\colon \boldsymbol{\beta}_1 \neq \mathbf{0}$$

It is also assumed that $\boldsymbol{\varepsilon} \sim N(\mathbf{0}, \sigma^2\mathbf{I})$ with *known* σ, and thus the *normal* likelihood would be used in arriving at a likelihood ratio test. Write out

$$\lambda(\boldsymbol{\beta}_1 \mid \boldsymbol{\beta}_2) = -2 \ln \frac{L(\hat{\boldsymbol{\beta}}_2)}{L(\hat{\boldsymbol{\beta}})}$$

where, of course, $L(\hat{\boldsymbol{\beta}}_2)$ is the maximized likelihood under the restricted model and $L(\hat{\boldsymbol{\beta}})$ is the maximized likelihood under the full model. How does the statistic $\lambda(\boldsymbol{\beta}_1 \mid \boldsymbol{\beta}_2)$ relate to the difference in the residual sums of squares for the full and reduced models?

7.10 In Section 7.6, which deals with the generalized linear models, the choice of appropriate link functions are discussed. The *canonical link* is given by

$$\mathbf{x}_i'\boldsymbol{\beta} = \theta_i$$

The use of the canonical link results in the identity link for the normal case, the log link for the Poisson case, and the logistic link for the binomial case; that is

$$\mathbf{x}_i'\boldsymbol{\beta} = \mu_i \qquad \text{(Normal case)}$$

$$\mathbf{x}_i'\boldsymbol{\beta} = \ln\left(\frac{\mu_i}{t_i}\right) \qquad \text{(Poisson)}$$

$$\mathbf{x}_i'\boldsymbol{\beta} = \ln\left(\frac{\mu_i}{1-\mu_i}\right) \qquad \text{(Binomial, ungrouped)}$$

Show that this is true, i.e., show that the canonical link for these three cases leads us to these three link functions.

7.11 Consider the Poisson regression discussed in Section 7.5. Suppose that the log link is used, producing the model (say, all $t_i = 1.0$)

$$\mu_i = e^{\mathbf{x}_i'\boldsymbol{\beta}} \qquad (i = 1, 2, \ldots, n)$$

Show that the maximum likelihood procedure for estimation of $\boldsymbol{\beta}$ results in the solution to

$$\mathbf{X}'\mathbf{e} = \mathbf{0}$$

which is given in Eq. (7.66). Since the canonical link is used, $\boldsymbol{\Delta}$ is the identity matrix here. Thus the estimation is given as the solution to $\hat{\boldsymbol{\beta}}$ in

$$\mathbf{X}'(\mathbf{y} - \hat{\mathbf{y}}) = \mathbf{0} \qquad \text{where} \quad \hat{y}_i = e^{\mathbf{x}_i'\hat{\boldsymbol{\beta}}}$$

In other words, we require the solution to $\hat{\boldsymbol{\beta}}$ of

$$\sum (y_i - \hat{y}_i)\mathbf{x}_i = \mathbf{0}$$

(*Comment*: Here you need merely to maximize the log likelihood for this Poisson case and show that this leads you to the above.

7.12 Do what you did in Exercise 7.11 for the binomial case, logistic link. That is, once again, the solution can be written

$$\mathbf{X}'(\mathbf{y} - \hat{\mathbf{y}}) = \mathbf{0}$$

where for the ungrouped case

$$\hat{y}_i = \frac{1}{1 + e^{-\mathbf{x}_i'\hat{\boldsymbol{\beta}}}}$$

7.13 Consider Table 7.10. Show how the canonical link function (and thus the model) is developed for the case of the gamma distribution and geometric distribution. (*Hint*: Put the density function in the form of the exponential family, and find $g(\theta_i)$ and μ_i. Then write $\theta_i = \mathbf{x}_i'\boldsymbol{\beta}$, and thus $\mathbf{x}_i'\boldsymbol{\beta} = s(\mu_i)$.

7.14 Consider the data of Exercise 5.18. Do a robust regression analysis using terms that are linear in the x's. Use IRWLS to compute M-estimators. Use a tuning constant of $r = 1.5$. Is there further diagnostic evidence here that observations 10, 16, and 17 are questionable? Explain.

7.15 In a study in collaboration with the Engineering Sciences and Mechanics department at Virginia Tech, personnel at the Statistics Consulting Center were called upon to analyze a data set dealing with cathodic debonding of elastomery metal bonds. A model relating the amount of debonding as a function of time (x_1), voltage (x_2), pH at time of bonding ($\mathbf{x}_3$), and temperature (x_4) was sought. The following data were collected.

y (cm)	x_1 (minutes)	x_2(volts)	x_3 (pH)	x_4 (°F)
2.240	24.00	1000	8.47	299
3.581	43.00	1000	8.55	299
5.131	84.00	1000	8.53	299
5.715	159.00	1000	8.60	297
0.889	3.79	1100	7.72	300
2.845	17.00	1100	7.83	299
6.147	37.00	1100	4.91	299
7.016	54.20	1100	3.79	299
7.747	58.00	1100	3.69	299
8.286	85.00	1100	3.70	298
9.321	89.00	1100	3.91	297
9.195	95.00	1100	3.57	296
1.895	1.85	1420	4.43	298
3.708	5.88	1420	3.67	298
3.467	7.02	1420	2.89	298
5.049	13.88	1420	2.56	298
4.846	17.19	1420	2.56	298
6.108	19.80	1420	2.84	298
7.137	33.80	1420	3.22	298
8.295	40.80	1420	3.54	298
1.168	0.83	1825	3.73	298
2.865	8.08	1825	3.45	295
3.912	21.08	1825	2.85	298
3.726	21.08	1825	2.85	298
4.521	7.00	1400	8.88	303
5.664	15.00	1400	8.88	303
7.620	27.00	1400	8.88	303
9.766	52.00	1400	8.88	303
12.497	65.00	1400	8.88	303
2.174	1.76	1310	5.46	313
8.153	7.90	1310	2.15	309
10.757	20.90	1310	2.55	312

(a) Compute a standard multiple linear regression equation.
(b) Use a Box–Cox power transformation on the response, and thus compute the appropriate transform y^{λ}.
(c) Does a confidence interval estimate on λ indicate whether or not there is a need to work with a transformed response rather than the natural metric y? Explain.
(d) If transformation is needed, fit a new regression using $y^{\hat{\lambda}}$ as the new response and indicate further evidence of the improved regression.

7.16 Consider the following data set in which nitrogen dioxide concentrations in parts per million are collected for 26 days in September, 1984 at a monitoring concentration in San Francisco Bay area. These data were taken from Chatterjee and Hadi (1988).

x_1 Windspeed (average mph)	x_2 Max. Temp. (°F)	x_3 Insolation (langleys per day)	NO^2
11.1	90	382	6
12.1	86	380	5
12.0	80	372	5
17.8	70	352	3
9.5	90	358	7
7.2	100	362	9
11.5	92	302	6
13.4	74	316	2
10.8	87	339	10
13.8	78	328	7
14.6	73	278	3
12.1	85	339	4
8.0	94	241	13
8.8	91	193	10
12.9	84	268	7
12.7	68	113	3
12.1	81	313	6
11.1	78	317	5
11.3	74	324	4
9.0	78	312	9
9.2	84	349	11
8.4	90	290	8
8.0	90	295	9
13.8	80	283	6
17.8	68	259	2

(a) Fit, with maximum likelihood, a multiple regression assuming gamma errors and identity link.
(b) Fit a model assuming gamma errors and reciprocal link (canonical link).
(c) Fit a model assuming gamma errors and a log link.
(d) What is your final conclusion concerning the appropriate model with our assumption of the error distribution?

7.17 Consider the data of Exercise 7.16. An alternative analysis might be to do standard least squares and use the Box–Cox procedure to estimate an appropriate transformation on the response. Fit a model using this procedure. Comment on the relative merits of the two model-fitting approaches.

REFERENCES FOR CHAPTER 7

1. Andrews, D.F. 1971. A note on the selection of data transformations. *Biometrika* 58: 249–254.
2. Andrews, D.F. 1974. A robust method for multiple linear regression. *Technometrics* 16: 523–531.
3. Andrews, D.F., P.J. Bickel, F.R. Hampel, P.J. Huber, W.H. Rogers, and J.W. Tukey. 1972. *Robust Estimates of Location: Survey and Advances.* Princeton, New Jersey: Princeton University Press.
4. Atkinson, A.C. 1983. Diagnostic regression for shifted power transformations. *Technometrics,* 25: 23–33.
5. Atkinson, A.C. 1985. *Plots, Transformations, and Regression.* Oxford: Clarendon Press.
6. Atkinson, A.C. 1982. Regression diagnostics, transformations and constructed variables. *Journal of the Royal Statistical Society, Series B* 44: 1–36.
7. Baker, R.J., and J.A. Nelder. 1978. *Generalized Linear Interactive Modeling (GLIM).* Release 3. Oxford: Numerical Algorithms Group.
8. Bartlett, M.S. 1947. The use of transformations. *Biometrics* 3: 39–52.
9. Beaton, A.E., and J.W. Tukey. 1974. The fitting of power series, meaning polynomials, illustrated on band spectroscopic data. *Technometrics* 16: 147–185.
10. Belsley, D.A., E. Kuh, and R. Welsch. 1980. *Regression Diagnostics: Identifying Influential Data and Sources of Collinearity.* New York: John Wiley & Sons.
11. Bement, T.R., and J.S. Williams. 1969. Variance of weighted regression estimators when sampling errors are independent and heteroscedastic. *Journal of the American Statistical Association,* 64: 1369–1382.
12. Berenblut, I.I., and G.I. Webb. 1973. A new test for autocorrelation errors in the linear regression model. *Journal of the Royal Statistical Society, Series B* 35: 33–50.
13. Berkson, J. 1950. Are there two regressions? *Journal of the American Statistical Association* 45: 164–180.
14. Bickel, P.J. 1975. One-step Huber estimates in the linear model. *Journal of the American Statistical Association* 70: 428–433.
15. Birch, J.B. 1980. Some convergence properties of iterated reweighted least squares in the location model. *Communications in Statistics* B9(4): 359–369.
16. Box, G.E.P., and D.R. Cox. 1964. An analysis of transformations (with discussion). *Journal of the Royal Statistical Society, Series B* 26: 211–246.
17. Box, G.E.P., and G.M. Jenkins. 1976. *Time Series Analysis: Forecasting and Control.* 2d ed. San Francisco: Holden-Day.
18. Box, G.E.P., and P.W. Tidwell. 1962. Transformation of the independent variables. *Technometrics* 4: 531–550.
19. Carroll, R.J., and D. Ruppert. 1985. Transformations in regression: A robust analysis. *Technometrics* 27: 1–12.

20. Chatterjee, S., and A.S. Hadi. 1988. *Sensitivity Analysis in Linear Regression.* New York: John Wiley & Sons.

21. Cochrane, D., and G.H. Orcutt. 1949. Application of least squares regression to relationships containing autocorrelated error terms. *Journal of the American Statistical Association* 44: 32–61.

22. Cox, D.R. 1970. *The Analysis of Binary Data.* London: Methuen.

23. Davies, R.B., and B. Hutton. 1975. The effects of errors in the independent variables in linear regression. *Biometrika* 62: 383–391.

24. Deaton, M.L., M.R. Reynolds, Jr., and R.H. Myers. 1983. Estimation and hypothesis testing in regression in the presence of nonhomogeneous error variances. *Communications in Statistics* B12(1): 45–66.

25. Draper, N.R., and D.R. Cox. 1969. On distributions and their transformations to normality. *Journal of the Royal Statistical Society, Series B* 31: 472–476.

26. Draper, N.R., and H. Smith. 1981. *Applied Regression Analysis.* 2d ed. New York: John Wiley & Sons.

27. Durbin, J. 1970. An alternative to the bounds test for testing for serial correlation in least squares regression. *Econometrica* 38: 422–429.

28. Durbin, J. 1969. Tests for serial correlation in regression analysis based on the periodogram of least squares residuals. *Biometrika* 56: 1–15.

29. Durbin, J., and G.S. Watson. 1950. Testing for serial correlation in least squares regression, I. *Biometrika* 37: 409–428.

30. Durbin, J., and G.S. Watson. 1951. Testing for serial correlation in least squares regression, II. *Biometrika* 38: 159–178.

31. Durbin, J., and G.S. Watson. 1971. Testing for serial correlation in least squares regression, III. *Biometrika* 58: 1–19.

32. Finney, D.J. 1947. The estimation from individual records of the relationship between dose and quantal response. *Biometrika* 34: 320–334.

33. Hampel, F.R. 1971. A general qualitative definition of robustness. *Annals of Mathematical Statistics* 42: 1887–1896.

34. Hampel, F.R. 1974. The influence curve and its role in robust estimation. *Journal of the American Statistical Association* 69: 383–393.

35. Hampel, F.R. 1978. Optimally bounding the gross-error-sensitivity and the influence of position in factor space. *1978 Proceedings of the Statistical Computing Section*, pp. 59–64. American Statistical Association, Washington, D.C.

36. Hedayat, A., and D.S. Robson. 1970. Independent stepwise residuals for testing homoscedasticity. *Journal of the American Statistical Association* 65: 1573–1581.

37. Hinkley, D.V. 1975. On power transformations to symmetry. *Biometrika* 62: 101–111.

38. Hoaglin, D.C., F. Mosteller, and J.W. Tukey. 1985. *Exploring Data Tables, Trends, and Shapes.* New York: John Wiley & Sons.

39. Hoaglin, D.C., F. Mosteller, and J.W. Tukey. 1983. *Understanding Robust and Exploratory Data Analysis.* New York: John Wiley & Sons.

40. Holland, P.W., and R.E. Welsch. 1977. Robust regression using iteratively reweighted least squares. *Communications in Statistics (Theory and Methods)* 6: 813–828.

41. Huber, P.J. 1977. Robust covariances, in *Statistical Decision Theory and Related Topics*, Vol. 2, edited by S.S. Gupta and D.S. Moore. New York: Academic Press, pp. 165–191.

42. Huber, P.J. 1964. Robust estimation of a location parameter. *Annals of Mathematical Statistics* 35: 73–101.

43. Huber, P.J. 1973. Robust regression: Asymptotics, conjectures, and Monte Carlo. *Annals of Statistics* 1: 799–821.

44. Huber, P.J. 1972. Robust statistics: A review. *Annals of Mathematical Statistics* 43: 1041–1067.

45. John, J.A., and N.R. Draper. 1980. An alternative family of transformations. *Applied Statistics* 29: 190–197.

46. Krasker, W.S. 1980. Estimation in linear regression models with disparate data points. *Econometrica* 48: 1333–1346.

47. Krasker, W.S., and R.E. Welsch. 1982. Efficient bounded-influence regression estimation. *Journal of the American Statistical Association* 77: 595–604.

48. Krasker, W.S., and R.E. Welsch. 1979. Efficient bounded-influence regression estimation using alternative definitions of sensitivity. Technical Report #3, Center for Computational Research in Economics and Management Science, Massachusetts Institute of Technology, Cambridge, Massachusetts.

49. Li, G. 1985. Robust regression, in *Exploring Data Tables, Trends, and Shapes*, edited by D. Hoaglin, F. Mosteller, and J. Tukey. New York: John Wiley & Sons. pp. 281–343.

50. Mandansky, A. 1959. The fitting of straight lines when both variables are subject to error. *Journal of the American Statistical Association* 54: 173–205.

51. McCullagh, P., and J.A. Nelder. 1983. *Generalized Linear Models*. New York: Chapman and Hall.

52. Mosteller, F., and J.W. Tukey. 1977. *Data Analysis and Regression*. Reading, Massachusetts: Addison-Wesley.

53. Nelder, J.A., and R.W.M. Wedderburn. 1972. Generalized linear models. *Journal of the Royal Statistical Society, Series A* 153: 370–384.

54. Pregibon, D. 1981. Logistic regression diagnostics. *Annals of Statistics*, vol. 9, no. 4: 705–724.

55. Rousseeuw, P.J., and A.M. Leroy. 1987. *Robust Regression and Outlier Detection*. New York: John Wiley & Sons.

56. SAS Institute Inc. 1983. *SUGI Supplemental Library Users Guide, 1983 Edition*. Cary, North Carolina: SAS Institute Inc.

57. Seber, G.A.F. 1977. *Linear Regression Analysis*. New York: John Wiley & Sons.

58. Stigler, S.M. 1977. Do robust estimators work with real data? *Annals of Statistics* 5: 1055–1098.

59. Stokes, M.E., and G.G. Koch. 1983. A macro for maximum likelihood fitting of log linear models to Poisson and multinomial counts. *Proceedings of Eighth Annual SAS Users Group International*. Cary, North Carolina: SAS Institute Inc. pp. 795–800.

60. Walker, S.H., and Duncan, D.B. 1967. Estimation of the probability of an event as a function of several independent variables. *Biometrika* 54, 167–179.

61. Walpole, R.E., and R.H. Myers. 1989. *Probability and Statistics for Engineers and Scientists*. 4th ed. New York: MacMillan, Inc.

62. Welsch, R.E. 1975. Confidence regions for robust regression. Paper presented at Statistical Computing Section Proceedings of the American Statistical Association, Washington, D.C.

63. Williams, J.S. 1967. The variance of weighted regression estimators. *Journal of the American Statistical Association* 62: 1290–1301.

CHAPTER 8

DETECTING AND COMBATING MULTICOLLINEARITY

In Chapter 3 we discussed the effects and hazards of building a model when the condition of multicollinearity is present among the regressor variables. Multicollinearity exists when the regressor variables are truly not independent and, rather, display redundant information. Thus any attempt to highlight the individual roles of the variables is badly clouded.

It is imperative that the potential users of least squares regression methodology not only understand the effects of multicollinearity, but learn to diagnose it. Alternative estimation procedures designed to combat collinearity settle into a category called *biased estimation* techniques. These methods represent a deviation from ordinary least squares. An economist who attempts to draw inferences from the sign or magnitude of a specific regression coefficient should be aware that least squares coefficients can be badly estimated in the presence of multicollinearity. An engineer who is interested in developing a linear prediction equation should be warned that even though a model fits his data quite well, multicollinearity may severely prohibit quality prediction.

8.1

MULTICOLLINEARITY DIAGNOSTICS

Much of the development in this chapter will generate quantities that serve as *multicollinearity diagnostics*. They allow one to evaluate the extent of the multicollinearity problem. Here we choose to illustrate these diagnostics with examples and to highlight other procedures that aid in detecting severe multicollinearity. The following represent formal diagnostic tools.

Simple Correlations Among the Regressor Variables

The analyst normally has access to the correlation matrix of the regressor variables, i.e., $\mathbf{X}^{*\prime}\mathbf{X}^{*}$, where the columns of the $\mathbf{X}^{*}$ matrix are centered and scaled as discussed in Section 3.8. These numbers, of course, indicate pairwise type correlations. However, we should clarify that multicollinearity, as the name implies, quite often involves associations among multiple regressor variables. As a result, the simple correlations themselves do not always underscore the extent of the problem. Many analysts constantly search for guideline values on the correlations, values above which one can assert that multicollinearity is severe. There are no definite guideline values on the simple correlations and, while they should be observed so that the analyst can see which *one-on-one* associations exist, they do not always indicate the actual nature or the extent of the multicollinearity.

Variance Inflation Factors

The VIFs represent the inflation that each regression coefficient experiences above ideal, i.e., above what would be experienced if the correlation matrix were an identity matrix. The variance inflation factor of the ith coefficient is defined in Eq. (3.34). It is easy to see that it involves the notion of multiple association. If R_i^2 in (3.34) is near unity, $(\text{VIF})_i$ will be quite large. This will occur if the ith regressor variable has a strong linear association with the remaining regressors. The VIFs represent a considerably more productive approach for detection than do the simple correlation values. They supply the user with an indication of which coefficients are adversely affected and to what extent. Though no rule of thumb on numerical values is foolproof, it is generally believed that if any VIF exceeds 10, there is reason for at least some concern; then one should consider variable deletion or an alternative to least squares estimation to combat the problem.

System of Eigenvalues of $\mathbf{X}'\mathbf{X}$

We noted in Chapter 3 that eigenvalues and eigenvectors of the correlation matrix play an important role in the multicollinearity that exists in a set of regression data. Indeed, the nearness to zero of the smallest eigenvalue is a measure of the strength of a linear dependency, while the elements of the associated normalized eigenvector display the *weights* on the corresponding regressor variables in the multicollinearity (see Section 3.8). Of course, the eigenvalues would all be unity if the variables define an orthogonal system so this provides a norm for the analyst. In addition, the *spectrum* of eigenvalues produces another diagnostic. Multicollinearity can be measured in terms of the ratio of the largest to the smallest eigenvalue, e.g., the quantity

$$\phi = \frac{\lambda_{\max}}{\lambda_{\min}}$$

which is called the *condition number* of the correlation matrix. Large values of ϕ are an indication of serious multicollinearity. An excessively large condition number is evidence that the regression coefficients are unstable (see Section 3.8), i.e., subject to major changes with small perturbations in the regressor data. Numerical rules of thumb regarding the eigenvalues and the spectrum of eigenvalues are also sought by analysts. When the condition number of the correlation matrix exceeds 1,000, one should be concerned about the effect of multicollinearity. As for the individual eigenvalues themselves, the number of eigenvalues near zero indicate the number of collinearities detected among the regressor variables. Of course, so-called reliable rules of thumb are not always reliable. As a result, it is difficult to attach a threshold value—a value below which a small eigenvalue indicates a serious collinearity. In fact, ratios of eigenvalues, i.e., ratios $\phi_j = \lambda_{\max}/\lambda_j$ are more reliable for diagnosing the impact of a dependency than the eigenvalue λ_j itself.

FURTHER COMMENTS CONCERNING DIAGNOSTICS

By now the reader should surmise that diagnosing multicollinearity involves consideration of many items, all of which should be an integral part of a computer printout in a regression analysis. The mathematically unsophisticated analyst may resist the use of the system of eigenvalues for diagnosis, but its usefulness cannot reasonably be denied. The diagnostic tools are designed to gauge the severity of collinearity and, of course, to determine if an alternative to least squares should be attempted. However, the analyst still cannot be certain if a biased estimation technique designed to combat multicollinearity (Section 8.4) will provide an improved estimation or prediction until the technique is attempted. Thus the diagnostics should not necessarily be viewed as indicators of the probable success of alternative estimation, but rather as indicators of the inefficiency of ordinary least squares.

Often the type of scaling used has an influence on the nature of the diagnostics. Recall that, in Chapter 3, we assumed that $\mathbf{X}'\mathbf{X}$ was in correlation form, with the correlations produced by the process of *centering* and *scaling* each regressor variable. The notation $\mathbf{X}^{*\prime}\mathbf{X}^*$ thus refers to the $k \times k$ matrix in which the constant term has been eliminated (see Eq. (3.31)). Alternatively, we can compute diagnostics using the $(k + 1)$ by $(k + 1)$ matrix $\mathbf{X}'\mathbf{X}$ for which each regressor has been scaled but *not centered*. In general, the eigenvalues will not be the same as in the case of centering. Data analysis computer packages vary in this approach when computing multicollinearity diagnostics. Centering each regressor variable will eliminate the capability of involving the constant term in the diagnostics. On the other hand, *scaling* and *centering* allows the objects of the diagnosis to be coefficients of *truly standardized variables*. In any case, it is important that the columns of the **X** matrix be of *unit length*, i.e., the sum of squares of the elements in a column be 1.0. Thus for uncentered regressors, the jth reading of the ith regressor is given by

$$\frac{x_{ij}}{\sqrt{\sum_{j=1}^{n} x_{ij}^2}} \tag{8.1}$$

At times the analyst may be at the mercy of whatever software package is available. However, there are some types of problems for which centering is necessary while, in other cases, diagnostics very well may be misleading if data are centered. In Section 8.3, we attempt to shed more light on this issue. Since the collinearity diagnostic results are dependent on whether or not the data are centered, the reader will better understand the issues involved after these diagnostics are thoroughly discussed and illustrated.

8.2 VARIANCE PROPORTIONS

The diagnostics presented in Section 8.1 are designed to indicate the strength of the linear dependencies and how much the variances of each regression coefficient is inflated above ideal. It should be emphasized that a serious multicollinearity does not deposit its effect on only one regression coefficient. Indeed, the appearance of a small eigenvalue of $\mathbf{X}'\mathbf{X}$ or $\mathbf{X}^{*\prime}\mathbf{X}^*$ implies that *any or all regression coefficients may be adversely affected*. It is often of interest to determine what proportion of the variance of each coefficient is attributed to each dependency. The model intercept is affected and may be evaluated in this type of analysis *as long as the regressor data are not centered*. Suppose we consider the eigenvalue decomposition of $\mathbf{X}'\mathbf{X}$ (scaled but not centered).

$$\mathbf{V}'(\mathbf{X}'\mathbf{X})\mathbf{V} = \begin{bmatrix} \lambda_0 & & & 0 \\ & \lambda_1 & & \\ & & \ddots & \\ 0 & & & \lambda_k \end{bmatrix} \tag{8.2}$$

This allows us to write the variance-covariance matrix $(\mathbf{X}'\mathbf{X})^{-1}$ as

$$(\mathbf{X}'\mathbf{X})^{-1} = [\mathbf{v}_0 \mathbf{v}_1 \cdots \mathbf{v}_k] \begin{bmatrix} 1/\lambda_0 & & & 0 \\ & 1/\lambda_1 & & \\ & & \ddots & \\ 0 & & & 1/\lambda_k \end{bmatrix} \begin{bmatrix} \mathbf{v}_0' \\ \mathbf{v}_1' \\ \vdots \\ \mathbf{v}_k' \end{bmatrix} \tag{8.3}$$

We use the notation $\lambda_0, \lambda_1, \ldots, \lambda_k$ to take into account one additional eigenvalue due to the increase in the dimension of $\mathbf{X}'\mathbf{X}$ when the intercept is involved. The variances of the coefficients (apart from σ^2) appear on the main diagonals of $(\mathbf{X}'\mathbf{X})^{-1}$. If we denote by v_{ij} the ith element in the eigenvector associated with λ_j with the eigenvectors appearing as columns of $\mathbf{V}$, then from Eq. (8.3) we can write

$$c_{ii} = \sum_{r=0}^{k} \frac{v_{ir}^2}{\lambda_r} \tag{8.4}$$

where $c_{ii} = \text{Var}(b_i)/\sigma^2$. From Eq. (8.4), it is easy to illustrate that a small eigenvalue deposits its influence, to some degree, on all variances. To quantify the extent, we can define

$$\boxed{p_{ji} = \frac{v_{ij}^2/\lambda_j}{c_{ii}}} \tag{8.5}$$

and interpret p_{ji} as the proportion of the variance of b_i, which is attributed to (or blamed on) the collinearity characterized by the eigenvalue λ_j. The variance proportions nicely complement the other diagnostics in assessing the effect of the linear dependencies. While numerical examples truly illustrate the usefulness of these variance proportions, the following is a qualitative description:

> A small eigenvalue (serious linear dependency), accompanied by a subset of regressors (at least two) with high variance proportions, represents a dependency involving the regressors in that subset, and the dependency is damaging to the precision of estimation of the coefficients in the subset.

Thus the total diagnosis involves several of the quantities together: the eigenvalue (or ratios) to assess the seriousness of a particular dependency, the variance proportions to signify what variables are involved in the dependency and to what extent, and the VIFs to aid in determining the damage to the individual coefficients.

Example 8.1 **Hospital Data**

Consider the Naval hospital data in Table 3.8. The variance inflation factors, first given in Chapter 3 and repeated here, indicate severe multicollinearity. Table 8.1 gives the eigenvalues and variance decomposition proportions for the hospital data. The variance proportion computations were made for uncentered data. Thus, six eigenvalues are reported.

Two of the eigenvalues represent serious or near serious dependencies. The smallest eigenvalue, 0.00002848, with $\phi_j = 182{,}607.1482$, reflects a dependency that is very damaging to coefficients of regressors x_1 and x_3, and, to a smaller extent, to the coefficient of x_4. Clearly, then, this dependency heavily involves these three regressors. One should recall from Chapter 3 that an extremely high simple correlation exists between x_1 and x_3. The impact of the second smallest eigenvalue (0.008215) is marginal since the eigenvalue ratio

$$\frac{5.2013}{0.0082153} = 633.12 < 1{,}000$$

We can interpret this dependency as one that affects β_5 and the intercept. However, its degree of importance does not match that of the dependency associated with the smallest eigenvalue. One should note that the estimated standard error of the intercept is 1071.36 man-hours, an indication of poor precision for the case of a problem such as this. Of course, it is quite likely that in this and many similar problems estimation of the constant term is not important. The response at the origin in the regressor variables essentially has little meaning.

In summary, then, a single dependency inflicts poor efficiency in the estimation of β_1, β_3, and β_4. A secondary, less important, dependency has an effect on the estimation of β_0 and β_5. Please note that the VIF associated with b_5 is only 4.2798.

TABLE 8.1 **Collinearity diagnostics for the hospital data**

Coefficient	Parameter Estimate	Estimated Standard Error	VIF (Regressors Centered)
b_0	1962.9482	1071.3616	
b_1	−15.8517	97.6530	9597.5708
b_2	0.0559	0.0213	7.9406
b_3	1.5896	3.0921	8933.0865
b_4	−4.2187	7.1766	23.2939
b_5	−394.3141	209.6395	4.2798

Variance proportions (regressors scaled, not centered)

Eigenvalue	ϕ_j	b_0	b_1	b_2	b_3	b_4	b_5
5.201286	1.0000	0.0005	0.0000	0.0023	0.0000	0.0007	0.0004
0.666629	7.8024	0.0138	0.0000	0.0105	0.0000	0.0017	0.0062
0.079094	65.7606	0.0259	0.0001	0.3787	0.0001	0.0121	0.0185
0.044747	116.2382	0.0089	0.0000	0.4635	0.0000	0.2940	0.0174
0.0082153	633.1248	0.8048	0.0004	0.1419	0.0007	0.2537	0.7574
0.00002848	182,607.1482	0.1460	0.9995	0.0031	0.9992	0.4378	0.2001

Condition Number = 182,607.1482

Example 8.2 ***Agriculture Production Data***

A second diagnostic illustration considers a typical agricultural production function involving some measure of production as the dependent variable and conventional regressor variables such as current expenses, land value, labor man-years, and annualized capital input, all on a per farm basis. Unconventional regressor variables often used include expenditures of land grant universities, expenditures for research, extension, and education. The data shown in Table 8.2 cover the years 1949 through 1979 for the State of Virginia. In addition to the conventional and unconventional variables, a rainfall variable (deviation from mean July rainfall) was used. A multiple regression was conducted with the purpose of determining the roles of the regressor variables. Often data of this type are infested with multicollinearity. Table 8.3 gives regression information including collinearity diagnostics. The diagnostics include the correlation matrix, eigenvalues of the correlation matrix, variance inflation factors, and variance decomposition proportions. We have included eigenvalue analysis for both centered and non-centered data. [The magnitudes of the unconventional regressors appear to be a bit unorthodox. They are computed by using a distributed lag structure (see Norton *et al.* (1984)).]

The multicollinearity diagnostics point to serious near linear dependencies. Before we discuss these results, we should indicate the analyst's knowledge about the role of the regressor variables in the production function. The following relate to what values of the coefficients are reasonable.

1. The conventional regressors (x_1, x_2, x_3, x_4) should have positive coefficients. Assuming *constant returns to scale*, the coefficients should sum approximately to 1.0.
2. The coefficients of the unconventional regressors should be smaller in magnitude than the conventional regressors. In addition, they should be negative.

The above represent conditions on the coefficients that allow for proper interpretation by the analyst. One can see that the fit of the model to the data is quite good (adjusted $R^2 = 0.9873$). However, for a problem in which interpretation of coefficients is crucial, the variance inflation factors are intolerably high, apart from the rainfall regressor. There are some linear dependencies (very small eigenvalues) that account for portions of the variances of several of the badly affected coefficients ($b_7, b_8, b_6, b_4, b_1, b_2$). It is interesting that the most severe dependency, characterized by the eigenvalue 0.0002298 (centered results) accounts for 96.81% of the variance of b_8 and 98.06% of the variance of b_7. The condition number is 29,108.71, revealing an unusually large spectrum in the eigenvalues. The OLS coefficients reveal several problems. The conventional regressors sum to a value that is uncomfortably larger than unity. This is likely to be a manifestation of a classical difficulty with OLS in the face of multicollinearity—*coefficients large in magnitude*. The signs of the research and education regressors (b_6, b_8) are positive, and therefore the coefficients are uninterpretable and unacceptable.

Included is a listing of the eigenvalues and variance proportions for noncentered data. It is instructive to see how the results compare to that of the centered data. Note that the eigenvalues are different than their counterparts for centered data. Note also, that in the noncentered case, the constant term is involved in one of the collinearities described by the serious eigenvalues. There are, perhaps, two damaging dependencies that surface in both analyses, according to the ϕ_j and the variance proportions. These two dependencies account for a large portion of the difficulty produced by the collinearity. The most serious dependency appears to heavily involve x_7 and x_8 in both centered and noncentered analyses. The second dependency involves x_4. However, in this situation, the diagnostics for noncentered data can produce what appear to be results that conflict with that of the centered diagnostics. One essential difference in this case centers around the involvement of the constant term. We emphasize that centering allows the estimate of the constant to be orthogonal to the estimates of the remaining coefficients. Whether or not its involvement is important depends on the analyst's interest in the intercept. In this case, coefficients *other than the intercept* are important. As a result, the focus should be on the *centered diagnostics*, which uncover a dependency involving x_1 and x_4. There will be additional discussion of centering and scaling in the next section.

TABLE 8.2 **Production and expenditure data for the State of Virginia**

Year	Production Log (Millions) (y)	Expenses Log (Millions) (x_1)	Capital Inputs Log (Millions) (x_2)	Log Man-Years (x_3)
1949	−5.8519	−6.5307	−8.2794	0.65915
1950	−5.7945	−6.4721	−8.1957	0.64107
1951	−5.7320	−6.3667	−8.1061	0.62840
1952	−5.6402	−6.2395	−7.8821	0.62910
1953	−5.6362	−6.1380	−7.7696	0.60627
1954	−5.4828	−6.0656	−7.7721	0.65957
1955	−5.3826	−5.9361	−7.6836	0.69576
1956	−5.3158	−5.8742	−7.6736	0.58642
1957	−5.4610	−5.8400	−7.6171	0.58618
1958	−5.2908	−5.7843	−7.6212	0.59382
1959	−5.3145	−5.7317	−7.5007	0.56240
1960	−5.2493	−5.7318	−7.4406	0.56839
1961	−5.2068	−5.6800	−7.3996	0.62328
1962	−5.1794	−5.6358	−7.3216	0.59072
1963	−5.2731	−5.6043	−7.2799	0.59035
1964	−5.0358	−5.5352	−7.2802	0.58990
1965	−5.0333	−5.4980	−7.2747	0.53535
1966	−5.1276	−5.4314	−7.1877	0.45323
1967	−4.9538	−5.3642	−7.1005	0.46824
1968	−5.0095	−5.3358	−7.0170	0.40742
1969	−4.9638	−5.3024	−6.9780	0.40449

TABLE 8.2 Continued

TABLE 8.2 Continued

Production and expenditure data for the State of Virginia

Year	Production Log (Millions) (y)	Expenses Log (Millions) (x_1)	Capital Inputs Log (Millions) (x_2)	Log Man-Years (x_3)
1970	−4.9044	−5.2630	−6.9072	0.32970
1971	−4.8976	−5.2330	−6.8526	0.33551
1972	−4.8451	−5.2190	−6.8440	0.35931
1973	−4.8697	−5.2066	−6.7915	0.39695
1974	−4.8540	−5.1529	−6.6067	0.17894
1975	−4.7772	−5.1144	−6.6749	0.19486
1976	−4.7574	−5.0340	−6.6052	0.13047
1977	−4.7081	−4.9250	−6.4522	−0.15286
1978	−4.5974	−4.8955	−6.5129	0.02989
1979	−4.5885	−4.8398	−6.3457	−0.08978

Year	Composite Value of Land, Log (Acres) (x_4)	Rainfall (x_5)	Log Expenditures for Research (x_6)	Expenditures for Extension (x_7)	Expenditures for Education (x_8)
1949	3.48567	1.30	−58.15	1783.68	9111.97
1950	3.48149	1.34	−84.47	1773.38	9060.97
1951	3.52176	−0.66	−113.38	1762.11	9012.99
1952	3.55833	−1.24	−145.01	1750.81	8966.80
1953	3.59757	−2.19	−178.08	1739.37	8923.86
1954	3.64740	−0.77	−211.77	1728.26	8883.57
1955	3.69940	−0.81	−245.44	1718.03	8845.19
1956	3.67984	2.11	−276.37	1707.17	8806.55
1957	3.71032	−2.47	−303.44	1696.00	8766.32
1958	3.72506	0.13	−329.01	1684.82	8724.07
1959	3.73144	1.49	−353.02	1673.70	8681.45
1960	3.74955	−0.18	−376.32	1661.90	8639.80
1961	3.77869	−1.20	−397.08	1650.22	8600.14
1962	3.79946	−0.09	−417.58	1638.72	8564.26
1963	3.82156	−2.16	−437.53	1627.15	8530.96
1964	3.84513	−0.02	−456.54	1615.65	8495.88
1965	3.87281	−0.09	−474.81	1605.25	8459.86
1966	3.91396	−1.11	−492.22	1595.64	8420.98
1967	3.91928	−0.40	−507.54	1586.36	8379.84
1968	3.92470	−0.75	−521.68	1577.43	8337.78
1969	3.93024	1.53	−532.39	1568.52	8294.21
1970	3.95533	0.73	−540.04	1559.27	8251.70
1971	3.98111	−0.56	−547.07	1549.94	8210.25
1972	4.00762	0.41	−552.93	1539.55	8170.04
1973	4.03490	−1.00	−557.53	1527.66	8132.14
1974	4.06300	−1.24	−561.21	1514.84	8092.94
1975	4.09197	1.89	−563.70	1501.19	8046.40
1976	4.12185	−1.55	−565.50	1486.13	7993.13

TABLE 8.2 Continued

Production and expenditure data for the State of Virginia

Year	Composite Value of Land, Log (Acres) (x_4)	Rainfall (x_5)	Log Expenditures for Research (x_6)	Expenditures for Extension (x_7)	Expenditures for Education (x_8)
1977	4.15271	−2.06	−567.24	1471.02	7934.00
1978	4.15155	0.11	−569.16	1456.93	7869.96
1979	4.16641	−0.62	−573.82	1443.32	7802.08

George W. Norton, Joseph D. Coffey, and E. Berrier Frye, "Estimating Returns to Agricultural Research, Extension, and Teaching at the State Level," *Southern Journal of Agricultural Economics* (July 1984): 121–128.

TABLE 8.3

Multicollinearity diagnostics and regression analysis for Virginia agricultural production data

	Regression Coefficients	VIF (Data Centered)	Standard Error	t	Prob > $\|t\|$
b_0	1.8278		4.0010	0.46	0.6523
b_1	0.9155	190.96	0.2144	4.27	0.0003
b_2	0.0378	198.45	0.1932	0.20	0.8467
b_3	0.3804	23.20	0.1525	2.49	0.0206
b_4	0.0118	230.31	0.5375	0.02	0.9827
b_5	0.0219	1.43	0.0070281	3.12	0.0050
b_6	0.0009878	111.60	0.0004615	2.14	0.0436
b_7	−0.0026121	2190.09	0.0033882	−0.77	0.4489
b_8	0.0003327	2082.11	0.0008822	0.38	0.7097

Standard regression analysis

Source	*SS*	df	*MS*	*F*
Regression	3.665990	8	0.458249	291.58
Residual	0.034575	22	0.001572	
Total	3.700565	30		

$R^2 = 0.9907$
Adjusted $R^2 = 0.9873$

Correlation matrix

	x_1	x_2	x_3	x_4	x_5	x_6	x_7	x_8
x_1	1	0.9906	−0.8437	0.9917	−0.0969	−0.9773	−0.9857	−0.9850
x_2	0.9906	1	−0.8817	0.9933	−0.1303	−0.9566	−0.9919	−0.9930
x_3	−0.8437	−0.8817	1	−0.8765	0.0992	0.7356	0.8948	0.8998
x_4	0.9917	0.9933	−0.8765	1	−0.1106	−0.9593	−0.9952	−0.9943
x_5	−0.0969	−0.1303	0.0992	−0.1106	1	0.0724	0.0851	0.0837
x_6	−0.9773	−0.9566	0.7356	−0.9593	0.0724	1	0.9513	0.9469
x_7	−0.9857	−0.9919	0.8948	−0.9952	0.0851	0.9513	1	0.9994
x_8	−0.9850	−0.9930	0.8998	−0.9943	0.0837	0.9469	0.9994	1

TABLE 8.3 Continued

TABLE 8.3 Continued

Multicollinearity diagnostics and regression analysis for Virginia agricultural production data

Eigenvalues of **X′X** *(regressors scaled but not centered) and variance decomposition proportions*

Eigenvalues	ϕ_j	b_0	b_1	b_2	b_3
7.727258	1.0000	0.0000	0.0000	0.0000	0.0001
0.930876	8.3011	0.0000	0.0000	0.0000	0.0001
0.304484	25.3782	0.0000	0.0000	0.0000	0.0142
0.037193	207.7596	0.0000	0.0001	0.0000	0.0905
0.00010694	72,258.5701	0.0018	0.0264	0.0288	0.5890
0.00004999	154,562.4997	0.0000	0.3678	0.0028	0.0287
0.00002843	271,846.0987	0.0009	0.2160	0.6191	0.0001
0.00000244	3,162,166.8164	0.9953	0.2656	0.0001	0.1634
0.00000059	13,117,195.6014	0.0019	0.1241	0.3491	0.1140

Eigenvalues	b_4	b_5	b_6	b_7	b_8
7.727258	0.0000	0.0009	0.0000	0.0000	0.0000
0.930876	0.0000	0.6857	0.0000	0.0000	0.0000
0.304484	0.0000	0.0167	0.0019	0.0000	0.0000
0.037193	0.0000	0.0011	0.0131	0.0000	0.0000
0.00010694	0.0708	0.0362	0.4627	0.0006	0.0001
0.00004999	0.0028	0.0171	0.0343	0.0111	0.0023
0.00002843	0.0081	0.1010	0.0173	0.0055	0.0006
0.00000244	0.7409	0.0523	0.1196	0.0171	0.0132
0.00000059	0.1773	0.0890	0.3510	0.9656	0.9839

Condition Number = 13,117,195.6014

Eigenvalues of **X′X** *(regressors scaled but not centered) and variance decomposition proportions*

Eigenvalues	ϕ_j	b_0	b_1	b_2	b_3
6.689917	1.0000	0.0000	0.0001	0.0001	0.0008
1.000000	6.6899	1.0000	0.0000	0.0000	0.0000
0.990778	6.7522	0.0000	0.0000	0.0000	0.0000
0.286900	23.3179	0.0000	0.0007	0.0000	0.1026
0.013931	480.2112	0.0000	0.0505	0.0000	0.4831
0.0089132	750.5590	0.0000	0.1234	0.1537	0.1478
0.0062952	1,062.6932	0.0000	0.1106	0.3971	0.0113
0.0030343	2,204.7508	0.0000	0.5734	0.1098	0.1144
0.0002298	29,108.7098	0.0000	9.1412	0.3393	0.1399

TABLE 8.3 Continued

Multicollinearity diagnostics and regression analysis for Virginia agricultural production data

Eigenvalues	b_4	b_5	b_6	b_7	b_8
6.689917	0.0001	0.0002	0.0002	0.0000	0.0000
1.000000	0.0000	0.0000	0.0000	0.0000	0.0000
0.990778	0.0000	0.7010	0.0000	0.0000	0.0000
0.286900	0.0001	0.0016	0.0085	0.0000	0.0000
0.013931	0.0314	0.0002	0.1499	0.0060	0.0066
0.0089132	0.0233	0.1050	0.2850	0.0065	0.0015
0.0062952	0.2280	0.0132	0.0123	0.0000	0.0023
0.0030343	0.5120	0.0975	0.1589	0.0069	0.0215
0.0002298	0.2052	0.0814	0.3852	0.9806	0.9681

Condition Number = 29,108.7098

Our use of eigenvalues, eigenvectors, VIFs, variance decomposition proportions, etc. results in a set of tools that are more complex than some analysts are accustomed to assimilating. It is important, however, that the user of analytical tools make maximum use of all diagnostic resources. After experience in diagnosis is gained in a particular field of application, one should become quite comfortable with these devices.

8.3

FURTHER TOPICS CONCERNING MULTICOLLINEARITY

Prediction

Consider the dilemma of the data analyst who is faced with the model-building task for which multicollinearity is a serious problem. An attempt has been made here to communicate the diagnostic information that can be used to determine the extent of deterioration of the least squares estimates in the case of severe multicollinearity. It is quite simple for the analyst to assess the influence of multicollinearity on the coefficients. Certainly the variance inflation factors or an inspection of the eigensystem reveals instability or intolerably large variances of the OLS estimates. However, the diagnostics do not extend easily to revealing the impact on prediction as we learned in Chapters 3 and 4. It is difficult to isolate any ideal single prediction norm and be able to say that prediction was damaged by collinearity. The reason for this is because the quality of the prediction, $\hat{y}(\mathbf{x}_0)$, *depends on where* $\mathbf{x}_0$ *is in the regressor space.* This statement carries particular import in the case of multicollinearity. If the fit of the least squares model is good, there are regions in $\mathbf{x}$ where prediction will be effective. However, the presence of serious multicollinearity will likely produce areas where prediction will be quite poor. Also, extrapolation in the cases of multicollinearity problems should be approached with extreme caution. Thus the diagnostic that might shed some light on the quality of prediction (during the model-building procedure), and thus influence one's judgment about the use of OLS, is the *prediction variance* (or standard error of prediction). Unfortunately, however, the analyst would need to determine and use specific $\mathbf{x}_0$ in the computation of the diagnostic. Recall that

$$\mathbf{x}_0 = \begin{bmatrix} 1 \\ x_{1,0} \\ x_{2,0} \\ \vdots \\ x_{k,0} \end{bmatrix}$$

is a vector that represents both the model and the levels of the k regressors at which prediction or extrapolation might be required. One would then compute, as a diagnostic,

$$\frac{\text{Var } \hat{y}(\mathbf{x}_0)}{\sigma^2} = \mathbf{x}_0'(\mathbf{X}'\mathbf{X})^{-1}\mathbf{x}_0$$

The reader should recall the role of prediction variance in the confidence interval development in Chapter 3 and the C_p statistic in Chapter 4. The initial value of 1.0 in $\mathbf{x}_0$ accounts for the constant term in the model. As far as $(\mathbf{X}'\mathbf{X})^{-1}$ is concerned, the regressor data may be centered and scaled, merely scaled, or neither. *The prediction variance does not change with either a scale or location change in the regressor variables.* Suppose that the regressor data are scaled but not centered, and thus $(\mathbf{X}'\mathbf{X})$ is a $(k+1)$ by $(k+1)$ matrix. Of course, the values $x_{1,0}, x_{2,0}, \ldots, x_{k,0}$ must be similarly scaled.

A variation in this approach is to use

$$s_{\hat{y}(\mathbf{x}_0)} = s\sqrt{\mathbf{x}_0'(\mathbf{X}'\mathbf{X})^{-1}\mathbf{x}_0}$$

with which the analyst may feel somewhat more comfortable since it can be used in a confidence bound interpretation. Some software regression packages allow for ease in the computation of $\mathbf{x}_0'(\mathbf{X}'\mathbf{X})^{-1}\mathbf{x}_0$ for regressor combinations that are not a part of the data set. Certainly, a thorough investigator, who has access to this quantity, can only become more informed by observing such a diagnostic for as many $\mathbf{x}_0$ as are practicable. This diagnostic will not allow for a "Yes" or "No" answer regarding prediction but will isolate areas where prediction is poor.

Intuition suggests that if the point $\mathbf{x}_0$ is consistent or *in line* with the multicollinearity as characterized by the data, prediction will not be severely affected. Thus the *fitted values* $\hat{y}(\mathbf{x}_i)$ $(i = 1, 2, \ldots, n)$ generally will not be poor estimates assuming the quality of fit is good. Similarly, any combinations, $\mathbf{x}_0$, in the data range that reflect similar linear dependencies as those experienced in the data are likely to yield reasonably good prediction.

Figure 3.3 illustrates this situation. The stability of the plane balanced on the picket fence is very poor in the direction that is perpendicular to the picket fence. As a result, we would expect precision of $\hat{y}$ to be best on the picket fence, corresponding, of course, to the $\hat{y}(\mathbf{x}_i)$, the fitted values. On the other hand, if $\mathbf{x}_0$ is a reasonable distance away from the fence in a perpendicular direction, $\hat{y}(\mathbf{x}_0)$ may have a large variance.

As a result, if *population multicollinearity* exists and is well depicted in the data, multicollinearity will not severely affect prediction at $\mathbf{x}_0$ in the data range. But the reader should not derive an undue amount of comfort from this. For in real-life problems, seldom is population multicollinearity well defined. In many cases, the dependencies in the data will reveal characteristics or the "personality" of *only that particular data set*; and if $\mathbf{x}_0$ for a future prediction does not have the same complexion as the data (with the x_0's moving together as they are in the data), prediction at $\mathbf{x}_0$ can be badly damaged. The real-life example that follows in Example 8.3 is a good illustration.

Suppose we have a point $\mathbf{x}_0$ at which one needs to predict. Suppose we denote the diagonal matrix by $\mathbf{\Lambda}$ (see Eq. (8.2)) with the eigenvalues of $\mathbf{X}'\mathbf{X}$ on the main diagonal. Then, we have

$$\frac{\text{Var } \hat{y}(\mathbf{x}_0)}{\sigma^2} = \mathbf{x}_0'\mathbf{V}\mathbf{\Lambda}^{-1}\mathbf{V}'\mathbf{x}_0 = \mathbf{z}_0'\mathbf{\Lambda}^{-1}\mathbf{z}_0 = \sum_{i=0}^{k} \frac{z_{0,i}^2}{\lambda_i}$$

where $\mathbf{z}_0 = \mathbf{V}'\mathbf{x}_0$. Now if multicollinearity is a serious problem, at least one $\lambda_i \cong 0$ resulting in a potentially large variance. However, the reader should recall from Chapter 3 that if $\lambda_i \cong 0$, then $\mathbf{X}\mathbf{v}_i \cong 0$ describes the near linear dependency characterized by λ_i. Suppose $z_{0,i}$ is the corresponding element of $\mathbf{z}_0$. The element $z_{0,i} = \mathbf{x}_0'\mathbf{v}_i$. Thus if the near linear dependency in $\mathbf{X}$ is characteristic of the elements in $\mathbf{x}_0$, then $z_{0,i} \cong 0$, thus countering the large value of $1/\lambda_i$. But, of course, if $\mathbf{x}_0'\mathbf{v}_i$ is not near zero, i.e., not consistent with the *sample multicollinearity*, the prediction variance will be large. This latter condition describes a situation in which $\mathbf{x}_0$ is not on or near the picket fence in Figure 3.3.

Example 8.3 ***Anesthesiology Service Data***

Using a real-life example, we illustrate how the presence of multicollinearity can cause deterioration of prediction in a model. We have learned that the reduction in prediction capability depends on the location $\mathbf{x}_0$ of the regressor levels. If $\mathbf{x}_0$ is a point that is somewhat remote from the "mainstream of multicollinearity" determined by the data on which the regression was based, prediction can be poor.

The data in Table 8.4 gives results of workload and man-hours associated with the anesthesiology service for 12 naval hospitals in the U.S. A regression analysis was conducted using monthly man-hours as the response, with x_1 (number of surgical cases), x_2 (eligible population ÷ 1000), and x_3 (number of operating rooms) as the regressor variables. The purpose of the regression exercise was to develop a prediction model to estimate manpower needs for similar existing or future hospitals. We used the least squares procedure to develop a regression equation in x_1, x_2, and x_3. The result is given by

$$\hat{y} = -176.282 + 1.547x_1 + 2.825x_2 + 124.993x_3$$

with $R^2 = 0.9858$ and $s = 163.740$. Also, please note the HAT diagonal values and the standard errors of prediction in Table 8.4.

Table 8.5 gives the regression diagnostics for these data. Note that the VIFs highlight problems with the coefficients of regressors x_1 and x_2, though the spectrum of eigenvalues indicates a condition number

$$\phi = \frac{3.698}{0.00835} = 442.8742$$

which is not large enough to signify a serious problem. The variance proportions suggest that a large portion of the VIFs associated with b_1 and b_2 are caused by a dependency involving x_1 and x_2.

Following the regression analysis, regressor data on six additional hospitals were obtained. In order to gain some insight into predictive capability of the model, values of Var $\hat{y}(\mathbf{x}_0)/\sigma^2 = \mathbf{x}_0'(\mathbf{X}'\mathbf{X})^{-1}\mathbf{x}_0$ were computed for the levels of workload for the six additional hospitals. Also, values of the standard error of prediction, $s\sqrt{\mathbf{x}_0'(\mathbf{X}'\mathbf{X})^{-1}\mathbf{x}_0}$, were computed and can be compared to the corresponding values for the hospitals in the data set. These results for the "new" hospitals appear in Table 8.6. The six additional hospitals are certainly typical of those for which the regression is to be used. Three, or perhaps four, of these hospitals represent points where the regression is not being extrapolated, and yet the standard errors of prediction are swelled considerably above those for the locations in the data set of twelve. In the case of hospitals 17 and 18, the demand on the regression involves extrapolation of x_1. In addition, however, both of these hospitals represent combinations of x_1 and x_2 that are not on the path defined by the collinearity in the data. As a result, the predictability is quantified by the relatively astronomical values of the standard error of prediction.

One very apparent conclusion comes through. The prediction variance (apart from σ^2) and the standard error of prediction for these hospitals illustrate that multicollinearity *can and does* reduce prediction capabilities of the model. Examples like the one used here are not rare. It provides emphasis for a recommendation made earlier in the text, namely that the standard error of prediction should be used as a guide for evaluating prediction capabilities, particularly when multicollinearity is a problem.

TABLE 8.4 **Hospital data for anesthesiology service**

Hospital	y (Man-hours)	x_1	x_2	x_3	$h_{ii} = \mathbf{x}_i'(\mathbf{X}'\mathbf{X})^{-1}\mathbf{x}_i$	$s_{\hat{y}(\mathbf{x})}$
1	304.37	89.00	25.50	4.00	0.1928	71.891
2	2616.32	513.00	194.30	11.00	0.4165	105.668
3	1139.12	231.00	83.70	4.00	0.2024	73.673
4	285.43	68.00	30.70	2.00	0.2658	84.416
5	1413.77	319.00	129.80	6.00	0.0917	49.585
6	1555.68	276.00	180.80	6.00	0.7478	141.597
7	383.78	82.00	43.40	4.00	0.1878	70.961
8	2174.27	427.00	165.20	10.00	0.3759	100.386
9	845.30	193.00	74.30	4.00	0.1510	63.629
10	1125.28	224.00	60.80	5.00	0.2486	81.635
11	3462.60	729.00	319.20	12.00	0.3530	97.280
12	3682.33	951.00	376.20	12.00	0.7668	143.379

Procedures and Analyses for Staffing Standards Development: Data/Regression Analysis Handbook (San Diego, California: Navy Manpower and Material Analysis Center, 1979).

TABLE 8.5 **Collinearity diagnostics for data of Table 8.4**

Coefficient	VIF
b_1	25.897
b_2	20.678
b_3	7.066

	Variance proportions (data not centered)			
Eigenvalue	b_0	b_1	b_2	b_3
3.698	0.0091	0.0010	0.0013	0.0021
0.2696	0.3958	0.0078	0.0106	0.0000
0.0237	0.5358	0.0111	0.1535	0.8708
0.00835	0.0593	0.9801	0.8347	0.1271

TABLE 8.6 **Standard errors of prediction for six additional hospitals**

Hospital	x_1	x_2	x_3	$\mathbf{x}_0'(\mathbf{X}'\mathbf{X})^{-1}\mathbf{x}_0$	$s_{\hat{y}(\mathbf{x})}$
13	309	40.0	5	1.2492	183.005
14	843	186.4	12	4.0126	327.996
15	670	70.8	6	7.1141	436.733
16	639	165.2	6	2.3649	251.803
17	1434	303.9	12	15.6798	648.373
18	1532	360.4	12	15.6100	646.929

MORE ON CENTERING AND SCALING DATA

The standard multiple regression model is given by

$$y = \beta_0 + \beta_1 x_1 + \beta_2 x_2 + \cdots + \beta_k x_k + \varepsilon \tag{8.6}$$

The process of centering and scaling allows for an alternative formulation

$$y = \beta'_0 + \beta'_1\left[\frac{x_1 - \bar{x}_1}{S_1}\right] + \beta'_2\left[\frac{x_2 - \bar{x}_2}{S_2}\right] + \cdots + \beta'_k\left[\frac{x_k - \bar{x}_k}{S_k}\right] + \varepsilon \tag{8.7}$$

This rewriting or reparameterization, as discussed in Chapter 3, has certain advantages and has long been recognized as a proper procedure. Strictly speaking, the two models are equivalent. The intercept, β'_0, for the centered and scaled model is related to β_0 by

$$\beta'_0 = \beta_0 + \beta'_1\frac{\bar{x}_1}{S_1} + \beta'_2\frac{\bar{x}_2}{S_2} + \cdots + \beta'_k\frac{\bar{x}_k}{S_k} \tag{8.8}$$

Strictly, aside from assessing collinearity, the best known advantage of centering and scaling is to eliminate rounding difficulties when computer storage or precision is low. However, the issues of substance to be discussed here include: (a) What does centering and scaling do to the *classical regression analysis*? (b) What is the effect on the collinearity diagnostics of centering?

Standard Regression Analysis

If a data set is used to fit the centered and scaled model of Eq. (8.7), we can obtain the estimated coefficients in the model of Eq. (8.6). The coefficients of the natural variables of (8.6) are obtained by dividing the ith estimated coefficient in the centered and scaled model by S_i. Thus the constant term β_0 is estimated from the estimate b'_0 by computing

$$b_0 = b'_0 - \frac{b'_1\bar{x}_1}{S_1} - \frac{b'_2\bar{x}_2}{S_2} - \cdots - \frac{b'_k\bar{x}_k}{S_k}$$

where the b'_i are estimates from the scaled and centered model of Eq. (8.7). Furthermore, $b'_0 = \bar{y}$, which is the mean of the y's. Thus, the analyst can always move from one model formulation to another regardless of which model was used for the analysis. But which model is most meaningful as far as reporting results is concerned, vis-à-vis t-test, standard errors of coefficients, etc.? First, any statistic related to the predicted response is the same for models of Eq. (8.6) and (8.7). This includes $\hat{y}$, the residuals, s^2, PRESS, C_p and $s_{\hat{y}}$. Thus if the fit is accomplished *solely for prediction*, the model formulation may be either (8.6) or (8.7). In addition, the t-tests on the regression coefficients (apart from the constant term) will be the same for the two formulations. The coefficient and the standard error of the ith regressor will deviate by a factor S_i, the scale constant. Thus the S_i cancels in the ratio, the t-statistic.

Suppose, however, the analyst's purpose is to *interpret regression coefficients* or to extract information regarding the roles of the individual regressor variables. It becomes clear from the model formulation that the coefficients, or rates of change, will not have the same interpretation for the two model formulations. As a result, the model formulation chosen often depends on the type of real-life situation. We shall consider an example to illustrate this.

Example 8.4 **Designed Experiment**

Suppose a chemist conducts an experiment in which two regressor variables, temperature (x_1) and reaction time (x_2) influence some response (y), say, the yield of an experiment. The researcher initially knows the interesting experimental region. Indeed, through *designing the experiment*, he or she selects the levels in that region. For example, the experiment may be the following:

x_1 (°F)	x_2 (hr)
200	0.5
250	0.5
200	0.75
250	0.75

The researcher has purposefully constructed a *rectangular* experimental region, and it is in this region where the function of the regressor variables must be studied. Now, if interest lies only in the *prediction* in this region, then either linear model, *natural* or *centered and scaled* is appropriate, and indeed, prediction results are the same for both. However, if we wish to draw interpretation from the constants being estimated, the intercept and two slope coefficients, then a selection must be made. In this case the chemist makes a preliminary judgment (based on educated opinion) that a change in 50°F is roughly equivalent in importance to a change in 0.25 hour. As a result, we can expect that the model definition will reflect the change. A meaningful regression coefficient should, then, be based on *standardized* variables, suitably defined so that a 0.25 hour change in reaction time is equivalent to a 50°F change in temperature. Then the centering and scaling results in the following:

Variable 1	Variable 2
−1/2	−1/2
1/2	−1/2
−1/2	1/2
1/2	1/2

or, written in a more standard manner

Variable 1	Variable 2
−1	−1
1	−1
−1	1
1	1

with centering being around $\bar{x}_1$ and $\bar{x}_2$ and the scale factors being 25 for x_1 and 0.125 for x_2. As a result, the coefficients are truly interpretable. The slopes b'_1 and b'_2 are rates of change of yield per *standardized* unit change in temperature and time respectively. These numbers are truly meaningful. Also, there will be no objective that evolves around β_0, the intercept in the natural variables model. No one is interested in the estimated yield at 0°F and 0 hour time. The true origin in the natural variables is uninteresting and of no practical importance. Alternatively, the intercept β'_0 in the centered and scaled model may be of considerable interest. The estimate $b'_0 = \bar{y}$ represents the estimated response in the *center* of the rectangular region, and it is very likely that this origin ((0, 0) in the standardized variables) is very important.

In Example 8.4, the results from a centered and scaled model are more interpretable than those in the natural variables model. However, there are certainly situations in which the coefficients of the natural variables are the only interpretable ones, and the only constant term that makes sense is that defined according to the natural variables, i.e., the value of the response when $x_1 = 0$, $x_2 = 0, \ldots, x_k = 0$, the *natural origin.* In the social sciences, particularly economics, this "structural interpretability" of the coefficients can be important. For example, in the application of Example 5.2 with the BOQ workload data, the increase in manpower *per unit workload* for each of the regressor variables is important. In addition the regressor averages $\bar{x}_1, \bar{x}_2, \ldots, \bar{x}_k$ may not be appropriate as an origin. Likely, there is nothing special or of interest scientifically about these levels. In other words, the consideration of centering around $\bar{x}_1, \bar{x}_2, \ldots, \bar{x}_k$ is derived *from the data, not the model.* If new sites were to be sampled, these averages would change. On the other hand, in Example 8.4, control of the levels allowed for a center that was structurally (or chemically) sensible, and, the scaling made sense due to prior considerations.

Centering and Collinearity Diagnostics

The foregoing, regarding the role of centering and scaling in the standard regression analysis, should be accepted by the reader as fairly rudimentary. However, in some individual practical situations, it will not be crystal clear whether the regression equation that is quoted should be in the form of natural variables of standardized regressors. One must often depend on the knowledge and experience of the subject matter scientist.

An issue, which is not as clear, is the role of standardization in collinearity diagnostics. In particular, there is some controversy regarding whether or not data should be centered prior to diagnosing collinearity. Thus, in what follows, let us assume that data have been scaled so that $\sum_{j=1}^{n} x_{ij}^2 = 1.0$ for $i = 1, 2, \ldots, k$. The two extreme views are "Data should be centered prior to diagnosing collinearity," and "Data should not be centered prior to diagnosing collinearity."

When regressor variables are centered, the resulting constant term, $b_0' = \bar{y}$ is clearly not damaged or affected at all by collinearity. This is nicely illustrated by considering the $\mathbf{X}'\mathbf{X}$ matrix. For centered and scaled data (in line with previous notation)

$$(\mathbf{X}'\mathbf{X}) = \begin{bmatrix} 1 & 0 & \cdots & 0 \\ 0 & & & \\ \vdots & & \mathbf{X}^{*\prime}\mathbf{X}^* & \\ 0 & & & \end{bmatrix}$$

where $\mathbf{X}^{*\prime}\mathbf{X}^*$ is the correlation matrix among the regressor variables. Thus $b_0' = \bar{y}$, which is independent of the regression coefficients of the standardized regressors. On the other hand, if collinearity is severe, the intercept of a noncentered model may be severely affected. This can be illustrated by merely considering that β_0 in Eq. (8.8) is a linear function of $\beta_1', \beta_2', \ldots, \beta_k'$, some of which are affected by the serious dependency or dependencies. If data are centered, the collinearity diagnostics *will not reflect impact on the constant term.* The variance decomposition proportions do not reveal any "damage" of the constant since centering has rendered the constant orthogonal to the other regressors. Thus a single or multiple variable "collinearity with the constant term" is not detectable. The term "collinearity with the constant term," perhaps confusing, actually implies that either some linear combination of the individual regressors is nearly constant or that a single regressor is nearly constant. The diagnostics revealed this situation in both the hospital data and the agricultural production data.

Centering may have a profound effect on the eigenvalues and on their ratios. These ratios are designed, as we indicated earlier, to determine the *sensitivity* of regression results to small changes or perturbations in the regressor data; but a difficulty may arise in interpretation of the ratios $\phi_j = \lambda_{max}/\lambda_j$ when one considers results from a centered data set as opposed to those from the same set uncentered. By centering, the analyst is defining a unit change in the x's differently than when one is not centering. As a result, the notion of "sensitivity to small changes" will be different for the two model formulations.

We have discussed the effect of regression centering on standard regression analysis and on collinearity diagnostics. Though they are treated separately, there are similarities. If there is little need for estimating the intercept (estimated response at zero values of the regressors), and if the regressor averages represent a location that is at or near a natural origin in the specific problem, then centering the regressors is very reasonable. In the case of diagnostics, if there is no need to study the efficiency with which the intercept (in the natural variables) is being estimated or, alternatively, the role of the intercept in the collinearity, then centering is not only permissible but necessary. In this latter case, collinearity of a single regressor with the constant term ceases to be important if the ranges covered by the regressor variable are typical and reasonable for the application in question.

Example 8.5 ***Coal-Cleansing Data***

Consider the coal data that was first presented in Example 5.3. The reader should recall that no collinearity discussion has evolved from that particular example. In fact, an inspection reveals that the data are from what appears to be a "finely tuned" experimental design. Indeed, the experimental array is that of a $\frac{1}{2}$ fraction of 2^3 factorial (see Box, Hunter and Hunter (1978)) with four additional replicates. One advantage in the design is the planned elimination of collinearity. This is easily illustrated when one observes the correlation matrix, the variance inflation factors, eigenvalues of the correlation matrix, and the condition number, all computed with data both centered and scaled. Table 8.7 shows this collinearity information along with collinearity diagnostics for the same data set not centered. Note that when the data are centered, there is no indication of collinearity, with everything appearing to be ideal. On the other hand, when the data are scaled but not centered, the diagnostics appear to reveal a different story. Relatively large variance inflation factors appear for all coefficients. One eigenvalue suggests difficulty with collinearity. Even the $\mathbf{X}'\mathbf{X}$ matrix (not the correlation matrix in this case) seems to paint a rather gloomy picture.

Now, which diagnostics are used in interpreting collinearity for the case of this data set? The rather depressing diagnostics for the noncentered data are strictly from a so-called "collinearity with the constant." This is displayed in the variance proportions. However, the estimate of the intercept in the noncentered model, i.e., the estimated response at (0, 0, 0) in the natural variables is totally unimportant. In fact, *zero* values for percent solids, pH, and flow rate mean nothing. The region of the experiment completely describes the areas for any future prediction. The center, $\bar{x}_1 = 2.0$, $\bar{x}_2 = 7.5$, and $\bar{x}_3 = 1593.33$ is an important centroid of the experimental region. In this case, centering the data, i.e., the act of eliminating the role of the intercept in the diagnostics, is essential. Thus the centered and scaled diagnostics should be employed.

TABLE 8.7 **Collinearity diagnostics for coal data of Example 5.3**

Centered and scaled

Coefficient	VIF
b_1	1.0
b_2	1.0
b_3	1.0

$$\mathbf{X}^{*\prime}\mathbf{X}^* \text{ (Correlation matrix)} = \begin{bmatrix} 1.0 & 0 & 0 \\ 0 & 1.0 & 0 \\ 0 & 0 & 1.0 \end{bmatrix}$$

Eigenvalues = 1.0, 1.0, 1.0
Condition Number = 1.0

Scaled, not centered

Coefficient	VIF
b_0	108.431
b_1	25.0
b_2	38.5
b_3	46.93

$$\mathbf{X}'\mathbf{X} = \begin{bmatrix} 1 & 0.9798 & 0.9869 & 0.9893 \\ & 1 & 0.9670 & 0.9693 \\ & & 1 & 0.9764 \\ & & & 1 \end{bmatrix}$$

	Variance proportions			
Eigenvalue	b_0	b_1	b_2	b_3
3.9344	0.0006	0.0025	0.0016	0.0014
0.03526	0.0035	0.8096	0.1392	0.0508
0.02360	0.0019	0.0166	0.5213	0.4620
0.006763	0.9940	0.1713	0.3379	0.4859

Condition Number = 581.77

8.4

ALTERNATIVES TO LEAST SQUARES IN CASES OF MULTICOLLINEARITY

There are many estimation procedures designed to combat multicollinearity, procedures that were developed to eliminate model instability and to reduce the variances of the regression coefficients. There is a certain amount of controversy surrounding their use, and thus the data analyst should not feel that he or she possesses a *carte blanche* to use the techniques at any time. Still, they can be a valuable part of the data analyst's repertoire, particularly when the subject matter field involves structures dealing with scientific variables that *inherently overlap in their influence* on a dependent response.

At the point in which the analyst has determined, by the use of the diagnostics, that multicollinearity is a problem, often a substantial benefit may be derived from an attempt to eliminate much of the multicollinearity *without* resorting to alternatives to least squares. The very presence of multicollinearity in the diagnostics suggests that, in the case of k regressor variables, the actual model-building exercise should involve fewer than k variables. In other words, there is not sufficient information in the regressor data to warrant modeling k regressors. As a result, the analyst often can eliminate or certainly reduce the effect of multi-collinearity by removing one or more regressors. Either the diagnostics or one's own knowledge of the phenomena being modeled might suggest which variables would be candidates for elimination. Clearly, of course, there may be a trade-off here. In a multiple regression, if x_1 and x_2 are highly correlated and x_2 is dropped from the model, some (perhaps nearly all) multicollinearity is eliminated. However, the analyst must be concerned about whether or not the quality of fit of the model has been severely compromised. The improvement in predictive capabilities of the model by variable elimination should be reflected in statistics such as PRESS, the sum of the absolute PRESS residuals, and, of course, the standard errors of prediction. In addition, the analyst should focus on the variance inflation factors for the remaining coefficients.

An alternative for reducing multicollinearity, but yet remaining with standard least squares estimation, is to try transformations on the regressor variables; these transformations often reduce the dimensionality of the regressor system while retaining some of the informational content of all regressors. For example, in the case of two regressors x_1 and x_2, which are highly correlated, redefining a variable $x_1 + x_2$, or perhaps forming ratios, might produce an effective result. Be cautious, however, in forming functions of regressor variables that do not make sense in the context of the problem. For example, the analyst should be extremely reluctant to add regressor variables that are measured in different units.

Example 8.6 **Hospital Data**

Recall the hospital data in Table 3.8. The data were used for purposes of illustration in both Chapters 3 and 4. In addition, the diagnostics shown in Table 8.1 revealed serious collinearity. This collinearity seemed to seriously hamper our model selection process in Chapter 4. Negative coefficients reside in the full model for three of the regressors, and two coefficients (b_1 and b_3) have VIFs of approximately 9,000.

It would be of interest to determine if a subset of the regressors can be chosen that enjoys little or only mild collinearity while maintaining good predictability. Several models were investigated in Example 4.5. Here we observe the variance inflation factors to determine if, indeed, the variable deletion that produced the competing prediction models also substantially reduced collinearity. The following is a synopsis of collinearity for several of the subset models that appear to have good prediction properties. We include the full model again for basis of comparison.

x_1, x_2, x_3, x_4, x_5

Coefficient	VIF	
b_1	9597.5708	
b_2	7.9406	Regression $\hat{y} = 1962.948 - 15.8517x_1 + 0.0559x_2$
b_3	8933.0865	$+ 1.5896x_3 - 4.2187x_4 - 394.314x_5$
b_4	23.2939	$R^2 = 0.9908$
b_5	4.2798	

x_1, x_5

Coefficient	VIF	
b_1	1.8199	Regression $\hat{y} = 2537.413 + 37.5336x_1 - 530.116x_5$
b_5	1.8199	$R^2 = 0.9840$

x_3, x_5

Coefficient	VIF	
b_3	1.8195	Regression $\hat{y} = 2585.520 + 1.2324x_3 - 530.933x_5$
b_5	1.8195	$R^2 = 0.9848$

x_1, x_2

Coefficient	VIF	
b_1	5.6605	Regression $\hat{y} = -96.9009 + 25.0211x_1 - 0.0752x_2$
b_2	5.6605	$R^2 = 0.9861$

x_2, x_3

Coefficient	VIF	
b_2	5.6471	Regression $\hat{y} = -68.3140 + 0.0749x_2 + 0.8229x_3$
b_3	5.6471	$R^2 = 0.9867$

x_1, x_3, x_5

Coefficient	VIF	
b_1	5207.441	Regression $\hat{y} = 2630.478 - 41.0544x_1 + 2.5795x_3 - 529.777x_5$
b_3	5206.324	$R^2 = 0.9850$
b_5	1.8199	

x_2, x_3, x_5

Coefficient	VIF	
b_2	7.7373	Regression $\hat{y} = 1523.389 + 0.0530x_2 + 0.9785x_3 - 320.951x_5$
b_3	11.2693	$R^2 = 0.9901$
b_5	2.4929	

x_1, x_2, x_5

Coefficient	VIF	
b_1	11.3214	Regression $\hat{y} = 1475.024 + 29.7316x_1 + 0.0534x_2 - 318.140x_5$
b_2	7.7714	$R^2 = 0.9894$
b_5	2.4985	

Model (x_1, x_3, x_5) is clearly unacceptable from a collinearity standpoint. This is not surprising due to the strong correlation between x_1 and x_3. The other models involving three regressors have collinearity that might be considered marginal while all of the two-variable models display only very mild collinearity. Apart from (x_1, x_3, x_5), all candidates studied here remain candidates because collinearity has been substantially reduced.

Note that variable x_5 always contains a negative regression coefficient. Thus, for interpretability of coefficients *and*, in addition, little or no collinearity, models (x_1, x_2) and (x_2, x_3) are strong competitors. In addition, Table 4.11 reveals that these two models perform well with regard to other performance criteria.

RIDGE REGRESSION

Ridge regression is one of the more popular, albeit controversial, estimation procedures for combating multicollinearity. The procedures discussed in this and subsequent sections fall into the category of *biased estimation techniques*. They are based on this notion: though ordinary least squares gives unbiased estimates and indeed enjoy the minimum variance of all linear unbiased estimators, there is no upper bound on the variance of the estimators and the presence of multicollinearity may produce large variances. As a result, one can visualize that, under the condition of multicollinearity, a huge price is paid for the unbiasedness property that one achieves by using ordinary least squares. Biased estimation is used to attain a substantial reduction in variance with an accompanied increase in stability of the regression coefficients. The coefficients become biased and, simply put, if one is successful, the reduction in variance is of greater magnitude than the bias induced in the estimators.

There are several ways of motivating ridge regression. Perhaps the most appealing approach is to visualize what technique offers a reduction or a "dampening" in the effect of the eigenvalues of $\mathbf{X}^{*\prime}\mathbf{X}^*$ (or $\mathbf{X}'\mathbf{X}$) on the regression results. We know from the development in Chapter 3 and in early sections of this chapter that the small eigenvalues produce the large variances. Suppose, for example, that we have a problem in which the correlation matrix is given by

$$(\mathbf{X}^{*\prime}\mathbf{X}^*) = \begin{bmatrix} 1.0 & 0.999 \\ 0.999 & 1.0 \end{bmatrix}$$

in a two-variable system. The eigenvalues are given by

$$\lambda_1 = 1.999 \qquad \lambda_2 = 0.001$$

It is clear that this produces a very inefficient estimation of the regression coefficients with the VIFs, i.e., the diagonals of the inverse $\mathbf{X}^{*\prime}\mathbf{X}^*$ being

$$(\text{VIF})_1 = 500.25$$
$$(\text{VIF})_2 = 500.25$$

Now, in this 2×2 case, we see that the eigenvalue λ_2 is small and the variance inflation factors are large because the correlation matrix does not have properties even resembling the desirable properties that $(\mathbf{X}^{*\prime}\mathbf{X}^*)$ enjoys in the orthogonal case, i.e., when the correlation matrix is given by

$$(\mathbf{X}^{*\prime}\mathbf{X}^*) = \begin{bmatrix} 1 & 0 \\ 0 & 1 \end{bmatrix}$$

In other words, for the case of collinearity, the diagonals *do not dominate* as in the case above. This nondominance of the diagonals, in general, causes at least one eigenvalue to be small. So, what can we do to $(\mathbf{X}^{*\prime}\mathbf{X}^*)$ to make it behave more like the orthogonal case? What can be done to increase the eigenvalues, decrease the determinant of the matrix, and hence decrease the elements of the inverse?

Suppose we consider replacing the matrix $(\mathbf{X}^{*\prime}\mathbf{X}^*)$ by the matrix $(\mathbf{X}^{*\prime}\mathbf{X}^* + k\mathbf{I})$, where k is a small positive quantity. To get an indication of what this accomplishes, suppose, for example, we use $k = 0$ in the previous illustration. Then we would replace the correlation matrix by the matrix

$$(\mathbf{X}^{*\prime}\mathbf{X}^* + k\mathbf{I}) = \begin{bmatrix} 1.1 & 0.999 \\ 0.999 & 1.1 \end{bmatrix}$$

Now, what does this do to the eigenvalues and the inverse elements? Since the $\mathbf{V}$

matrix diagonalizes $(\mathbf{X}^{*\prime}\mathbf{X}^*)$, it also diagonalizes $(\mathbf{X}^{*\prime}\mathbf{X}^* + k\mathbf{I})$. Thus

$$\mathbf{V}'(\mathbf{X}^{*\prime}\mathbf{X}^* + k\mathbf{I})\mathbf{V} = \begin{bmatrix} \lambda_1 + k & 0 & \cdots\cdots & 0 \\ 0 & \lambda_2 + k & 0 \quad \cdots & 0 \\ \vdots & 0 & \ddots & \vdots \\ \vdots & \vdots & & \vdots \\ 0 & 0 & \cdots\cdots & \lambda_k + k \end{bmatrix}$$

Thus the eigenvalues of the new matrix $(\mathbf{X}^{*\prime}\mathbf{X}^* + k\mathbf{I})$ are $\lambda_i + k$ for $i = 1, 2, \ldots, k$. Adding k to the main diagonal effectively replaces λ_i by $\lambda_i + k$. Thus, in the previous illustration, the eigenvalues are replaced by 2.099 and 0.101. In addition, the inverse matrix is given by

$$\begin{bmatrix} 1.1 & 0.999 \\ 0.999 & 1.1 \end{bmatrix}^{-1} = \begin{bmatrix} 5.1887 & -4.71229 \\ -4.71229 & 5.1887 \end{bmatrix}$$

As a result, the damaging effect of $\lambda_2 = 0.001$ is countered, and the difficulties created by large elements on the inverse (including large variance inflation factors) are alleviated.

The reader should not consider the foregoing as representing total justification for ridge regression. Rather, consider it as a simple *eye opener* regarding what major changes occur when one merely adds a small positive constant to the main diagonal of a poorly conditioned correlation matrix. In the following pages, we link this development to the estimators, which we commonly refer to as the *ridge regression* estimators.

The ridge regression estimator of the coefficient $\boldsymbol{\beta}$ is found by solving for $\mathbf{b}_R$ in the system of equations

$$\boxed{(\mathbf{X}'\mathbf{X} + k\mathbf{I})\mathbf{b}_R = \mathbf{X}'\mathbf{y}} \tag{8.9}$$

where $k \geqslant 0$ is often referred to as a *shrinkage parameter*. The solution, of course, is given by

$$\mathbf{b}_R = (\mathbf{X}'\mathbf{X} + k\mathbf{I})^{-1}\mathbf{X}'\mathbf{y} \tag{8.10}$$

There are various procedures for choosing the shrinkage parameter k. One must visualize $(\mathbf{X}'\mathbf{X})$ as being the result of either scaling or centering and scaling the regressor variables. In the case of centering, $\mathbf{X}'\mathbf{X}$ becomes $\mathbf{X}^{*\prime}\mathbf{X}^*$, the correlation matrix. A fairly simple study of the properties of the ridge estimator in (8.10) reveals the role of k in moderating the variance of the estimators. Perhaps the most dramatic illustration of the impact of small eigenvalues on the variances of the least squares coefficients is the expression for $\sum_i(\text{Var } b_i)/\sigma^2$ given in Eq. (3.37). In the case of the ridge regression estimator, the equivalent property is given by (see Exercise 8.5)

$$\sum_i \frac{\text{Var } b_{i,R}}{\sigma^2} = \sum_i \frac{\lambda_i}{(\lambda_i + k)^2} \tag{8.11}$$

For example, in the case of three regressor variables with $\lambda_1 = 2.985$, $\lambda_2 = 0.01$, and $\lambda_3 = 0.005$, least squares estimation gives

$$\frac{\sum_{i=1}^{3} \text{Var } b_i}{\sigma^2} = \sum_{i=1}^{3} \frac{1}{\lambda_i}$$
$$= .3350 + 100 + 200$$
$$= 300.3350$$

If ridge regression with say $k = 0.10$ is used, the sum of the variances is given by

$$\sum_{i=1}^{3} \frac{\lambda_i}{(\lambda_i + k)^2} \cong 2.3$$

It is clear that when multicollinearity is severe, i.e., when there is at least one near zero eigenvalue, much improvement in variance, and thus coefficient stability, can be experienced. Equation (8.11) emphasizes a point made earlier in this section, namely, that the k in ridge regression moderates the damaging impact of the small eigenvalues that result from the collinearity.

The bias that results for a selection of $k > 0$ is best quantified by observing an expression for $\sum_{i=1}^{k} (\text{Bias } b_{i,R})^2 = \sum_{i=1}^{k} [E(b_{i,R}) - \beta_i]^2$, the sum of the squared biases of the regression coefficients. This expression is given by (see Hoerl and Kennard (1970(a))

$$\sum_{i=1}^{k} [E(b_{i,R}) - \beta_i]^2 = k^2 \boldsymbol{\beta}'[\mathbf{X}'\mathbf{X} + k\mathbf{I}]^{-2}\boldsymbol{\beta} \tag{8.12}$$

Thus we can expect that the procedure of ridge regression would be successful if a k is chosen so that the variance reduction is greater than the bias term given in Eq. (8.12). There is no assurance that this can be done because the analyst will never know what the bias is. A close study of the variance contribution given in (8.11) suggests that the analyst has a better chance for improvement, in a sense of variance reduction, when collinearity is severe, i.e., when at least one $\lambda_i \cong 0$. Equation (8.12) involves unknown coefficients in $\boldsymbol{\beta}$; thus any attempt at calculating the bias can be misleading. The choice of k belongs to the analyst, of course, and a parameter value should be chosen where results show strong evidence that improvements in the estimates are being experienced. These improvements often take the form of evidence that the estimates are more stable or that prediction is improved. The reader should be aware that there are many procedures in the literature for choosing k. In this text we point out but a few and provide illustrations. The user of ridge regression must decide if the technique has sufficient merit to use in his or her field.

Notice that, in the foregoing, we have chosen to use $\sum_{i=1}^{k}$ in the summation notation as opposed to $\sum_{i=1}^{p}$ or $\sum_{i=0}^{k}$. No inconsistency in notation is intended here. We are still assuming k regressor variables with $p = k + 1$, when the constant term, β_0, is part of the parameter set. Strictly speaking, which summation applies depends on whether one centers and scales, or merely scales. If centering is accomplished, the resulting term, β'_0, in Eq. (8.7) is not affected by collinearity, and indeed, its least squares estimate is merely $\bar{y} = \sum_{i=1}^{n} y_i/n$. So, the assessment of the damage of collinearity and, indeed, the moderating influence of ridge regression are best illustrated through an expression (given in Eq. (8.11)) for the sum of the variances of the coefficients. Thus when ridge regression is performed on centered data, the matrix involved in (8.9) and (8.10) is $(\mathbf{X}^{*\prime}\mathbf{X}^{*} + k\mathbf{I})$ and is merely the $k \times k$ correlation matrix with the shrinkage parameter k added to the main diagonal. The resulting ridge regression estimators (which we must call the $b'_{i,R}$ in the notation of Eq. (8.7)) are coefficients of centered and scaled variables, and the estimate of β'_0 is $\bar{y}$. We then compute the coefficients of the natural variables (if necessary) in model (8.6) as follows:

$$b_{i,R} = \frac{b'_{i,R}}{S_i} \qquad (i = 1, 2, \ldots, k)$$

The constant term of model (8.6) is estimated by (see (8.8))

$$b_{0,R} = b'_{0,R} - \frac{b'_{1,R}\bar{x}_1}{S_1} - \frac{b'_{2,R}\bar{x}_2}{S_2} - \cdots - \frac{b'_{k,R}\bar{x}_k}{S_k}$$

Choice of k

The *ridge trace* is a very pragmatic procedure for choosing the shrinkage parameter (see Hoerl and Kennard (1970(b)). The analyst simply allows k to increase until *stability* is indicated in all coefficients. Quite often a plot of the coefficients against k pictorially displays the trace and helps the analyst make a decision regarding the appropriate value of k. It should be emphasized that stability does not imply that the regression coefficients have converged. For values of k close to zero, multicollinearity will cause rapid changes in the coefficients. These quick changes occur in an interval of k in which one expects coefficient variances to be inflated. As k grows, variances reduce, and coefficients become more stable. A value for k is chosen at the point for which the coefficients are no longer changing rapidly. Quite often the analyst's use of the ridge trace procedure involves viewing plots of coefficients of standardized, i.e., centered and scaled, regressors. However, at times the analyst has a better notion of stability or interpretability of coefficients by observing the coefficients of the natural variables. Such is the case in Example 8.7.

Example 8.7 ***Agriculture Production Data***

Consider Example 8.2 in which a production function was fit to the data of Table 8.2. The regression coefficients using OLS produced results that were difficult to accept and uninterpretable by the analyst. A series of ridge regression coefficients were computed. Table 8.8 contains the results. The numbers reveal coefficients of the natural variables. The resulting SS_{Res} is shown for each k. Table 8.8

clearly illustrates the instability of the OLS coefficients. Notice how rapidly several of the coefficients change as k moves only slightly away from zero. Coefficient estimation, rather than prediction, is important here. The reader should recall that the coefficients of x_1, x_2, x_3, and x_4 should sum approximately to 1.0. For fairly small values of k, the coefficients of the conventional variables (in particular, variables x_1 and x_3) have reduced in magnitude to the point where they are interpretable. Also, the coefficients of the unconventional regressors all become negative for a fairly small value of k. As a result, they too become interpretable. Note, of course, that SS_{Res} increases with an increasing k. Stability and reasonableness are criteria for choice of k in this case. Instability, as depicted in this example, will appear with considerable regularity when multicollinearity is severe. The value of k chosen by the researcher in this case was 0.02. At this point all coefficients have "settled down," a condition manifesting a leveling in the variances of the coefficients.

TABLE 8.8 **Values of ridge regression for agricultural production data**

k	$b_{0,R}$	$b_{1,R}$	$b_{2,R}$	$b_{3,R}$	$b_{4,R}$	$b_{5,R}$
0.00	1.8278	0.915484	0.037779	0.380389	0.011813	0.0219163
0.01	−2.3852	0.379918	0.091081	0.157677	0.389306	0.0213710
0.02*	−2.6932	0.285443	0.100162	0.108384	0.360036	0.0207249
0.03	−2.7891	0.243342	0.101967	0.081723	0.339508	0.0202563
0.04	−2.8336	0.219170	0.102275	0.063643	0.325695	0.0198859
0.05	−2.8590	0.203322	0.102146	0.049895	0.315827	0.0195708
0.06	−2.8753	0.192031	0.101871	0.038717	0.308389	0.0192900
0.07	−2.8867	0.183514	0.101546	0.029237	0.302540	0.0190322
0.08	−2.8953	0.176816	0.101210	0.020968	0.297785	0.0187908
0.09	−2.9020	0.171378	0.100875	0.013612	0.293816	0.0185619
0.10	−2.9076	0.166851	0.100550	0.006974	0.290429	0.0183428
0.11	−2.9123	0.163006	0.100235	0.000921	0.287489	0.0181317

k	$b_{6,R}$	$b_{7,R}$	$b_{8,R}$	SS_{Res}
0.00	0.00098783	−0.0026121	0.00033270	0.0015716
0.01	0.00000748	−0.0003424	−0.00011648	0.0021074
0.02*	−0.00017543	−0.0003826	−0.00011976	0.0023646
0.03	−0.00025674	−0.0004106	−0.00012245	0.0025090
0.04	−0.00030137	−0.0004289	−0.00012428	0.0026059
0.05	−0.00032864	−0.0004414	−0.00012556	0.0026792
0.06	−0.00034638	−0.0004503	−0.00012649	0.0027395
0.07	−0.00035836	−0.0004570	−0.00012718	0.0027921
0.08	−0.00036662	−0.0004621	−0.00012771	0.0028400
0.09	−0.00037238	−0.0004661	−0.00012811	0.0028848
0.10	−0.00037637	−0.0004692	−0.00012843	0.0029276
0.11	−0.00037909	−0.0004718	−0.00012867	0.0029690

Perhaps the most positive aspect of the ridge trace, or the consideration of stability, is the fact that it is practical and data-dependent. It is not based on conceptual criteria, but rather on the analyst's own notions concerning what stability actually means. Of course, the latter may be interpreted as a non-virtue of the procedure simply because what is regarded as stability may be somewhat subjective. The choice of k may be somewhat more arbitrary than would be the case where the parameter is determined by a more exact recipe.

Use of Prediction Criteria for Choice of k

Many of the procedures for choice of k are designed to select values that are in the interval on k for which there is an apparent improvement in the estimation of the coefficients. As a result, it is often more important to base the selection of k on a criterion that is more directly descriptive of prediction performance. Several criteria have been proposed and will be discussed and illustrated here. We shall initially consider a C_p-like statistic that is based on the same variance-bias type trade-off, discussed in Chapter 4. The statistic, very much analogous to that given in Eq. (4.20), is

$$C_k = \frac{SS_{\text{Res},k}}{\hat{\sigma}^2} - n + 2 + 2\,\text{tr}[\mathbf{H}_k] \qquad \textbf{(8.13)}$$

where $\mathbf{H}_k = [\mathbf{X}^*(\mathbf{X}^{*\prime}\mathbf{X}^* + k\mathbf{I})^{-1}\mathbf{X}^{*\prime}]$, $SS_{\text{Res},k}$ is the residual sum of squares using ridge regression, and $\text{tr}[\mathbf{H}_k]$ is the trace of $\mathbf{H}_k$.

In Eq. (8.13), $\mathbf{X}^*$ and $\mathbf{X}^{*\prime}\mathbf{X}^*$ do not reflect the constant term; i.e., if there are k regressor variables, $\mathbf{X}^*$ is an $n \times k$, and $\mathbf{X}^{*\prime}\mathbf{X}^*$ is the correlation matrix among the k regressor variables. Notice that $\mathbf{H}_k$ plays the same role as the HAT matrix in ordinary least squares. In fact, if one observes Eq. (4.20) for the C_p statistic and replaces p by $1 + \text{tr}\,\mathbf{H}_k$, the result is identical to that given in Eq. (8.13). The statistic $\hat{\sigma}^2$ comes from the residual mean square from OLS estimation. Details on the development of Eq. (8.13) are very interesting and appear in Appendix B.10.

Procedurally, the use of the C_k statistic may involve a simple plotting of C_k against k with the use of the k-value for which C_k is minimized. See Example 8.8.

Since prediction can be an extremely important criterion for a choice of k, it seems only natural to consider cross validation in some sense. A PRESS-like statistic of the type

$$\text{PR(Ridge)} = \sum_{i=1}^{n} \left[\frac{e_{i,k}}{1 - \frac{1}{n} - h_{ii,k}} \right]^2 \qquad \textbf{(8.14)}$$

may be considered. Here $e_{i,k}$ is merely the ith residual for a specific value of k, and $h_{ii,k}$ is the ith diagonal element of $\mathbf{H}_k$. Again, the value of k is chosen so as to minimize PR(Ridge). A plot of PR(Ridge) against k will be sufficient and informative. The reader should notice the resemblance between Eq. (8.14) and the PRESS statistic for OLS in Chapter 4. Merely replace the HAT diagonal by $(1/n) + h_{ii,k}$. This comes from the matrix $\mathbf{H}_k$ that plays, in the case of ridge regression, a role somewhat similar to that of the HAT matrix. Now, for pragmatic reasons, the expression in Eq. (8.14) cannot always be used. When we center and scale the data, the setting aside of a data point changes the centering and scaling constants and thus changes all the regressor observations. As a result, Eq. (8.14) is only an approximation to the true PRESS that would be obtained if we delete observations one at a time, with ridge regression recomputed *from scratch* each time.

There are situations in which we can feel very comfortable about using Eq. (8.14). Yet, in other situations, misleading results can occur. In fact, as one would suspect, the use of PR(Ridge) is quite good where:

1. Sample size is not small.
2. Sample data contain no high leverage observations, i.e. no large HAT diagonals.

True numerical guidelines on items 1 and 2 are difficult. Example 8.8 will illustrate.

If the data set is moderate or small in size or if high leverage observations do appear, we can still make use of the PRESS criterion but without exploiting the shortcut "deletion formula" that motivates Eq. (8.14). In fact, the actual computer elimination of data points, one at a time with ridge regression computed repeatedly, is a relatively practical approach for moderate or small samples. In fact, there is a SAS PROC IML program (1987) for accomplishing this regression and producing a plot of PRESS against k. Here, of course, one truly computes

$$\text{PRESS}, k = \sum_{i=1}^{n} e_{i,-i,k}^2$$

where $e_{i,-i,k}$ is the ith PRESS residual for ridge regression. The $e_{i,-i,k}$ involve $\hat{y}_{i,-i,k}$. To accomplish this prediction at the ith point for a particular k, we eliminate the ith point, recompute the centered and scaled regressor data, recompute the ridge regression, and thus compute $\hat{y}_{i,-i,k}$.

A third criterion, which represents a prediction approach, or rather a cross validation approach, is the *generalized cross validation* (GCV) given by

$$\text{GCV} = \frac{\sum_{i=1}^{n} e_{i,k}^2}{\{n - [1 + \text{tr}(\mathbf{H}_k)]\}^2} = \frac{SS_{\text{Res},k}}{\{n - [1 + \text{tr}(\mathbf{H}_k)]\}^2} \tag{8.15}$$

The value "1" in $1 + \text{tr}(\mathbf{H}_k)$ accounts for the fact that the role of the constant term is not involved in $\mathbf{H}_k$. For further details regarding the philosophy behind GCV, the reader is referred to Wahba *et al.* (1979). This statistic is a norm that is in the same spirit of prediction as PRESS. It is easy to observe the resemblance between the GCV in (8.15) and the statistic in (8.14). Again, the appropriate procedure is to choose k so as to minimize GCV. Simple plotting against k will suffice and can be very enlightening.

Further Comments Concerning the Prediction Methods for Choosing k (Role of H_k)

By now, the reader should understand the relationship between C_k and the corresponding C_p criterion in OLS. In addition, the text shows how closely PR(Ridge) resembles the PRESS criterion in OLS. The quantity that draws the distinction is the $\mathbf{H}_k$ matrix. We indicated earlier that this matrix is the counterpart to the HAT matrix in ridge regression. In fact, we can view the quantity $\text{tr}(\mathbf{H}_k)$ as being very similar conceptually to regression degrees of freedom. Recall in the OLS case that

$$\text{tr}[\mathbf{X}^*(\mathbf{X}^{*\prime}\mathbf{X}^*)^{-1}\mathbf{X}^{*\prime}] = k$$

where k is the number of regressors, i.e., the regression degrees of freedom. Thus the role of $\mathbf{H}_k$ in the C_k criterion is very simple. The p in the C_p statistic in Eq. (4.20) is replaced by $1 + \text{tr}(\mathbf{H}_k)$ where the latter represents *constant term plus* ***effective*** *regression degrees of freedom* or put another way, *constant term plus effective number of regressors after the redundancy has been removed by ridge regression.* This, of course, makes the effective error degrees of freedom equal to $n - (1 + \text{tr}(\mathbf{H}_k))$. This *effective degrees of freedom* notion will subsequently be exploited in yet another very simple method for choosing k.

It is instructive, however, to take note of the role of $\text{tr}(\mathbf{H}_k)$ in PR(Ridge), C_k, and GCV. We know that

$$\begin{aligned} \text{tr}(\mathbf{H}_k) &= \text{tr}[\mathbf{X}^*(\mathbf{X}^{*\prime}\mathbf{X}^* + k\mathbf{I})^{-1}\mathbf{X}^{*\prime}] \\ &= \text{tr}[\mathbf{X}^{*\prime}\mathbf{X}^*(\mathbf{X}^{*\prime}\mathbf{X}^* + k\mathbf{I})^{-1}] \\ &= \sum_{i=1}^{k} \frac{\lambda_i}{(\lambda_i + k)} \end{aligned}$$

As in the case of $\sum_{i=1}^{k} \text{Var}\, b_{i,R}$, if there is serious collinearity, $\text{tr}(\mathbf{H}_k)$ will experience a sharp decrease followed by a "leveling off." In the cases of C_k and GCV, this will produce a decrease in the criterion followed by an increase, which is an inevitable result of $SS_{\text{Res},k}$ becoming larger. As a result, these two criteria are a result of a trade-off between the variance reduction manifested by the behavior of $\mathbf{H}_k$ and the bias induced, the latter resulting in the eventual inflation of $SS_{\text{Res},k}$.

In the foregoing discussion, we focused on ridge regression procedures that emphasize prediction, the philosophy reflecting less concern for individual coefficients and more concentration on the predictor $\hat{y}$. Certainly the ridge trace method concerns itself with stability of the coefficients. Also there are other methods used for choosing the shrinkage parameter that revolve around quality estimation of the coefficients. In certain fields of application, interpretation of a system through the sign and magnitude of one or more regression coefficients is crucial. Thus in many cases, the analyst may be tempted to reject least squares because it results in signs or magnitudes of regression coefficients that make little or no sense. The natural approach then is to select a value of k, through, say, ridge trace, for which the coefficients result in reasonable values. This approach is different from the philosophy of determining k solely on the basis of a prediction criterion. There are many volumes written on philosophies for choosing k. For most of these procedures, stability and estimation of coefficients is paramount. For an extensive review into various methods of choosing k, the reader is referred to Draper *et al.* (1979) and Hocking (1976).

Nonstochastic Choices of k and the df*-Trace*

There are certain advantages to using a value of the shrinkage parameter that is not a function of the response data. In the previous developments that deal with variances (and thus standard errors) of the ridge regression estimator, we assume that the constant k is *not a random variable*. However, the methods discussed thus far for choosing k have all exploited the y-data. As a result, the ridge trace, the use of C_k, PR(Ridge), and GCV are termed "stochastic methods" for choosing k, with the choice of k being a random variable. There is a case for choosing k as a function only of the regressor data; therefore the choice of k is determined by the nature of the collinearity itself. In this case, k is not a random variable. As a result, variance inflation factors, namely the diagonal elements of the matrix

$$\frac{\text{Var}\,\mathbf{b}_R}{\sigma^2} = (\mathbf{X}^{*\prime}\mathbf{X}^* + k\mathbf{I})^{-1}(\mathbf{X}^{*\prime}\mathbf{X}^*)(\mathbf{X}^{*\prime}\mathbf{X}^* + k\mathbf{I})^{-1}$$

can be used to reflect the true variances of the coefficients. Thus the standard errors of the regression coefficients would be given by the square root of the diagonal elements (multiplied by the root error mean square). For the stochastic choices of k, the preceding expression is only an approximation of the variance-covariance matrix of the ridge estimators, and in specific cases, very little can be said about the quality of the approximation.

One simple nonstochastic choice of k is to let k grow until the variance inflation factors of all coefficients are reduced sufficiently. As an example, one may wish all VIFs to be below the value 10.0. Since the variance inflation factors do not depend on the y-observations, this decision procedure for choice of k is nonstochastic.

A second, perhaps more appealing, method is called the df-trace criterion (see Tripp (1983)). This criterion centers around the matrix $\mathbf{H}_k$. The reader should recall what an important role $\mathbf{H}_k$ played in the C_k, PR(Ridge), and GCV criteria for choosing k. The criterion suggested here is based on

$$\text{df} = \text{tr}(\mathbf{H}_k)$$

The "df" implies *degrees of freedom*, namely, the effective regression degrees of freedom as described previously. The procedure involves plotting df against k with a view toward choosing k where df stabilizes. This methodology is much like the ridge trace in which the k is chosen where all coefficients stabilize.

The df-trace criterion is a sound approach in which k is allowed to grow until there is a stabilizing or a "settling" in the effective regression degrees of freedom or the *effective rank of the data matrix*. The analyst should keep in mind that df is a nonstochastic procedure for choosing k, based only on the structure of the collinearity; this procedure is not at all divorced from the prediction based norms, i.e., C_k and GCV.

The typical picture of the df-trace is illustrated in Figure 8.1. Example 8.8 provides a realistic description of the use of df-trace.

FIGURE 8.1 **A typical df-trace and choice of k**

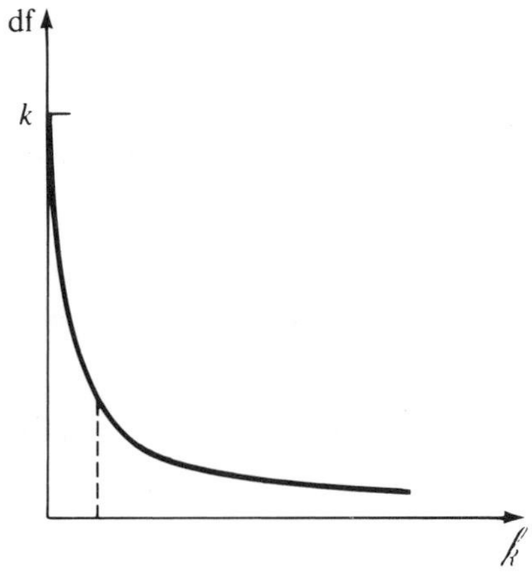

Example 8.8 ***Hospital Data***

Consider again the hospital data in Table 3.8. Suppose it is important to predict the response, in man-hours, from a regression using all five variables. Initially consider Table 8.9, which presents ridge coefficient values in the natural regressors for k from 0 to 0.24. As in Example 8.7, note how obviously unstable some of the OLS coefficients are, particularly coefficients of regressors x_1, x_3, x_4, and x_5. Relatively large changes occur in coefficients between $k = 0$ and $k = 0.01$.

The PRESS statistic (not PR(Ridge)) was computed for each value of k, and results are given in the table. In addition, the plot in Figure 8.2 illustrates how predictive capability of the model improves with increasing k, with a "leveling off" at k of approximately 0.23. The appropriate ridge coefficients are given by

$$b_{0,R} = -475.56$$
$$b_{1,R} = 8.8293$$
$$b_{2,R} = 0.0632$$
$$b_{3,R} = 0.2940$$
$$b_{4,R} = 9.3502$$
$$b_{5,R} = 116.12$$

These coefficients would appear preferable, from the prediction standpoint, to the OLS estimates *if we need to use all five regressors.* Of course, as another alternative here, we could compute multicollinearity diagnostics for a subset and possibly use ridge regression on a subset of regressors.

In this example, ridge regression produced a change from negative to positive in the coefficient of regressors x_1, x_4, and x_5.

The PRESS computations were accomplished here by use of the SAS PROC IML program, referenced previously in this section. To compare the coefficient estimates obtained using the PRESS concept with another prediction criteria, a plot was constructed in which C_k was plotted in the same manner. Figure 8.3 reveals this plot. The resulting "optimum" k-values and coefficients are given in Table 8.10. Notice that, as in the case of the PRESS criterion, C_k drops substantially for $k > 0$ and then rises. The "optimal" k-value and the coefficients are indicated in Table 8.10. Here $k = 0.005$ is the selection, resulting in a considerably more conservative application of ridge regression than in the case of the PRESS criterion. The analyst may be dissatisfied by the fact that the two regression coefficients, b_4 and b_5, are negative for this particular k-value. It is also interesting to note that PRESS and C_k do not give the same results. As mentioned earlier, this is expected when high leverage observations are prevalent, which they are in this data set.

The df-trace criterion was applied to the same hospital data. The value of $\text{tr}(\mathbf{H}_k)$ was computed for a grid of values of k. Table 8.11 reveals some numerical results for k from 0 to 0.2. Notice how much $\text{tr}(\mathbf{H}_k)$ drops between $k = 0$ and $k = 0.01$, followed by a more modest taper thereafter. As a result, it would seem reasonable to investigate a finer grid of k-values close to $k = 0$. This was done and the plot in Figure 8.4 reveals the relationship for k on the interval $[0, 0.001]$. Certainly $\text{tr}(\mathbf{H}_k)$ has stabilized before $k = 0.0004$. Now, in order to illustrate how this reflects stability in coefficients, consider the information in Table 8.12, reflecting the coefficient values (in the natural variables) for k in the interval $[0, 0.0004]$. The $\text{tr}(\mathbf{H}_k)$ appears to be a reasonable composite criterion for reflecting stability in the regression coefficients.

From the examples produced here for choosing k in the case of the hospital data, we see that methods with even slightly differing philosophies will not result in the same procedure for choosing k. The PRESS criterion produced the least conserva-

TABLE 8.9 **Ridge coefficient values for hospital data**

k	$b_{0,R}$	$b_{1,R}$	$b_{2,R}$	$b_{3,R}$	$b_{4,R}$	$b_{5,R}$	PRESS
0.00	1962.95	−15.8517	0.0559	1.5896	−4.2187	−394.31	32,195,222
0.01	1515.07	14.5765	0.0600	0.5104	−2.1732	−312.71	26,277,673
0.02	1122.83	13.5101	0.0621	0.4664	0.2488	−236.20	24,129,646
0.03	839.55	12.7104	0.0634	0.4358	1.9882	−180.25	22,522,091
0.04	624.89	12.0993	0.0643	0.4130	3.2949	−137.25	21,271,926
0.05	456.27	11.6180	0.0648	0.3951	4.3098	−102.94	20,272,861
0.06	320.08	11.2286	0.0652	0.3808	5.1188	−74.75	19,458,508
0.07	207.65	10.9066	0.0653	0.3690	5.7768	−51.05	18,785,094
0.08	113.17	10.6353	0.0654	0.3591	6.3209	−30.75	18,222,422
0.09	32.61	10.4031	0.0654	0.3507	6.7768	−13.07	17,748,879
0.10	−36.91	10.2016	0.0654	0.3434	7.1632	2.50	17,348,509
0.11	−97.52	10.0247	0.0653	0.3370	7.4937	16.39	17,009,230
0.12	−150.81	9.8679	0.0652	0.3313	7.7787	28.88	16,721,694
0.13	−198.00	9.7276	0.0651	0.3262	8.0261	40.21	16,478,532
0.14	−240.04	9.6010	0.0649	0.3216	8.2422	50.56	16,273,841
0.15	−277.70	9.4860	0.0648	0.3175	8.4319	60.07	16,102,829
0.16	−311.58	9.3808	0.0646	0.3137	8.5990	68.85	15,961,551
0.17	−342.18	9.2841	0.0644	0.3103	8.7469	77.00	15,846,729
0.18	−369.91	9.1948	0.0642	0.3071	8.8782	84.59	15,755,610
0.19	−395.10	9.1118	0.0640	0.3041	8.9950	91.69	15,685,862
0.20	−418.03	9.0343	0.0638	0.3013	9.0992	98.35	15,635,491
0.21	−438.96	8.9618	0.0636	0.2987	9.1922	104.62	15,602,781
0.22	−458.07	8.8936	0.0634	0.2963	9.2755	110.53	15,586,248
0.23	−475.56	8.8293	0.0632	0.2940	9.3502	116.12	15,584,595
0.24	−491.56	8.7684	0.0630	0.2919	9.4170	121.42	15,596,687

tive (largest k) biased estimation of the coefficients, whereas the df-trace resulted in the smallest k. An intermediate value of k was the result of using the C_k criterion, which is prediction oriented but much influenced by $\text{tr}(\mathbf{H}_k)$. Again, the difference between the results obtained using PRESS and other criteria will be most pronounced when the data contain high leverage points. As a result, the analyst must remember the purpose of the regression. If interpretation of coefficients is important, the ridge trace or df-trace approach is important. Also, other methods should be investigated. If prediction is important, GCV, C_k, and PRESS should be attempted. (See Exercise 8.8.)

We have not exhausted all of the various methods for choosing the shrinkage parameter k. Indeed, a complete discussion would fill volumes. We have attempted to expose the reader to a few procedures with the distinct philosophies represented. We emphasize that there is never an assurance that the method used will be the best available for a specific data set. The analyst may wish to use more than one procedure for ridge regression when multicollinearity diagnostics suggest that a problem exists and no simple variable deletion can be used to eliminate the

FIGURE 8.2 **Plot of PRESS against k; hospital data of Table 3.8**

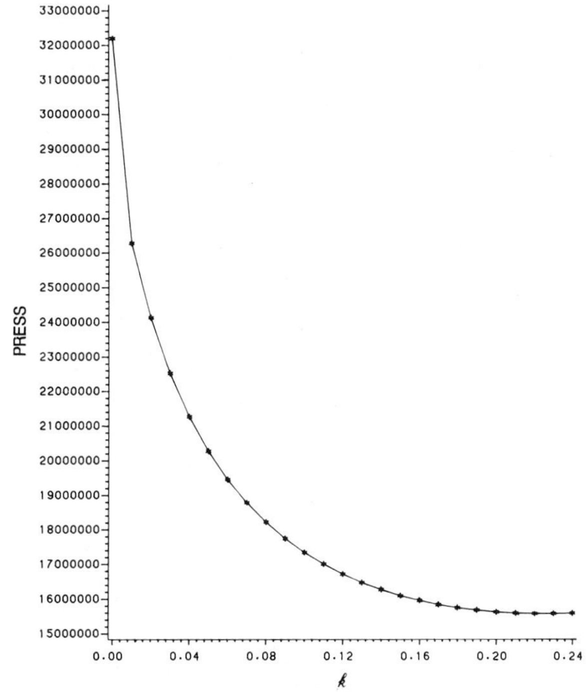

FIGURE 8.3 **Plot of C_k against k; hospital data of Table 3.8**

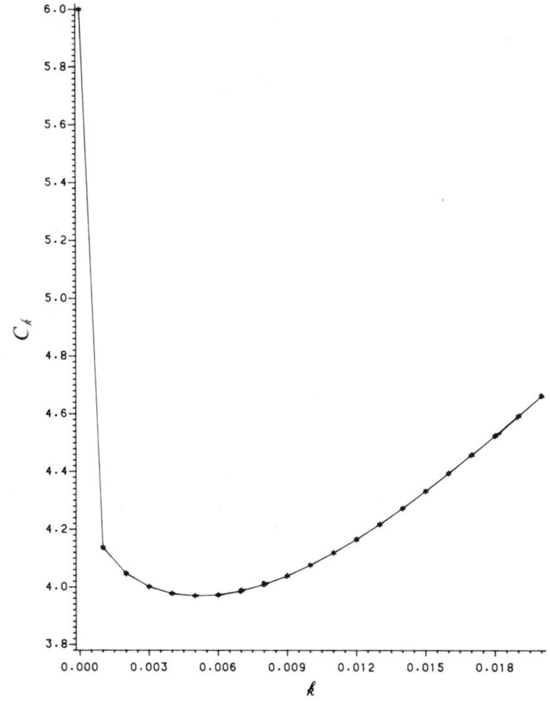

TABLE 8.10 **Values of C_k and the regression coefficients for various values of k (hospital data)**

k	C_k	b_0	b_1	b_2	b_3	b_4	b_5
0.000	6.0000	1962.95	−15.852	0.0559	1.5896	−4.2187	−394.31
0.001	4.1372	2020.49	14.604	0.0568	0.6112	−5.2711	−410.11
0.002	4.0468	1956.48	15.211	0.0572	0.5786	−4.8927	−397.91
0.003	4.0014	1893.33	15.299	0.0576	0.5630	−4.5076	−385.79
0.004	3.9775	1832.45	15.258	0.0580	0.5521	−4.1334	−374.07
0.005*	3.9689	1773.99	15.170	0.0584	0.5432	−3.7731	−362.80
0.006	3.9724	1717.91	15.061	0.0588	0.5354	−3.4269	−351.97
0.007	3.9862	1664.10	14.943	0.0591	0.5284	−3.0944	−341.57
0.008	4.0088	1612.43	14.821	0.0594	0.5220	−2.7751	−331.58
0.009	4.0390	1562.79	14.698	0.0597	0.5160	−2.4682	−321.96
0.010	4.0757	1515.07	14.577	0.0600	0.5104	−2.1732	−312.71
0.011	4.1182	1469.16	14.457	0.0602	0.5050	−1.8893	−303.79
0.012	4.1656	1424.96	14.340	0.0605	0.5000	−1.6161	−295.20
0.013	4.2174	1382.37	14.226	0.0607	0.4951	−1.3529	−286.92
0.014	4.2729	1341.30	14.114	0.0610	0.4905	−1.0992	−278.92
0.015	4.3318	1301.69	14.006	0.0612	0.4861	−0.8545	−271.20
0.016	4.3935	1263.45	13.901	0.0614	0.4818	−0.6184	−263.73
0.017	4.4576	1226.50	13.799	0.0616	0.4777	−0.3904	−256.51
0.018	4.5239	1190.79	13.700	0.0618	0.4738	−0.1701	−249.52
0.019	4.5919	1156.25	13.604	0.0620	0.4700	0.0428	−242.75
0.020	4.6615	1122.83	13.510	0.0621	0.4664	0.2488	−236.20

FIGURE 8.4 **df-trace on the hospital data of Table 3.8**

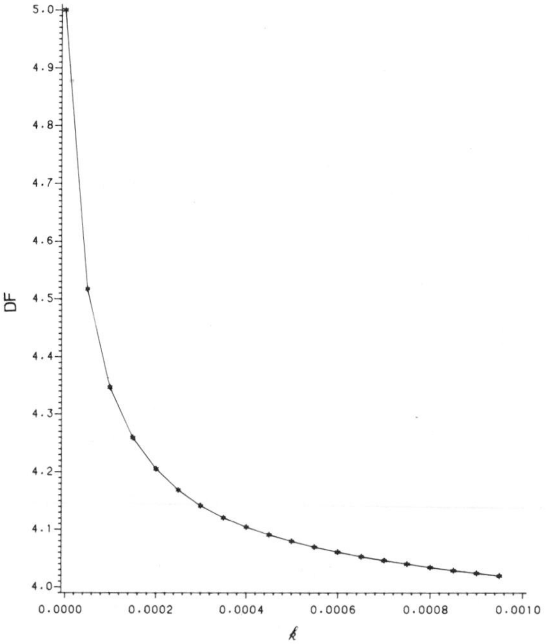

TABLE 8.11 **Values of tr($\mathbf{H}_k$) (hospital data)**

k	tr($\mathbf{H}_k$)	k	tr($\mathbf{H}_k$)
0.00	5.0000	0.11	2.5661
0.01	3.6955	0.12	2.5145
0.02	3.4650	0.13	2.4671
0.03	3.2867	0.14	2.4233
0.04	3.1427	0.15	2.3827
0.05	3.0227	0.16	2.3447
0.06	2.9206	0.17	2.3092
0.07	2.8320	0.18	2.2758
0.08	2.7541	0.19	2.2443
0.09	2.6848	0.20	2.2146
0.10	2.6225		

TABLE 8.12 **Coefficient values (hospital data)**

k	$b_{0,R}$	$b_{1,R}$	$b_{2,R}$	$b_{3,R}$	$b_{4,R}$	$b_{5,R}$
0.00000	1962.95	−15.8517	0.0559	1.5896	−4.2187	−394.31
0.00005	2024.31	−0.3193	0.0561	1.0971	−4.9516	−408.51
0.00010	2043.48	5.1181	0.0562	0.9242	−5.1941	−413.04
0.00015	2051.49	7.8846	0.0563	0.8359	−5.3067	−415.01
0.00020	2054.94	9.5575	0.0563	0.7822	−5.3663	−415.93
0.00025	2056.08	10.6767	0.0564	0.7461	−5.3990	−416.33
0.00030	2055.90	11.4767	0.0564	0.7201	−5.4163	−416.42
0.00035	2054.89	12.0761	0.0564	0.7004	−5.4238	−416.32
0.00040	2053.33	12.5411	0.0565	0.6849	−5.4249	−416.09

difficulty. The goal, then, is for one to gain the experience necessary to determine which procedure or procedures should receive more attention. One may view the various procedures as a collection of methods, each of which produces different sets of biased estimates. The user can certainly gain some comfort from the fact that if collinearity is seen from the diagnostics as being damaging, all of the methods from which one chooses may be preferable to OLS. Although, clearly, ridge regression should not be used to solve all model-fitting problems involving multicollinearity, enough positive evidence about ridge regression exists to suggest that it should be a part of any model builder's arsenal of techniques.

Example 8.9 ***Tobacco Data***

In the previous illustrations with the hospital data set, df-trace, PRESS, and the C_k statistic were used to compute k and the resulting coefficients. The three criteria produced quite different results. We alluded to the fact that the presence of high leverage data points was responsible, at least in part. This example provides the reader with one more experience in observing collinearity diagnostics and gives a case in which very little leverage is carried by any single data point.

Consider the following data set in which 30 tobacco blends were made. The goal was to develop a linear equation that relates the percentage concentration of four important components to a response that measures the amount of heat given off by the tobacco during the smoking process. Table 8.13 supplies the data. Table 8.14 reveals the collinearity diagnostics. Also note that the HAT diagonals reveal that no single point carries a disproportionate amount of leverage.

The coefficients of regressors x_1, x_3, and x_4 appear to be badly damaged by the collinearity. In fact, the coefficient of x_2 is not totally immune. The eigenvalue spectrum (notice the ratios, the ϕ_j values) indicates that there are dependencies involving regressors x_1, x_3, and x_4 simultaneously. In addition, regressor x_2 is involved in "collinearity with the constant." One can appreciate this by noticing that x_2 varies very little in the data set. The eigenvalues and variance proportions are for the data that are scaled but not centered. Of course, the constant in this problem is not particularly crucial because it was felt that the region of the data is the region in which the regression would be used for any future prediction. In addition, the origin, $x_1 = 0$, $x_2 = 0$, $x_3 = 0$, $x_4 = 0$, is of no practical importance at all.

The PR(Ridge) and PRESS statistic were computed, with the latter representing the *exact* computation and the former representing the approximation using Eq. (8.14). In this case, the agreement is quite good. The value of $k = 0.004$ results in the smallest PRESS for both methods of computation. The results, along with the values of the coefficients, the C_k statistic, and $\text{tr}(\mathbf{H}_k)$ appear in Table 8.15. Figures 8.5 and 8.6 contain, respectively, plots of C_k against k and the df-trace, containing $\text{tr}(\mathbf{H}_k)$ plotted against k. From the PRESS information, C_k plot, and df-trace, it appears that if one were to use ridge regression in this data set, a value of k from 0.002 to 0.004 would be appropriate.

TABLE 8.13 **Tobacco data**

Sample	x_1 (%)	x_2 (%)	x_3 (%)	x_4 (%)	y	h_{ii}
1	5.5	4.0	9.55	13.25	527.91	0.1824
2	6.2	4.3	11.10	15.32	518.29	0.2239
3	7.7	5.2	12.84	17.41	549.56	0.0809
4	8.5	5.3	13.32	18.08	738.06	0.2580
5	11.0	6.3	17.84	24.16	704.82	0.0873
6	11.5	6.5	18.57	24.29	697.94	0.1176
7	13.0	7.2	21.96	27.29	826.86	0.1129
8	15.0	7.6	25.87	31.32	998.18	0.1771
9	16.2	7.8	26.82	34.62	1040.22	0.1923
10	16.9	8.7	27.89	36.03	1040.46	0.2076

TABLE 8.13 Continued

Tobacco data

Sample	x_1 (%)	x_2 (%)	x_3 (%)	x_4 (%)	y	h_{ii}
11	14.1	7.2	23.99	28.48	803.26	0.2235
12	17.5	8.8	29.61	36.88	1009.51	0.2029
13	15.0	7.5	25.80	31.41	916.44	0.1710
14	6.3	4.8	11.49	15.59	394.23	0.2245
15	9.2	5.4	14.68	19.69	583.20	0.1936
16	11.5	6.5	19.10	25.40	744.81	0.0609
17	12.0	7.0	19.60	26.39	825.93	0.1769
18	16.8	8.4	27.16	36.49	1070.88	0.3067
19	14.2	7.4	23.95	28.84	840.91	0.1718
20	17.1	8.3	27.94	35.40	991.58	0.1277
21	11.9	6.9	19.56	26.10	767.40	0.1291
22	13.1	7.1	22.05	27.40	807.18	0.0763
23	14.3	7.5	23.91	29.03	857.15	0.1712
24	7.2	5.0	12.14	16.13	526.05	0.0950
25	6.7	4.9	11.91	15.98	495.89	0.1421
26	5.6	4.1	9.56	13.34	476.38	0.1532
27	7.1	5.0	11.98	16.09	520.82	0.0943
28	17.0	8.0	27.77	35.45	1066.99	0.2194
29	16.1	7.6	26.99	34.16	1020.25	0.2591
30	6.3	4.6	11.26	15.52	494.59	0.1608

TABLE 8.14

Collinearity diagnostics for tobacco data

Coefficient	Estimate	VIF	Standard Error
b_0	311.7818		99.5390
b_1	77.6248	324.1412	38.3609
b_2	−75.1377	45.1728	40.1006
b_3	−26.5371	173.2577	17.1727
b_4	21.8326	138.1753	12.9226

$R^2 = 0.9572$

Eigenvalue	*Variance proportions (regressors scaled, not centered)* ϕ_j	b_0	b_1	b_2	b_3	b_4
4.92150343	1.0000	0.0003	0.0000	0.0000	0.0000	0.0000
0.07679247	64.0884	0.0701	0.0005	0.0001	0.0008	0.0005
0.00089756	5,483.2091	0.7059	0.0049	0.8341	0.1222	0.0003
0.00058578	8,401.6673	0.0074	0.0005	0.1131	0.2988	0.6993
0.00022076	22,293.0880	0.2163	0.9941	0.0526	0.5782	0.2999

TABLE 8.15 Values of ridge criteria for tobacco data

k	PR (Ridge)	$\mathrm{tr}(\mathbf{H}_k)$	PRESS	C_k	$b_{0,R}$	$b_{1,R}$	$b_{2,R}$	$b_{3,R}$	$b_{4,R}$
0.000	87848.1	4.0000	87848.1	5.0000	311.78	77.6262	−75.1379	−26.5373	21.8321
0.001	82287.0	3.4922	82259.1	4.3410	282.14	58.2332	−64.6811	−17.265	22.1139
0.002	80462.4	3.1759	80367.6	4.2534	263.70	48.1895	−57.1178	−11.9549	21.4347
0.003	79841.5	2.9503	79690.5	4.3273	250.22	42.0198	−51.0765	−8.4518	20.5558
0.004	79723.3	2.7771	79527.8	4.4593	239.51	37.8248	−46.0104	−5.9503	19.6793
0.005	79843.4	2.6379	79612.6	4.6132	230.57	34.7731	−41.6399	−4.0706	18.8639
0.006	80083.5	2.5226	79824.7	4.7740	222.90	32.4436	−37.7992	−2.6068	18.1228
0.007	80384.5	2.4247	80103.2	4.9345	216.16	30.6003	−34.3793	−1.4360	17.4547
0.008	80714.9	2.3402	80415.5	5.0914	210.17	29.1005	−31.3038	−0.4797	16.8531
0.009	81056.9	2.2663	80742.9	5.2430	204.78	27.8529	−28.5164	0.3146	16.3105
0.010	81400.3	2.2009	81074.6	5.3887	199.90	26.7964	−25.9738	0.9836	15.8196

FIGURE 8.5 Plot of C_k against k; tobacco data of Table 8.13

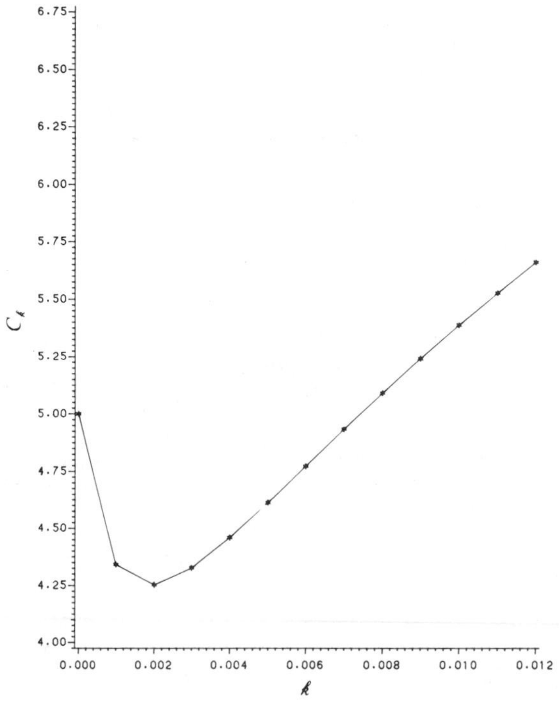

FIGURE 8.6 **df-trace on the tobacco data of Table 8.13**

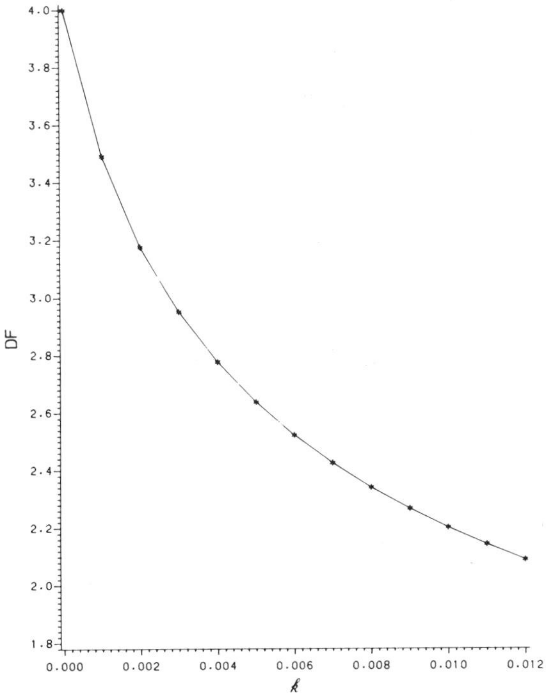

PRINCIPAL COMPONENTS REGRESSION

Principal components regression represents another biased estimation technique for combating multicollinearity. With this method, we perform least squares estimation on a set of artificial variables called the *principal components* of the correlation matrix. Based on the nature of the analysis, we eliminate a certain number of the principal components to effect a substantial reduction in variance. The method varies somewhat in philosophy from ridge regression but, like ridge, gives biased estimates; when used successfully, this method results in estimation and prediction that is superior to OLS. Principal components are orthogonal to each other, so that it becomes quite easy to *attribute a specific amount of variance* to each.

Consider the matrix of normalized eigenvectors associated with the eigenvalues $\lambda_1, \lambda_2, \ldots, \lambda_k$ of $\mathbf{X}^{*\prime}\mathbf{X}^*$ (correlation form). We know that $\mathbf{VV}' = \mathbf{I}$ since $\mathbf{V}$ is an orthogonal matrix. Hence we can write the original regression model in the form

$$\mathbf{y} = \beta_0 \mathbf{1} + \mathbf{X}^*\mathbf{VV}'\boldsymbol{\beta} + \boldsymbol{\varepsilon} \quad \textbf{(8.16)}$$

$$\mathbf{y} = \beta_0 \mathbf{1} + \mathbf{Z}\boldsymbol{\alpha} + \boldsymbol{\varepsilon} \quad \textbf{(8.17)}$$

where $\mathbf{Z} = \mathbf{X}^*\mathbf{V}$ and $\boldsymbol{\alpha} = \mathbf{V}'\boldsymbol{\beta}$. $\mathbf{Z}$ is an $n \times k$ matrix and $\boldsymbol{\alpha}$ is a $k \times 1$ vector of new coefficients $\alpha_1, \alpha_2, \ldots, \alpha_k$. We can visualize the columns of $\mathbf{Z}$ (typical element z_{ij}) as representing readings on k *new variables*, the *principal components*. It is easy to see that the components are orthogonal to each other. We have

$$\begin{aligned}\mathbf{Z}'\mathbf{Z} &= (\mathbf{X}^*\mathbf{V})'(\mathbf{X}^*\mathbf{V}) \\ &= \mathbf{V}'\mathbf{X}^{*\prime}\mathbf{X}^*\mathbf{V} \\ &= \operatorname{diag}(\lambda_1, \lambda_2, \ldots, \lambda_k)\end{aligned} \tag{8.18}$$

So, if regression is performed on the z's via the model in Eq. (8.17), the variances of the coefficients (the diagonal elements of $(\mathbf{Z}'\mathbf{Z})^{-1}$ apart from σ^2) are the reciprocals of eigenvalues. That is,

$$\frac{\operatorname{Var}(\hat{\alpha}_j)}{\sigma^2} = \frac{1}{\lambda_j} \qquad (j = 1, 2, \ldots, k) \tag{8.19}$$

Note that the $\hat{\alpha}$'s are, indeed, least squares estimators. If all of the principal components are retained in the regression model, then all that has been accomplished by the transformation is essentially a *rotation of the regressor variables*. Even though the new variables are orthogonal, the same magnitude of variance (due to the ill-conditioning in $\mathbf{X}'\mathbf{X}$) is retained. In a sense, the total variance has merely been redistributed. If multicollinearity is severe, there will be at least one small eigenvalue. An elimination of one (at least one) principal component, that associated with the small eigenvalue, may *substantially* reduce the total variance in the model and thus produce an appreciably improved prediction equation.

What Are Principal Components?

The rotation of the regressor variables that produces the z's, the principal components, essentially allows for a new set of coefficients, the α's; these α's are defined so that we can directly attribute the variance of an estimator $\hat{\alpha}_j$ to a specific linear dependency. This is apparent from Eq. (8.19). Graphically, Figure 8.7 will allow the potential user of principal components regression to better understand what the z's actually are.

Clearly, the data indicate a strong association between x_1 and x_2. Now, this dependency will deposit its effect on the estimates of both β_1 and β_2. But consider the z-coordinate system. In fact, consider

$$\mathbf{Z} = \mathbf{X}^*\mathbf{V}$$

with a specific column of $\mathbf{Z}$ being given by

$$\mathbf{z}_j = \mathbf{X}^*\mathbf{v}_j \tag{8.20}$$

The elements in $\mathbf{z}_j$ are the data measured on the z_j axis, where for our illustration, $j = 1, 2$. The "variation" in the resulting z_j values is given by

$$\mathbf{z}_j'\mathbf{z}_j = \mathbf{v}_j'(\mathbf{X}^{*\prime}\mathbf{X}^*)\mathbf{v}_j = \lambda_j \qquad (j = 1, 2)$$

Thus the regression on the principal components for this case would involve

$$\mathbf{Z}'\mathbf{Z} = \begin{bmatrix} \lambda_1 & 0 \\ 0 & \lambda_2 \end{bmatrix} \tag{8.21}$$

So, from Figure 8.7, λ_1 is the large eigenvalue, and λ_2 is small. The variance of coefficient $\hat{\alpha}_2$ is given by $\sigma^2(1/\lambda_2)$. The small variation in the $\mathbf{z}_2$ direction is responsible for the large variance in $\hat{\alpha}_2$. The variation in the data lies principally along the z_1 axis. Thus a large λ_1 allows $\text{Var}(\hat{\alpha}_1)$ to be relatively unaffected by the dependency between x_1 and x_2. From the foregoing illustration, the reader can understand that the transformation to the z's allows the linear dependencies, characterized by small eigenvalues, to be focused sharply on a small number of coefficients. This allows for ease in decision making regarding which z's are eliminated. Clearly in the case of Figure 8.7, the data have assumed the direction of z_1, and thus the principal component z_2 contributes essentially nothing to the regression. Thus z_2 would be eliminated, thereby reducing the variance contributed by $\hat{\alpha}_2$. So principal components regression is nothing more than variable screening in a regression on the principal components.

FIGURE 8.7 **Principal components for $k = 2$**

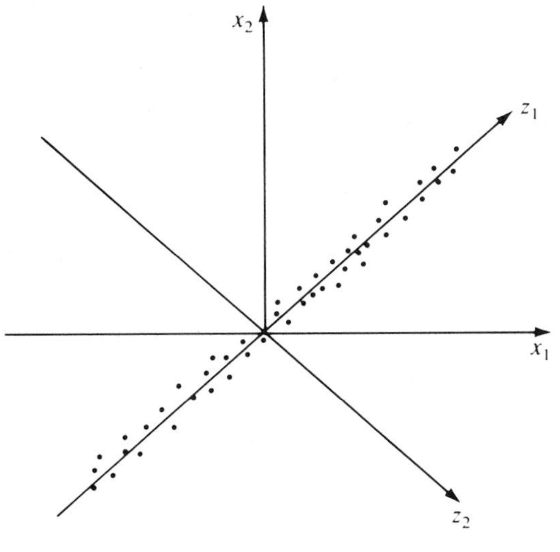

How Many Principal Components Are Eliminated?

The philosophy of principal component (pc) regression very much resembles the philosophy of least squares variable screening in general. Least squares estimation is conducted on the components, and if a component is eliminated, the resulting estimators of the coefficients of the original variables, the x's, are biased. Since variance producing components are eliminated, variance is reduced, as in the case of least squares variable screening. The difficulty arises in the decision of how many (if any) components we should eliminate. The analyst has at his or her disposal the ordinary type of least squares criteria, s^2, PRESS, sum of absolute PRESS residuals, C_p, and the width of the confidence intervals on $E(y)$. These seem like natural criteria on which decisions should be based. Example 8.10 contains an illustration.

Transformation Back to Original Variables

Objections to principal components regression are quite often the result of the artificiality of the principal components themselves. Without a doubt, if principal components regression is used successfully, the analyst can expect the resulting model in the original variables to improve. Of course, computed statistics such as s^2, PRESS, etc. apply to the model that is transformed back to the original standardized variables. Suppose, for example, with k variables and hence k principal components, $r < k$ components are eliminated. From Eq. (8.17), with the retention of all components, we can write $\boldsymbol{\alpha} = \mathbf{V}'\boldsymbol{\beta}$, and hence

$$\boldsymbol{\beta} = \mathbf{V}\boldsymbol{\alpha} \tag{8.22}$$

Clearly then, if we eliminate the last r components, the least squares estimators of the regression coefficients for all k parameters (deleting principal components does not imply deletion of any of the original regressors) are given by

$$\mathbf{b}_{\text{pc}} = \begin{bmatrix} b_{1,\text{pc}} \\ b_{2,\text{pc}} \\ \vdots \\ b_{k,\text{pc}} \end{bmatrix} = [\mathbf{v}_1 \mathbf{v}_2 \cdots \mathbf{v}_{k-r}] \begin{bmatrix} \hat{\alpha}_1 \\ \hat{\alpha}_2 \\ \vdots \\ \hat{\alpha}_{k-r} \end{bmatrix} \tag{8.23}$$

Thus elimination of r principal components is tantamount to elimination of r eigenvectors and, of course, r of the α's. We are still assuming that the x's are centered and scaled so the constant term in the transformed model is $\bar{y}$.

Bias in Principal Components Coefficients

Suppose we consider the principal components procedure with r principal components eliminated and s components retained, where $s + r = k$. Also suppose we consider the matrix $\mathbf{V} = [\mathbf{v}_1 \mathbf{v}_2 \cdots \mathbf{v}_k]$ of normalized eigenvectors of $\mathbf{X}^{*\prime}\mathbf{X}^*$ partitioned into

$$\mathbf{V} = [\mathbf{V}_r \,\vdots\, \mathbf{V}_s]$$

and similarly consider the matrix $\boldsymbol{\Lambda}$ to be a diagonal matrix of eigenvalues of $\mathbf{X}^{*\prime}\mathbf{X}^*$. We partition $\boldsymbol{\Lambda}$ as

$$\boldsymbol{\Lambda} = \begin{bmatrix} \Lambda_r & 0 \\ 0 & \Lambda_s \end{bmatrix}$$

where $\mathbf{\Lambda}_r$ and $\mathbf{\Lambda}_s$ are diagonal matrices, with $\mathbf{\Lambda}_r$ containing the eigenvalues associated with the eliminated components. Since $\mathbf{V}'(\mathbf{X}^{*\prime}\mathbf{X}^*)\mathbf{V} = \mathbf{Z}'\mathbf{Z} = \mathbf{\Lambda}$, the least squares estimates of the α's can be written

$$\hat{\boldsymbol{\alpha}} = (\mathbf{Z}'\mathbf{Z})^{-1}\mathbf{Z}'\mathbf{y} = \mathbf{\Lambda}^{-1}\mathbf{V}'\mathbf{X}^{*\prime}\mathbf{y} \tag{8.24}$$

which implies that the estimator for the α's that are retained is given by

$$\hat{\boldsymbol{\alpha}}_s = \mathbf{\Lambda}_s^{-1}\mathbf{V}_s'\mathbf{X}^{*\prime}\mathbf{y}$$

As indicated earlier, we must view principal components regression as standard least squares model-building on the principal components. Since the principal components are orthogonal, we can use Eq. (4.7) to show that $\hat{\boldsymbol{\alpha}}_s$ is an unbiased estimator for $\boldsymbol{\alpha}_s$. Consider now Eq. (8.23) for $\mathbf{b}_{\text{pc}}$. We can write

$$\mathbf{b}_{\text{pc}} = \mathbf{V}_s\hat{\boldsymbol{\alpha}}_s \tag{8.25}$$

Thus

$$E(\mathbf{b}_{\text{pc}}) = \mathbf{V}_s\boldsymbol{\alpha}_s = \mathbf{V}_s\mathbf{V}_s'\boldsymbol{\beta}$$

Since $\mathbf{V}\mathbf{V}' = \mathbf{I} = \mathbf{V}_r\mathbf{V}_r' + \mathbf{V}_s\mathbf{V}_s'$,

$$\begin{aligned} E(\mathbf{b}_{\text{pc}}) &= [\mathbf{I} - \mathbf{V}_r\mathbf{V}_r']\boldsymbol{\beta} \\ &= \boldsymbol{\beta} - \mathbf{V}_r\mathbf{V}_r'\boldsymbol{\beta} \\ &= \boldsymbol{\beta} - \mathbf{V}_r\boldsymbol{\alpha}_r \end{aligned}$$

Thus the estimators of the p regression coefficients are biased by the quantity $\mathbf{V}_r\boldsymbol{\alpha}_r$, with $\boldsymbol{\alpha}_r$ being the vector of principal components that have been eliminated.

Variance in Principal Components Coefficients

As one would expect, the elimination of principal components results in a decrease in the variances of the regression coefficients in $\mathbf{b}_{\text{pc}}$. The magnitude of this decrease, as in the case of ridge regression, depends on the extent of the multicollinearity involved. It is relatively easy to determine, analytically, what the variance reduction is. If all components are retained, $\mathbf{b}_{\text{pc}}$ reduces to ordinary least squares, and hence

$$\begin{aligned} \frac{\text{Var}\,\mathbf{b}}{\sigma^2} &= (\mathbf{X}^{*\prime}\mathbf{X}^*)^{-1} \\ &= \mathbf{V}\mathbf{\Lambda}^{-1}\mathbf{V}' \\ &= \mathbf{V}_r\mathbf{\Lambda}_r^{-1}\mathbf{V}_r' + \mathbf{V}_s\mathbf{\Lambda}_s^{-1}\mathbf{V}_s' \end{aligned} \tag{8.26}$$

From Eq. (8.25), and using the fact that $\text{Var}\,\hat{\boldsymbol{\alpha}}_s = \mathbf{\Lambda}_s^{-1}$, we have the variance-covariance matrix

$$\frac{\text{Var}\,\mathbf{b}_{\text{pc}}}{\sigma^2} = \mathbf{V}_s\mathbf{\Lambda}_s^{-1}\mathbf{V}_s' \tag{8.27}$$

Thus the difference in the variance-covariance matrix for the OLS estimator and the principal components estimator is the quantity $\mathbf{V}_r\mathbf{\Lambda}_r^{-1}\mathbf{V}_r'$. The diagonal elements of this matrix are merely weighted sums of the *reciprocals* of the eigenvalues associated with the eliminated principal components. As a result, if the ignored principal components are associated with small eigenvalues, one may expect a substantial variance reduction.

Example 8.10 ***Principal Components Example***

Consider the following data set:

y	x_1	x_2	x_3	x_4
17.6	8.8	2589	83.1	158.2
10.9	8.5	1186	24.2	96.2
9.2	7.7	291	4.5	31.8
16.2	4.9	1276	9.1	95.0
10.1	9.6	6633	158.2	407.2
11.7	10.0	12125	132.2	404.6
17.9	11.5	36717	501.5	1180.6
21.1	11.6	43319	904.0	1807.5
14.7	11.2	10530	227.6	470.0
7.7	10.7	3931	66.6	151.4
8.4	10.0	1536	43.4	93.8
32.8	6.8	61400	1253.0	3293.4

A least squares regression gives the prediction equation

$$\hat{y} = 21.971979 - 1.277560x_1 + 0.00015026x_2 + 0.015533x_3 - 0.002854x_4$$

with $SS_{\text{Res}} = 63.17323$, $s^2 = 9.025$, PRESS $= 670.303$, and $R^2 = 0.8857$. The following are multicollinearity diagnostics (variance decomposition proportions and eigenvalues).

Eigenvalue	Portion Intercept	Portion b_1	Portion b_2	Portion b_3	Portion b_4
4.022	0.0009	0.0009	0.0010	0.0004	0.0004
0.932897	0.0077	0.0073	0.0019	0.0010	0.0010
0.030779	0.2296	0.1932	0.1141	0.0001	0.0476
0.010341	0.1778	0.1989	0.8239	0.2644	0.0061
0.003661	0.5839	0.5998	0.0592	0.7340	0.9448

Condition Number = 1,098.6069

These eigenvalues of $\mathbf{X}'\mathbf{X}$ are for a scaled $\mathbf{X}$ matrix (not centered). The variance inflation factors are 2.080, 34.723, 79.155, and 82.967 for the coefficients b_1, b_2, b_3, and b_4, respectively. The correlation matrix is given by

$$\mathbf{X}^{*\prime}\mathbf{X}^{*} = \begin{bmatrix} 1 & 0.13141 & 0.08008 & -0.01470 \\ 0.13141 & 1 & 0.98161 & 0.97364 \\ 0.08008 & 0.98161 & 1 & 0.98871 \\ -0.01470 & 0.97364 & 0.98871 & 1 \end{bmatrix}$$

The eigenvalues of the correlation matrix[1] are given by $\lambda_1 = 2.9692$, $\lambda_2 = 1.00464$, $\lambda_3 = 0.019438$, and $\lambda_4 = 0.0067049$ and the resulting matrix of eigenvectors is as follows:

$$\mathbf{V} = \begin{bmatrix} -0.05768 & -0.99267 & -0.07732 & 0.07280 \\ -0.57623 & -0.03408 & 0.81466 & -0.05591 \\ -0.57817 & 0.01934 & -0.45462 & -0.67725 \\ -0.57476 & 0.11432 & -0.35167 & 0.73000 \end{bmatrix}$$

Three regression coefficients, b_2, b_3, and b_4, appear to be affected by collinearity. The simple correlations indicate that x_2, x_3, and x_4 are involved in pairwise correlations. The variance decomposition proportions (noncentered diagnostics) indicate one or perhaps two collinearities that are causing difficulties. One collinearity (eigenvalue = 0.003661) is responsible for a substantial portion of the variance of b_3 and b_4, while a second dependency (eigenvalue = 0.010341) is responsible for a majority (82.39%) of the variance of b_2 and a modest portion of the variance of b_3. If principal components regression is successful, it will involve the elimination of one or perhaps both of the principal components associated with the two dependencies.

The $\mathbf{Z}$ matrix of principal components is found by $\mathbf{Z} = \mathbf{X}^*\mathbf{V}$, where $\mathbf{X}^*$ is centered and scaled. $\mathbf{X}^*$ is a 12×4 matrix without the column of ones, i.e., with the model shown in Eq. (8.16). The principal components are columns of

$$\mathbf{Z} = \begin{bmatrix} z_1 & z_2 & z_3 & z_4 \\ 0.29068 & 0.05460 & -0.02032 & -0.01064 \\ 0.34169 & 0.09626 & -0.00699 & 0.00326 \\ 0.37607 & 0.21148 & 0.00506 & -0.00908 \\ 0.37836 & 0.62388 & 0.04049 & -0.02813 \\ 0.17280 & -0.05490 & -0.03332 & 0.01247 \\ 0.13435 & -0.11679 & 0.03738 & 0.02485 \\ -0.38528 & -0.31647 & 0.10644 & 0.00762 \\ -0.72778 & -0.30655 & -0.02028 & -0.06018 \\ 0.08479 & -0.28827 & -0.03513 & 0.00534 \\ 0.27142 & -0.22518 & -0.01950 & 0.01554 \\ 0.31803 & -0.12368 & -0.02620 & 0.00882 \\ -1.25514 & 0.44561 & -0.02761 & 0.03011 \end{bmatrix}$$

[1]Note how the eigenvalues for the case of centering and scaling differ from those that occur when only scaling is used.

The fourth column of **Z** is the principal component associated with the smallest eigenvalue. Notice the small variation in the z_{4i}.

The application of principal components regression involves the removal, initially, of z_4 with the response y being regressed against the remaining components. This regression gives the following coefficients of the **z**'s:

$$\hat{\alpha}_1 = -12.0121$$
$$\hat{\alpha}_2 = 7.57925$$
$$\hat{\alpha}_3 = 2.79077$$

Ordinary least squares procedures are applied to this regression with the computation of the residual sum of squares and PRESS statistic as discussed in Chapters 3 and 4. The results are

$$SS_{\text{Res}} = 66.4176$$
$$s^2 = 8.302$$
$$\text{PRESS} = 116.007$$

The regression with the reduced number of components has resulted in a slight increase in the residual SS but a substantial reduction in PRESS from 670.303 to 116.007 and a reduction in s^2 from 9.025 to 8.302. The improvement in the model is illustrated by the following PRESS residuals.

	Before Deletion	After Deletion
Observation	PRESS Residual	PRESS Residual
1	6.6604	6.8011
2	−0.5697	−0.6516
3	−3.6473	−3.3503
4	1.5293	2.6342
5	−2.3068	−2.5677
6	−0.2959	−0.9254
7	3.8907	2.7974
8	−17.3040	−0.1931
9	4.2692	4.0924
10	−2.2844	−2.6002
11	−1.7473	−1.9560
12	16.2328	−2.9183

Equation (8.23) can then be used to determine the estimates of the coefficients in terms of the centered and scaled regressors; the constant term for the regression is $\bar{y}$. This is followed by a transformation to the coefficients of the natural variables as described in Section 8.3. The results are given by

$$b_{1,\text{pc}} = -1.04099$$
$$b_{2,\text{pc}} = 0.000132085$$
$$b_{3,\text{pc}} = 0.00436507$$
$$b_{4,\text{pc}} = 0.00209076$$

The constant term obtained through this transformation is $b_{0,\text{pc}} = 19.849$.

The deletion of a single principal component seems to have produced a regression that is somewhat superior to the OLS regression; the deletion of another principal component is tempting. The principal component that was eliminated was associated with a dependency, which was damaging to coefficients of regressors x_3 and x_4. The other infected coefficient is b_2. From the variance decomposition proportions, it would seem that the deletion of another principal component might be at least marginally effective. Such an analysis does slightly reduce PRESS (to a value of 100.001 with $SS_{\text{Res}} = 66.569$ and $s^2 = 7.396$). We shall not show the details here. Obviously, continuing the deletion of principal components will result in eventual deterioration of the regression.

What Is the Appropriate Order of Elimination of Principal Components?

Through our work with Example 8.10, we have implied that the strategy of elimination of principal components should be to begin by discarding the principal component associated with the smallest eigenvalue. Often this will be the appropriate plan of action, the rationale being that this is the least informative component and thus deposits the largest amount of variance. However, a more acceptable strategy is to treat the principal component reduction as if it were a standard variable screening problem (which it actually is). Now, of course, the principal components, as regressors, are orthogonal, and thus a reasonable set of statistics to dictate the order of reduction are the t-statistics given by

$$t = \frac{\hat{\alpha}_j}{s_{\hat{\alpha}j}} = \frac{\hat{\alpha}_j\sqrt{\lambda_j}}{s}$$

In other words, the t-values should be rank ordered and components be considered for elimination beginning with the *smallest t-value*, in magnitude. Obviously many times the rank order of t-values will be that which is the descending order of the eigenvalues, but this is not necessarily the case.

EXERCISES FOR CHAPTER 8

8.1 Thirty firms were chosen for a study of the effect of several factors on firm return on assets, y, for 1982. The data are as follows:

Firm	y	x_1	x_2	x_3	x_4	x_5	x_6	x_7
1	14.5	3.6	20.6	34.3	44.0	13.6	14.1	15.8
2	11.5	4.7	−35.9	4.8	−29.2	9.0	7.3	7.2
3	10.3	6.9	22.3	26.8	27.1	11.8	8.5	14.3
4	12.8	1.7	36.3	76.8	51.5	14.4	13.6	15.9
5	18.1	2.6	25.2	51.2	473.1	17.2	11.5	12.0
6	24.9	4.4	22.1	33.7	77.6	22.9	4.0	−1.6
7	26.8	4.1	20.4	37.4	112.3	23.6	15.0	13.9
8	22.9	3.2	78.3	57.0	−2.8	24.8	12.5	16.0
9	11.9	8.2	37.2	41.0	65.5	14.3	7.3	7.2
10	11.0	12.3	128.7	32.2	62.3	15.3	8.5	14.3
11	18.1	5.6	50.7	63.3	98.3	27.7	8.5	14.3
12	22.6	1.8	10.6	30.1	95.1	23.3	14.5	12.2
13	15.2	3.9	29.0	33.1	16.3	16.0	10.5	8.9
14	18.4	3.8	−238.5	52.2	69.5	36.0	12.5	12.4
15	10.7	4.8	−45.1	2.6	−9.9	4.7	7.3	7.2
16	7.1	6.5	−21.3	20.8	88.4	6.4	9.6	6.4
17	22.2	5.7	2.4	19.6	62.4	18.5	14.0	14.7
18	8.9	4.6	52.5	16.5	49.6	9.7	14.0	12.2
19	15.0	11.2	−50.7	−9.2	135.4	12.2	12.3	8.4
20	−2.3	41.1	−83.8	−35.5	−80.2	2.7	1.4	−18.3
21	12.1	5.6	−2.4	5.1	−5.1	14.1	14.9	14.1
22	4.0	8.5	8.3	24.3	−61.8	−2.3	6.2	−2.9
23	4.4	4.1	−7.3	35.2	169.4	9.8	11.6	7.5
24	18.1	2.6	143.5	103.8	111.8	18.3	12.5	16.0
25	16.2	3.1	36.2	47.0	60.4	17.3	11.6	7.5
26	24.3	5.0	−15.5	7.4	114.7	24.2	10.5	5.8
27	16.5	7.2	56.1	113.5	96.1	12.0	12.3	8.4
28	13.5	3.0	34.2	28.3	−7.2	10.5	13.5	18.3
29	14.3	4.1	−6.1	29.0	−10.3	10.8	14.9	14.1
30	13.4	2.1	−23.7	10.3	−78.6	20.7	16.0	16.1

Data analyzed for the Department of Business Administration by the Statistical Consulting Center, Virginia Polytechnic Institute and State University, Blacksburg, Virginia, 1984.

The description of the variables are given by

x_1: Inventory turnover
x_2: Operating income growth
x_3: Sales growth
x_4: Capital spending growth
x_5: Operating income/total assets
x_6: Industry return on assets
x_7: Industry sales growth
y: Firm return on assets for 1982

Use computed values for VIFs, condition indices, eigenvalues of the correlation matrix, and variance decomposition proportions to diagnose multicollinearity. Discuss the collinearity problem.

8.2 Consider the executive compensation data of Exercise 3.11.

(a) Compute variance inflation factors, eigenvalues of the correlation matrix, and variance decomposition proportions.

(b) Use ridge regression for estimation of the coefficients. Use ℓ on the basis of the ridge trace, i.e., on the basis of stability of coefficients.

(c) Use ridge regression with the ℓ-value determined by considering PR(Ridge).

(d) Estimate the coefficients using ridge regression with ℓ chosen by considering the C_ℓ criterion.

8.3 Consider the results of Exercise 8.2 with the executive compensation data.

(a) Use principal components to estimate coefficients. Make an appropriate choice regarding the number of components to be eliminated.

(b) Which biased estimation procedure appears to be more appropriate: ridge regression or principal components regression?

8.4 Refer to the heat transfer data in Exercise 4.4.

(a) Compute variance inflation factors on the regression coefficients.

(b) Compute the eigenvalues of the correlation matrix.

(c) Compute the variance decomposition proportions, and discuss the multicollinearity. What regressor variables are involved in dependencies that are damaging to the regression coefficients?

(d) Use principal components regression to estimate regression coefficients. Use the PRESS statistic as the criterion on which to base your decision as to how many components should be eliminated.

8.5 For the ridge regression estimator of Eq. (8.10), show that

$$\sum_{i=1}^{k} \frac{\text{Var } b_{i,R}}{\sigma^2} = \sum_{i=1}^{k} \frac{\lambda_i}{(\lambda_i + \ell)^2}$$

where the λ_i are the eigenvalues of the $(\mathbf{X}^{*\prime}\mathbf{X}^*)$ matrix (correlation form). (*Hint*: Write the variance-covariance matrix of $\mathbf{b}_R$.

$$\frac{\text{Var } \mathbf{b}_R}{\sigma^2} = (\mathbf{X}^{*\prime}\mathbf{X}^* + \ell\mathbf{I})^{-1}\mathbf{X}^{*\prime}\mathbf{X}^*(\mathbf{X}^{*\prime}\mathbf{X}^* + \ell\mathbf{I})^{-1}$$

Then diagonalize $\mathbf{X}^{*\prime}\mathbf{X}^*$ and $(\mathbf{X}^{*\prime}\mathbf{X}^* + \ell\mathbf{I})$ with the orthogonal matrix $\mathbf{V}$. See Appendix A.3.)

8.6 Show that the expression for PR(Ridge) in Eq. (8.14) is the PRESS statistic in the case of ridge regression for the case in which scaling is not a function of the data. (*Hint*: Use a development nearly identical to that in Appendix B.4.)

8.7 Show that in the case of ridge regression, the variance inflation factors, which are the diagonal elements of

$$(\mathbf{X}^{*\prime}\mathbf{X}^* + \ell\mathbf{I})^{-1}(\mathbf{X}^{*\prime}\mathbf{X}^*)(\mathbf{X}^{*\prime}\mathbf{X}^* + \ell\mathbf{I})^{-1}$$

decrease with an increasing ℓ.

8.8 Use the GCV criterion to compute the k-value for ridge regression and the resulting coefficients for the case of the hospital data discussed in this chapter and listed in Table 3.8.

8.9 In Section 3.8, Eq. (3.38) relates $\sum_{i=1}^{n} \text{Var } b_i$ to the eigenvalues of the $\mathbf{X}'\mathbf{X}$ (or $\mathbf{X}^{*\prime}\mathbf{X}^*$) matrix for multiple linear regression. In fact, we learned that

$$\sum_{i=1}^{n} \frac{\text{Var } b_i}{\sigma^2} = \sum_{i=1}^{k} \frac{1}{\lambda_i}$$

This directly gives us

$$E(\mathbf{b}'\mathbf{b}) = \boldsymbol{\beta}'\boldsymbol{\beta} + \sigma^2 \sum \frac{1}{\lambda_i}$$

As a result, we say that a symptom of collinearity is that the coefficients are too large in magnitude. Show that the ridge regression estimator of Eq. (8.10) is obtained by minimizing the residual sum of squares, i.e., minimizing $(\mathbf{y} - \mathbf{X}\mathbf{b}_R)'(\mathbf{y} - \mathbf{X}\mathbf{b}_R)$ subject to the restriction that, say, $\mathbf{b}_R'\mathbf{b}_R = \rho$. That is, ridge regression is motivated by the need to reduce the size of the coefficients.

8.10 Prove the result of Eq. (8.12). It will be helpful to write $\mathbf{b}_R$ in terms of $\mathbf{b}$, the least squares estimator. Begin with

$$[I + k(\mathbf{X}^{*\prime}\mathbf{X}^*)^{-1}]\mathbf{b}_R = \mathbf{b}$$

8.11 Show that $\sum_{i=1}^{k} (\text{Var } b_{i,R})/\sigma^2$ decreases monatonically with an increase in k.

8.12 Show that $\sum_{i=1}^{k} (\text{bias } b_{i,R})^2/\sigma^2$ increases monatonically with an increase in k. You may benefit by starting with Eq. (8.12) and use the eigenvalue decomposition.

REFERENCES FOR CHAPTER 8

1. Alldredge, J.R., and N.S. Gilb. 1976. Ridge regression: an annotated bibliography. *International Statistical Review* 44: 355–360.
2. Box, G.E.P., W.G. Hunter, and J.S. Hunter. 1978. *Statistics for Experimenters.* New York: John Wiley.
3. Brown, P.J. 1977. Centering and scaling in ridge regression. *Technometrics* 19: 35–36.
4. Conniffe, D., and J. Stone. 1973. A critical view of ridge regression. *Statist.* 22: 181–187.
5. Draper, N.R., and R.C. Van Nostrand. 1979. Ridge regression and James Stein estimators: Review and comments. *Technometrics* 21: 451–466.
6. Gunst, R.F., and R.L. Mason. 1977. Biased estimation in regression: Evaluation using mean squared error. *Journal of the American Statistical Association* 72: 616–628.
7. Gunst, R.F., and R.L. Mason. 1980. *Regression Analysis and Its Applications: A Data Oriented Approach.* New York: Marcel Dekker.
8. Hawkins, D.M. 1973. On the investigation of alternative regressions by principal component analysis. *Applied Statistics* 22: 275–286.
9. Hemmerle, W.J. 1975. An explicit solution for generalized ridge regression. *Technometrics* 17: 309–314.

10. Hocking, R.R. 1976. The analysis and selection of variables in linear regression. *Biometrics* 32: 1–51.

11. Hocking, R.R., and O.J. Pendleton. 1983. The regression dilemma. *Communications in Statistics* A12(5): 497–527.

12. Hocking, R.R., F.M. Speed, and M.J. Lynn. 1976. A class of biased estimators in linear regression. *Technometrics* 18: 425–438.

13. Hoerl, A.E. 1962. Application of ridge analysis to regression problems. *Chemical Engineering Progress* 58: 54–59.

14. Hoerl, A.E., and R.W. Kennard. 1975. A note on a power generalization of ridge regression. *Technometrics* 17: 269.

15. Hoerl, A.E., and R.W. Kennard. 1970(b). Ridge regression: applications to nonorthogonal problems. *Technometrics* 12: 69–82.

16. Hoerl, A.E., and R.W. Kennard. 1970(a). Ridge regression: biased estimation for nonorthogonal problems. *Technometrics* 12: 55–67.

17. Hoerl, A.E., R.W. Kennard, and K.F. Baldwin. 1975. Ridge regression: some simulations. *Communications in Statistics* 4: 105–123.

18. Mallows, C.L. 1973. Some comments on C_p. *Technometrics* 15: 661–675.

19. Marquardt, D.W., and R.D. Snee. 1975. Ridge regression in practice. *American Statistician* 29: 3–20.

20. Mayer, L.S., and T.A. Willke. 1973. On biased estimation in linear models. *Technometrics* 15: 497–508.

21. Norton, G.W., J.D. Coffey, and B.E. Frye. 1984. Estimating returns to agricultural research, extension, and teaching at the state level. *Southern Journal of Agricultural Economics* (July): 121–128.

22. Obenchain, R.L. 1977. Classical *F*-tests and confidence regions for ridge regression. *Technometrics* 19: 429–439.

23. Obenchain, R.L. 1978. Good and optimal ridge estimators. *Annals of Statistics* 6: 1111–1121.

24. SAS Institute Inc. 1987. *SAS Views: SAS Principles of Regression Analysis, 1984 Edition.* Cary, North Carolina: SAS Institute Inc.

25. Snee, R.D. 1973. Some aspects of nonorthogonal data analysis, Part I. Developing prediction equations. *Journal of Quality Technology* 5: 67–79.

26. Tripp, R.E. 1983. Non-stochastic ridge regression and effective rank of the regressors matrix. Unpublished doctoral dissertation, Department of Statistics, Virginia Polytechnic Institute and State University, Blacksburg, Virginia.

27. Vinod, H.D. 1976. Application of new ridge regression methods to a study of Bell system scaled economies. *Journal of American Statistical Association* 71: 835–841.

28. Wahba, G., G.H. Golub, and C.G. Heath. 1979. Generalized cross validation as a method for choosing a good ridge parameter. *Technometrics* 21: 215–223.

29. Webster, J.T., R.F. Gunst, and R.L. Mason. 1974. Latent root regression analysis. *Technometrics* 16: 513–522.

30. White, J.W., and R.F. Gunst. 1979. Latent root regression: large sample analysis. *Technometrics* 21: 481–488.

CHAPTER 9

NONLINEAR REGRESSION

In all of the previous chapters, the prevailing assumption regarding model description is that the structure is *linear in the model coefficients.* In many areas of the physical, chemical, engineering, and biological sciences, knowledge about the experimental situation suggests the use of a less empirical, more theoretically based, *nonlinear* model.

There are many examples of nonlinear models. We offer but a few here as illustrations:

$$y = \alpha e^{\beta x} + \varepsilon \tag{9.1}$$

$$y = \frac{\alpha}{1 + \exp(-(\beta_1 x_1 + \beta_2 x_2 + \cdots + \beta_k x_k))} + \varepsilon \tag{9.2}$$

$$y = \alpha + \beta_1 x_1^{\gamma_1} + \beta_2 x_2^{\gamma_2} + \cdots + \beta_k x_k^{\gamma_k} + \varepsilon \tag{9.3}$$

$$y = \alpha \exp[-\beta_1 e^{-\beta_2 x}] + \varepsilon \tag{9.4}$$

Note that, in all of the models of Eq. (9.1)–(9.4), at least one of the parameters enters the model in a nonlinear way. As in the case of linear models, the analyst's initial task is to estimate the parameters involved. Again, the technique of least squares is used extensively.

9.1

NONLINEAR LEAST SQUARES

The development of the least squares estimators for a nonlinear model brings about complications not encountered in the case of the linear model. This is easily illustrated in the case of the exponential regression model of Eq. (9.1). Given a set of data (y_i, x_i) for $i = 1, 2, \ldots, n$, the estimators of α and β are found by minimizing

$$SS_{\text{Res}} = \sum_{i=1}^{n} (y_i - \alpha e^{\beta x_i})^2 \tag{9.5}$$

We differentiate the result of (9.5) with respect to α and β and set each derivative to zero. This yields the following equations:

$$\sum_{i=1}^{n} (y_i - \hat{\alpha} e^{\hat{\beta} x_i})(-e^{\hat{\beta} x_i}) = 0 \tag{9.6}$$

$$\sum_{i=1}^{n} (y_i - \hat{\alpha} e^{\hat{\beta} x_i})(-\hat{\alpha} e^{\hat{\beta} x_i} \cdot x_i) = 0 \tag{9.7}$$

Unlike the least squares estimating equations given by (3.4), Eq. (9.6) and (9.7) are *nonlinear in the parameter estimators* $\hat{\alpha}$ *and* $\hat{\beta}$. Thus we cannot compute estimates by elementary matrix algebra. Some type of iterative process must be used.

The literature is quite rich in algorithms for minimization of the residual sum of squares in nonlinear model situations. In addition, there are many regression computer packages available that contain at least one nonlinear estimation method. In a later section, we give the details of some of these methods. However, any analyst who embarks on a nonlinear estimation exercise should be made aware of what is known (or unknown) about the properties of the estimators.

9.2

PROPERTIES OF THE LEAST SQUARES ESTIMATORS

In Chapter 3 details were given regarding the performance of the least squares estimators for the general *linear* model. In Chapter 7 these performance characteristics were reviewed as we discussed methods that apply when conditions are not ideal and thus when assumptions do not hold. It was noted that when the ε_i are normal and independent with homogeneous variance, the estimators achieve minimum variance of all unbiased estimators; if the Gaussian assumption is relaxed, the estimators achieve minimum variance of all linear unbiased estimators. The basis for these properties is the linearity in the parameters in the model of Eq. (3.2). Unfortunately, in the case of the nonlinear model, none of these properties are possessed by the least squares estimators. Consider as a general formulation, the model

$$y_i = f(\mathbf{x}_i, \boldsymbol{\theta}) + \varepsilon_i \qquad (i = 1, 2, \ldots, n) \tag{9.8}$$

where $\boldsymbol{\theta}$ is a vector containing p parameters and $n > p$. We assume further, of course, that f is nonlinear in $\boldsymbol{\theta}' = [\theta_1, \theta_2, \ldots, \theta_p]$. Suppose we call the vector $\hat{\boldsymbol{\theta}}$ the estimator of $\boldsymbol{\theta}$ that minimizes

$$SS_{\text{Res}} = \sum_{i=1}^{n} [y_i - f(\mathbf{x}_i, \hat{\boldsymbol{\theta}})]^2 \tag{9.9}$$

Suppose we also make the assumptions that the ε_i are normal and independent with mean zero and common variance σ^2. We do know that $\hat{\boldsymbol{\theta}}$ is a maximum likelihood estimator of $\boldsymbol{\theta}$. (See Appendix B.8.) However, under these circumstances, one cannot make any general statements about the properties of the estimators *except* for large samples. In other words, the only properties are *asymptotic properties*. The estimators in $\hat{\boldsymbol{\theta}}$ are not unbiased in general, but they are unbiased and minimum variance estimators *in the limit*. That is, the unbiasedness and minimum variance properties are only approached as the sample size grows large. As a result, for a specific nonlinear model and a specific sample size, nothing can truly be stated regarding the properties of the estimators. There are asymptotic variance-covariance results that we can use to obtain approximate confidence intervals and to construct t-statistics on the parameters. Details and illustrations will be presented in Section 9.3.

In Sections 9.3 and 9.4, we give details of methods that allow us to do the computations involved in finding the least squares estimates. We learned by the example in Section 9.1 that direct computation of the estimates cannot be accomplished as they are in the case of linear regression. However, computer algorithms are available in many commercial software packages. These methods vary somewhat in technique and philosphy, but several are quite successful in a wide variety of applications.

9.3

THE GAUSS–NEWTON PROCEDURE FOR FINDING ESTIMATES

The method most often used in software computing algorithms for finding the least squares estimator $\hat{\boldsymbol{\theta}}$ in a nonlinear model is the *Gauss–Newton* procedure. See Bard (1974), Draper and Smith (1981), Kennedy and Gentle (1980), and Bates and Watts (1988).

There are also many modifications of the Gauss–Newton method, and some will be illustrated in this chapter. Essentially, the procedure is iterative and requires *starting value* estimates for the parameters. We shall denote these estimates by the vector $\boldsymbol{\theta}_0' = (\theta_{1,0}, \theta_{2,0}, \ldots, \theta_{p,0})$. In the attempt to find the value of $\boldsymbol{\theta}$ that minimizes the residual sum of squares in Eq. (9.9), we first expand the nonlinear function in (9.8) in a Taylor series (see Appendix B.9) around $\boldsymbol{\theta} = \boldsymbol{\theta}_0$ and retain only linear terms. Thus

$$f(\mathbf{x}_i, \boldsymbol{\theta}) \cong f(\mathbf{x}_i, \boldsymbol{\theta}_0) + (\theta_1 - \theta_{1,0})\left[\frac{\partial f(\mathbf{x}_i, \boldsymbol{\theta})}{\partial \theta_1}\right]_{\boldsymbol{\theta}=\boldsymbol{\theta}_0} + (\theta_2 - \theta_{2,0})\left[\frac{\partial f(\mathbf{x}_i, \boldsymbol{\theta})}{\partial \theta_2}\right]_{\boldsymbol{\theta}=\boldsymbol{\theta}_0} + \cdots + (\theta_p - \theta_{p,0})\left[\frac{\partial f(\mathbf{x}_i, \boldsymbol{\theta})}{\partial \theta_p}\right]_{\boldsymbol{\theta}=\boldsymbol{\theta}_0} \qquad (i = 1, 2, \ldots, n) \tag{9.10}$$

Equation (9.10) represents what is essentially a linearization of the nonlinear form $f(\mathbf{x}_i, \boldsymbol{\theta})$ in (9.8). The reader may view (9.10) as a linear approximation in a neighborhood of the starting values.

An inspection of the linearization in (9.10) reveals that it is of the form

$$\boxed{f(\mathbf{x}_i, \boldsymbol{\theta}) - f(\mathbf{x}_i, \boldsymbol{\theta}_0) \cong \gamma_1 w_{1i} + \gamma_2 w_{2i} + \cdots + \gamma_p w_{pi} \qquad (i = 1, 2, \ldots, n)} \tag{9.11}$$

where

$$w_{ji} = \left[\frac{\partial f(\mathbf{x}_i, \boldsymbol{\theta})}{\partial \theta_j}\right]_{\boldsymbol{\theta}=\boldsymbol{\theta}_0}$$

represents the derivative of the nonlinear function with respect to the jth parameter. Here the derivative is evaluated at all the starting values, and

$$\gamma_j = \theta_j - \theta_{j,0}$$

We can then consider the left-hand side of Eq. (9.11) as the residual $y_i - f(\mathbf{x}_i, \boldsymbol{\theta}_0)$ where the parameters are replaced by starting values. The w_{ji} are known, and play the role of regressor variables in a linear regression, while γ_j, the difference between a parameter value and the starting value, plays the role of a regression coefficient. As a result, the Gauss–Newton procedure builds on the *linear regression* structure

$$y_i - f(\mathbf{x}_i, \boldsymbol{\theta}_0) = \gamma_1 w_{1i} + \gamma_2 w_{2i} + \cdots + \gamma_p w_{pi} + \varepsilon_i \qquad (i = 1, 2, \ldots, n) \tag{9.12}$$

The procedure essentially involves a linear regression analysis on the model of Eq. (9.12). Of course, the analyst cannot work with the model of (9.12) directly in a one-step operation. Estimation of the γ_j does produce estimates of the θ_j, which can be viewed as an improvement on the initial guess $\boldsymbol{\theta}_0$. This operation changes focus from a neighborhood of $\boldsymbol{\theta}_0$ to a new neighborhood around the improved estimate of $\boldsymbol{\theta}$. This operation continues; the exact iterative process is as follows:

1. Estimate $\gamma_1, \gamma_2, \ldots, \gamma_p$ in the model of (9.12) by linear least squares. We shall denote these *first iteration* estimates by $\hat{\gamma}_{1,1}, \hat{\gamma}_{2,1}, \ldots, \hat{\gamma}_{p,1}$.

2. Compute $\hat{\theta}_{j,1} = \theta_{j,0} + \hat{\gamma}_{j,1} (j = 1, 2, \ldots, p)$. At this point, we do not have our final estimates. Rather, $\hat{\theta}_{1,1}, \hat{\theta}_{2,1}, \ldots, \hat{\theta}_{p,1}$ represent first iteration values.
3. The $\hat{\theta}$ values from step 2 replace the starting values in the model in (9.12).
4. Return to the first step, and compute $\hat{\gamma}_{1,2}, \hat{\gamma}_{2,2}, \ldots, \hat{\gamma}_{p,2}$ and eventually $\hat{\theta}_{1,2}, \hat{\theta}_{2,2}, \ldots, \hat{\theta}_{p,2}$.
5. Continue the process until convergence is reached. Convergence implies that after, say, r iterations, the residual sum of squares and the parameter estimates are no longer changing.

At each iteration, the $\hat{\gamma}_j$'s represent "increments" that are added to the estimates from the previous iteration according to step 2. If convergence is achieved, these increments become negligible. As a result, the Gauss–Newton procedure represents a succession of linear regressions in which the residuals are regressed against the w's, i.e., the derivatives evaluated at the current estimates. At each step, the estimates (the $\hat{\theta}_j$'s) become updated. Minimum residual sum of squares is attained when convergence is achieved in the estimates. Formally, then, we can state that the estimate vector at the sth iteration, $\hat{\boldsymbol{\theta}}_s = (\hat{\theta}_{1,s}, \hat{\theta}_{2,s}, \ldots, \hat{\theta}_{p,s})$, is related to that of iteration $(s-1)$ by

$$\begin{aligned} \hat{\boldsymbol{\theta}}_s &= \hat{\boldsymbol{\theta}}_{s-1} + (\mathbf{W}'_{s-1}\mathbf{W}_{s-1})^{-1}\mathbf{W}'_{s-1}[\mathbf{y} - \mathbf{f}(\hat{\boldsymbol{\theta}}_{s-1})] \\ &= \hat{\boldsymbol{\theta}}_{s-1} + \hat{\boldsymbol{\gamma}}_{s-1} \end{aligned} \tag{9.13}$$

Here $\mathbf{W}_{s-1}$ is an $n \times p$ matrix whose (i, j) element is $[\partial f(\mathbf{x}_i, \boldsymbol{\theta})/\partial \theta_j]_{\boldsymbol{\theta} = \hat{\boldsymbol{\theta}}_{s-1}}$, and $[\mathbf{y} - \mathbf{f}(\hat{\boldsymbol{\theta}}_{s-1})]$ represents the *vector* of residuals in which the function, i.e., the n-dimensional vector containing the $f(\mathbf{x}_i, \boldsymbol{\theta}_{s-1})$, is evaluated at $\hat{\boldsymbol{\theta}}_{s-1}$.

Any inference on the coefficients that follows Gauss–Newton nonlinear estimation is based on the *asymptotic* variance-covariance matrix of the regression coefficients. There is a nonlinear analog to the variance-covariance matrix $\sigma^2(\mathbf{X}'\mathbf{X})^{-1}$ in the linear case. The estimate of the *asymptotic* variance-covariance matrix of $\hat{\boldsymbol{\theta}}$ is given by

$$\widehat{\text{Var}}(\hat{\boldsymbol{\theta}}) = s^2(\mathbf{W}'\mathbf{W})^{-1} \tag{9.14}$$

Here $\mathbf{W}$ is the matrix of partial derivatives discussed previously, *evaluated at the least squares estimates* obtained from the final iteration, and s^2 is the familiar residual mean square given by

$$s^2 = \sum_{i=1}^{n} \frac{[y_i - f(\mathbf{x}_i, \hat{\boldsymbol{\theta}})]^2}{n - p}$$

The use of the result in Eq. (9.14) will be illustrated in Examples 9.1 and 9.2.

DIFFICULTIES WITH THE GAUSS–NEWTON PROCEDURE

The user may experience some difficulties with the Gauss–Newton procedure as described here. The drawbacks evolve from the fact that the incremental changes, namely the γ's as described previously, can be estimated very poorly in some problems. The result is that the convergence may be very slow with a large number of iterations being required. In some cases, wrong signs may occur on the $\hat{\gamma}$'s, and then the procedure will move in the wrong direction. Indeed, the method may not converge at all with the residual sum of squares continuing to increase.

There are several modifications of the Gauss–Newton procedure that are designed to alleviate some of the difficulties. Most commercial regression computer packages that do nonlinear regression provide not merely the Gauss–Newton procedure but modifications which generally bring about improvements. Some of these improvements are designed to directly reduce the calculated incremental changes in the regression coefficients when the results indicate that reduction is needed. In other cases, the structure of these increments is changed in a subtle but major way.

USE OF FRACTIONAL INCREMENTS

In many real-life instances, the ordinary Gauss–Newton procedure can be improved by simply allowing the "jump size," i.e., the $\hat{\gamma}_j$'s, at each iteration to be reduced in magnitude. The goal is not to allow the computed increments at a specific iteration to increase SS_{Res}. To this end, we present the following reasonable strategy:

1. Use $(\mathbf{W}'_{s-1}\mathbf{W}_{s-1})^{-1}\mathbf{W}'_{s-1}[\mathbf{y} - \mathbf{f}(\hat{\boldsymbol{\theta}}_{s-1})] = \hat{\boldsymbol{\gamma}}_{s-1}$ to compute the standard Gauss–Newton increment vector for the sth iteration ($s = 1, 2, \ldots$).
2. Compute $\hat{\boldsymbol{\theta}}_s = \hat{\boldsymbol{\theta}}_{s-1} + \hat{\boldsymbol{\gamma}}_{s-1}$ as the Gauss–Newton procedure suggests.
3. If $SS_{\text{Res},s} < SS_{\text{Res},s-1}$, continue to the next iteration using $\hat{\boldsymbol{\theta}}_s$.
4. If $SS_{\text{Res},s} > SS_{\text{Res},s-1}$, go back to step 2; use $\hat{\boldsymbol{\gamma}}_{s-1}/2$ as the vector of increments.

The halving may be used, say, 10 times during an iteration if necessary. Generally, if the increment is halved approximately 10 times and a decrease in SS_{Res} is not experienced, the procedure should be terminated.

Though the methodology based on fractional increments is truly a strategic modification of the classic Gauss–Newton linearization, *Gauss–Newton* is often used as terminology to describe the basic Gauss–Newton procedure *with* the halving mechanism. Some computer packages automatically invoke the halving procedure in what they call the Gauss–Newton method. This modification does improve the Gauss–Newton estimation procedure, and the user should be aware of whether or not it is being used.

Example 9.1 ***Mining Excavation Data***

A major problem associated with many mining projects is subsidence, or sinking of the ground above the excavation. The mining engineer needs to control the amount and distribution of this subsidence. This will ensure that structures on the surface survive the excavation.

There are several factors that affect the amount and nature of the subsidence. Among these are the depth of the mine and the width of the excavation. An important variable, which aids in characterizing the condition, is known as the angle of draw, y. It is defined as the angle between the perpendicular at the edge of the excavation and the line that connects the same edge of excavation with the point on the surface for which there is zero subsidence. Engineers generally feel that the angle of draw should relate well to the ratio of the width (w) of the excavation to the depth (d) of the mine. It also is known that any relationship involved is nonlinear. The following is a data set involving y, w, and d for mining excavations in West Virginia.

Observation	w (ft)	d (ft)	y (degrees)
1	610	550	33.6
2	450	500	22.3
3	450	520	22.0
4	430	740	18.7
5	410	800	20.2
6	500	230	31.0
7	500	235	30.0
8	500	240	32.0
9	450	600	26.6
10	450	650	15.1
11	480	230	30.0
12	475	1400	13.5
13	485	615	26.8
14	474	515	25.0
15	485	700	20.4
16	600	750	15.0

Data were collected by the Mining Engineering Department and analyzed by the Statistical Consulting Center, Virginia Polytechnic Institute and State University, Blacksburg, Virginia, 1982.

A nonlinear model was required, which depicts the *growth* of subsidence with an increase in the *ratio* w/d and a "leveling" of the growth at some w/d value. A growth model of the type

$$y_i = \alpha\left[1 - \exp\left(-\beta\left(\frac{w}{d}\right)_i\right)\right] + \varepsilon_i \qquad (i = 1, 2, \ldots, 16)$$

was attempted. This model is a variation of the Mitcherlich Law, which will be discussed in Section 9.5. The NLIN procedure in SAS (1985) was used to estimate the parameters. The Gauss–Newton option was chosen. In SAS, the Gauss–Newton method involves the halving procedure, which we discussed in this section. Starting values can be chosen by taking into account the nature of the model. As w/d grows large, y approaches α. Thus a value for α_0 of 35 seems reasonable from the data. In addition, a reasonable starting value for β can be determined by considering the nonrandom portion of the model, namely

$$y = \alpha\left[1 - \exp\left(-\beta\left(\frac{w}{d}\right)\right)\right]$$

A crude estimate of β, and thus a starting value, is achieved by taking the negative of the slope of the graph of $\ln(1 - (y/\alpha_0))$ against w/d. From this exercise, a value of $\beta_0 = 1.0$ was adopted. The reader should refer to Section 9.6 for more on selection of starting values. The SAS Gauss–Newton method converged in four iterations. Table 9.1 illustrates the values of the parameter estimate at each iteration. Convergence was achieved at $\hat{\alpha} = 32.4644$ and $\hat{\beta} = 1.5112$ with a resulting residual sum of squares of 204.6331.

The error mean square, used in computing approximate standard errors of $\hat{\alpha}$ and $\hat{\beta}$, is given by

$$s^2 = \frac{SS_{\text{Res}}}{n-2} = \frac{204.6331}{14} = 14.6167$$

Of course, much of the inference regarding tests and confidence intervals on α and β evolve from the asymptotic variance-covariance matrix determined from Eq. (9.14). Here, the **W** matrix of derivatives, evaluated at the parameter estimates, must be computed and $s^2(\mathbf{W}'\mathbf{W})^{-1}$ evaluated. This result is given by

$$s^2(\mathbf{W}'\mathbf{W})^{-1} = \begin{bmatrix} 7.0109 & -0.6918 \\ -0.6918 & 0.0887 \end{bmatrix}$$

The asymptotic standard errors of $\hat{\alpha}$ and $\hat{\beta}$, retrieved by computing the square roots of the diagonal elements of $s^2(\mathbf{W}'\mathbf{W})^{-1}$, are $s_{\hat{\alpha}} = 2.6478$ and $s_{\hat{\beta}} = 0.2978$.

Asymptotic confidence intervals on the parameters are a standard part of most nonlinear commercial software packages. They are based on the use of the t-distribution. In this case, there are 14 residual degrees of freedom, and hence the 95% confidence intervals are of the form $\hat{\alpha} \pm t_{.025,14}s_{\hat{\alpha}}$ and $\hat{\beta} \pm t_{.025,14}s_{\hat{\beta}}$. The numerical results are given by,

	Lower	Upper
α	26.7854	38.1433
β	0.8725	2.1498

TABLE 9.1 **Information from the four Gauss–Newton iterations for mining subsidence data**

Iteration	$\hat{\alpha}$	$\hat{\beta}$	SS_{Res}
0	35.0000	1.0000	345.1943
1	30.3500	1.4872	252.2746
2	32.4579	1.5134	204.6350
3	32.4648	1.5111	204.6331
4	32.4644	1.5111	204.6331

FIGURE 9.1 **y = angle of draw in degrees; w/d = width of excavation to depth of mine; mining excavation data fit to nonlinear growth model**

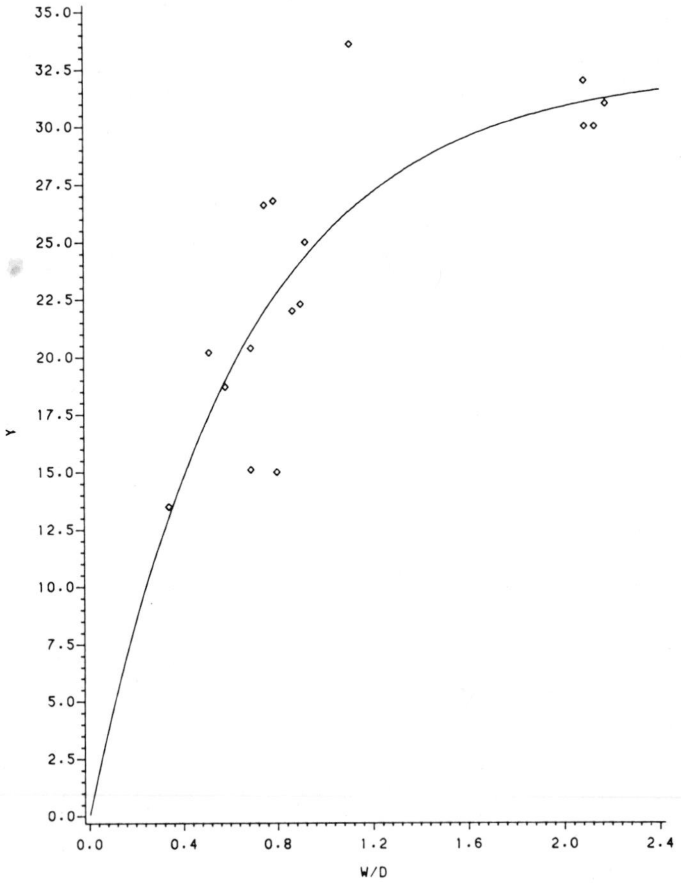

We offer the following table of residuals so that the reader gains insight into the quality of fit.

Observation	y (angle)	$\hat{y}$	$y - \hat{y}$
1	33.6	26.3896	7.2104
2	22.3	24.1323	−1.8323
3	22.0	23.6848	−1.6848
4	18.7	18.9731	−0.2731
5	20.2	17.4999	2.7001
6	31.0	31.2490	−0.2490
7	30.0	31.1610	−1.1610
8	32.0	31.0707	0.9293
9	26.6	22.0125	4.5875
10	15.1	21.0603	−5.9603
11	30.0	31.0783	−1.0783
12	13.5	13.0224	0.4776
13	26.8	22.6049	4.1951
14	25.0	24.3851	0.6149
15	20.4	21.0698	−0.6698
16	15.0	22.7731	−7.7731

The plot of the data along with the estimated growth model is given in Figure 9.1.

9.4

OTHER MODIFICATIONS OF THE GAUSS–NEWTON PROCEDURE

Difficulties with the Gauss–Newton procedure surface most often in functions that behave *very much unlike linear* functions. For some details, see Ratkowsky (1983). In addition, the quality of starting values plays a major role. While the halving procedure described previously certainly improves the method, there are other important modifications of the Gauss–Newton procedure. Hartley (1961) suggested a modification that is guaranteed to converge and does not suffer from the difficulties we often experience with the ordinary Gauss–Newton method. This alteration again shrinks the increment size according to the behavior of the SS_{Res} at each iteration. A second and very popular modification was developed by Marquardt (1963). The Marquardt procedure is based on a revision of Eq. (9.13) for computing the vector of incremental changes. In this method, the structure of the vector of increments for the sth iteration is given by the solution $\hat{\boldsymbol{\gamma}}_s$ to

$$(\mathbf{W}_s'\mathbf{W}_s + \lambda\mathbf{I}_p)\hat{\boldsymbol{\gamma}}_s = \mathbf{W}_s'[\mathbf{y} - \mathbf{f}(\hat{\boldsymbol{\theta}}_s)] \qquad (\lambda > 0) \tag{9.15}$$

The information in Marquardt's paper provides some geometric insight into why the use of the constant, λ, given by Eq. (9.15), can improve convergence. The reader who is familiar with multicollinearity and biased estimation in linear regression should understand the similarity between (9.15) and the *ridge regression* (see Chapter 8) estimator of regression coefficients. The Marquardt procedure may be viewed as a ridge regression approach to the improvement of the estimators of the incremental changes at each iteration. The reader should come to understand why this ridge regression approach is reasonable. The Gauss–Newton method calculates the incremental changes as a result of ordinary least squares; since the regressor variables are derivatives of the same function, ridge regression is a natural solution to inevitable multicollinearity. Thus the λ value of Eq. (9.15) is essentially the k value in ridge regression. Scaling the columns of the $\mathbf{W}_s$ matrix is accomplished in order to produce unity elements on the diagonal of $\mathbf{W}_s'\mathbf{W}_s$.

COMPUTATION OF λ

The procedure for the computation of λ quite naturally invites attention. It may vary depending on the computer software package. However, *to assure convergence*, the following inequality must hold:

$$SS_{\text{Res}}(\hat{\boldsymbol{\theta}}_{s+1}) < SS_{\text{Res}}(\hat{\boldsymbol{\theta}}_s) \qquad (s = 0, 1, 2, \ldots) \tag{9.16}$$

Marquardt suggested a type of trial and error evaluation to find a value that reduces SS_{Res} at each iteration. A small value of λ should be used when conditions suggest that the ordinary Gauss–Newton procedure will converge satisfactorily. We should use relatively large values of λ only when necessary to satisfy (9.16).

There is certainly no unique algorithm or strategy to use in the computation of λ at each iteration. One possible method, currently used by SAS in the NLIN procedure begins with $\lambda = 10^{-8}$. A series of trial and error computations is done at each iteration, with λ continually being multiplied by 10 until (9.16) is satisfied. The process also involves reduction by a factor of 10 at each iteration as long as (9.16) is satisfied. The design is to keep λ as small as possible while assuring that we experience an improvement in SS_{Res} at each iteration.

Example 9.2 ***Chemical Kinetic Data***

In the field of ecology, the relationship between the concentration of available dissolved organic substrate and the rate of uptake (velocity) of that substrate by heterotrophic microbial communities has been described by the Michaelis–Menten Equation (1913). This equation has been used for many decades as a standard model for estimating kinetic parameters. The form of the equation to be used here is as follows.

$$y = \frac{V_{max}}{k + x}$$

where

$$\begin{aligned} y &= \text{velocity of uptake, } \mu\text{g/l/hr} \\ V_{max} &= \text{maximum velocity} \\ k &= \text{transport constant, } \mu\text{g/l} \\ x &= \text{concentration of substrate, } \mu\text{g/l} \end{aligned}$$

In this example, we use Michaelis–Menten kinetics to build a model to estimate k and V_{max} and to be able to predict velocity of uptake from data involving a glucose type substrate. The substrate is added to a sediment sample, and the sample is incubated. The data collected include the velocity (y) and the concentration (x). Data are as follows:

Observation	y	x
1	0.0773895	0.417
2	0.0688714	0.417
3	0.0819351	0.417
4	0.0737034	0.833
5	0.0738753	0.833
6	0.0712396	0.833
7	0.0650420	1.670
8	0.0547667	1.670
9	0.0497128	3.750
10	0.0642727	3.750
11	0.0613005	6.250
12	0.0643576	6.250
13	0.0393892	6.250

Data analyzed for the Department of Biology by the Statistical Consulting Center, Virginia Polytechnic Institute and State University, Blacksburg, Virginia, 1983.

The statistical model is written

$$y_i = \frac{V_{max}}{k + x_i} + \varepsilon_i \qquad (i = 1, 2, \ldots, 13)$$

The kinetic data were fit to the above model. The Marquardt option of the NLIN procedure in SAS was used. Convergence was achieved in eight iterations. The starting values chosen were $V_{max,0} = 0.5$ and $k_0 = 17$. The information from each iteration is as follows:

Iteration	V_{max}	k	SS_{Res}
0	0.50000000	17.00000000	0.02113709
1	0.75394344	7.50632545	0.00381199
2	0.70786719	8.53308100	0.00099829
3	0.81417449	10.34075292	0.00084822
4	0.97064721	12.67814397	0.00080054
5	1.03769475	13.68324303	0.00079634
6	1.04825994	13.84202905	0.00079625
7	1.04914440	13.85524467	0.00079625
8	1.04920879	13.85620450	0.00079625

Starting values here can easily be determined by linearizing the Michaelis–Menten Equation. For example, if we were to plot $1/y$ against x, a reasonable $V_{max,0}$ is the reciprocal of the slope. In addition, k_0 can be computed as the product of the intercept of that plot and $V_{max,0}$. Using this method, the analyst obtains $V_{max,0} \cong 1.0$ and $k_0 \cong 14$. We chose *not* to use these starting values in order to illustrate relative insensitivity to starting values in this case. The asymptotic variance-covariance matrix is

$$s^2(\mathbf{W}'\mathbf{W})^{-1} = \begin{bmatrix} 0.114473 & 1.685088 \\ 1.685088 & 25.117753 \end{bmatrix}$$

with $s^2 = 7.239 \times 10^{-5}$. From this matrix, we can easily verify that the standard errors of $\hat{V}_{max}$ and $\hat{k}$, the estimators, are

$$s_{\hat{V}_{max}} = 0.3383 \qquad s_{\hat{k}} = 5.0118$$

9.5

SOME SPECIAL CLASSES OF NONLINEAR MODELS

There are certain fields of application in which nonlinear modeling is rarely successful. Success is enjoyed most often in areas where the theory or knowledge of the subject suggests a model, as in Examples 9.1 and 9.2. The pioneers of nonlinear modeling applications were analysts working in scientific fields where mathematical formulae indicate how factors relate to one another. In the latter case, there is no need to resort to more empirical linear models. We emphasize that one should not adopt a specific nonlinear model without a reasonable understanding of its implications. Many analysts unsuccessfully try nonlinear model building simply because they cannot find a linear model to fit the data in question.

Many nonlinear models fall into categories that are designed for specific situations. In each category, there are several models that have been used successfully by analysts in various fields. In the following subsections, we discuss some of these categories of models and give the reader some insight into appropriate applications. The best known category is that of *growth models*, which are used to describe how something grows with an increase in a certain independent variable (quite often, time). The classic fields of application are biology, forestry, and zoology where organisms and plants grow over time. Economics, manpower applications, and even reliability engineering represent other fields where growth models apply. In addition, growth models have received attention in chemotherapy research. See Carter *et al.* (1983).

The Logistic Growth Model

Logistic regression received considerable attention in Chapter 7 where we focused on methods of handling the (0, 1) response. In the case of *logistic growth*, a measurable quantity y varies with some quantity x according to the model

$$y = \frac{\alpha}{1 + \beta \exp(-kx)} + \varepsilon \tag{9.17}$$

Note that the form of the growth model in (9.17) is similar to the model discussed in Chapter 7 for handling binary type responses. In the case of the logistic model, the parameters take on a special meaning. For $x = 0$, $y = \alpha/(1 + \beta)$, and thus this quantity can be considered the level of y at time zero, or zero level of application. On the other hand, the important parameter α is the *limiting growth*, i.e., the value that y approaches as x grows larger. The values of β and k must be positive so that the logistic function can be interpretable. The plot of y against x for the logistic of (9.17) is S-shaped. Figures 9.2 and 9.3 provide a graphical illustration of the appearance of the logistic function for various k and β.

In some applications of the logistic model, the kx portion in the exponent is replaced by the more general linear structure $\beta_0 + \beta_1 x_1 + \beta_2 x_2 + \cdots$ or by a polynomial in a single regressor structure. (See Carter *et al.* (1983).)

The Gompertz Growth Model

The Gompertz model is another S-shaped functional form that is applicable in growth situations. The form of the model is given by

$$y = \alpha \exp[-\beta e^{-kx}] + \varepsilon \tag{9.18}$$

Notice that the form of the Gompertz model is a double exponential. Again, the parameter α is limiting growth. For $x = 0$, $y = \alpha e^{-\beta}$.

The Richards Growth Model

The Richards model (see Richards (1959)) is a simple variation of the logistic model in which an additional parameter is adopted. The functional form

$$y = \frac{\alpha}{[1 + \beta e^{-kx}]^{1/\delta}} + \varepsilon \tag{9.19}$$

is used.

FIGURE 9.2 Graph of logistic growth model for $\alpha = 10$, $\beta = 2$

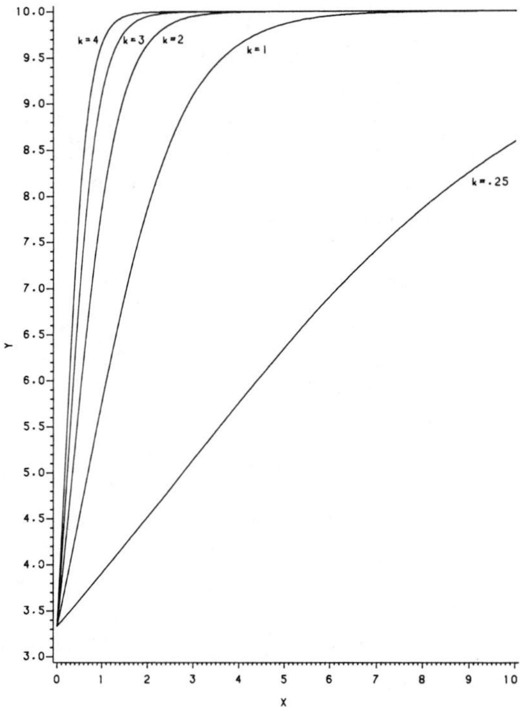

FIGURE 9.3 Graph of logistic growth model for $\alpha = 10$, $\beta = 5$

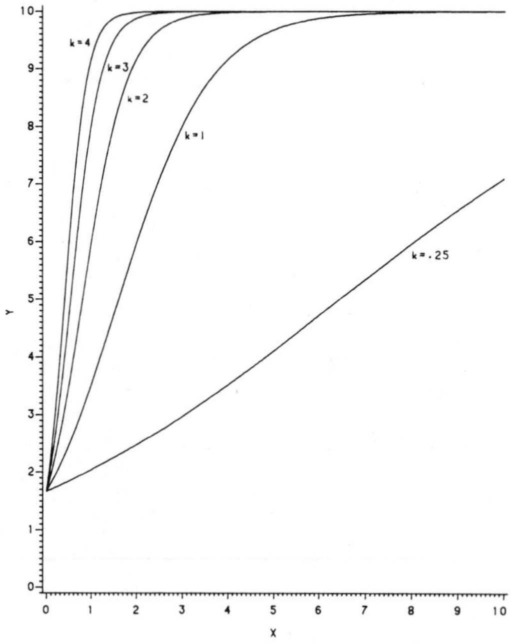

The Weibull Growth Model

The Weibull growth model is also a candidate when the data set involves a growth mechanism. The functional form is given by

$$y = \alpha - \beta \exp[-\gamma x^{\delta}] + \varepsilon \tag{9.20}$$

In this case, the growth at $x = 0$ is $\alpha - \beta$, while again the growth approaches its maximum $y = \alpha$ as $x \to \infty$.

Though the four models in (9.17)–(9.20) give the reader a rich set of functional forms, they certainly do not exhaust the entire family of growth models. For other forms and further discussion, see Ratkowsky (1983) and Draper and Smith (1981).

The Mitcherlich Law

We shift our focus somewhat away from growth models and turn toward a class of models useful in agriculture, chemistry, and the engineering sciences. Here the interest is in y, the yield of some mechanism, such as a crop yield or a chemical reaction, and x, the regressor variable that provides the impetus for the increasing yield; the variable x may be rate of fertilizer application, time, etc. The mechanism certainly has similarities to the growth situation. Most evolve from the model known as Mitcherlich's Law. See Phillips and Campbell (1968). The model is written

$$y = \alpha - \beta\gamma^{x} + \varepsilon \tag{9.21}$$

The model in (9.21) is also used in growth situations. There is no point of inflection (the point where $\partial^2 y/\partial x^2$ is zero) as in the case of the logistic or Gompertz model; thus it is not S-shaped. There are many different reparameterizations of the model of (9.21) that carry the same name. A few are given below:

$$y = \alpha - \beta e^{-\gamma x} \tag{9.22}$$

$$y = \alpha - e^{-(\beta + \gamma x)} \tag{9.23}$$

$$y = \alpha[1 - e^{-\beta x}] \tag{9.24}$$

$$y = e^{\alpha} - \beta\gamma^{x} \tag{9.25}$$

The model of Eq. (9.24) was illustrated in Example 9.1. It is also often called the MacArthur–Wilson Growth Equation. It is not our intention to do an exhaustive presentation of types of nonlinear models. For a more thorough treatment, see Ratkowsky (1983).

9.6

FURTHER CONSIDERATIONS IN NONLINEAR REGRESSION

In this section we discuss some issues that often puzzle many users of nonlinear regression. There are several troublesome loose ends that deal with starting values, standard statistical inference, and *transformations to linearize.*

STARTING VALUES ON PARAMETERS

The use of nonlinear regression methodology presumes the presence of initial estimates, which are the values in the vector $\boldsymbol{\theta}_0$. Obviously if $\boldsymbol{\theta}_0$ is relatively close to the least squares estimate $\hat{\boldsymbol{\theta}}$, convergence difficulties will be minimal. In the case of the unmodified Gauss–Newton procedure, good starting values are vital. The modifications (i.e., reduced increments, Hartley Modification, and Marquardt Modification) render the procedure less sensitive to starting values. However, even with the modifications, the need for good starting values cannot be ignored. Poor starting values may result in convergence to a *local minimum* of the SS_{Res} function. Indeed, it is quite possible that the analyst would not even be aware of this convergence to the "wrong values."

In some instances, the experienced analyst will know of approximate parameter values that serve as starting values. Certainly in the case of the growth functions, either the parameters or the ratios of the parameters have obvious interpretations, and if much is known about the system and the model, starting values can be found. In many cases, reasonable starting values can be determined from the data set. A simple example of this is the exponential model of Eq. (9.1). Consider the deterministic portion of the model, namely $y = \alpha e^{\beta x}$. A linearized form is found by taking natural logs as follows:

$$\ln y = \ln \alpha + \beta x \tag{9.26}$$

As a result, a linear regression of $\ln y$ against x can easily be accomplished, with the antilog of the intercept used as a starting value for α and the slope of the regression used as a starting value for β.

Similar *tricks* can be used for other models. For example, in the case of the logistic model of Eq. (9.17), a value close to $y_{\max}$ in the data can be used as α_0, the starting value on α. Then a simple regression of $\ln((\alpha_0/y) - 1)$ against x can be accomplished, with the antilog of the intercept of that regression being used as β_0 and the slope being $-k_0$. If a linear regression exercise appears a bit time consuming, a simple plotting of $\ln((\alpha_0/y) - 1)$ against x can produce "eyeball" values of the slope and intercept, and starting values can be captured accordingly.

In the case of the Gompertz growth model of Eq. (9.18), initial estimates are achieved quite easily, and the procedure is obvious after a little rearranging. Again, a value close to $y_{\max}$ produces an initial value α_0. By working with the deterministic portion of the model, we discover that

$$\ln\left[-\ln\frac{y}{\alpha_0}\right] = \ln\beta - kx$$

and thus a linear regression (or a plotting) of $\ln[-\ln(y/\alpha_0)]$ against x can be accomplished, with β_0 being the antilog of the intercept and $-k_0$ being the slope.

Similar procedures can be used for obtaining starting values in the case of other nonlinear models. At times the complexity of the nonlinear structure can cause difficulty. There are computer software algorithms that use grid search routines, which compute starting values from crude ranges suggested by the user. (See *SAS User's Guide: Statistics* (1985).)

STANDARD STATISTICAL INFERENCE

We pointed out earlier that the user of nonlinear regression does not have the luxury of making exact statistical inferences, e.g., tests and confidence intervals on model parameters, confidence intervals on mean response and prediction intervals, etc. For finite samples, approximations exist that depend on the computed value of the asymptotic variance-covariance matrix of the parameter estimates. Examples 9.1 and 9.2 illustrate the role of that matrix. Many analysts are discouraged from using nonlinear regression because some of the procedures that one takes for granted in linear regression are either not available or are questionable in nonlinear regression.

It is important for the user to understand what is available in nonlinear regression in terms of standard techniques. We have used the $(\mathbf{W}'\mathbf{W})^{-1}$ matrix to play the role of $(\mathbf{X}'\mathbf{X})^{-1}$ in obtaining standard errors of parameter estimates. Strictly speaking, the $\mathbf{W}$ matrix is that which is *available to us*, but it serves as an *estimate* of the matrix W where

$$W = \begin{bmatrix} \dfrac{\partial f(\mathbf{x}_1, \boldsymbol{\theta})}{\partial \theta_1} & \dfrac{\partial f(\mathbf{x}_1, \boldsymbol{\theta})}{\partial \theta_2} & \cdots & \dfrac{\partial f(\mathbf{x}_1, \boldsymbol{\theta})}{\partial \theta_p} \\ \dfrac{\partial f(\mathbf{x}_2, \boldsymbol{\theta})}{\partial \theta_1} & \dfrac{\partial f(\mathbf{x}_2, \boldsymbol{\theta})}{\partial \theta_2} & \cdots & \dfrac{\partial f(\mathbf{x}_2, \boldsymbol{\theta})}{\partial \theta_p} \\ \vdots & \vdots & & \vdots \\ \dfrac{\partial f(\mathbf{x}_n, \boldsymbol{\theta})}{\partial \theta_1} & \dfrac{\partial f(\mathbf{x}_n, \boldsymbol{\theta})}{\partial \theta_2} & \cdots & \dfrac{\partial f(\mathbf{x}_n, \boldsymbol{\theta})}{\partial \theta_p} \end{bmatrix}$$

and, the derivatives are evaluated at the *true parameters*. To formalize what is known about the least squares estimators $\hat{\boldsymbol{\theta}}$ in a nonlinear model, we can write that if ε, the vector of model errors in (9.8), is normally distributed with mean $\mathbf{0}$ and variance covariance matrix $\sigma^2\mathbf{I}$, then

> $\hat{\boldsymbol{\theta}}$ *is approximately normal with mean* $\boldsymbol{\theta}$ *and variance-covariance matrix* $\sigma^2(W'W)^{-1}$

Actually, $\hat{\boldsymbol{\theta}}$ is asymptotically normal even in cases where $\boldsymbol{\varepsilon}$ is not normal. For details regarding this theory, the reader is referred to Gallant (1987). As a result, we referred to $(\mathbf{W}'\mathbf{W})^{-1}$ in (9.14) as the *estimated asymptotic* variance-covariance matrix. The "estimate" comes from the use of s^2 instead of σ^2 and $\mathbf{W}$ instead of W. Similar asymptotic statements can be made about the residual sum of squares. The student of statistics knows that for linear regression, $SS_{\mathbf{Res}}/\sigma^2$ has a χ^2-distribution with $n-p$ degrees of freedom. In nonlinear regression,

> *$SS_{\mathbf{Res}}/\sigma^2$ has approximately a χ^2-distribution with $n-p$ degrees of freedom and $\hat{\boldsymbol{\theta}}$, the least squares estimator, and $SS_{\mathbf{Res}}$ are independent*

The foregoing information allows for approximate procedures that are very much like the standard methodology used in linear regression. For example, approximate t-tests can be constructed (though the t-statistics often do not appear in standard statistical packages) on coefficients by merely computing

$$t = \frac{\hat{\theta}_j - \theta_{j,0}}{s_{\hat{\theta}_j}} \tag{9.27}$$

where $\hat{\theta}_j$ is the least squares estimator of θ_j, $s_{\hat{\theta}_j}$ is the standard error of $\hat{\theta}_j$ as in Example 9.1, and $\theta_{j,0}$ is the value of θ_j under the hypothesis

$$H_0: \theta_j = \theta_{j,0}$$
$$H_1: \theta_j \neq \theta_{j,0}$$

The statistic in Eq. (9.27) is approximately t_{n-p-1} under H_0. Obviously, either one- or two-tailed hypotheses can be accommodated.

NONLINEAR FUNCTIONS IN $\hat{\theta}$; CONFIDENCE LIMITS AND PREDICTION LIMITS

More often than not, the user of nonlinear regression is interested in functions of the parameter estimates that are nonlinear in the elements of $\hat{\boldsymbol{\theta}}$. Obvious applications are confidence intervals on mean response and prediction limits on a new observed response. These procedures for linear regression were covered in detail in Chapters 2 and 3. Actually the role of the $\mathbf{W}$ matrix in this type of computation is as one would expect. Indeed, the computational procedure follows along lines very similar to that of linear regression. Suppose we address the notion of any nonlinear function $g(\boldsymbol{\theta})$. Now, for the estimate $g(\hat{\boldsymbol{\theta}})$,

> *$g(\hat{\boldsymbol{\theta}})$ is approximately normal with approximate mean and variance given by $g(\boldsymbol{\theta})$ and $\sigma^2\mathbf{u}'(W'W)^{-1}\mathbf{u}$ respectively. Here, the vector $\mathbf{u}' = [\partial g(\boldsymbol{\theta})/\partial\theta_1, \partial g(\boldsymbol{\theta})/\partial\theta_2, \ldots, \partial g(\boldsymbol{\theta})/\partial\theta_p]$*

The reader should now visualize applying this result to, say, confidence limits on mean response. Keep in mind that $\mathbf{u}'$, the vector of derivatives in this application, would become

$$w_0' = \left[\frac{\partial f(\boldsymbol{\theta}, \mathbf{x}_0)}{\partial \theta_1}, \frac{\partial f(\boldsymbol{\theta}, \mathbf{x}_0)}{\partial \theta_2}, \ldots, \frac{\partial f(\boldsymbol{\theta}, \mathbf{x}_0)}{\partial \theta_p}\right]$$

in the situation where one wishes to predict at an arbitrary location $\mathbf{x}_0$ in the regressors. Obviously the user must substitute estimates in for parameters in w_0. As a result, an *estimate* of the approximate *standard error of a predicted response* $f(\mathbf{x}_0, \hat{\boldsymbol{\theta}})$ is given by

$$s_{f(\mathbf{x}_0, \hat{\boldsymbol{\theta}})} = s w_0'(\mathbf{W}'\mathbf{W})^{-1} w_0$$

where w_0 is a vector of derivatives evaluated at the location $\mathbf{x}_0$ and with parameter estimates substituted. Of course, $\mathbf{W}$ contains parameter estimates. Notice the structure similar to the standard error of a predicted value in multiple *linear* regression. Thus approximate $100(1-\alpha)\%$ confidence intervals on the mean response at an arbitrary location $\mathbf{x}_0$ is given by

$$f(\mathbf{x}_0, \hat{\boldsymbol{\theta}}) \pm t_{\alpha/2, n-p-1} s \sqrt{w_0'(\mathbf{W}'\mathbf{W})^{-1} w_0}$$

and of course, approximate $100(1-\alpha)\%$ prediction limits on a new observation at $\mathbf{x}_0$ is given by

$$f(\mathbf{x}_0, \hat{\boldsymbol{\theta}}) \pm t_{\alpha/2, n-p-1} s \sqrt{1 + w_0'(\mathbf{W}'\mathbf{W})^{-1} w_0}$$

The confidence limits and prediction limits are certainly important when the analyst is interested in using the nonlinear regression function for prediction. They may also be used in comparing the performance of different fitted nonlinear models. Clearly if prediction is important, the model that produces the "tighter" confidence limits is one that should be given serious consideration. Also, of course, one should consider the use of the error mean square, s^2, in comparing competing models.

The general nature of the residuals can also provide useful information for determining the most appropriate model from a set of competing nonlinear models. Diagnostics similar to those discussed in Chapter 6 can be generated for nonlinear models, though there is no easy access to these diagnostics in commercial computer packages as given for the case of linear models. For the Gauss–Newton procedure (or modifications), the $\mathbf{W}'\mathbf{W}$ matrix, evaluated at the solutions to the parameter estimates, plays the role of $\mathbf{X}'\mathbf{X}$. As a result, HAT-diagonal values, studentized residuals, and all diagnostics derived from them can be computed.

9.7

WHY NOT TRANSFORM DATA TO LINEARIZE?

The data analyst is likely to consider simplifying the estimation procedure by performing transformations on the model to induce linearity in a model that is inherently nonlinear. It seems like a reasonable approach, and indeed in the numerical Examples 9.1 and 9.2, it was suggested as a method to *locate starting values.* In Section 9.6, we gave other illustrations in which starting values were found by linearizing first. However, the obvious question arises: "Why not linearize and then estimate parameters with a linear model?" The concepts that must be understood to answer this question often go unappreciated. A classic example is the exponential model form of Eq. (9.1). By taking natural logarithms, we linearize the deterministic part of the model as we did in Eq. (9.26). As a result, consider the alternative model

$$\ln y = \ln \alpha + \beta x_i + \varepsilon_i \tag{9.28}$$

Use of Eq. (9.28) surely eliminates the complexity of the nonlinear approach, and a simple linear regression can be applied directly by regressing $\ln y$ against x. This is not an isolated example. There are many situations where a transformation linearizes (or simplifies) the model. The Michaelis–Menten model for the chemical kinetic illustration of Example 9.2 is another case in point. The original model is

$$y_i = \frac{V_{\max}}{k + x_i} + \varepsilon_i \tag{9.29}$$

But we can simplify estimation by dealing with $1/y$. Hence we *might* postulate a model of the type

$$\frac{1}{y_i} = \frac{k}{V_{\max}} + \left(\frac{1}{V_{\max}}\right) x_i + \varepsilon_i \tag{9.30}$$

Then we can perform a simple linear regression by regressing $1/y$ against x, and we can estimate the parameters $V_{\max}$ and k accordingly.

The least squares procedure on the linearized form as in (9.28) or (9.30) will not produce the same answers for the parameter estimates as in the case of the nonlinear analog. For the nonlinear model (Eq. (9.1) or (9.29)), least squares implies the minimization of the sum of squares of residuals on y. But, in the model forms of (9.28) and (9.30), we are minimizing the sum of squares of residuals on a *transform* of y, $\ln y$ in the case of (9.28) and $1/y$ in the case of (9.30). Which is correct? Is it strictly incorrect to linearize? The answer lies in the issue of which model assumptions are correct. There are two separate models and model philosophies. For the model of Eq. (9.1), the error structure is *additive.* If one assumes this is reasonable, merely taking logs does not produce the model of (9.28). Thus (9.1) and (9.28) are different, with (9.28) being a result of taking logs of an exponential model with a *multiplicative error structure.* A similar distinction can be made between models (9.29) and (9.30).

We often make certain assumptions on the ε_i, the model errors, namely that they are normal, independent, with mean 0 and common variance σ^2. If a commitment is made to these error assumptions in, say, (9.29), it makes little sense to transform to (9.30) with an entirely different error structure (and hence different assumptions) required. Thus the issue becomes whether to assume that well-behaved model errors accompany the original (nonlinear) model or the transformed one.

Another difficulty is derived from the nature of the parameters that result from the transformed model. In nearly all cases in which a transformation results in a linear model, complications arise in the properties of the estimators of the parameters. An example is produced by models (9.29) and (9.30). The transformation to (9.30) allows a linear regression to be performed. However, even if the usual assumptions on the model error in (9.30) hold, we have the displeasing property of minimum variance of all linear unbiased estimators of $1/V_{max}$ and k/V_{max}, not k and V_{max}. Thus often we do not gain a clear advantage by linearizing, as far as the properties of the estimators are concerned.

From the foregoing, one dominating argument remains intact. Linearization of a nonlinear model does not produce an equivalent model. There are really two models with differing error structures. If transformation to linearize is accomplished, the user must realize the consequences of having inherited: (i) an unreasonable error structure, and (ii) poorer properties of the estimators of the parameters that *really count,* i.e., the interpretable ones. A related discussion appeared in Section 7.3.

Models exist that are known to be reasonably stable under the contrasting two assumptions on the error term. That is, the nonlinear estimators tend to have properties similar to those found under transformation. For example, the logistic is known to have this stability property. For an excellent comparison regarding competing models, see Ratkowsky (1983).

EXERCISES FOR CHAPTER 9

9.1 In a study to develop the growth behavior for protozoa colonization in a particular lake, an experiment was conducted in which 15 sponges were placed in a lake and 3 sponges at a time were gathered. Then the number of protozoa were counted at 1, 3, 6, 15, and 21 days. In this case, the MacArthur–Wilson Equation was used to describe the growth mechanism. The model is given by

$$y = S_{eq}(1 - e^{-g_0 t})$$

where

- y: Total protozoa on the sponge
- S_{eq}: Species equilibrium constant
- g_0: Parameter that measures how quickly growth rises
- t: Time, number of days

The data are as follows:

Observation	Day	y (Total Protozoa)
1	1	17
2	1	21
3	1	16
4	3	30
5	3	25
6	3	25
7	6	33
8	6	31
9	6	32
10	15	34
11	15	33
12	15	33
13	21	39
14	21	35
15	21	36

Data analyzed for the Department of Biology by the Statistical Consulting Center, Virginia Polytechnic Institute and State University, Blacksburg, Virginia, 1983.

(a) Estimate S_{eq} and g_0 using nonlinear regression. Supply your own starting values.
(b) Give estimated standard errors of the parameter estimates.

9.2 An investigation was made to study age and growth characteristics of selected freshwater mussel species in Southwest Virginia. For a particular type of mussel, age and length were measured for 20 females with the following results:

Observation	Length (in.)	Age (yr)	Observation	Length (in.)	Age (yr)
1	29.2	4	11	36.9	8
2	28.6	4	12	40.2	8
3	29.4	5	13	39.2	9
4	33.0	5	14	40.6	9
5	28.2	6	15	35.2	10
6	33.9	6	16	43.3	10
7	33.1	6	17	42.3	13
8	33.2	7	18	41.4	13
9	31.4	7	19	45.2	14
10	37.8	7	20	47.1	18

Data analyzed for the Department of Fisheries and Wildlife by the Statistical Consulting Center, Virginia Polytechnic Institute and State University, Blacksburg, Virginia, 1983.

(a) Use nonlinear regression to fit the model

$$y_i = \alpha - \exp[-(\beta + \gamma x_i)] + \varepsilon_i \qquad (i = 1, 2, \ldots, 20)$$

where y_i is the length and x_i the age, respectively for the ith mussel. Supply your own starting values.

(b) Give an estimate of the variance-covariance matrix of the parameter estimates α, β, and γ.

9.3 The following data were collected on specific gravity and spectrophotometer analysis for 26 mixtures of NG (nitroglycerine), TA (triacetin) and 2 NDPA (2-nitrodiphenylamine).

Mixture	x_1 (% NG)	x_2 (% TA)	x_3 (% 2 NDPA)	d (Specific Gravity)
1	79.98	19.85	0	1.4774
2	80.06	18.91	1.00	1.4807
3	80.10	16.87	3.00	1.4829
4	77.61	22.36	0	1.4664
5	77.60	21.38	1.00	1.4677
6	77.63	20.35	2.00	1.4686
7	77.34	19.65	2.99	1.4684
8	75.02	24.96	0	1.4524
9	75.03	23.95	1.00	1.4537
10	74.99	22.99	2.00	1.4549
11	74.98	22.00	3.00	1.4565
12	72.50	27.47	0	1.4410
13	72.50	26.48	1.00	1.4414
14	72.50	25.48	2.00	1.4426
15	72.49	24.49	3.00	1.4438
16	69.98	29.99	0	1.4279
17	69.98	29.00	1.00	1.4287
18	69.99	27.99	2.00	1.4291
19	69.99	26.99	3.00	1.4301
20	67.51	32.47	0	1.4157
21	67.50	31.47	1.00	1.4172
22	67.48	30.50	2.00	1.4183
23	67.49	29.49	3.00	1.4188
24	64.98	34.00	1.00	1.4042
25	64.98	33.00	2.00	1.4060
26	64.99	31.99	3.00	1.4068

Raymond H. Myers, "Methods for Estimating the Composition of a Three Component Liquid Mixture," *Technometrics*, vol. 6, no. 4 (November 1964): 343–356.

There is a need to estimate "activity" coefficients from the model

$$d_i = \frac{1}{\alpha_1^* x_{1i} + \alpha_2^* x_{2i} + \alpha_3^* x_{3i}} + \varepsilon_i$$

The quantities α_1^*, α_2^*, and α_3^* are ratios of activity coefficients to the individual specific gravity of the NG, TA, and 2 NDPA respectively.
(a) Compute starting values for the parameters α_1^*, α_2^*, and α_3^*.
(b) Use nonlinear regression to estimate α_1^*, α_2^*, and α_3^*.
(c) Compute the ordinary residuals of the nonlinear regression.

9.4 Consider the surgical services data of Example 7.4.
(a) Fit the hyperbola

$$y_i = \frac{x_i}{\alpha + \beta x_i} + \varepsilon_i \qquad (i = 1, 2, \ldots, 15)$$

using nonlinear regression.
(b) Form the residuals from the nonlinear fit in part (a). Compare the residuals to those given for the transformed hyperbola given in Example 7.4. Which seems to be more appropriate: the nonlinear fit, or the transformation described in Example 7.4?

9.5 Consider the kinetic data of Example 9.2. Fit a linear regression of $1/y$ against x, and estimate $V_{\max}$ and k. Use the fit to compute residuals on y. Comment on the comparison between the nonlinear fit and the linear transformation. Which is preferable?

9.6 Consider the magnesium uptake data of Exercise 3.20. The data are taken under two different treatments. Consider the following exponential model

$$y_i = \alpha_1 e^{\beta_1 x_i} + \varepsilon_i \qquad (i = 1, 2, \ldots, 28)$$
$$y_i = \alpha_2 e^{\beta_2 x_i} + \varepsilon_i \qquad (i = 29, 30, \ldots, 56)$$

That is, two separate exponential decay models are assumed for each treatment.
(a) Use data plotting to arrive at appropriate starting values for α_1, β_1, α_2, and β_2.
(b) Use nonlinear least squares to estimate parameters of the nonlinear regression. Also compute approximate standard errors.
(c) Is there evidence (exploratory or formal) that a single model might be appropriate for both treatments? Explain.

9.7 In Chapter 7 considerable attention was placed on logistic and Poisson regression. In addition, a generalization to the so-called generalized linear models was discussed. Please refer to this material and also to Exercise 7.10. When the canonical link function is used, maximum likelihood estimation of parameters in the generalized linear model is equivalent to minimization of

$$\sum_{i=1}^{n} \left[\frac{(y_i - \hat{y}_i)^2}{\sigma_i^2} \right]$$

In cases other than the normal error case, the fitted model is nonlinear in the parameters. For example, in the logistic regression case

$$\hat{y}_i = \frac{1}{1 + e^{-\mathbf{x}_i' \boldsymbol{\beta}}} \qquad (i = 1, 2, \ldots, n)$$

This equivalence between maximum likelihood and *weighted* nonlinear least squares allows for parameter estimation using nonlinear regression algorithms. Many nonlinear software algorithms allow for use of weights, where weights are reestimated at each iteration. As a result, the procedure becomes nonlinear *iterative reweighted* least squares (IRWLS). (The NLIN procedure in SAS allows for weighting.)

Consider Example 7.10. Using the Poisson error assumption with the canonical link function, fit the model

$$E(y) = e^{\beta_0 + \beta_1 x_1 + \beta_2 x_2 + \beta_4 x_4}$$

using weighted nonlinear regression. For starting values, use the following:

$$\beta_{0,0} = -3.00$$
$$\beta_{1,0} = -0.001$$
$$\beta_{2,0} = 0.1$$
$$\beta_{4,0} = 0.0$$

REFERENCES FOR CHAPTER 9

1. Amer, F.A., and W.T. Williams. 1957. Leaf-area growth in Perlargonium zonale. *Annals of Botany* 21: 339–342.
2. Bard, Y. 1974. *Nonlinear Parameter Estimation.* New York: Academic Press.
3. Bates, D.M., and D.G. Watts. 1988. *Nonlinear Regression Analysis and Its Applications.* New York: John Wiley & Sons.
4. Bates, D.M., and D.G. Watts. 1980. Relative curvature measures of nonlinearity. *Journal of the Royal Statistical Society, Series B* 42: 1–16, discussion pp. 16–25.
5. Beale, E.M.L. 1960. Confidence regions in non-linear estimation. *Journal of the Royal Statistical Society, Series B* 22: 41–76.
6. Beck, J.V., and K.J. Arnold. 1977. *Parameter Estimation in Engineering and Science.* New York: Wiley.
7. Bertalanffy, L. von. 1957. Quantitative laws in metabolism and growth. *Biological Revelation Quarterly* 32: 218–231.
8. Bertalanffy, L. von. 1941. Stoffwecheseltypen und Wachstumstypen. *Biol. Zentralbl.* 61: 510–532.
9. Bliss, C.I., and A.T. James. 1966. Fitting the rectangular hyperbola. *Biometrics* 22: 573–602.
10. Bowden, D.C., and R.K. Steinhorst. 1973. Tolerance bands for growth curves. *Biometrics* 29: 361–371.
11. Box, G.E.P. 1958. Use of statistical methods in the elucidation of basic mechanism. *Bulletin of International Statistical Institute* 36: 215–222.
12. Box, G.E.P., and W.G. Hunter. 1965. The experimental study of physical mechanisms. *Technometrics* 7: 23–42.
13. Box, G.E.P., and W.G. Hunter. 1962. A useful method for model building. *Technometrics* 4: 301–318.
14. Box, G.E.P., and H.L. Lucas. 1959. Design of experiments in non-linear situations. *Biometrika* 46: 77–90.
15. Brand, R.J., D.E. Pinnock, and K.L. Jackson. 1973. Large sample confidence bands for the logistic response curve and its inverse. *American Statistician* 33: 157–160.
16. Carter, W.H., Jr., G.L. Wampler, and D.M. Stablein. 1983. *Regression Analysis of Survival Data in Cancer Chemotherapy.* New York: Marcel Dekker.

17. Causton, D.R. 1969. A computer program for fitting the Richards function. *Biometrics* 25: 401–409, see also p. 779 (1969).

18. Colquhoun, D. 1969. A comparison of estimators for a two-parameter hyperbola. *Applied Statistics* 18: 130–140.

19. Cornell, R.G. 1962. A method for fitting linear combinations of exponentials. *Biometrics* 18: 104–113.

20. Cornell, R.G., and J.A. Speakman. 1967. Estimation for a simple exponential model. *Biometrics* 23: 717–737.

21. Curry, H.B. 1944. The method of steepest descent for non-linear minimization problems. *Applied Mathematics Quarterly* 2: 258–261.

22. Draper, N.R., and H. Smith. 1981. *Applied Regression Analysis*. 2nd ed. New York: John Wiley & Sons.

23. Fletcher, R., and M.J.D. Powell. 1963. A rapidly convergent descent method for minimization. *Computer Journal* 6: 163–168.

24. Foss, S.D. 1969. A method for obtaining initial estimates of the parameters in exponential curve fitting. *Biometrics* 25: 580–584.

25. Gallant, A.R. 1987. *Nonlinear Statistical Models*. New York: John Wiley & Sons.

26. Gallant, A.R. 1968. A note on the measurement of cost/quantity relationships in the aircraft industry. *Journal of the American Statistical Association* 63: 1252–1274.

27. Gallant, A.R. 1975. Testing a subset of the parameters of a non-linear regression model. *Journal of the American Statistical Association* 70: 927–932.

28. Gallucci, V.F., and T.J. Quinn II. 1979. Reparameterizing, fitting, and testing a simple growth model. *Transactions of American Fisheries Society* 108: 14–25.

29. Grizzle, J.E., and D.M. Allen. 1969. Analysis of growth and dose response curves. *Biometrics* 25: 357–381.

30. Halperin, M. 1962. Confidence interval estimation in non-linear regression. Program 360, Applied Mathematics Department SRRC-RR-62-28, Sperry-Rand Research Center.

31. Hartley, H.O. 1948. The estimation of parameters by "internal least squares." *Biometrika* 35: 32–45.

32. Hartley, H.O. 1961. The modified Gauss–Newton method for the fitting of nonlinear regression functions by least squares. *Technometrics* 3: 269–280.

33. Harville, D.A. 1973. Fitting partially linear models by weighted least squares. *Technometrics* 15: 638–650.

34. Hunter, W.G., and R. Mezaki. 1964. A model-building technique for chemical engineering kinetics. *Journal of American Institute of Chemical Engineering* 10: 315–322.

35. Hunter, W.G., and A.M. Reiner. 1965. Designs for discriminating between two rival models. *Technometrics* 7: 307–323.

36. Kennedy, W.J., and J.E. Gentle. 1980. *Statistical Computing*. New York: Marcel Dekker.

37. Llewelyn, F.W.M. 1968. The log (−log) transformation in the analysis of fruit retention records. *Biometrics* 24: 627–638.

38. Marquardt, D.W. 1963. An algorithm for least squares estimation of nonlinear parameters. *Journal of the Society of Industrial and Applied Mathematics* 2: 431–441.

39. Marquardt, D.W. 1959. Solution of non-linear chemical engineering models. *Chemical Engineering Progress* 55: 65–70.

40. Marquardt, D.W., R.G. Bennett, and E.J. Burrell. 1961. Least-squares analysis of electron paramagnetic resonance spectra. *Journal of Molecular Spectroscopy* 7: 269–279.

41. Michaelis, L., and M.L. Menten. 1913. Die Kinetik der Invertinwirkung. *Biochemische Zeitschrift* 49: 333–369.

42. Mitchell, A.F.S. 1968. Exponential regression with corelated observations. *Biometrika* 55: 149–162.

43. Moore, R.H., and R.K. Ziegler. 1967. The use of nonlinear regression methods for analyzing sensitivity and quantal response data. *Biometrics* 23: 563–566.

44. Oliver, F.R. 1970. Estimating the exponential growth function by direct least squares. *Applied Statistics* 19: 92–100.

45. Oliver, F.R. 1964. Methods of estimating the logistic growth function. *Applied Statistics* 13: 57–66.

46. Olsson, D.M. 1972. Fitting two widely useful nonlinear models. *Journal of Quality Technology* 4: 113–117.

47. Olsson, D.M. 1974. A sequential simplex program for solving minimization problems. *Journal of Quality Technology* 6: 53–57.

48. Phillips, B.F., and N.A. Campbell. 1968. A new method of fitting the Von Bertalanffy Growth Curve using data on the whelk. *Dicathais, Growth* 32: 317–329.

49. Ratkowsky, D.A. 1983. *Nonlinear Regression Modeling*. New York: Marcel Dekker.

50. Richards, F.J. 1959. A flexible growth function for empirical use. *Journal of Experimental Biology* 10: 290–300.

51. Riffenburgh, R.H. 1966. On growth parameter estimation for early life stages. *Biometrics* 22: 162–178.

52. SAS Institute Inc. 1985. *SAS User's Guide: Statistics, Version 5*. Cary, North Carolina: SAS Institute Inc.

53. Smith, H., and S.D. Dubey. 1964. Some reliability problems in the chemical industry. *Industrial Quality Control* 22: 64–70.

54. Sprent, P. 1968. Linear relationships in growth and size studies. *Biometrics* 24: 639–656.

55. Turner, M.E., B.A. Blumenstein, and J.L. Sebaugh. 1969. A generalization of the logistic law of growth. *Biometrics* 25: 577–580.

56. Wilson, A.L., and A.W. Douglas. 1969. A note on nonlinear curve fitting. *American Statistician* 23: 37–38.

57. Ziegel, E.R., and J.W. Gorman. 1980. Kinetic modelling with multiresponse data. *Technometrics* 22: 139–151.

APPENDIX A

SOME SPECIAL CONCEPTS IN MATRIX ALGEBRA

In this appendix, we review a few fundamental notions in matrix algebra. Certain definitions and theorems that are used extensively are presented. We begin with a presentation of the matrix manipulation necessary for obtaining solutions to linear equations.

A.1 SOLUTIONS TO SIMULTANEOUS LINEAR EQUATIONS

In Section 3.2, we introduced the general linear model of the form

$$\mathbf{y} = \mathbf{X}\boldsymbol{\beta} + \boldsymbol{\varepsilon}$$

The solution to the least squares normal equations given by Eq. (3.4) involves solving systems of simultaneous linear equations. Consider a set of linear equations

$$\begin{aligned} a_{11}b_1 + a_{12}b_2 + \cdots + a_{1p}b_p &= g_1 \\ a_{21}b_1 + a_{22}b_2 + \cdots + a_{2p}b_p &= g_2 \\ \vdots \qquad \vdots \qquad\qquad \vdots \quad &\quad \vdots \\ a_{p1}b_1 + a_{p2}b_2 + \cdots + a_{pp}b_p &= g_p \end{aligned}$$

We can write this system of equations as

$$\mathbf{Ab} = \mathbf{g} \tag{A.1}$$

where

$$\mathbf{A} = \begin{bmatrix} a_{11} & a_{12} & \cdots & a_{1p} \\ a_{21} & a_{22} & \cdots & a_{2p} \\ \vdots & \vdots & & \vdots \\ a_{p1} & a_{p2} & \cdots & a_{pp} \end{bmatrix} \qquad \mathbf{b} = \begin{bmatrix} b_1 \\ b_2 \\ \vdots \\ b_p \end{bmatrix} \qquad \mathbf{g} = \begin{bmatrix} g_1 \\ g_2 \\ \vdots \\ g_p \end{bmatrix}$$

A unique solution $\mathbf{b}$ to the set of simultaneous equations will exist when the matrix $\mathbf{A}$ is *nonsingular*, i.e., when there exists a $p \times p$ matrix $\mathbf{A}^{-1}$ such that

$$\mathbf{AA}^{-1} = \mathbf{A}^{-1}\mathbf{A} = \mathbf{I}_p$$

where $\mathbf{I}_p$ is the identity matrix of order p. As a result, the solution to Eq. (A.1) is produced by premultiplying both sides by $\mathbf{A}^{-1}$. As a result,

$$\begin{aligned} \mathbf{A}^{-1}\mathbf{Ab} &= \mathbf{A}^{-1}\mathbf{g} \\ \mathbf{Ib} &= \mathbf{A}^{-1}\mathbf{g} \\ \mathbf{b} &= \mathbf{A}^{-1}\mathbf{g} \end{aligned}$$

The matrix $\mathbf{A}^{-1}$ is called the *inverse* of $\mathbf{A}$. In the regression application of Chapter 3, the solution is given by the least squares estimators in $\mathbf{b}$ of Eq. (3.4), namely

$$\mathbf{b} = (\mathbf{X}'\mathbf{X})^{-1}\mathbf{X}'\mathbf{y}$$

The matrix $\mathbf{X}'\mathbf{X}$ plays the role of the $\mathbf{A}$ matrix in Eq. (A.1). The $\mathbf{X}'\mathbf{X}$ matrix is a *symmetric* array that contains sums of squares and sums of cross products. The diagonal elements of $\mathbf{X}'\mathbf{X}$ contain sums of squares and the off-diagonal elements contain sums of cross products. For example, consider the following:

$$\mathbf{X} = \begin{bmatrix} 1 & x_{11} & x_{21} & x_{31} \\ 1 & x_{12} & x_{22} & x_{32} \\ 1 & x_{13} & x_{23} & x_{33} \\ \vdots & \vdots & \vdots & \vdots \\ 1 & x_{1n} & x_{2n} & x_{3n} \end{bmatrix}$$

The $\mathbf{X}'\mathbf{X}$ matrix is given by

$$\mathbf{X}'\mathbf{X} = \begin{bmatrix} n & \sum_{i=1}^{n} x_{1i} & \sum_{i=1}^{n} x_{2i} & \sum_{i=1}^{n} x_{3i} \\ & \sum_{i=1}^{n} x_{1i}^2 & \sum_{i=1}^{n} x_{1i}x_{2i} & \sum_{i=1}^{n} x_{1i}x_{3i} \\ & & \sum_{i=1}^{n} x_{2i}^2 & \sum_{i=1}^{n} x_{2i}x_{3i} \\ \text{Symmetric} & & & \sum_{i=1}^{n} x_{3i}^2 \end{bmatrix}$$

For the reader who is unfamiliar with fundamental matrix algebra, Section A.3 will provide further insight into the nature of $\mathbf{X}'\mathbf{X}$, and how its characteristics can result in serious collinearity problems.

A.2

THE QUADRATIC FORM

One of the more important matrix manipulations in regression is the *quadratic form*. The concept is quite simple. Suppose we have a column vector $\mathbf{z}$ (n elements) and an $n \times n$ symmetric matrix $\mathbf{A}$ with typical element a_{ij}. Then the scalar quantity

$$\mathbf{z}'\mathbf{A}\mathbf{z} = \sum_{i=1}^{n} a_{ii} z_i^2 + 2 \sum_{\substack{i=1 \\ i<j}}^{n} \sum_{j=1}^{n} a_{ij} z_i z_j$$

is called a *quadratic form* in $\mathbf{z}$ with matrix $\mathbf{A}$.

A *positive definite quadratic form* is one that is greater than zero for all $\mathbf{z} \neq \mathbf{0}$. The notion of positive definiteness transfers to the matrix $\mathbf{A}$. That is, a positive definite matrix $\mathbf{A}$ is one for which $\mathbf{z}'\mathbf{A}\mathbf{z} > 0$ for all $\mathbf{z} \neq \mathbf{0}$. A *positive semi-definite matrix* is one for which $\mathbf{z}'\mathbf{A}\mathbf{z} \geqslant 0$ for all $\mathbf{z}$, but $\mathbf{z}'\mathbf{A}\mathbf{z} = 0$ for some $\mathbf{z} \neq \mathbf{0}$.

In the text there are numerous applications of the quadratic form. In Section 3.4, attention is focused on the *extra sum of squares* principle, used when one tests a hypothesis concerning a subset of regression coefficients. Equation (3.13) expresses the regression sum of squares explained by the subset of terms $\mathbf{X}_1\boldsymbol{\beta}_1$ in the presence of a second subset $X_2\boldsymbol{\beta}_2$. The resulting explained regression is given by a quadratic form with vector $\mathbf{y}$ and matrix $\mathbf{X}(\mathbf{X}'\mathbf{X})^{-1}\mathbf{X}' - \mathbf{X}_2(\mathbf{X}_2'\mathbf{X}_2)^{-1}\mathbf{X}_2'$.

We can view the genesis of the quadratic form in Eq. (3.13) from the viewpoint of the regression sum of squares for the full model. Consider the general linear model

$$\mathbf{y} = \mathbf{X}\boldsymbol{\beta} + \boldsymbol{\varepsilon}$$

with

$$\mathbf{X} = \begin{bmatrix} 1 & x_{11} & \cdots & x_{k1} \\ 1 & x_{12} & \cdots & x_{k2} \\ \vdots & \vdots & & \vdots \\ 1 & x_{1n} & \cdots & x_{kn} \end{bmatrix} \qquad \boldsymbol{\beta} = \begin{bmatrix} \beta_0 \\ \beta_1 \\ \vdots \\ \beta_k \end{bmatrix}$$

and $n > k+1$. The regression sum of squares $\sum_{i=1}^{n} (\hat{y}_i - \bar{y})^2$, is discussed in Chapter 3. This quantity is the k degrees of freedom sum of squares, which accounts for the variation explained by the regressors $x_1, x_2, \ldots, x_k$. The $k+1$ regression sum of squares explained by the regressors *and the constant term* is given by

$$\sum_{i=1}^{n} \hat{y}_i^2 = (\mathbf{Xb})'(\mathbf{Xb})$$

That is,

$$\begin{aligned} \sum_{i=1}^{n} \hat{y}_i^2 &= \mathbf{b}'\mathbf{X}'\mathbf{X}\mathbf{b} \\ &= \mathbf{y}'\mathbf{X}(\mathbf{X}'\mathbf{X})^{-1}\mathbf{X}'\mathbf{X}(\mathbf{X}'\mathbf{X})^{-1}\mathbf{X}'\mathbf{y} \\ &= \mathbf{y}'\mathbf{X}(\mathbf{X}'\mathbf{X})^{-1}\mathbf{X}'\mathbf{y} \end{aligned}$$

As a result, the expression in Eq. (3.13), with the regression explained by the regressors in $\mathbf{X}_1$ *in the presence of the regressors in* $\mathbf{X}_2$, is given by the difference between the regression sum of squares in the full model, $\mathbf{y} = \mathbf{X}\boldsymbol{\beta} + \boldsymbol{\varepsilon}$ and the reduced model, which contains only the regressors in $\mathbf{X}_2$. Thus $R(\boldsymbol{\beta}_1 | \boldsymbol{\beta}_2)$ is the difference between the two corresponding quadratic forms: namely

$$\begin{aligned} R(\boldsymbol{\beta}_1 | \boldsymbol{\beta}_2) &= \mathbf{y}'\mathbf{X}(\mathbf{X}'\mathbf{X})^{-1}\mathbf{X}'\mathbf{y} - \mathbf{y}'\mathbf{X}_2(\mathbf{X}_2'\mathbf{X}_2)^{-1}\mathbf{X}_2'\mathbf{y} \\ &= \mathbf{y}'[\mathbf{X}(\mathbf{X}'\mathbf{X})^{-1}\mathbf{X}' - \mathbf{X}_2(\mathbf{X}_2'\mathbf{X}_2)^{-1}\mathbf{X}_2']\mathbf{y} \end{aligned}$$

Another very important application of the quadratic form appears in the computation of the *prediction variance*, a notation which appears throughout the text. The first usage arises in Section 3.5. Equation (3.20) gives an expression for the variance of the fitted (or predicted) value of y at a location $\mathbf{x}_0$. The *variance of this prediction* is given by $\sigma^2 \cdot \mathbf{x}_0'(\mathbf{X}'\mathbf{X})^{-1}\mathbf{x}_0$, a quadratic form in the vector $\mathbf{x}_0$ with matrix $(\mathbf{X}'\mathbf{X})^{-1}$. A close inspection of this particular quadratic form reveals that

$$\frac{\operatorname{Var}(\hat{y} | \mathbf{x}_0)}{\sigma^2} = \operatorname{Var} b_0 + \sum_{j=1}^{k} x_{j,0}^2 \operatorname{Var} b_j + 2 \sum_{\substack{j=0 \\ j<\ell}}^{k} \sum_{\ell=0}^{k} x_{j,0} x_{\ell,0} \operatorname{Cov}(b_j, b_\ell)$$

where $x_{0,0}$ is taken to be unity. One can easily see that the above expanded writing of the prediction variance takes into account, through $(\mathbf{X}'\mathbf{X})^{-1}$, variances and covariances of all coefficients.

Another obvious use of the quadratic form and, indeed, a rather special case of the prediction variance, is the HAT diagonal, which is introduced in Section 3.9

and discussed throughout the text. If the location x_0 in question happens to be one of the data points, then $\mathbf{x}'_0$ becomes $\mathbf{x}'_i$, a row of the $\mathbf{X}$ matrix. Then the quadratic form $\mathbf{x}'_i(\mathbf{X}'\mathbf{X})^{-1}\mathbf{x}_i$ is the diagonal of the HAT matrix $\mathbf{X}(\mathbf{X}'\mathbf{X})^{-1}\mathbf{X}'$ introduced in Eq. (3.39).

A.3

EIGENVALUES AND EIGENVECTORS

The notion of eigenvalues are prominent in the discussion of multicollinearity in Chapters 3 and 8. The eigenvalue-vector concept was first introduced in Section 3.8. Here we give definitions and also provide some extra insight into why eigenvalues play such an important role in diagnosing and quantifying the extent of multicollinearity.

Consider a $k \times k$ symmetric matrix $\mathbf{A}$. The eigenvalues $\lambda_1, \lambda_2, \ldots, \lambda_k$ of the matrix $\mathbf{A}$ are given by the solutions to the determinantal equation

$$|\mathbf{A} - \lambda\mathbf{I}| = 0 \qquad \textbf{(A.2)}$$

Associated with the ith eigenvalue, λ_i, is an *eigenvector* defined by the solution $\mathbf{v}_i$ to

$$(\mathbf{A} - \lambda_i\mathbf{I})\mathbf{v}_i = 0$$

When $\mathbf{A}$ is symmetric, the eigenvalues will all be real.

While there are many applications of eigenvalues and eigenvectors in the theory of statistics, we shall concern ourselves with collinearity. As we indicated in Section 3.8, the matrix $\mathbf{V}$, whose columns are the associated normalized eigenvectors, can be used to diagonalize $\mathbf{A}$. That is,

$$\mathbf{V}'\mathbf{A}\mathbf{V} = \text{diag}\,(\lambda_1, \lambda_2, \ldots, \lambda_k) \qquad \textbf{(A.3)}$$

In addition, $\mathbf{V}$ is an orthogonal matrix; i.e., $\mathbf{V}'\mathbf{V} = \mathbf{V}\mathbf{V}' = \mathbf{I}_p$. Now, our concern is with the case where $\mathbf{A} = (\mathbf{X}^{*\prime}\mathbf{X}^*)$, the correlation matrix described in Chapter 3. If the correlation matrix is diagonal, i.e., there are no linear associations among the regressor variables, then all the eigenvalues will be unity. This is the ideal case. If the correlation matrix is near-singular, i.e., there is at least one near dependency among the columns of $\mathbf{X}^*$, then the determinant of $\mathbf{X}^{*\prime}\mathbf{X}^*$ will be near zero. The eigenvalues have an important relationship with the determinant; consider Eq. (A.3) with $\mathbf{A} = (\mathbf{X}^{*\prime}\mathbf{X}^*)$. Since $\mathbf{V}$ is orthogonal, $|\mathbf{V}| = 1.0$, $|\mathbf{V}'| = 1.0$, and

$$\begin{aligned} |\mathbf{X}^{*\prime}\mathbf{X}^*| &= |\mathbf{V}| \prod_{i=1}^{k} \lambda_i |\mathbf{V}'| \\ &= \prod_{i=1}^{k} \lambda_i \end{aligned} \qquad \textbf{(A.4)}$$

Another important result relates the sum of the eigenvalues of the correlation matrix to the dimension of the matrix. A well-known result from matrix algebra (see Graybill (1976)) is tr (**AB**) = tr (**BA**), given that **A** and **B** are matrices conformable to the multiplication **AB** and **BA**. As a result,

$$\text{tr}(\mathbf{V}'\mathbf{A}\mathbf{V}) = \text{tr}\,\mathbf{V}\mathbf{V}'\mathbf{A}$$

Now, since $\mathbf{V}\mathbf{V}' = \mathbf{I}$,

$$\text{tr}(\mathbf{V}'\mathbf{A}\mathbf{V}) = \text{tr}(\mathbf{A})$$

But, from Eq. (A.3),

$$\text{tr}(\mathbf{A}) = \sum_{i=1}^{k} \lambda_i$$

Thus if **A** is a correlation matrix,

$$\sum_{i=1}^{k} \lambda_i = k \tag{A.5}$$

where k is the *dimension* of the correlation matrix.

It is clear from the foregoing that if the correlation matrix is rendered near singular through multicollinearity, at least *one eigenvalue is near zero*. That is to be compared with the ideal case in which no collinearity implies all eigenvalues are 1.0.

There is yet another development that reveals how near zero eigenvalues of the correlation matrix present serious problems in a regression situation. Consider Eq. (3.35) and (3.36) in Section 3.8. It is obvious from Eq. (3.36) that even one near zero eigenvalue produces an undesirable value for the sum of the variances of the regression coefficients. Equation (3.36) is easily verified by considering Eq. (A.3), with **A** again being the correlation matrix $\mathbf{X}^{*\prime}\mathbf{X}^*$. We have, then

$$\mathbf{X}^{*\prime}\mathbf{X}^* = \mathbf{V}\begin{bmatrix} \lambda_1 & & & 0 \\ & \lambda_2 & & \\ & & \ddots & \\ 0 & & & \lambda_k \end{bmatrix}\mathbf{V}'$$

For the orthogonal matrix **V**, its inverse is equal to its transpose. We can then take the inverse of both sides to obtain

$$(\mathbf{X}^{*\prime}\mathbf{X}^*)^{-1} = \mathbf{V}\begin{bmatrix} 1/\lambda_1 & & & 0 \\ & 1/\lambda_2 & & \\ & & \ddots & \\ 0 & & & 1/\lambda_k \end{bmatrix}\mathbf{V}'$$

The sum of the variances of the regression coefficients (apart from σ^2) is, then, the trace of $(\mathbf{X}^{*\prime}\mathbf{X}^{*})^{-1}$.

$$\operatorname{tr}(\mathbf{X}^{*\prime}\mathbf{X}^{*})^{-1} = \operatorname{tr}\mathbf{V}'\mathbf{V}\begin{bmatrix} 1/\lambda_2 & & & 0 \\ & 1/\lambda_2 & & \\ & & \ddots & \\ 0 & & & 1/\lambda_k \end{bmatrix}$$

$$= \operatorname{tr}\begin{bmatrix} 1/\lambda_1 & & & 0 \\ & 1/\lambda_2 & & \\ & & \ddots & \\ 0 & & & 1/\lambda_k \end{bmatrix}$$

$$= \sum_{i=1}^{k} \frac{1}{\lambda_i}$$

which is the result in Eq. (3.36).

A.4

THE INVERSE OF A PARTITIONED MATRIX

In Section 4.1, we presented the bias in the error mean square associated with an *underspecified* model. Much of the material in Chapter 4 evolved from the presumption that an underspecified modeling setting can be structured by assuming that one postulates the model

$$\mathbf{y} = \mathbf{X}_1\boldsymbol{\beta}_1 + \boldsymbol{\varepsilon}^* \tag{A.6}$$

while the true model is

$$\mathbf{y} = \mathbf{X}_1\boldsymbol{\beta}_1 + \mathbf{X}_2\boldsymbol{\beta}_2 + \boldsymbol{\varepsilon} \tag{A.7}$$

with m parameters and $m > p$. Here,

$$\boldsymbol{\beta} = \begin{bmatrix} \boldsymbol{\beta}_1 \\ \cdots \\ \boldsymbol{\beta}_2 \end{bmatrix}$$

The expected value of the error mean square for the underspecified model is given in Eq. (4.1). As indicated in the text, this bias is given by

$$\frac{1}{n-p}\boldsymbol{\beta}_2'[\mathbf{X}_2'\mathbf{X}_2 - \mathbf{X}_2'\mathbf{X}_1(\mathbf{X}_1'\mathbf{X}_1)^{-1}\mathbf{X}_1'\mathbf{X}_2]\boldsymbol{\beta}_2$$

which is a *standardized form* of the ignored coefficients, i.e., those in $\boldsymbol{\beta}_2$. The implication here is that the variance-covariance matrix of $\mathbf{b}_2$ (if one were to fit the complete model of Eq. (A.7)) is given by

$$\text{Var}(\mathbf{b}_2) = \sigma^2(\mathbf{X}_2'\mathbf{X}_2 - \mathbf{X}_2'\mathbf{X}_1(\mathbf{X}_1'\mathbf{X}_1)^{-1}\mathbf{X}_1'\mathbf{X}_2)^{-1}$$

We can easily verify this by considering the inverse of the partitioned matrix

$$\mathbf{X}'\mathbf{X} = \begin{bmatrix} \mathbf{X}_1'\mathbf{X}_1 & \mathbf{X}_1'\mathbf{X}_2 \\ \mathbf{X}_2'\mathbf{X}_1 & \mathbf{X}_2'\mathbf{X}_2 \end{bmatrix} \tag{A.8}$$

Now, $\mathbf{X}_1'\mathbf{X}_1$ is a $p \times p$ square symmetric matrix, and $\mathbf{X}_2'\mathbf{X}_2$ is symmetric of dimension $(m-p) \times (m-p)$. From Chapter 3 we recall that the variance-covariance matrix of $\mathbf{b}$ is given by

$$\text{Var } \mathbf{b} = \sigma^2(\mathbf{X}'\mathbf{X})^{-1}$$

As a result, apart from σ^2, the variance-covariance matrix of $\mathbf{b}_1$ is the $(p \times p)$ square matrix in the upper left-hand corner of the *inverse* of the matrix in (A.8), and $\text{Var}(\mathbf{b}_2)$ is the square symmetric matrix in the bottom right-hand corner of the same inverse. In Graybill (1976) the partitions of the inverse of $\mathbf{X}'\mathbf{X}$ are presented. They are given by

$$(\mathbf{X}'\mathbf{X})^{-1} = \begin{bmatrix} (\mathbf{C}_{11})^{-1} & -(\mathbf{X}_1'\mathbf{X}_1)^{-1}\mathbf{X}_1'\mathbf{X}_2\mathbf{C}_{22}^{-1} \\ -\mathbf{C}_{22}^{-1}\mathbf{X}_2'\mathbf{X}_1(\mathbf{X}_1'\mathbf{X}_1)^{-1} & (\mathbf{C}_{22})^{-1} \end{bmatrix} \tag{A.9}$$

where

$$\begin{aligned} \mathbf{C}_{11} &= (\mathbf{X}_1'\mathbf{X}_1 - \mathbf{X}_1'\mathbf{X}_2(\mathbf{X}_2'\mathbf{X}_2)^{-1}\mathbf{X}_2'\mathbf{X}_1) \\ \mathbf{C}_{22} &= (\mathbf{X}_2'\mathbf{X}_2 - \mathbf{X}_2'\mathbf{X}_1(\mathbf{X}_1'\mathbf{X}_1)^{-1}\mathbf{X}_1'\mathbf{X}_2) \end{aligned} \tag{A.10}$$

The reader is encouraged to verify that $(\mathbf{X}'\mathbf{X})(\mathbf{X}'\mathbf{X})^{-1} = \mathbf{I}$.

A.5

SHERMAN–MORRISON–WOODBURY THEOREM

The result in this section serves as the basis for modern *single data point* diagnostics discussed in Chapter 6 as well as the PRESS statistic presented in Chapter 4. Essentially it offers an ease in computation of important regression statistics for the case in which the ith data point is removed or *set aside*. Here we give the fundamental result in a very general form. In Appendix B, the result is used to explain the development of certain important diagnostic tools.

Consider a square nonsingular matrix $\mathbf{A}$, which is $p \times p$, and a p-dimensional column vector $\mathbf{z}$. In our application, $\mathbf{A}$ is the $\mathbf{X}'\mathbf{X}$ matrix. The vector $\mathbf{z}'$ is the ith row of the $\mathbf{X}$ matrix. Thus $(\mathbf{A} - \mathbf{z}\mathbf{z}')$ becomes the $\mathbf{X}'\mathbf{X}$ matrix with the ith data point not involved. The theorem states (see Rao (1973))

$$(\mathbf{A} - \mathbf{z}\mathbf{z}')^{-1} = \mathbf{A}^{-1} + \frac{\mathbf{A}^{-1}\mathbf{z}\mathbf{z}'\mathbf{A}^{-1}}{1 - \mathbf{z}'\mathbf{A}^{-1}\mathbf{z}} \tag{A.11}$$

We can prove this result by merely demonstrating that multiplication of the right-hand side by $\mathbf{A}-\mathbf{zz}'$ gives the identity matrix.

$$\left[\mathbf{A}^{-1}+\frac{\mathbf{A}^{-1}\mathbf{zz}'\mathbf{A}^{-1}}{1-\mathbf{z}'\mathbf{A}^{-1}\mathbf{z}}\right][\mathbf{A}-\mathbf{zz}']$$

$$=\mathbf{I}+\frac{\mathbf{A}^{-1}\mathbf{zz}'}{1-\mathbf{z}'\mathbf{A}^{-1}\mathbf{z}}-\mathbf{A}^{-1}\mathbf{zz}'-\frac{\mathbf{A}^{-1}\mathbf{zz}'}{1-\mathbf{z}'\mathbf{A}^{-1}\mathbf{z}}\mathbf{A}^{-1}\mathbf{zz}'$$

$$=\mathbf{I}+\frac{\mathbf{A}^{-1}\mathbf{zz}'-\mathbf{A}^{-1}\mathbf{zz}'(1-\mathbf{z}'\mathbf{A}^{-1}\mathbf{z})-\mathbf{A}^{-1}\mathbf{z}(\mathbf{z}'\mathbf{A}^{-1}\mathbf{z})\mathbf{z}'}{1-\mathbf{z}'\mathbf{A}^{-1}\mathbf{z}}$$

$$=\mathbf{I}+\frac{\mathbf{A}^{-1}\mathbf{zz}'-\mathbf{A}^{-1}\mathbf{zz}'+\mathbf{A}^{-1}\mathbf{zz}'(\mathbf{z}'\mathbf{A}^{-1}\mathbf{z})-\mathbf{A}^{-1}\mathbf{zz}'(\mathbf{z}'\mathbf{A}^{-1}\mathbf{z})}{1-\mathbf{z}'\mathbf{A}^{-1}\mathbf{z}}$$

$$=\mathbf{I}$$

REFERENCES FOR APPENDIX A

1. Graybill, F.A. 1976. *Theory and Application of the Linear Model.* Boston, Massachusetts: Duxbury.
2. Rao, C.R. 1973. *Linear Statistical Inference and Its Applications.* 2d ed. p. 33 New York: John Wiley.

APPENDIX B

SOME SPECIAL MANIPULATIONS

In this appendix we offer details in the development of several important statistical results and concepts that appear in the text. The reader who is well-versed in matrix algebra will be able to follow these developments.

B.1

UNBIASEDNESS OF THE RESIDUAL MEAN SQUARE

Here we verify the unbiasedness of the estimator s^2 for the parameter σ^2. We begin with the residual mean square as formulated in Section 3.2.

$$s^2 = \frac{(\mathbf{y} - \mathbf{Xb})'(\mathbf{y} - \mathbf{Xb})}{n - p} \tag{B.1}$$

Using the least squares normal equations $(\mathbf{X}'\mathbf{X})\mathbf{b} = \mathbf{X}'\mathbf{y}$,

$$s^2 = \frac{\mathbf{y}'\mathbf{y} - \mathbf{b}'\mathbf{X}'\mathbf{y}}{n - p}$$

The numerator of the above equation can be viewed as the difference between the total sum of squares, $\mathbf{y}'\mathbf{y}$, and the *regression* sum of squares

$$\mathbf{b}'\mathbf{X}'\mathbf{y} = \mathbf{y}'\mathbf{X}(\mathbf{X}'\mathbf{X})^{-1}\mathbf{X}'\mathbf{y}$$

Thus, Eq. (B.1) becomes

$$s^2 = \frac{\mathbf{y}'\mathbf{y} - \mathbf{y}'\mathbf{X}(\mathbf{X}'\mathbf{X})^{-1}\mathbf{X}'\mathbf{y}}{n-p}$$

Now, the residual sum of squares in the numerator can be structured as a quadratic form in $\mathbf{y}$ as illustrated:

$$s^2 = \frac{\mathbf{y}'[\mathbf{I} - \mathbf{X}(\mathbf{X}'\mathbf{X})^{-1}\mathbf{X}']\mathbf{y}}{n-p} \tag{B.2}$$

At this point, we make use of a theorem (see Graybill (1976)) on the expected value of a quadratic form. Given a random vector $\mathbf{y}$ with mean $E(\mathbf{y}) = \boldsymbol{\mu}$ and variance-covariance matrix $\text{Var}(\mathbf{y}) = \sigma^2\mathbf{I}$,

$$E(\mathbf{y}'\mathbf{A}\mathbf{y}) = \sigma^2 \text{tr}(\mathbf{A}) + \boldsymbol{\mu}'\mathbf{A}\boldsymbol{\mu} \tag{B.3}$$

We can use this theorem to find the expected value of the quadratic form in Eq. (B.2). We have

$$\begin{aligned} E(s^2)(n-p) &= \sigma^2 \text{tr}[\mathbf{I} - \mathbf{X}(\mathbf{X}'\mathbf{X})^{-1}\mathbf{X}'] + (\mathbf{X}\boldsymbol{\beta})'[\mathbf{I} - \mathbf{X}(\mathbf{X}'\mathbf{X})^{-1}\mathbf{X}'](\mathbf{X}\boldsymbol{\beta}) \\ &= \sigma^2(n-p) + \boldsymbol{\beta}'\mathbf{X}'\mathbf{X}\boldsymbol{\beta} - \boldsymbol{\beta}'\mathbf{X}'\mathbf{X}(\mathbf{X}'\mathbf{X})^{-1}\mathbf{X}'\mathbf{X}\boldsymbol{\beta} \\ &= \sigma^2(n-p) \end{aligned}$$

Here we have used the fact that $\text{tr}\ \mathbf{X}(\mathbf{X}'\mathbf{X})^{-1}\mathbf{X}' = \text{tr}(\mathbf{X}'\mathbf{X})(\mathbf{X}'\mathbf{X})^{-1} = p$. We can now write

$$E(\mathbf{s}^2) = \sigma^2$$

and thus s^2 is an unbiased estimator for σ^2.

B.2

EXPECTED VALUE OF RESIDUAL SUM OF SQUARES AND MEAN SQUARE FOR AN UNDERSPECIFIED MODEL

In Section 3.6 and in Section 4.1, we focused attention on the *bias* induced in s^2 when the model fit by the analyst is underspecified. That is, a "short" model

$$\mathbf{y} = \mathbf{X}_1\boldsymbol{\beta}_1 + \boldsymbol{\varepsilon}^* \qquad (p \text{ parameters})$$

is fit, and the true model is

$$\mathbf{y} = \mathbf{X}_1\boldsymbol{\beta}_1 + \mathbf{X}_2\boldsymbol{\beta}_2 + \boldsymbol{\varepsilon} \qquad (m \text{ parameters})$$

where, of course, $m > p$. We attempt, then, to determine how the residual mean square for the incorrect model, i.e., the p-term model, is inflated by the *extra* parameters $\boldsymbol{\beta}_2$ of the m-term model. To obtain the $E(s^2)$ for the p-term model, we use a result identical to that used in Section B.1, namely the expected value of a quadratic form. Recall from Section B.1 that if a random vector $\mathbf{y}$ has mean $\boldsymbol{\mu}$ and variance-covariance matrix $\sigma^2\mathbf{I}$, and $\mathbf{y}'\mathbf{A}\mathbf{y}$ represents a quadratic form in y, then

$$E(\mathbf{y}'\mathbf{A}\mathbf{y}) = \sigma^2\text{tr}(\mathbf{A}) + \boldsymbol{\mu}'\mathbf{A}\boldsymbol{\mu}$$

In the present situation, the error mean square for the underfitted model is given by

$$\frac{1}{n-p}\mathbf{y}'[\mathbf{I} - \mathbf{X}_1(\mathbf{X}_1'\mathbf{X}_1)^{-1}\mathbf{X}_1']\mathbf{y}$$

Note how this resembles the form of the residual mean square for the full model. The matrix $\mathbf{X}_1$ replaces the matrix $\mathbf{X}$. If we apply the expectation operator,

$$E\{\mathbf{y}'[\mathbf{I} - \mathbf{X}_1(\mathbf{X}_1'\mathbf{X}_1)^{-1}\mathbf{X}_1']\mathbf{y}\} = \sigma^2\text{tr}[\mathbf{I} - \mathbf{X}_1(\mathbf{X}_1'\mathbf{X}_1)^{-1}\mathbf{X}_1'] + E(\mathbf{y})'[\mathbf{I} - \mathbf{X}_1(\mathbf{X}_1'\mathbf{X}_1)^{-1}\mathbf{X}_1']E(\mathbf{y}) \quad \textbf{(B.4)}$$

Now,

$$\text{tr}[\mathbf{I} - \mathbf{X}_1(\mathbf{X}_1'\mathbf{X}_1)^{-1}\mathbf{X}_1'] = n - p$$

with

$$\text{tr}[\mathbf{X}_1(\mathbf{X}_1'\mathbf{X}_1)^{-1}\mathbf{X}_1'] = \text{tr}[(\mathbf{X}_1'\mathbf{X}_1) \times (\mathbf{X}_1'\mathbf{X}_1)^{-1}] = p$$

Then we evaluate the second term on the right-hand side of Eq. (B.4). We know that since the "true" model has $E(\mathbf{y}) = \mathbf{X}_1\boldsymbol{\beta}_1 + \mathbf{X}_2\boldsymbol{\beta}_2$, Eq. (B.4) becomes

$$E\{\mathbf{y}'[\mathbf{I} - \mathbf{X}_1(\mathbf{X}_1'\mathbf{X}_1)^{-1}\mathbf{X}_1']\mathbf{y}\} = \sigma^2(n-p) + (\mathbf{X}_1\boldsymbol{\beta}_1 + \mathbf{X}_2\boldsymbol{\beta}_2)' \times [\mathbf{I} - \mathbf{X}_1(\mathbf{X}_1'\mathbf{X}_1)^{-1}\mathbf{X}_1'](\mathbf{X}_1\boldsymbol{\beta}_1 + \mathbf{X}_2\boldsymbol{\beta}_2)$$

Now, the quadratic form on the right-hand side simply reduces to

$$(\mathbf{X}_2\boldsymbol{\beta}_2)'[\mathbf{I} - \mathbf{X}_1(\mathbf{X}_1'\mathbf{X}_1)^{-1}\mathbf{X}_1'](\mathbf{X}_2\boldsymbol{\beta}_2)$$

which can be written $\boldsymbol{\beta}_2'[\mathbf{X}_2'\mathbf{X}_2 - \mathbf{X}_2'\mathbf{X}_1(\mathbf{X}_1'\mathbf{X}_1)^{-1}\mathbf{X}_1'\mathbf{X}_2]\boldsymbol{\beta}_2$. As a result, we have

$$E(SS_{\text{Res}}) = \sigma^2(n-p) + \boldsymbol{\beta}_2'[\mathbf{X}_2'\mathbf{X}_2 - \mathbf{X}_2'\mathbf{X}_1(\mathbf{X}_1'\mathbf{X}_1)^{-1}\mathbf{X}_1'\mathbf{X}_2]\boldsymbol{\beta}_2$$

and hence

$$E(s_p^2) = \sigma^2 + \frac{1}{n-p}\boldsymbol{\beta}_2'[\mathbf{X}_2'\mathbf{X}_2 - \mathbf{X}_2'\mathbf{X}_1(\mathbf{X}_1'\mathbf{X}_1)^{-1}\mathbf{X}_1'\mathbf{X}_2]\boldsymbol{\beta}_2$$

This is the result presented in Eq. (3.27) and (4.1).

B.3

THE MAXIMUM LIKELIHOOD ESTIMATOR

In Section 3.3 there was considerable emphasis on the least squares procedure under the condition that the errors are Gaussian. We concluded that, in the case of the linear regression model, the performance of the least squares estimator is better under Gaussian errors than in the nonideal situation where the errors are not normally distributed. We also mentioned that under normality, the least squares estimator is the *maximum likelihood estimator* of the vector, $\boldsymbol{\beta}$, of regression coefficients. Other references to maximum likelihood are given in Chapters 7 and 9.

The development of the properties of the maximum likelihood procedure is beyond the scope of this text. For details see Kendall and Stuart (1973) or Graybill (1976). However, for the student who does have some familiarity with the Gaussian density function and the notion of *likelihood*, we can easily verify that, in the case of normal errors, the least squares estimator is also the maximum likelihood estimator.

Consider the general linear model of Eq. (3.2), namely

$$\mathbf{y} = \mathbf{X}\boldsymbol{\beta} + \boldsymbol{\varepsilon}$$

We make the usual assumptions that $\boldsymbol{\varepsilon} \sim N(\mathbf{0}, \sigma^2\mathbf{I})$. The normal error *density function* for ε_i is as follows:

$$f(\varepsilon_i) = \frac{1}{\sqrt{2\pi}\,\sigma} \exp\left\{-\frac{1}{2\sigma^2}\varepsilon_i^2\right\} \qquad \begin{matrix}(-\infty < \varepsilon_i < \infty) \\ (i = 1, 2, \ldots, n)\end{matrix} \tag{B.5}$$

The *likelihood* is given by the joint density of $\varepsilon_1, \varepsilon_2, \ldots, \varepsilon_n$, which is $\prod_{i=1}^{n} f(\varepsilon_i)$. From Eq. (B.5) the likelihood is given by

$$\prod_{i=1}^{n} f(\varepsilon_i) = \frac{1}{(2\pi)^{n/2}\sigma^n} \exp\left\{-\frac{1}{2\sigma^2}(\mathbf{y} - \mathbf{X}\boldsymbol{\beta})'(\mathbf{y} - \mathbf{X}\mathbf{b})\right\}$$

It is convenient to work with the natural log of the likelihood. As a result, we seek to find $\mathbf{b}$ that maximizes

$$\ln \prod_{i=1}^{n} f(\varepsilon_i) = -\frac{n}{2}\log 2\pi - n\log\sigma - \frac{1}{2\sigma^2}(\mathbf{y} - \mathbf{X}\boldsymbol{\beta})'(\mathbf{y} - \mathbf{X}\boldsymbol{\beta}) \tag{B.6}$$

It is clear that the log likelihood is maximized when the term

$$(\mathbf{y} - \mathbf{X}\boldsymbol{\beta})'(\mathbf{y} - \mathbf{X}\boldsymbol{\beta})$$

which is the residual sum of squares, is minimized. Thus the maximum likelihood estimator of $\boldsymbol{\beta}$ under normal errors is equivalent to the least squares estimator given by

$$\mathbf{b} = (\mathbf{X}'\mathbf{X})^{-1}\mathbf{X}'\mathbf{y}$$

In Chapter 7, we presented the notion of generalized least squares. As a special case, weighted least squares is suggested for consideration when we know that the error variances are not equal. The emphasis here is on weighted regression. It is stated that if the variance-covariance matrix, $\mathbf{V}$, of the errors is known, the generalized least squares estimator $\boldsymbol{\beta}^*$, given in Eq. (7.4), is also a maximum likelihood estimator. Of course, in the weighted least squares case,

$$\mathbf{V} = \text{diag}(\sigma_1^2, \sigma_2^2, \ldots, \sigma_n^2)$$

The development of $\boldsymbol{\beta}^*$ as a least squares estimator follows along lines similar to that used in the ordinary least squares case. The likelihood for the more general case is given by (see Graybill (1976) or Seber (1977))

$$f(\varepsilon_i, \varepsilon_2, \ldots, \varepsilon_n) = \frac{1}{(2\pi)^{n/2}|\mathbf{V}|^{1/2}} \exp\left\{-\frac{1}{2}(\mathbf{y} - \mathbf{X}\boldsymbol{\beta})'\mathbf{V}^{-1}(\mathbf{y} - \mathbf{X}\mathbf{b})\right\} \qquad \begin{matrix}(-\infty < \varepsilon_i < \infty) \\ (i = 1, 2, \ldots, n)\end{matrix}$$

Here we use the $f(\varepsilon_1, \varepsilon_2, \ldots, \varepsilon_n)$ notation to denote *joint probability density function.* As before, maximization of $f(\varepsilon_1, \varepsilon_2, \ldots, \varepsilon_n)$ implies minimization of the exponential portion. Thus the maximum likelihood estimator for $\boldsymbol{\beta}$ is $\boldsymbol{\beta}^*$ for which $SS_{\text{Res},\mathbf{V}}$ in Eq. (7.5) is minimized. This, of course, results in the estimator

$$\boldsymbol{\beta}^* = (\mathbf{X}'\mathbf{V}^{-1}\mathbf{X})^{-1}\mathbf{X}'\mathbf{V}^{-1}\mathbf{y}$$

which is the generalized least squares estimator given by Eq. (7.4).

B.4

DEVELOPMENT OF THE PRESS STATISTIC

The expression for the ith PRESS residual given by Eq. (4.6) provides for computation of $y_i - \hat{y}_{i,-i}$ without actually eliminating the ith point. We can verify the equation by using the Sherman–Morrison–Woodbury Theorem developed in Section A.5. Suppose we let $\mathbf{X}'\mathbf{X}$ play the role of $\mathbf{A}$ and $\mathbf{x}_i'$ (the ith row of $\mathbf{X}$) play the role of $\mathbf{z}'$. Then $(\mathbf{X}'\mathbf{X} - \mathbf{x}_i\mathbf{x}_i')$ is the $\mathbf{X}'\mathbf{X}$ matrix when the ith data point is "set aside." The matrix $(\mathbf{X}'\mathbf{X} - \mathbf{x}_i\mathbf{x}_i')$ is $\mathbf{X}'\mathbf{X}$ reduced by

$$\mathbf{x}_i\mathbf{x}_i' = \begin{bmatrix} 1 & x_{1i} & x_{2i} & \cdots & x_{ki} \\ & x_{1i}^2 & x_{1i}x_{2i} & \cdots & x_{1i}x_{ki} \\ & & x_{2i}^2 & & x_{2i}x_{ki} \\ & & & \ddots & \vdots \\ & & & & x_{ki}^2 \end{bmatrix}$$

As a result,

$$(\mathbf{X}'\mathbf{X} - \mathbf{x}_i\mathbf{x}_i') = \begin{bmatrix} n-1 & \sum_{j\neq i} x_{1j} & \sum_{j\neq i} x_{2j} & \cdots & \sum_{j\neq i} x_{kj} \\ & \sum_{j\neq i} x_{1j}^2 & \sum_{j\neq i} x_{1j}x_{2j} & \cdots & \sum_{j\neq i} x_{1j}x_{kj} \\ & & \sum_{j\neq i} x_{2j}^2 & \cdots & \sum_{j\neq i} x_{2j}x_{kj} \\ & & & \ddots & \vdots \\ & & & & \sum_{j\neq i} x_{kj}^2 \end{bmatrix}$$

which is the $\mathbf{X}'\mathbf{X}$ without the use of the ith data point. The notation used is $(\mathbf{X}'_{-i}\mathbf{X}_{-i})$. From Eq. (A.11), we have

$$(\mathbf{X}'_{-i}\mathbf{X}_{-i})^{-1} = (\mathbf{X}'\mathbf{X})^{-1} + \frac{(\mathbf{X}'\mathbf{X})^{-1}\mathbf{x}_i\mathbf{x}_i'(\mathbf{X}'\mathbf{X})^{-1}}{1-h_{ii}} \tag{B.7}$$

Equation (B.7) is fundamental in developing diagnostic criteria used in Chapter 6. The PRESS residual is given by

$$e_{i,-i} = y_i - \mathbf{x}_i'\mathbf{b}_{-i}$$

where $\mathbf{b}_{-i}$ is the vector of coefficients computed with the ith data point set aside. From (B.7),

$$\begin{aligned} e_{i,-i} &= y_i - \mathbf{x}_i'\left[(\mathbf{X}'\mathbf{X})^{-1} + \frac{(\mathbf{X}'\mathbf{X})^{-1}\mathbf{x}_i\mathbf{x}_i'(\mathbf{X}'\mathbf{X})^{-1}}{1-h_{ii}}\right]\mathbf{X}'_{-i}\mathbf{y}_{-i} \\ &= y_i - \mathbf{x}_i'(\mathbf{X}'\mathbf{X})^{-1}\mathbf{X}'_{-i}\mathbf{y}_{-i} - \frac{h_{ii}\mathbf{x}_i'(\mathbf{X}'\mathbf{X})^{-1}\mathbf{X}'_{-i}\mathbf{y}_{-i}}{1-h_{ii}} \\ &= \frac{(1-h_{ii})y_i - (1-h_{ii})\mathbf{x}_i'(\mathbf{X}'\mathbf{X})^{-1}\mathbf{X}'_{-i}\mathbf{y}_{-i} - h_{ii}\mathbf{x}_i'(\mathbf{X}'\mathbf{X})^{-1}\mathbf{X}'_{-i}\mathbf{y}_{-i}}{1-h_{ii}} \\ &= \frac{(1-h_{ii})y_i - \mathbf{x}_i'(\mathbf{X}'\mathbf{X})^{-1}\mathbf{X}'_{-i}\mathbf{y}_{-i}}{1-h_{ii}} \end{aligned}$$

Now $\mathbf{X}'_{-i}\mathbf{y}_{-i} + \mathbf{x}_i y_i = \mathbf{X}'\mathbf{y}$. Thus we can write the ith PRESS residual

$$e_{i,-i} = \frac{(1-h_{ii})y_i - \mathbf{x}_i'(\mathbf{X}'\mathbf{X})^{-1}(\mathbf{X}'\mathbf{y} - \mathbf{x}_i y_i)}{1-h_{ii}}$$

Now, $\mathbf{x}_i'(\mathbf{X}'\mathbf{X})^{-1}\mathbf{X}'\mathbf{y} = \hat{y}_i$; so we have

$$\begin{aligned} e_{i,-i} &= \frac{(1-h_{ii})y_i - \hat{y}_i + h_{ii}y_i}{1-h_{ii}} \\ &= \frac{y_i - \hat{y}}{1-h_{ii}} \\ &= \frac{e_i}{1-h_{ii}} \end{aligned}$$

B.5

COMPUTATION OF s_{-i}

In Eq. (5.6), an expression was given for the estimate of the residual standard deviation, computed with the ith observation set aside. The statistic s_{-i} is used in the formation of the R-student statistic. The s_{-i} computation illustrates another use of the Sherman–Morrison–Woodbury Theorem to determine an important statistic with a row of the $\mathbf{X}$ matrix eliminated.

We begin with Eq. (B.7), which relates $(\mathbf{X}'_{-i}\mathbf{X}_{-i})^{-1}$ to $(\mathbf{X}'\mathbf{X})^{-1}$. If we multiply both sides by $\mathbf{X}'\mathbf{y} - \mathbf{x}_i y_i$, we obtain

$$\mathbf{b}_{-i} = \mathbf{b} - (\mathbf{X}'\mathbf{X})^{-1}\mathbf{x}_i y_i + \frac{(\mathbf{X}'\mathbf{X})^{-1}\mathbf{x}_i\mathbf{x}_i'(\mathbf{X}'\mathbf{X})^{-1}(\mathbf{X}'\mathbf{y} - \mathbf{x}_i y_i)}{1 - h_{ii}}$$

Collecting terms and simplifying yields the very useful equation,

$$\mathbf{b} - \mathbf{b}_{-i} = \frac{(\mathbf{X}'\mathbf{X})^{-1}\mathbf{x}_i e_i}{1 - h_{ii}} \qquad \text{(B.8)}$$

Now, we can write

$$(n - p - 1)s_{-i}^2 = \sum_{j \neq i} (y_j - \mathbf{x}_j'\mathbf{b}_{-i})^2 \qquad \text{(B.9)}$$

As a result, since

$$\mathbf{b}_{-i} = \mathbf{b} - \frac{(\mathbf{X}'\mathbf{X})^{-1}\mathbf{x}_i e_i}{1 - h_{ii}}$$

we can write

$$\begin{aligned}\sum_{j \neq i} (y_j - \mathbf{x}_j'\mathbf{b}_{-i})^2 &= \sum_{j=1}^{n} \left(y_j - \mathbf{x}_j'\mathbf{b} + \frac{\mathbf{x}_j'(\mathbf{X}'\mathbf{X})^{-1}\mathbf{x}_i e_i}{1 - h_{ii}}\right)^2 - \left(y_i - \mathbf{x}_i'\mathbf{b} + \frac{h_{ii}e_i}{1 - h_{ii}}\right)^2 \\ &= \sum_{j=1}^{n} \left(e_j + \frac{h_{ij}e_i}{1 - h_{ii}}\right)^2 - \frac{e_i^2}{(1 - h_{ii})^2}\end{aligned}$$

Now we can make use of an interesting relationship between the HAT matrix and the residuals. By expanding the term

$$\sum_j \left(e_j + \frac{h_{ij}e_i}{1 - h_{ii}}\right)^2$$

we obtain

$$\sum_{j=1}^{n} e_j^2 + \frac{2e_i}{1 - h_{ii}} \sum_{j=1}^{n} e_j h_{ij} + \frac{e_i^2}{(1 - h_{ii})^2} \sum_{j=1}^{n} h_{ij}^2$$

Since $\mathbf{H}\mathbf{y} = \mathbf{H}\hat{\mathbf{y}}$, then $\sum_{j=1}^{n} e_j h_{ij} = 0$. In addition, since $\mathbf{H}^2 = \mathbf{H}$, $\sum_{j=1}^{n} h_{ij}^2 = h_{ii}$. As a result, $\sum_{j \neq i}(y_j - \mathbf{x}_j'\mathbf{b}_{-i})^2$ can be written

$$\begin{aligned}(n-p-1)s_{-i}^2 &= \sum_{j=1}^{n} e_j^2 + \frac{h_{ii}e_i^2}{(1-h_{ii})^2} - \frac{e_i^2}{(1-h_{ii})^2} \\ &= \sum_{j=1}^{n} e_j^2 - \frac{e_i^2}{1-h_{ii}} \\ &= (n-p)s^2 - \frac{e_i^2}{1-h_{ii}}\end{aligned}$$

Finally, we obtain Eq. (5.6); namely

$$s_{-i} = \sqrt{\frac{(n-p)s^2 - [e_i^2/(1-h_{ii})]}{n-p-1}}$$

B.6

DOMINANCE OF A RESIDUAL BY THE CORRESPONDING MODEL ERROR

In Section 5.7, we studied residuals in order to recover evidence concerning the normality of the ε_i. Equation (5.14) was used to illustrate that e_i is not merely a function of the corresponding ε_i, but rather a linear combination of all of the model errors. Repeating Eq. (5.14), we have

$$e_i = \varepsilon_i - \sum_{j=1}^{n} h_{ij}\varepsilon_j$$

Now, if the sample size n grows large while p, the number of parameters, remains constant, the h_{ij} will tend toward zero, resulting in ε_i dominating. In addition, the variance of $\sum_{j=1}^{n} h_{ij}\varepsilon_j$ is $\sigma^2(\sum_{j=1}^{n} h_{ij}^2) = \sigma^2 h_{ii}$. If $n \gg p$, h_{ii} will be small in comparison to 1.0; thus $\text{Var}(e_i) \cong \sigma^2$. As a result, one can ignore the $\sum_{j=1}^{n} h_{ij}\varepsilon_j$ portion in these circumstances and presume that the information carried in e_i is essentially that of ε_i. Clearly, then, assessment of normality is more accurate in the case where the sample size is much larger than the number of parameters in the model.

B.7

COMPUTATION OF INFLUENCE DIAGNOSTICS

In this section, we show the details that allow for computational ease in obtaining the influence diagnostic tools discussed in Chapter 6. The statistics DFFITS, DFBETAS, and Cook's D are developed. In each case we make use of results already recovered from the Sherman–Morrison–Woodbury Theorem.

DFFITS

Consider the *definitive* form of DFFITS given by Eq. (6.4). The reader should reinspect the very useful formula of Eq. (B.8) derived from the Sherman–Morrison–Woodbury Theorem. Equation (B.8) gives the expression for $\mathbf{b} - \mathbf{b}_{-i}$, where $\mathbf{b}_{-i}$ is the vector of least squares estimators obtained without the use of the ith data point. We have

$$\mathbf{b} - \mathbf{b}_{-i} = \frac{(\mathbf{X}'\mathbf{X})^{-1}\mathbf{x}_i e_i}{1 - h_{ii}}$$

Now, if we merely multiply both sides by $\mathbf{x}_i'$, we have

$$\hat{y}_i - \hat{y}_{i,-i} = \frac{h_{ii} e_i}{1 - h_{ii}}$$

We now merely standardize $\hat{y}_i - \hat{y}_{i,-i}$ to obtain DFFITS.

$$\frac{\hat{y}_i - \hat{y}_{i,-i}}{s_{-i}\sqrt{h_{ii}}} = \left[\frac{e_i h_{ii}}{1 - h_{ii}}\right] \frac{1}{s_{-i}\sqrt{h_{ii}}} = \frac{e_i}{s_{-i}\sqrt{1 - h_{ii}}} \left(\frac{h_{ii}}{1 - h_{ii}}\right)^{1/2}$$

and thus the standardized difference in fitted values is

$$(\text{DFFITS})_i = (R\text{-student})_i \left[\frac{h_{ii}}{1 - h_{ii}}\right]^{1/2}$$

as given by Eq. (6.5).

COOK'S D

Cook's D is easily developed from the expression for $\mathbf{b} - \mathbf{b}_{-i}$ in Eq. (B.8). The definitive form, given by Eq. (6.8) is the quantity

$$D_i = \frac{(\mathbf{b} - \mathbf{b}_{-i})'(\mathbf{X}'\mathbf{X})(\mathbf{b} - \mathbf{b}_{-i})}{ps^2}$$

Making use of (B.8), we have

$$D_i = \frac{\mathbf{x}_i'(\mathbf{X}'\mathbf{X})^{-1}(\mathbf{X}'\mathbf{X})(\mathbf{X}'\mathbf{X})^{-1}\mathbf{x}_i e_i^2}{(1 - h_{ii})^2 ps^2} = \left(\frac{e_i^2}{(1 - h_{ii})^2}\right)\left(\frac{h_{ii}}{ps^2}\right)$$

Finally, we have

$$D_i = \left(\frac{r_i^2}{p}\right)\left(\frac{h_{ii}}{1 - h_{ii}}\right)$$

as given by Eq. (6.9).

DFBETAS

The DFBETAS diagnostic, with definitive form given by Eq. (6.6), is the jth element of $\mathbf{b} - \mathbf{b}_{-i}$ given in Eq. (B.8). Before standardizing,

$$b_j - b_{j,-i} = \frac{r_{j,i} e_i}{1 - h_{ii}} \tag{B.10}$$

By standardizing, we divide (B.10) by $s_{-i}\sqrt{c_{jj}}$, where c_{jj} is the jth diagonal element of $(\mathbf{X}'\mathbf{X})^{-1}$. The c_{jj} element is merely the scalar $\mathbf{r}_j'\mathbf{r}_j$ since $\mathbf{R}'\mathbf{R} = (\mathbf{X}'\mathbf{X})^{-1}$. As a result,

$$\frac{b_j - b_{j,-i}}{s_{-i}\sqrt{c_{jj}}} = \left(\frac{r_{j,i}}{\sqrt{\mathbf{r}_j'\mathbf{r}_j}}\right)\left(\frac{e_i}{s_{-i}(1 - h_{ii})}\right)$$

which is given in Eq. (6.7).

B.8

MAXIMUM LIKELIHOOD ESTIMATOR IN THE NONLINEAR MODEL

The least squares estimator in the case of a nonlinear model involves the minimization of SS_{Res} given by Eq. (9.9). If the model errors are normal, independent, with common variance, σ^2, the resulting estimator is a *maximum likelihood estimator*. Recall that the notion of maximum likelihood was discussed in Appendix B.3 in the case of the linear model. This can easily be carried over to the nonlinear case. If we consider the model notation of Eq. (9.8), then the likelihood, or joint density of $\varepsilon_1, \varepsilon_2, \ldots, \varepsilon_n$ is given by

$$\prod_{i=1}^{n} f(\varepsilon_i) = \frac{1}{(2\pi)^{n/2}\sigma^n} \exp\left\{-\frac{1}{2\sigma^2} \textstyle\sum_{i=1}^{n} [y_i - f(\mathbf{x}_i, \boldsymbol{\theta})]^2\right\}$$

As a result, the maximum likelihood estimator is that value $\hat{\boldsymbol{\theta}}$ for $\boldsymbol{\theta}$ that maximize $\prod_{i=1}^{n} f(\varepsilon_i)$. As in the case of the linear model, the likelihood is maximized when the exponent is minimized. As a result, the *least squares* estimator, with a minimization of $\sum_{i=1}^{n} [y_i - f(\mathbf{x}_i, \hat{\boldsymbol{\theta}})]^2$ is also the maximum likelihood estimator.

B.9

TAYLOR SERIES

A Taylor series approximation is presented in Section 9.3 as a basis for the Gauss–Newton nonlinear estimation procedure. The Taylor series expansion as outlined in Eq. (9.10) is a natural mechanism since it is *linear* in the parameters involved. As a result, the Gauss–Newton procedure involves estimation of coefficients in the linearized version of $f(\mathbf{x}, \boldsymbol{\theta})$ and a continual updating of the estimates through an iterative procedure.

The Taylor series expansion is a standard analytical device for approximating functions. The use in this application is rather special in that it is truncated after only linear terms. More generally, suppose we have a function $y = f(z_1, z_2, \ldots, z_p)$. The Taylor series expansion, representing a local approximation in the neighborhood of $\mathbf{z} = \mathbf{z}_0$; i.e., $(z_1, z_2, \ldots, z_p) = (z_{1,0}, z_{2,0}, \ldots, z_{p,0})$, is given by

$$f(\mathbf{z}) = f(\mathbf{z}_0) + \sum_{i=1}^{p} (z_i - z_{i,0}) \left[\frac{\partial f}{\partial z_i}\right]_{\mathbf{z}=\mathbf{z}_0} + \sum_{i=1}^{p} \frac{(z_i - z_{i,0})^2}{2!} \left[\frac{\partial^2 f}{\partial z_i^2}\right]_{\mathbf{z}=\mathbf{z}_0} + \sum_{i<j}\sum (z_i - z_{i,0})(z_i - z_{j,0}) \left[\frac{\partial^2 f}{\partial z_i \partial z_j}\right]_{\mathbf{z}=\mathbf{z}_0} + \cdots \tag{B.11}$$

A simple example is the expansion of e^z around $z = 0$; we have

$$e^z = e^0 + e^0 z + \frac{e^0 z^2}{2!} + \frac{e^0 z^3}{3!} + \cdots$$
$$= 1 + z + \frac{z^2}{2!} + \frac{z^3}{3!} + \cdots$$

In the Gauss–Newton type application, the θ's are the z's of Eq. (B.11), and the expansion is truncated after the linear terms. The expansion is done around the values $(\theta_{1,0}, \theta_{2,0}, \ldots, \theta_{p,0})$, the starting values of the parameters.

B.10

DEVELOPMENT OF THE C_k-STATISTIC

The C_k-statistic is discussed in Section 8.4. An expansion is given in Eq. (8.13). As in the case of the C_p statistic, the intention of C_k is to estimate the quantity

$$\sum_{i=1}^{n} \text{Var}\ \hat{y}_{i,R} + \sum_{i=1}^{n} [\text{Bias}\ \hat{y}_{i,R}]^2 \tag{B.12}$$

where $\hat{y}_{i,R}$ is the prediction or fitted value at $\mathbf{x}_i$ with the use of ridge regression. Namely,

$$\hat{y}_{i,R} = \mathbf{x}_i' \mathbf{b}_R$$

The estimators $b'_{0,R}, b'_{1,R}, \ldots, b'_{k,R}$, which are the coefficients of the centered and scaled regressors, can be formulated as the solution to

$$\begin{bmatrix} n & 0 & 0 & \cdots & 0 \\ 0 & & & & \\ 0 & & (\mathbf{X}^{*\prime}\mathbf{X}^* + k\mathbf{I}) & & \\ \vdots & & & & \\ 0 & & & & \end{bmatrix} \begin{bmatrix} b'_{0,R} \\ b'_{1,R} \\ \vdots \\ \vdots \\ b'_{k,R} \end{bmatrix} = \mathbf{X}'\mathbf{y} \tag{B.13}$$

where

$$\mathbf{X} = \begin{bmatrix} 1 & \\ 1 & \\ \vdots & (\mathbf{X}^*) \\ 1 & \end{bmatrix}$$

At the point $\mathbf{x}_i$ (centered and scaled), the variance of the prediction $\hat{y}_{i,R}$ is given by

$$\frac{\text{Var } \hat{y}_{i,R}}{\sigma^2} = \mathbf{x}_i' \begin{bmatrix} n & 0 & \cdots & 0 \\ 0 & & & \\ \vdots & & (\mathbf{X}^{*\prime}\mathbf{X}^* + k\mathbf{I}) & \\ 0 & & & \end{bmatrix}^{-1} (\mathbf{X}'\mathbf{X}) \begin{bmatrix} n & 0 & \cdots & 0 \\ 0 & & & \\ \vdots & & (\mathbf{X}^{*\prime}\mathbf{X}^* + k\mathbf{I}) & \\ 0 & & & \end{bmatrix}^{-1} \mathbf{x}_i$$

We can now write $(\sum \text{Var } \hat{y}_{i,R}/\sigma^2)$ as follows:

$$\frac{\sum_{i=1}^{n} \text{Var } \hat{y}_{i,R}}{\sigma^2} = \text{tr}\left[\mathbf{X} \begin{bmatrix} n & 0 & \cdots & 0 \\ 0 & & & \\ \vdots & & (\mathbf{X}^{*\prime}\mathbf{X}^* + k\mathbf{I}) & \\ 0 & & & \end{bmatrix}^{-1} (\mathbf{X}'\mathbf{X}) \begin{bmatrix} n & 0 & \cdots & 0 \\ 0 & & & \\ \vdots & & (\mathbf{X}^{*\prime}\mathbf{X}^* + k\mathbf{I}) & \\ 0 & & & \end{bmatrix}^{-1} \mathbf{X}' \right]$$

$$= \text{tr}\left[\mathbf{X} \begin{bmatrix} n & 0 & \cdots & 0 \\ 0 & & & \\ \vdots & & (\mathbf{X}^{*\prime}\mathbf{X}^* + k\mathbf{I}) & \\ 0 & & & \end{bmatrix}^{-1} \mathbf{X}' \right]^2$$

$$= \text{tr}[\mathbf{A}_k]^2 \tag{B.14}$$

Now, let us consider $\sum_{i=1}^{n} [(\text{Bias } \hat{y}_{i,R})^2/\sigma^2]$. As in the case of the C_p-statistic, the bias portion is estimated from the residual sum of squares. We learned in Chapter 8 that the coefficients are biased in the case of ridge regression. The bias in the fitted values, the $\hat{y}_{i,R}$, inflate the residual sum of squares. We can write

$$SS_{\text{Res},k} = (\mathbf{y} - \mathbf{X}\mathbf{b}_R)'(\mathbf{y} - \mathbf{X}\mathbf{b}_R) = \mathbf{y}'[\mathbf{I} - \mathbf{A}_k]^2\mathbf{y}$$

We can now use the theorem given by Eq. (B.3) on the expected value of a quadratic form to obtain

$$\begin{aligned} E\{\mathbf{y}'[\mathbf{I} - \mathbf{A}_k]^2\mathbf{y}\} &= \sigma^2 \text{ tr}[\mathbf{I} - \mathbf{A}_k]^2 + (\mathbf{X}\boldsymbol{\beta})'[\mathbf{I} - \mathbf{A}_k]^2(\mathbf{X}\boldsymbol{\beta}) \\ &= \sigma^2 \text{ tr}[\mathbf{I} - \mathbf{A}_k]^2 + \boldsymbol{\beta}'\mathbf{X}'[\mathbf{I} - \mathbf{A}_k]^2\mathbf{X}\boldsymbol{\beta} \end{aligned}$$

The bias portion of the *total error* in Eq. (B.12) is given by

$$\begin{aligned} \sum_{i=1}^{n} (\text{Bias } \hat{y}_{i,R})^2 &= E(\mathbf{X}\boldsymbol{\beta} - \mathbf{X}\mathbf{b}_R)'E(\mathbf{X}\boldsymbol{\beta} - \mathbf{X}\mathbf{b}_R) \\ &= [\mathbf{X}\boldsymbol{\beta} - \mathbf{X}E(\mathbf{b}_R)]'[\mathbf{X}\boldsymbol{\beta} - \mathbf{X}E(\mathbf{b}_R)] \\ &= [\mathbf{X}\boldsymbol{\beta} - \mathbf{A}_k\mathbf{X}\boldsymbol{\beta}]'[\mathbf{X}\boldsymbol{\beta} - \mathbf{A}_k\mathbf{X}\boldsymbol{\beta}] \\ &= (\mathbf{X}\boldsymbol{\beta})'[\mathbf{I} - \mathbf{A}_k]^2(\mathbf{X}\boldsymbol{\beta}) \end{aligned}$$

Thus, an unbiased estimator for $\boldsymbol{\beta}'\mathbf{X}'[\mathbf{I}-\mathbf{A}_k]^2\mathbf{X}\boldsymbol{\beta}$, the sum of the squared biases of $\hat{y}_{i,R}$, is given by

$$\sum_{i=1}^{n} (\text{Bias } \hat{y}_{i,R})^2 = SS_{\text{Res},k} - \sigma^2 \operatorname{tr}(\mathbf{I}-\mathbf{A}_k)^2$$

Thus, the estimator of the expression in Eq. (B.12) is given by

$$C_k = \operatorname{tr}(\mathbf{A}_k)^2 + \frac{SS_{\text{Res},k}}{\sigma^2} - \operatorname{tr}(\mathbf{I}-\mathbf{A}_k)^2$$

$$= \frac{SS_{\text{Res},k}}{\sigma^2} - n + 2\operatorname{tr}(\mathbf{A}_k)$$

From the definition of $\mathbf{A}_k$ and the matrix

$$\mathbf{H}_k = \mathbf{X}^*(\mathbf{X}^{*\prime}\mathbf{X}^* + k\mathbf{I}]^{-1}\mathbf{X}^{*\prime}$$

$$\operatorname{tr}(\mathbf{A}_k) = \operatorname{tr}(\mathbf{H}_k) + 1$$

Thus, using $\hat{\sigma}^2$ (from OLS) in place of σ^2,

$$C_k = \frac{SS_{\text{Res},k}}{\hat{\sigma}^2} - n + 2 + 2\operatorname{tr}(\mathbf{H}_k)$$

REFERENCES FOR APPENDIX B

1. Graybill, F.A. 1976. *Theory and Application of the Linear Model*. Boston, Massachusetts: Duxbury.
2. Kendall, M.G., and A. Stuart, 1973. *The Advanced Theory of Statistics* vol. 2, *Inference and Relationship*. 3d ed. New York: Hafner.
3. Seber, G.A.F. 1977. *Linear Regression Analysis*. New York: John Wiley & Sons.

APPENDIX C

STATISTICAL TABLES

TABLE C.1

Cumulative standard normal p.d.f.
Entries are $\int_{-\infty}^{N_\alpha} n(z: 0, 1)\, dz = 1 - \alpha$

N_α	.00	.01	.02	.03	.04	.05	.06	.07	.08	.09
.0	.5000	.5040	.5080	.5120	.5160	.5199	.5239	.5279	.5319	.5359
.1	.5398	.5438	.5478	.5517	.5557	.5596	.5636	.5675	.5714	.5753
.2	.5793	.5832	.5871	.5910	.5948	.5987	.6026	.6064	.6103	.6141
.3	.6179	.6217	.6255	.6293	.6331	.6368	.6406	.6443	.6480	.6517
.4	.6554	.6591	.6628	.6664	.6700	.6736	.6772	.6808	.6844	.6879
.5	.6915	.6950	.6985	.7019	.7054	.7088	.7123	.7157	.7190	.7224
.6	.7257	.7291	.7324	.7357	.7389	.7422	.7454	.7486	.7517	.7549
.7	.7580	.7611	.7642	.7673	.7704	.7734	.7764	.7794	.7823	.7852
.8	.7881	.7910	.7939	.7967	.7995	.8023	.8051	.8078	.8106	.8133
.9	.8159	.8186	.8212	.8238	.8264	.8289	.8315	.8340	.8365	.8389
1.0	.8413	.8438	.8461	.8485	.8508	.8531	.8554	.8577	.8599	.8621
1.1	.8643	.8665	.8686	.8708	.8729	.8749	.8770	.8790	.8810	.8830
1.2	.8849	.8869	.8888	.8907	.8925	.8944	.8962	.8980	.8997	.9015
1.3	.9032	.9049	.9066	.9082	.9099	.9115	.9131	.9147	.9162	.9177
1.4	.9192	.9207	.9222	.9236	.9251	.9265	.9279	.9292	.9306	.9319
1.5	.9332	.9345	.9357	.9370	.9382	.9394	.9406	.9418	.9429	.9441
1.6	.9452	.9463	.9474	.9484	.9495	.9505	.9515	.9525	.9535	.9545
1.7	.9554	.9564	.9573	.9582	.9591	.9599	.9608	.9616	.9625	.9633
1.8	.9641	.9649	.9656	.9664	.9671	.9678	.9686	.9693	.9699	.9706
1.9	.9713	.9719	.9726	.9732	.9738	.9744	.9750	.9756	.9761	.9767
2.0	.9772	.9778	.9783	.9788	.9793	.9798	.9803	.9808	.9812	.9817
2.1	.9821	.9826	.9830	.9834	.9838	.9842	.9846	.9850	.9854	.9857
2.2	.9861	.9864	.9868	.9871	.9875	.9878	.9881	.9884	.9887	.9890
2.3	.9893	.9896	.9898	.9901	.9904	.9906	.9909	.9911	.9913	.9916
2.4	.9918	.9920	.9922	.9925	.9927	.9929	.9931	.9932	.9934	.9936
2.5	.9938	.9940	.9941	.9943	.9945	.9946	.9948	.9949	.9951	.9952
2.6	.9953	.9955	.9956	.9957	.9959	.9960	.9961	.9962	.9963	.9964
2.7	.9965	.9966	.9967	.9968	.9969	.9970	.9971	.9972	.9973	.9974
2.8	.9974	.9975	.9976	.9977	.9977	.9978	.9979	.9979	.9980	.9981
2.9	.9981	.9982	.9982	.9983	.9984	.9984	.9985	.9985	.9986	.9986
3.0	.9987	.9987	.9987	.9988	.9988	.9989	.9989	.9989	.9990	.9990
3.1	.9990	.9991	.9991	.9991	.9992	.9992	.9992	.9992	.9993	.9993
3.2	.9993	.9993	.9994	.9994	.9994	.9994	.9994	.9995	.9995	.9995
3.3	.9995	.9995	.9995	.9996	.9996	.9996	.9996	.9996	.9996	.9997
3.4	.9997	.9997	.9997	.9997	.9997	.9997	.9997	.9997	.9997	.9998

N_α	1.282	1.645	1.960	2.326	2.576	3.090	3.291	3.891	4.417
$1 - \alpha$	.90	.95	.975	.99	.995	.999	.9995	.99995	.999995

Franklin A. Graybill, *Theory and Application of the Linear Model.* (Boston, Massachusetts: Duxbury, 1976), p. 651.

TABLE C.2 **Percentage points of the student's *t*-distribution**

ν	$\alpha^{a} = 0.4$	0.25	0.1	0.05	0.025	0.01	0.005	0.0025	0.001	0.0005
1	0.325	1.000	3.078	6.314	12.706	31.821	63.657	127.32	318.31	636.62
2	0.289	0.816	1.886	2.920	4.303	6.965	9.925	14.089	22.327	31.598
3	0.277	0.765	1.638	2.353	3.182	4.541	5.841	7.453	10.214	12.924
4	0.271	0.741	1.533	2.132	2.776	3.747	4.604	5.598	7.173	8.610
5	0.267	0.727	1.476	2.015	2.571	3.365	4.032	4.773	5.893	6.869
6	0.265	0.718	1.440	1.943	2.447	3.143	3.707	4.317	5.208	5.959
7	0.263	0.711	1.415	1.895	2.365	2.998	3.499	4.029	4.785	5.408
8	0.262	0.706	1.397	1.860	2.306	2.896	3.355	3.833	4.501	5.041
9	0.261	0.703	1.383	1.833	2.262	2.821	3.250	3.690	4.297	4.781
10	0.260	0.700	1.372	1.812	2.228	2.764	3.169	3.581	4.144	4.587
11	0.260	0.697	1.363	1.796	2.201	2.718	3.106	3.497	4.025	4.437
12	0.259	0.695	1.356	1.782	2.179	2.681	3.055	3.428	3.930	4.318
13	0.259	0.694	1.350	1.771	2.160	2.650	3.012	3.372	3.852	4.221
14	0.258	0.692	1.345	1.761	2.145	2.624	2.977	3.326	3.787	4.140
15	0.258	0.691	1.341	1.753	2.131	2.602	2.947	3.286	3.733	4.073
16	0.258	0.690	1.337	1.746	2.120	2.583	2.921	3.252	3.686	4.015
17	0.257	0.689	1.333	1.740	2.110	2.567	2.898	3.222	3.646	3.965
18	0.257	0.688	1.330	1.734	2.101	2.552	2.878	3.197	3.610	3.922
19	0.257	0.688	1.328	1.729	2.093	2.539	2.861	3.174	3.579	3.883
20	0.257	0.687	1.325	1.725	2.086	2.528	2.845	3.153	3.552	3.850
21	0.257	0.686	1.323	1.721	2.080	2.518	2.831	3.135	3.527	3.819
22	0.256	0.686	1.321	1.717	2.074	2.508	2.819	3.119	3.505	3.792
23	0.256	0.685	1.319	1.714	2.069	2.500	2.807	3.104	3.485	3.767
24	0.256	0.685	1.318	1.711	2.064	2.492	2.797	3.091	3.467	3.745
25	0.256	0.684	1.316	1.708	2.060	2.485	2.787	3.078	3.450	3.725
26	0.256	0.684	1.315	1.706	2.056	2.479	2.779	3.067	3.435	3.707
27	0.256	0.684	1.314	1.703	2.052	2.473	2.771	3.057	3.421	3.690
28	0.256	0.683	1.313	1.701	2.048	2.467	2.763	3.047	3.408	3.674
29	0.256	0.683	1.311	1.699	2.045	2.462	2.756	3.038	3.396	3.659
30	0.256	0.683	1.310	1.697	2.042	2.457	2.750	3.030	3.385	3.646
40	0.255	0.681	1.303	1.684	2.021	2.423	2.704	2.971	3.307	3.551
60	0.254	0.679	1.296	1.671	2.000	2.390	2.660	2.915	3.232	3.460
120	0.254	0.677	1.289	1.658	1.980	2.358	2.617	2.860	3.160	3.373
∞	0.253	0.674	1.282	1.645	1.960	2.326	2.576	2.807	3.090	3.291

[a] The quantity α is the upper-tail area of the distribution for ν degrees of freedom.

TABLE C.3

Percentage points of the *F*-distribution with degrees of freedom ν_1 and ν_2
Upper 10% points

ν_2 \ ν_1	1	2	3	4	5	6	7	8	9	10	12	15	20	24	30	40	60	120	∞
1	39.86	49.50	53.59	55.83	57.24	58.20	58.91	59.44	59.86	60.19	60.71	61.22	61.74	62.00	62.26	62.53	62.79	63.06	63.33
2	8.53	9.00	9.16	9.24	9.29	9.33	9.35	9.37	9.38	9.39	9.41	9.42	9.44	9.45	9.46	9.47	9.47	9.48	9.49
3	5.54	5.46	5.39	5.34	5.31	5.28	5.27	5.25	5.24	5.23	5.22	5.20	5.18	5.18	5.17	5.16	5.15	5.14	5.13
4	4.54	4.32	4.19	4.11	4.05	4.01	3.98	3.95	3.94	3.92	3.90	3.87	3.84	3.83	3.82	3.80	3.79	3.78	3.76
5	4.06	3.78	3.62	3.52	3.45	3.40	3.37	3.34	3.32	3.30	3.27	3.24	3.21	3.19	3.17	3.16	3.14	3.12	3.10
6	3.78	3.46	3.29	3.18	3.11	3.05	3.01	2.98	2.96	2.94	2.90	2.87	2.84	2.82	2.80	2.78	2.76	2.74	2.72
7	3.59	3.26	3.07	2.96	2.88	2.83	2.78	2.75	2.72	2.70	2.67	2.63	2.59	2.58	2.56	2.54	2.51	2.49	2.47
8	3.46	3.11	2.92	2.81	2.73	2.67	2.62	2.59	2.56	2.54	2.50	2.46	2.42	2.40	2.38	2.36	2.34	2.32	2.29
9	3.36	3.01	2.81	2.69	2.61	2.55	2.51	2.47	2.44	2.42	2.38	2.34	2.30	2.28	2.25	2.23	2.21	2.18	2.16
10	3.29	2.92	2.73	2.61	2.52	2.46	2.41	2.38	2.35	2.32	2.28	2.24	2.20	2.18	2.16	2.13	2.11	2.08	2.06
11	3.23	2.86	2.66	2.54	2.45	2.39	2.34	2.30	2.27	2.25	2.21	2.17	2.12	2.10	2.08	2.05	2.03	2.00	1.97
12	3.18	2.81	2.61	2.48	2.39	2.33	2.28	2.24	2.21	2.19	2.15	2.10	2.06	2.04	2.01	1.99	1.96	1.93	1.90
13	3.14	2.76	2.56	2.43	2.35	2.28	2.23	2.20	2.16	2.14	2.10	2.05	2.01	1.98	1.96	1.93	1.90	1.88	1.85
14	3.10	2.73	2.52	2.39	2.31	2.24	2.19	2.15	2.12	2.10	2.05	2.01	1.96	1.94	1.91	1.89	1.86	1.83	1.80
15	3.07	2.70	2.49	2.36	2.27	2.21	2.16	2.12	2.09	2.06	2.02	1.97	1.92	1.90	1.87	1.85	1.82	1.79	1.76
16	3.05	2.67	2.46	2.33	2.24	2.18	2.13	2.09	2.06	2.03	1.99	1.94	1.89	1.87	1.84	1.81	1.78	1.75	1.72
17	3.03	2.64	2.44	2.31	2.22	2.15	2.10	2.06	2.03	2.00	1.96	1.91	1.86	1.84	1.81	1.78	1.75	1.72	1.69
18	3.01	2.62	2.42	2.29	2.20	2.13	2.08	2.04	2.00	1.98	1.93	1.89	1.84	1.81	1.78	1.75	1.72	1.69	1.66
19	2.99	2.61	2.40	2.27	2.18	2.11	2.06	2.02	1.98	1.96	1.91	1.86	1.81	1.79	1.76	1.73	1.70	1.67	1.63
20	2.97	2.59	2.38	2.25	2.16	2.09	2.04	2.00	1.96	1.94	1.89	1.84	1.79	1.77	1.74	1.71	1.68	1.64	1.61
21	2.96	2.57	2.36	2.23	2.14	2.08	2.02	1.98	1.95	1.92	1.87	1.83	1.78	1.75	1.72	1.69	1.66	1.62	1.59
22	2.95	2.56	2.35	2.22	2.13	2.06	2.01	1.97	1.93	1.90	1.86	1.81	1.76	1.73	1.70	1.67	1.64	1.60	1.57
23	2.94	2.55	2.34	2.21	2.11	2.05	1.99	1.95	1.92	1.89	1.84	1.80	1.74	1.72	1.69	1.66	1.62	1.59	1.55
24	2.93	2.54	2.33	2.19	2.10	2.04	1.98	1.94	1.91	1.88	1.83	1.78	1.73	1.70	1.67	1.64	1.61	1.57	1.53
25	2.92	2.53	2.32	2.18	2.09	2.02	1.97	1.93	1.89	1.87	1.82	1.77	1.72	1.69	1.66	1.63	1.59	1.56	1.52
26	2.91	2.52	2.31	2.17	2.08	2.01	1.96	1.92	1.88	1.86	1.81	1.76	1.71	1.68	1.65	1.61	1.58	1.54	1.50
27	2.90	2.51	2.30	2.17	2.07	2.00	1.95	1.91	1.87	1.85	1.80	1.75	1.70	1.67	1.64	1.60	1.57	1.53	1.49
28	2.89	2.50	2.29	2.16	2.06	2.00	1.94	1.90	1.87	1.84	1.79	1.74	1.69	1.66	1.63	1.59	1.56	1.52	1.48
29	2.89	2.50	2.28	2.15	2.06	1.99	1.93	1.89	1.86	1.83	1.78	1.73	1.68	1.65	1.62	1.58	1.55	1.51	1.47
30	2.88	2.49	2.28	2.14	2.05	1.98	1.93	1.88	1.85	1.82	1.77	1.72	1.67	1.64	1.61	1.57	1.54	1.50	1.46
40	2.84	2.44	2.23	2.09	2.00	1.93	1.87	1.83	1.79	1.76	1.71	1.66	1.61	1.57	1.54	1.51	1.47	1.42	1.38
60	2.79	2.39	2.18	2.04	1.95	1.87	1.82	1.77	1.74	1.71	1.66	1.60	1.54	1.51	1.48	1.44	1.40	1.35	1.29
120	2.75	2.35	2.13	1.99	1.90	1.82	1.77	1.72	1.68	1.65	1.60	1.55	1.48	1.45	1.41	1.37	1.32	1.26	1.19
∞	2.71	2.30	2.08	1.94	1.85	1.77	1.72	1.67	1.63	1.60	1.55	1.49	1.42	1.38	1.34	1.30	1.24	1.17	1.00

TABLE C.3 Continued

TABLE C.3 Continued

Percentage points of the *F*-distribution with degrees of freedom ν_1 and ν_2
Upper 5% points

ν_2 \ ν_1	1	2	3	4	5	6	7	8	9	10	12	15	20	24	30	40	60	120	∞
1	161.4	199.5	215.7	224.6	230.2	234.0	236.8	238.9	240.5	241.9	243.9	245.9	248.0	249.1	250.1	251.1	252.2	253.3	254.3
2	18.51	19.00	19.16	19.25	19.30	19.33	19.35	19.37	19.38	19.40	19.41	19.43	19.45	19.45	19.46	19.47	19.48	19.49	19.50
3	10.13	9.55	9.28	9.12	9.01	8.94	8.89	8.85	8.81	8.79	8.74	8.70	8.66	8.64	8.62	8.59	8.57	8.55	8.53
4	7.71	6.94	6.59	6.39	6.26	6.16	6.09	6.04	6.00	5.96	5.91	5.86	5.80	5.77	5.75	5.72	5.69	5.66	5.63
5	6.61	5.79	5.41	5.19	5.05	4.95	4.88	4.82	4.77	4.74	4.68	4.62	4.56	4.53	4.50	4.46	4.43	4.40	4.36
6	5.99	5.14	4.76	4.53	4.39	4.28	4.21	4.15	4.10	4.06	4.00	3.94	3.87	3.84	3.81	3.77	3.74	3.70	3.67
7	5.59	4.74	4.35	4.12	3.97	3.87	3.79	3.73	3.68	3.64	3.57	3.51	3.44	3.41	3.38	3.34	3.30	3.27	3.23
8	5.32	4.46	4.07	3.84	3.69	3.58	3.50	3.44	3.39	3.35	3.28	3.22	3.15	3.12	3.08	3.04	3.01	2.97	2.93
9	5.12	4.26	3.86	3.63	3.48	3.37	3.29	3.23	3.18	3.14	3.07	3.01	2.94	2.90	2.86	2.83	2.79	2.75	2.71
10	4.96	4.10	3.71	3.48	3.33	3.22	3.14	3.07	3.02	2.98	2.91	2.85	2.77	2.74	2.70	2.66	2.62	2.58	2.54
11	4.84	3.98	3.59	3.36	3.20	3.09	3.01	2.95	2.90	2.85	2.79	2.72	2.65	2.61	2.57	2.53	2.49	2.45	2.40
12	4.75	3.89	3.49	3.26	3.11	3.00	2.91	2.85	2.80	2.75	2.69	2.62	2.54	2.51	2.47	2.43	2.38	2.34	2.30
13	4.67	3.81	3.41	3.18	3.03	2.92	2.83	2.77	2.71	2.67	2.60	2.53	2.46	2.42	2.38	2.34	2.30	2.25	2.21
14	4.60	3.74	3.34	3.11	2.96	2.85	2.76	2.70	2.65	2.60	2.53	2.46	2.39	2.35	2.31	2.27	2.22	2.18	2.13
15	4.54	3.68	3.29	3.06	2.90	2.79	2.71	2.64	2.59	2.54	2.48	2.40	2.33	2.29	2.25	2.20	2.16	2.11	2.07
16	4.49	3.63	3.24	3.01	2.85	2.74	2.66	2.59	2.54	2.49	2.42	2.35	2.28	2.24	2.19	2.15	2.11	2.06	2.01
17	4.45	3.59	3.20	2.96	2.81	2.70	2.61	2.55	2.49	2.45	2.38	2.31	2.23	2.19	2.15	2.10	2.06	2.01	1.96
18	4.41	3.55	3.16	2.93	2.77	2.66	2.58	2.51	2.46	2.41	2.34	2.27	2.19	2.15	2.11	2.06	2.02	1.97	1.92
19	4.38	3.52	3.13	2.90	2.74	2.63	2.54	2.48	2.42	2.38	2.31	2.23	2.16	2.11	2.07	2.03	1.98	1.93	1.88
20	4.35	3.49	3.10	2.87	2.71	2.60	2.51	2.45	2.39	2.35	2.28	2.20	2.12	2.08	2.04	1.99	1.95	1.90	1.84
21	4.32	3.47	3.07	2.84	2.68	2.57	2.49	2.42	2.37	2.32	2.25	2.18	2.10	2.05	2.01	1.96	1.92	1.87	1.81
22	4.30	3.44	3.05	2.82	2.66	2.55	2.46	2.40	2.34	2.30	2.23	2.15	2.07	2.03	1.98	1.94	1.89	1.84	1.78
23	4.28	3.42	3.03	2.80	2.64	2.53	2.44	2.37	2.32	2.27	2.20	2.13	2.05	2.01	1.96	1.91	1.86	1.81	1.76
24	4.26	3.40	3.01	2.78	2.62	2.51	2.42	2.36	2.30	2.25	2.18	2.11	2.03	1.98	1.94	1.89	1.84	1.79	1.73
25	4.24	3.39	2.99	2.76	2.60	2.49	2.40	2.34	2.28	2.24	2.16	2.09	2.01	1.96	1.92	1.87	1.82	1.77	1.71
26	4.23	3.37	2.98	2.74	2.59	2.47	2.39	2.32	2.27	2.22	2.15	2.07	1.99	1.95	1.90	1.85	1.80	1.75	1.69
27	4.21	3.35	2.96	2.73	2.57	2.46	2.37	2.31	2.25	2.20	2.13	2.06	1.97	1.93	1.88	1.84	1.79	1.73	1.67
28	4.20	3.34	2.95	2.71	2.56	2.45	2.36	2.29	2.24	2.19	2.12	2.04	1.96	1.91	1.87	1.82	1.77	1.71	1.65
29	4.18	3.33	2.93	2.70	2.55	2.43	2.35	2.28	2.22	2.18	2.10	2.03	1.94	1.90	1.85	1.81	1.75	1.70	1.64
30	4.17	3.32	2.92	2.69	2.53	2.42	2.33	2.27	2.21	2.16	2.09	2.01	1.93	1.89	1.84	1.79	1.74	1.68	1.62
40	4.08	3.23	2.84	2.61	2.45	2.34	2.25	2.18	2.12	2.08	2.00	1.92	1.84	1.79	1.74	1.69	1.64	1.58	1.51
60	4.00	3.15	2.76	2.53	2.37	2.25	2.17	2.10	2.04	1.99	1.92	1.84	1.75	1.70	1.65	1.59	1.53	1.47	1.39
120	3.92	3.07	2.68	2.45	2.29	2.17	2.09	2.02	1.96	1.91	1.83	1.75	1.66	1.61	1.55	1.50	1.43	1.35	1.25
∞	3.84	3.00	2.60	2.37	2.21	2.10	2.01	1.94	1.88	1.83	1.75	1.67	1.57	1.52	1.46	1.39	1.32	1.22	1.00

TABLE C.3 Continued

Percentage points of the *F*-distribution with degrees of freedom ν_1 and ν_2
Upper 2.5% points

ν_2 \ ν_1	1	2	3	4	5	6	7	8	9	10	12	15	20	24	30	40	60	120	∞
1	647.8	799.5	864.2	899.6	921.8	937.1	948.2	956.7	963.3	968.6	976.7	984.9	993.1	997.2	1001	1006	1010	1014	1018
2	38.51	39.00	39.17	39.25	39.30	39.33	39.36	39.37	39.39	39.40	39.41	39.43	39.45	39.46	39.46	39.47	39.48	39.49	39.50
3	17.44	16.04	15.44	15.10	14.88	14.73	14.62	14.54	14.47	14.42	14.34	14.25	14.17	14.12	14.08	14.04	13.99	13.95	13.90
4	12.22	10.65	9.98	9.60	9.36	9.20	9.07	8.98	8.90	8.84	8.75	8.66	8.56	8.51	8.46	8.41	8.36	8.31	8.26
5	10.01	8.43	7.76	7.39	7.15	6.98	6.85	6.76	6.68	6.62	6.52	6.43	6.33	6.28	6.23	6.18	6.12	6.07	6.02
6	8.81	7.26	6.60	6.23	5.99	5.82	5.70	5.60	5.52	5.46	5.37	5.27	5.17	5.12	5.07	5.01	4.96	4.90	4.85
7	8.07	6.54	5.89	5.52	5.29	5.12	4.99	4.90	4.82	4.76	4.67	4.57	4.47	4.42	4.36	4.31	4.25	4.20	4.14
8	7.57	6.06	5.42	5.05	4.82	4.65	4.53	4.43	4.36	4.30	4.20	4.10	4.00	3.95	3.89	3.84	3.78	3.73	3.67
9	7.21	5.71	5.08	4.72	4.48	4.32	4.20	4.10	4.03	3.96	3.87	3.77	3.67	3.61	3.56	3.51	3.45	3.39	3.33
10	6.94	5.46	4.83	4.47	4.24	4.07	3.95	3.85	3.78	3.72	3.62	3.52	3.42	3.37	3.31	3.26	3.20	3.14	3.08
11	6.72	5.26	4.63	4.28	4.04	3.88	3.76	3.66	3.59	3.53	3.43	3.33	3.23	3.17	3.12	3.06	3.00	2.94	2.88
12	6.55	5.10	4.47	4.12	3.89	3.73	3.61	3.51	3.44	3.37	3.28	3.18	3.07	3.02	2.96	2.91	2.85	2.79	2.72
13	6.41	4.97	4.35	4.00	3.77	3.60	3.48	3.39	3.31	3.25	3.15	3.05	2.95	2.89	2.84	2.78	2.72	2.66	2.60
14	6.30	4.86	4.24	3.89	3.66	3.50	3.38	3.29	3.21	3.15	3.05	2.95	2.84	2.79	2.73	2.67	2.61	2.55	2.49
15	6.20	4.77	4.15	3.80	3.58	3.41	3.29	3.20	3.12	3.06	2.96	2.86	2.76	2.70	2.64	2.59	2.52	2.46	2.40
16	6.12	4.69	4.08	3.73	3.50	3.34	3.22	3.12	3.05	2.99	2.89	2.79	2.68	2.63	2.57	2.51	2.45	2.38	2.32
17	6.04	4.62	4.01	3.66	3.44	3.28	3.16	3.06	2.98	2.92	2.82	2.72	2.62	2.56	2.50	2.44	2.38	2.32	2.25
18	5.98	4.56	3.95	3.61	3.38	3.22	3.10	3.01	2.93	2.87	2.77	2.67	2.56	2.50	2.44	2.38	2.32	2.26	2.19
19	5.92	4.51	3.90	3.56	3.33	3.17	3.05	2.96	2.88	2.82	2.72	2.62	2.51	2.45	2.39	2.33	2.27	2.20	2.13
20	5.87	4.46	3.86	3.51	3.29	3.13	3.01	2.91	2.84	2.77	2.68	2.57	2.46	2.41	2.35	2.29	2.22	2.16	2.09
21	5.83	4.42	3.82	3.48	3.25	3.09	2.97	2.87	2.80	2.73	2.64	2.53	2.42	2.37	2.31	2.25	2.18	2.11	2.04
22	5.79	4.38	3.78	3.44	3.22	3.05	2.93	2.84	2.76	2.70	2.60	2.50	2.39	2.33	2.27	2.21	2.14	2.08	2.00
23	5.75	4.35	3.75	3.41	3.18	3.02	2.90	2.81	2.73	2.67	2.57	2.47	2.36	2.30	2.24	2.18	2.11	2.04	1.97
24	5.72	4.32	3.72	3.38	3.15	2.99	2.87	2.78	2.70	2.64	2.54	2.44	2.33	2.27	2.21	2.15	2.08	2.01	1.94
25	5.69	4.29	3.69	3.35	3.13	2.97	2.85	2.75	2.68	2.61	2.51	2.41	2.30	2.24	2.18	2.12	2.05	1.98	1.91
26	5.66	4.27	3.67	3.33	3.10	2.94	2.82	2.73	2.65	2.59	2.49	2.39	2.28	2.22	2.16	2.09	2.03	1.95	1.88
27	5.63	4.24	3.65	3.31	3.08	2.92	2.80	2.71	2.63	2.57	2.47	2.36	2.25	2.19	2.13	2.07	2.00	1.93	1.85
28	5.61	4.22	3.63	3.29	3.06	2.90	2.78	2.69	2.61	2.55	2.45	2.34	2.23	2.17	2.11	2.05	1.98	1.91	1.83
29	5.59	4.20	3.61	3.27	3.04	2.88	2.76	2.67	2.59	2.53	2.43	2.32	2.21	2.15	2.09	2.03	1.96	1.89	1.81
30	5.57	4.18	3.59	3.25	3.03	2.87	2.75	2.65	2.57	2.51	2.41	2.31	2.20	2.14	2.07	2.01	1.94	1.87	1.79
40	5.42	4.05	3.46	3.13	2.90	2.74	2.62	2.53	2.45	2.39	2.29	2.18	2.07	2.01	1.94	1.88	1.80	1.72	1.64
60	5.29	3.93	3.34	3.01	2.79	2.63	2.51	2.41	2.33	2.27	2.17	2.06	1.94	1.88	1.82	1.74	1.67	1.58	1.48
120	5.15	3.80	3.23	2.89	2.67	2.52	2.39	2.30	2.22	2.16	2.05	1.94	1.82	1.76	1.69	1.61	1.53	1.43	1.31
∞	5.02	3.69	3.12	2.79	2.57	2.41	2.29	2.19	2.11	2.05	1.94	1.83	1.71	1.64	1.57	1.48	1.39	1.27	1.00

TABLE C.3 Continued

Percentage points of the *F*-distribution with degrees of freedom ν_1 and ν_2
Upper 1% points

ν_2 \ ν_1	1	2	3	4	5	6	7	8	9	10	12	15	20	24	30	40	60	120	∞
1	4052	4999.5	5403	5625	5764	5859	5928	5981	6022	6056	6106	6157	6209	6235	6261	6287	6313	6339	6366
2	98.50	99.00	99.17	99.25	99.30	99.33	99.36	99.37	99.39	99.40	99.42	99.43	99.45	99.46	99.47	99.47	99.48	99.49	99.50
3	34.12	30.82	29.46	28.71	28.24	27.91	27.67	27.49	27.35	27.23	27.05	26.87	26.69	26.60	26.50	26.41	26.32	26.22	26.13
4	21.20	18.00	16.69	15.98	15.52	15.21	14.98	14.80	14.66	14.55	14.37	14.20	14.02	13.93	13.84	13.75	13.65	13.56	13.46
5	16.26	13.27	12.06	11.39	10.97	10.67	10.46	10.29	10.16	10.05	9.89	9.72	9.55	9.47	9.38	9.29	9.20	9.11	9.02
6	13.75	10.92	9.78	9.15	8.75	8.47	8.26	8.10	7.98	7.87	7.72	7.56	7.40	7.31	7.23	7.14	7.06	6.97	6.88
7	12.25	9.55	8.45	7.85	7.46	7.19	6.99	6.84	6.72	6.62	6.47	6.31	6.16	6.07	5.99	5.91	5.82	5.74	5.65
8	11.26	8.65	7.59	7.01	6.63	6.37	6.18	6.03	5.91	5.81	5.67	5.52	5.36	5.28	5.20	5.12	5.03	4.95	4.86
9	10.56	8.02	6.99	6.42	6.06	5.80	5.61	5.47	5.35	5.26	5.11	4.96	4.81	4.73	4.65	4.57	4.48	4.40	4.31
10	10.04	7.56	6.55	5.99	5.64	5.39	5.20	5.06	4.94	4.85	4.71	4.56	4.41	4.33	4.25	4.17	4.08	4.00	3.91
11	9.65	7.21	6.22	5.67	5.32	5.07	4.89	4.74	4.63	4.54	4.40	4.25	4.10	4.02	3.94	3.86	3.78	3.69	3.60
12	9.33	6.93	5.95	5.41	5.06	4.82	4.64	4.50	4.39	4.30	4.16	4.01	3.86	3.78	3.70	3.62	3.54	3.45	3.36
13	9.07	6.70	5.74	5.21	4.86	4.62	4.44	4.30	4.19	4.10	3.96	3.82	3.66	3.59	3.51	3.43	3.34	3.25	3.17
14	8.86	6.51	5.56	5.04	4.69	4.46	4.28	4.14	4.03	3.94	3.80	3.66	3.51	3.43	3.35	3.27	3.18	3.09	3.00
15	8.68	6.36	5.42	4.89	4.56	4.32	4.14	4.00	3.89	3.80	3.67	3.52	3.37	3.29	3.21	3.13	3.05	2.96	2.87
16	8.53	6.23	5.29	4.77	4.44	4.20	4.03	3.89	3.78	3.69	3.55	3.41	3.26	3.18	3.10	3.02	2.93	2.84	2.75
17	8.40	6.11	5.18	4.67	4.34	4.10	3.93	3.79	3.68	3.59	3.46	3.31	3.16	3.08	3.00	2.92	2.83	2.75	2.65
18	8.29	6.01	5.09	4.58	4.25	4.01	3.84	3.71	3.60	3.51	3.37	3.23	3.08	3.00	2.92	2.84	2.75	2.66	2.57
19	8.18	5.93	5.01	4.50	4.17	3.94	3.77	3.63	3.52	3.43	3.30	3.15	3.00	2.92	2.84	2.76	2.67	2.58	2.49
20	8.10	5.85	4.94	4.43	4.10	3.87	3.70	3.56	3.46	3.37	3.23	3.09	2.94	2.86	2.78	2.69	2.61	2.52	2.42
21	8.02	5.78	4.87	4.37	4.04	3.81	3.64	3.51	3.40	3.31	3.17	3.03	2.88	2.80	2.72	2.64	2.55	2.46	2.36
22	7.95	5.72	4.82	4.31	3.99	3.76	3.59	3.45	3.35	3.26	3.12	2.98	2.83	2.75	2.67	2.58	2.50	2.40	2.31
23	7.88	5.66	4.76	4.26	3.94	3.71	3.54	3.41	3.30	3.21	3.07	2.93	2.78	2.70	2.62	2.54	2.45	2.35	2.26
24	7.82	5.61	4.72	4.22	3.90	3.67	3.50	3.36	3.26	3.17	3.03	2.89	2.74	2.66	2.58	2.49	2.40	2.31	2.21
25	7.77	5.57	4.68	4.18	3.85	3.63	3.46	3.32	3.22	3.13	2.99	2.85	2.70	2.62	2.54	2.45	2.36	2.27	2.17
26	7.72	5.53	4.64	4.14	3.82	3.59	3.42	3.29	3.18	3.09	2.96	2.81	2.66	2.58	2.50	2.42	2.33	2.23	2.13
27	7.68	5.49	4.60	4.11	3.78	3.56	3.39	3.26	3.15	3.06	2.93	2.78	2.63	2.55	2.47	2.38	2.29	2.20	2.10
28	7.64	5.45	4.57	4.07	3.75	3.53	3.36	3.23	3.12	3.03	2.90	2.75	2.60	2.52	2.44	2.35	2.26	2.17	2.06
29	7.60	5.42	4.54	4.04	3.73	3.50	3.33	3.20	3.09	3.00	2.87	2.73	2.57	2.49	2.41	2.33	2.23	2.14	2.03
30	7.56	5.39	4.51	4.02	3.70	3.47	3.30	3.17	3.07	2.98	2.84	2.70	2.55	2.47	2.39	2.30	2.21	2.11	2.01
40	7.31	5.18	4.31	3.83	3.51	3.29	3.12	2.99	2.89	2.80	2.66	2.52	2.37	2.29	2.20	2.11	2.02	1.92	1.80
60	7.08	4.98	4.13	3.65	3.34	3.12	2.95	2.82	2.72	2.63	2.50	2.35	2.20	2.12	2.03	1.94	1.84	1.73	1.60
120	6.85	4.79	3.95	3.48	3.17	2.96	2.79	2.66	2.56	2.47	2.34	2.19	2.03	1.95	1.86	1.76	1.66	1.53	1.38
∞	6.63	4.61	3.78	3.32	3.02	2.80	2.64	2.51	2.41	2.32	2.18	2.04	1.88	1.79	1.70	1.59	1.47	1.32	1.00

TABLE C.4
Critical values for the outlier test using the *R*-student statistic*

	$\alpha = 0.05$																	
n \ p	1	2	3	4	5	6	7	8	9	10	11	12	13	14	15	20	25	30
6	4.85	6.23	10.89	76.39														
7	4.38	5.07	6.58	11.77	89.12													
8	4.12	4.53	5.26	6.90	12.59	101.9												
9	3.95	4.22	4.66	5.44	7.18	13.36	114.6											
10	3.83	4.03	4.32	4.77	5.60	7.45	14.09	127.3										
11	3.75	3.90	4.10	4.40	4.88	5.75	7.70	14.78	140.1									
12	3.69	3.81	3.96	4.17	4.49	4.98	5.89	7.94	15.44	152.8								
13	3.65	3.74	3.86	4.02	4.24	4.56	5.08	6.02	8.16	16.08	165.5							
14	3.61	3.69	3.79	3.91	4.07	4.30	4.63	5.16	6.14	8.37	16.69	178.2						
15	3.58	3.65	3.73	3.83	3.95	4.12	4.36	4.70	5.25	6.25	8.58	17.28	191.0					
16	3.56	3.62	3.68	3.77	3.87	4.00	4.17	4.41	4.76	5.33	6.36	8.77	17.85	203.7				
17	3.54	3.59	3.65	3.72	3.80	3.90	4.04	4.21	4.46	4.82	5.40	6.47	8.95	18.40	216.4			
18	3.53	3.57	3.62	3.68	3.75	3.83	3.94	4.08	4.26	4.51	4.88	5.47	6.57	9.13	18.93			
19	3.52	3.56	3.60	3.65	3.71	3.78	3.86	3.97	4.11	4.30	4.55	4.93	5.54	6.67	9.30			
20	3.51	3.54	3.58	3.62	3.67	3.73	3.81	3.89	4.00	4.15	4.33	4.59	4.98	5.60	6.76			
21	3.50	3.53	3.57	3.60	3.65	3.70	3.76	3.83	3.92	4.03	4.18	4.37	4.64	5.03	5.67			
22	3.50	3.52	3.55	3.59	3.63	3.67	3.72	3.78	3.86	3.95	4.06	4.21	4.40	4.68	5.08	280.1		
23	3.49	3.52	3.54	3.57	3.61	3.65	3.69	3.75	3.81	3.88	3.98	4.09	4.24	4.44	4.71	21.41		
24	3.49	3.51	3.53	3.56	3.59	3.63	3.67	3.71	3.77	3.83	3.91	4.00	4.12	4.27	4.47	10.07		
25	3.48	3.50	3.53	3.55	3.58	3.61	3.65	3.69	3.73	3.79	3.85	3.93	4.02	4.14	4.30	7.17		
26	3.48	3.50	3.52	3.54	3.57	3.60	3.63	3.66	3.70	3.75	3.81	3.87	3.95	4.05	4.17	5.95		
27	3.48	3.50	3.52	3.54	3.56	3.58	3.61	3.65	3.68	3.72	3.77	3.83	3.89	3.97	4.07	5.29	343.8	
28	3.48	3.50	3.51	3.53	3.55	3.58	3.60	3.63	3.66	3.70	3.74	3.79	3.84	3.91	3.99	4.88	23.63	
29	3.48	3.49	3.51	3.53	3.55	3.57	3.59	3.62	3.64	3.68	3.71	3.76	3.81	3.86	3.93	4.61	10.74	
30	3.48	3.49	3.51	3.52	3.54	3.56	3.58	3.60	3.63	3.66	3.69	3.73	3.77	3.82	3.88	4.42	7.53	
31	3.48	3.49	3.50	3.52	3.54	3.55	3.57	3.59	3.62	3.64	3.67	3.71	3.74	3.79	3.84	4.28	6.18	
32	3.48	3.49	3.50	3.52	3.53	3.55	3.57	3.59	3.61	3.63	3.66	3.69	3.72	3.76	3.80	4.17	5.47	407.4
33	3.48	3.49	3.50	3.52	3.53	3.54	3.56	3.58	3.60	3.62	3.64	3.67	3.70	3.74	3.77	4.08	5.03	25.66
34	3.48	3.49	3.50	3.51	3.53	3.54	3.56	3.57	3.59	3.61	3.63	3.66	3.68	3.71	3.74	4.01	4.74	11.34
35	3.48	3.49	3.50	3.51	3.52	3.54	3.55	3.57	3.58	3.60	3.62	3.64	3.67	3.70	3.73	3.96	4.53	7.84
36	3.48	3.49	3.50	3.51	3.52	3.54	3.55	3.56	3.58	3.60	3.61	3.63	3.66	3.68	3.71	3.91	4.37	6.39
37	3.48	3.49	3.50	3.51	3.52	3.53	3.55	3.56	3.57	3.59	3.61	3.62	3.65	3.67	3.69	3.87	4.26	5.62
38	3.48	3.49	3.50	3.51	3.52	3.53	3.54	3.56	3.57	3.58	3.60	3.62	3.64	3.66	3.68	3.84	4.16	5.16
39	3.49	3.49	3.50	3.51	3.52	3.53	3.54	3.55	3.57	3.58	3.59	3.61	3.63	3.65	3.67	3.81	4.09	4.84
40	3.49	3.49	3.50	3.51	3.52	3.53	3.54	3.55	3.56	3.58	3.59	3.60	3.62	3.64	3.66	3.79	4.03	4.62
50	3.51	3.51	3.51	3.52	3.53	3.53	3.54	3.54	3.55	3.56	3.57	3.57	3.58	3.59	3.60	3.66	3.75	3.88
60	3.53	3.53	3.53	3.54	3.54	3.54	3.55	3.55	3.56	3.56	3.57	3.57	3.58	3.58	3.59	3.62	3.67	3.73
70	3.55	3.55	3.55	3.55	3.56	3.56	3.56	3.56	3.57	3.57	3.57	3.58	3.58	3.59	3.59	3.61	3.64	3.67
80	3.57	3.57	3.57	3.57	3.57	3.58	3.58	3.58	3.58	3.58	3.59	3.59	3.59	3.60	3.60	3.61	3.63	3.66
90	3.58	3.59	3.59	3.59	3.59	3.59	3.59	3.60	3.60	3.60	3.60	3.60	3.60	3.61	3.61	3.62	3.63	3.65
100	3.60	3.60	3.60	3.60	3.61	3.61	3.61	3.61	3.61	3.61	3.61	3.62	3.62	3.62	3.62	3.63	3.64	3.65
200	3.73	3.73	3.73	3.73	3.73	3.73	3.73	3.73	3.73	3.73	3.73	3.73	3.73	3.73	3.74	3.74	3.74	3.74
300	3.81	3.81	3.81	3.81	3.81	3.81	3.81	3.81	3.81	3.81	3.82	3.82	3.82	3.82	3.82	3.82	3.82	3.82
400	3.87	3.87	3.87	3.87	3.87	3.87	3.87	3.88	3.88	3.88	3.88	3.88	3.88	3.88	3.88	3.88	3.88	3.88
500	3.92	3.92	3.92	3.92	3.92	3.92	3.92	3.92	3.92	3.92	3.92	3.92	3.92	3.92	3.92	3.92	3.92	3.92

n is the sample size. *p* is the total number of parameters in the model.

*Used with permission. Sanford Weisberg, *Applied Linear Regression.* (New York: John Wiley and Sons, 1980), pp. 264–267.

TABLE C.4 Continued

TABLE C.4 Continued

Critical values for the outlier test using the *R*-student statistic*

n \ p ($\alpha = 0.01$)	1	2	3	4	5	6	7	8	9	10	11	12	13	14	15	20	25	30
6	7.53	10.87	24.46	382.0														
7	6.35	7.84	11.45	26.43	445.6													
8	5.71	6.54	8.12	11.98	28.26	509.3												
9	5.31	5.84	6.71	8.38	12.47	29.97	573.0											
10	5.04	5.41	5.96	6.87	8.61	12.92	31.60	636.6										
11	4.85	5.12	5.50	6.07	7.01	8.83	13.35	33.14	700.3									
12	4.71	4.91	5.19	5.58	6.17	7.15	9.03	13.75	34.62	763.9								
13	4.60	4.76	4.97	5.25	5.66	6.26	7.27	9.22	14.12	36.03	827.6							
14	4.51	4.64	4.81	5.02	5.32	5.73	6.35	7.39	9.40	14.48	37.40	891.3						
15	4.44	4.55	4.68	4.85	5.08	5.37	5.80	6.43	7.50	9.57	14.82	38.71	954.9					
16	4.38	4.48	4.59	4.72	4.90	5.12	5.43	5.86	6.51	7.60	9.73	15.15	39.98					
17	4.34	4.41	4.51	4.62	4.76	4.94	5.17	5.48	5.92	6.59	7.70	9.88	15.46	41.21				
18	4.30	4.36	4.44	4.54	4.66	4.80	4.98	5.21	5.53	5.98	6.66	7.80	10.03	15.76	42.41			
19	4.26	4.32	4.39	4.47	4.57	4.69	4.83	5.01	5.25	5.57	6.03	6.72	7.89	10.17	16.05			
20	4.23	4.29	4.35	4.42	4.50	4.60	4.72	4.86	5.05	5.29	5.62	6.08	6.79	7.98	10.31			
21	4.21	4.26	4.31	4.37	4.44	4.52	4.62	4.74	4.89	5.08	5.33	5.66	6.13	6.85	8.06			
22	4.19	4.23	4.28	4.33	4.39	4.46	4.55	4.65	4.77	4.92	5.11	5.36	5.70	6.18	6.91			
23	4.17	4.21	4.25	4.30	4.35	4.41	4.49	4.57	4.67	4.80	4.95	5.14	5.40	5.74	6.22	47.94		
24	4.15	4.19	4.22	4.27	4.32	4.37	4.43	4.51	4.59	4.70	4.82	4.98	5.17	5.43	5.78	17.36		
25	4.14	4.17	4.20	4.24	4.28	4.33	4.39	4.45	4.53	4.62	4.72	4.85	5.00	5.20	5.46	10.92		
26	4.12	4.15	4.18	4.22	4.26	4.30	4.35	4.41	4.47	4.55	4.64	4.74	4.87	5.03	5.23	8.43		
27	4.11	4.14	4.17	4.20	4.24	4.27	4.32	4.37	4.43	4.49	4.57	4.66	4.76	4.89	5.05	7.17		
28	4.10	4.13	4.15	4.18	4.21	4.25	4.29	4.33	4.38	4.44	4.51	4.59	4.68	4.78	4.91	6.43	52.90	
29	4.09	4.12	4.14	4.17	4.20	4.23	4.26	4.30	4.35	4.40	4.46	4.53	4.60	4.69	4.80	5.94	18.50	
30	4.09	4.11	4.13	4.15	4.18	4.21	4.24	4.28	4.32	4.36	4.42	4.47	4.54	4.62	4.71	5.60	11.44	
31	4.08	4.10	4.12	4.14	4.17	4.19	4.22	4.26	4.29	4.33	4.38	4.43	4.49	4.56	4.64	5.35	8.75	
32	4.07	4.09	4.11	4.13	4.15	4.18	4.21	4.24	4.27	4.31	4.35	4.39	4.45	4.50	4.57	5.16	7.40	
33	4.07	4.08	4.10	4.12	4.14	4.17	4.19	4.22	4.25	4.28	4.32	4.36	4.41	4.46	4.52	5.01	6.60	57.43
34	4.06	4.08	4.09	4.11	4.13	4.15	4.18	4.20	4.23	4.26	4.29	4.33	4.37	4.42	4.47	4.89	6.09	19.51
35	4.06	4.07	4.09	4.11	4.12	4.14	4.16	4.19	4.21	4.24	4.27	4.31	4.34	4.39	4.43	4.79	5.72	11.90
36	4.05	4.07	4.08	4.10	4.12	4.13	4.15	4.18	4.20	4.22	4.25	4.28	4.32	4.36	4.40	4.71	5.46	9.03
37	4.05	4.06	4.08	4.09	4.11	4.13	4.14	4.16	4.19	4.21	4.24	4.26	4.29	4.33	4.37	4.64	5.26	7.60
38	4.05	4.06	4.07	4.09	4.10	4.12	4.13	4.15	4.17	4.20	4.22	4.25	4.27	4.31	4.34	4.59	5.10	6.76
39	4.04	4.06	4.07	4.08	4.10	4.11	4.13	4.14	4.16	4.18	4.21	4.23	4.26	4.28	4.32	4.54	4.97	6.21
40	4.04	4.05	4.06	4.08	4.09	4.10	4.12	4.14	4.15	4.17	4.19	4.22	4.24	4.27	4.29	4.49	4.87	5.83
50	4.03	4.03	4.04	4.05	4.06	4.07	4.07	4.08	4.09	4.10	4.12	4.13	4.14	4.15	4.17	4.25	4.38	4.59
60	4.03	4.03	4.04	4.04	4.05	4.05	4.06	4.06	4.07	4.08	4.08	4.09	4.10	4.11	4.12	4.17	4.23	4.32
70	4.03	4.03	4.04	4.04	4.05	4.05	4.05	4.06	4.06	4.07	4.07	4.08	4.08	4.09	4.09	4.13	4.17	4.22
80	4.04	4.04	4.04	4.05	4.05	4.05	4.06	4.06	4.06	4.07	4.07	4.07	4.08	4.08	4.09	4.11	4.13	4.17
90	4.05	4.05	4.05	4.05	4.06	4.06	4.06	4.06	4.07	4.07	4.07	4.07	4.08	4.08	4.08	4.10	4.12	4.14
100	4.06	4.06	4.06	4.06	4.06	4.07	4.07	4.07	4.07	4.07	4.08	4.08	4.08	4.08	4.09	4.10	4.11	4.13
200	4.15	4.15	4.15	4.15	4.15	4.15	4.15	4.15	4.15	4.15	4.15	4.15	4.15	4.15	4.15	4.16	4.16	4.16
300	4.21	4.21	4.21	4.21	4.21	4.21	4.22	4.22	4.22	4.22	4.22	4.22	4.22	4.22	4.22	4.22	4.22	4.22
400	4.26	4.27	4.27	4.27	4.27	4.27	4.27	4.27	4.27	4.27	4.27	4.27	4.27	4.27	4.27	4.27	4.27	4.27
500	4.31	4.31	4.31	4.31	4.31	4.31	4.31	4.31	4.31	4.31	4.31	4.31	4.31	4.31	4.31	4.31	4.31	4.31

n is the sample size, p is the total number of parameters in the model.

*Used with permission. Sanford Weisberg, *Applied Linear Regression.* (New York: John Wiley and Sons, 1980), pp. 264–267.

TABLE C.5 **Plots of rankits for testing for normality in residuals***

	n									
i	1	2	3	4	5	6	7	8	9	10
1	0	−0.56	−0.85	−1.03	−1.16	−1.27	−1.35	−1.42	−1.49	−1.54
2		0.56	0.00	−0.30	−0.50	−0.64	−0.76	−0.85	−0.93	−1.00
3			0.85	0.30	0.00	−0.20	−0.35	−0.47	−0.57	−0.66
4				1.03	0.50	0.20	0.00	−0.15	−0.27	−0.38
5					1.16	0.64	0.35	0.15	0.00	−0.12
6						1.27	0.76	0.47	0.27	0.12

	n									
i	11	12	13	14	15	16	17	18	19	20
1	−1.59	−1.63	−1.67	−1.70	−1.74	−1.77	−1.79	−1.82	−1.84	−1.87
2	−1.06	−1.12	−1.16	−1.21	−1.25	−1.28	−1.32	−1.35	−1.38	−1.41
3	−0.73	−0.79	−0.85	−0.90	−0.95	−0.99	−1.03	−1.07	−1.10	−1.13
4	−0.46	−0.54	−0.60	−0.66	−0.71	−0.76	−0.81	−0.85	−0.89	−0.92
5	−0.22	−0.31	−0.39	−0.46	−0.52	−0.57	−0.62	−0.66	−0.71	−0.75
6	0.00	−0.10	−0.19	−0.27	−0.34	−0.40	−0.45	−0.50	−0.55	−0.59
7	0.22	0.10	0.00	−0.09	−0.17	−0.23	−0.30	−0.35	−0.40	−0.45
8	0.46	0.31	0.19	0.09	0.00	−0.08	−0.15	−0.21	−0.26	−0.31
9	0.73	0.54	0.39	0.27	0.17	0.08	0.00	−0.07	−0.13	−0.19
10	1.06	0.79	0.60	0.46	0.34	0.23	0.15	0.07	0.00	−0.06

*Plots not shown are found by symmetry.

Used with permission, Sanford Weisberg, *Applied Linear Regression.* (New York: John Wiley and Sons, 1980), p. 268.

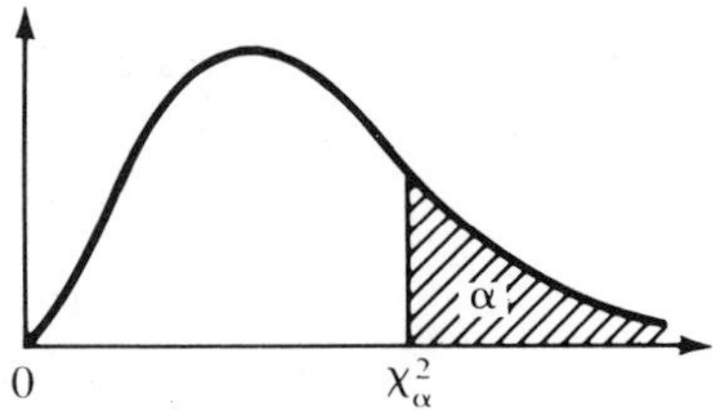

TABLE C.6 **Critical values of the χ-square distribution**

df	α							
ν	0.995	0.99	0.975	0.95	0.05	0.025	0.01	0.005
1	0.0^4393	0.0^3157	0.0^3982	0.0^2393	3.841	5.024	6.635	7.879
2	0.0100	0.0201	0.0506	0.103	5.991	7.378	9.210	10.597
3	0.0717	0.115	0.216	0.352	7.815	9.348	11.345	12.838
4	0.207	0.297	0.484	0.711	9.488	11.143	13.277	14.860
5	0.412	0.554	0.831	1.145	11.070	12.832	15.086	16.750
6	0.676	0.872	1.237	1.635	12.592	14.449	16.812	18.548
7	0.989	1.239	1.690	2.167	14.067	16.013	18.475	20.278
8	1.344	1.646	2.180	2.733	15.507	17.535	20.090	21.955
9	1.735	2.088	2.700	3.325	16.919	19.023	21.666	23.589
10	2.156	2.558	3.247	3.940	18.307	20.483	23.209	25.188
11	2.603	3.053	3.816	4.575	19.675	21.920	24.725	26.757
12	3.074	3.571	4.404	5.226	21.026	23.337	26.217	28.300
13	3.565	4.107	5.009	5.892	22.362	24.736	27.688	29.819
14	4.075	4.660	5.629	6.571	23.685	26.119	29.141	31.319
15	4.601	5.229	6.262	7.261	24.996	27.488	30.578	32.801
16	5.142	5.812	6.908	7.962	26.296	28.845	32.000	34.267
17	5.697	6.408	7.564	8.672	27.587	30.191	33.409	35.718
18	6.265	7.015	8.231	9.390	28.869	31.526	34.805	37.156
19	6.844	7.633	8.907	10.117	30.144	32.852	36.191	38.582
20	7.434	8.260	9.591	10.851	31.410	34.170	37.566	39.997
21	8.034	8.897	10.283	11.591	32.671	35.479	38.932	41.401
22	8.643	9.542	10.982	12.338	33.924	36.781	40.289	42.796
23	9.260	10.196	11.689	13.091	35.172	38.076	41.638	44.181
24	9.886	10.856	12.401	13.848	36.415	39.364	42.980	45.558
25	10.520	11.524	13.120	14.611	37.652	40.646	44.314	46.928
26	11.160	12.198	13.844	15.379	38.885	41.923	45.642	48.290
27	11.808	12.879	14.573	16.151	40.113	43.194	46.963	49.645
28	12.461	13.565	15.308	16.928	41.337	44.461	48.278	50.993
29	13.121	14.256	16.047	17.708	42.557	45.722	49.588	52.336
30	13.787	14.953	16.791	18.493	43.773	46.979	50.892	53.672

TABLE C.7 **Critical values of the Durbin–Watson statistic**

		k = Number of regressors (excluding the intercept)									
		1		2		3		4		5	
Sample Size	Significance Level	d_L	d_U	d_L	d_U	d_L	d_U	d_L	d_U	d_L	d_U
	.01	.81	1.07	.70	1.25	.59	1.46	.49	1.70	.39	1.96
15	.025	.95	1.23	.83	1.40	.71	1.61	.59	1.84	.48	2.09
	.05	1.08	1.36	.95	1.54	.82	1.75	.69	1.97	.56	2.21
	.01	.95	1.15	.86	1.27	.77	1.41	.63	1.57	.60	1.74
20	.025	1.08	1.28	.99	1.41	.89	1.55	.79	1.70	.70	1.87
	.05	1.20	1.41	1.10	1.54	1.00	1.68	.90	1.83	.79	1.99
	.01	1.05	1.21	.98	1.30	.90	1.41	.83	1.52	.75	1.65
25	.025	1.13	1.34	1.10	1.43	1.02	1.54	.94	1.65	.86	1.77
	.05	1.20	1.45	1.21	1.55	1.12	1.66	1.04	1.77	.95	1.89
	.01	1.13	1.26	1.07	1.34	1.01	1.42	.94	1.51	.88	1.61
30	.025	1.25	1.38	1.18	1.46	1.12	1.54	1.05	1.63	.98	1.73
	.05	1.35	1.49	1.28	1.57	1.21	1.65	1.14	1.74	1.07	1.83
	.01	1.25	1.34	1.20	1.40	1.15	1.46	1.10	1.52	1.05	1.58
40	.025	1.35	1.45	1.30	1.51	1.25	1.57	1.20	1.63	1.15	1.69
	.05	1.44	1.54	1.39	1.60	1.34	1.66	1.29	1.72	1.23	1.79
	.01	1.32	1.40	1.28	1.45	1.24	1.49	1.20	1.54	1.16	1.59
50	.025	1.42	1.50	1.38	1.54	1.34	1.59	1.30	1.64	1.26	1.69
	.05	1.50	1.59	1.46	1.63	1.42	1.67	1.38	1.72	1.34	1.77
	.01	1.38	1.45	1.35	1.48	1.32	1.52	1.28	.156	1.25	1.60
60	.025	1.47	1.54	1.44	1.57	1.40	1.61	1.37	1.65	1.33	1.69
	.05	1.55	1.62	1.51	1.65	1.48	1.69	1.44	1.73	1.41	1.77
	.01	1.47	1.52	1.44	1.54	1.42	1.57	1.39	1.60	1.36	1.62
80	.025	1.54	1.59	1.52	1.62	1.49	1.65	1.47	1.67	1.44	1.70
	.05	1.61	1.66	1.59	1.69	1.56	1.72	1.53	1.74	1.51	1.77
	.01	1.52	1.56	1.50	1.58	1.48	1.60	1.45	1.63	1.44	1.65
100	.025	1.59	1.63	1.57	1.65	1.55	1.67	1.53	1.70	1.51	1.72
	.05	1.65	1.69	1.63	1.72	1.61	1.74	1.59	1.76	1.57	1.78

INDEX